Ecology of Insects

CONCEPTS AND APPLICATIONS

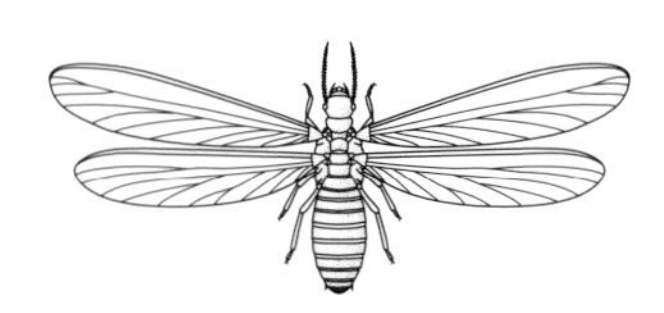

MARTIN R.SPEIGHT
Department of Zoology, University of Oxford
South Parks Road, Oxford
and St Anne's College, Woodstock Road
Oxford

MARK D.HUNTER
Institute of Ecology
University of Georgia
Athens, Georgia
USA

ALLAN D. WATT
Institute of Terrestrial Ecology
Edinburgh Research Station
Penicuik, Midlothian
Scotland

b

Blackwell Science

Editorial Offices:
Osney Mead, Oxford OX2 0EL
25 John Street, London WC1N 2BL
23 Ainslie Place, Edinburgh EH3 6AJ
350 Main Street, Malden
MA 02148 5018, USA
54 University Street, Carlton
Victoria 3053, Australia
10, rue Casimir Delavigne
75006 Paris, France

Other Editorial Offices:
Blackwell Wissenschafts-Verlag GmbH
Kurfürstendamm 57
10707 Berlin, Germany

Blackwell Science KK
MG Kodenmacho Building
7–10 Kodenmacho Nihombashi
Chuo-ku, Tokyo 104, Japan

First published 1999

Set by Excel Typesetters Co., Hong Kong
Printed and bound in Great Britain
at the Alden Press Ltd, Oxford and
Northampton

A catalogue record for this title
is available from the British Library

ISBN 0-86542-745-3

Library of Congress
Cataloging-in-publication Data

Speight, Martin R.
Ecology of insects: concepts and
applications/ Martin R. Speight,
Mark D. Hunter, Allan D. Watt.
p cm
Includes bibliographical references
(p.) and index.
ISBN 0-86542-745-3
1. Insects—Ecology. I. Hunter,
Mark D. II. Watt, Allan D. III. Title.
QL496.4.S66 1999
595.717—dc21 99-11441
CIP

DISTRIBUTORS
Marston Book Services Ltd
PO Box 269
Abingdon, Oxon OX14 4YN
(*Orders*: Tel: 01235 465500
Fax: 01235 465555)

USA
Blackwell Science, Inc.
Commerce Place
350 Main Street
Malden, MA 02148 5018
(*Orders*: Tel: 800 759 6102
781 388 8250
Fax: 781 388 8255)

Canada
Login Brothers Book Company
324 Saulteaux Crescent
Winnipeg, Manitoba R3J 3T2
(*Orders*: Tel: 204 837 2987)

Australia
Blackwell Science Pty Ltd
54 University Street
Carlton, Victoria 3053
(*Orders*: Tel: 3 9347 0300
Fax: 3 9347 5001)

For further information on
Blackwell Science, visit our website:
www.blackwell-science.com

Contents

Foreword

Today, it is not necessary to stress the significance of ecology or its place in the world. The authors of this volume have provided a succinct overview of the subject: both its fundamental principles and the application of these to matters affecting human welfare. They have limited their canvas to insects, but these provide excellent models. Without in any way decrying other branches of the subject, one can point out that in so many respects insects provide the mid-point optimum on the spectrum between the study of mammals and birds on the one hand and that of microbes on the other. They may be seen with the naked eye, yet their generation times are sufficiently short and the sizes of their populations sufficiently large to permit the quantitative study of population dynamics on time and spatial scales appropriate for students—undergraduate and postgraduate.

Insects have other claims on our attention. The vast number of species make them the major contributor to biodiversity in the great majority of habitats other than the sea. From a more anthropocentric view, one must also recognize that they are our major competitors taking about 10% of the food that we grow and infecting one in six of the world's population with a pathogen. All these facets are explored in this volume.

The authors bring a diversity of experience to the task of encapsulating the essence of the subject in a compact book. A glance at the colour plate illustrations shows the range of insects and habitats that have interested the principal author. From their hands-on experience in the field the authors bring a freshness to even the most mathematical of concepts. Not surprisingly the text emphasizes the truth that 'it all starts with good observations in the field'. As this book shows, as well as providing a good exemplar of a key subject for today's world, insect ecology is also fun.

T.R.E. Southwood
October 1998

Acknowledgements

A large number of friends and colleagues have helped us during the writing of this book—so many in fact that we hesitate to list them in case we miss someone out! With this risk in mind, we would firstly like to thank all those who read parts of the early drafts and suggested changes and additions, or provided photographs and other graphics. Apart from anonymous referees, we would like to thank Dave Thompson, Chris Conlon, Andy Foggo, Dave Kelly, Tristram Wyatt, David Rogers, Paul Embden, Ross Wylie, Alissa Salmore, Kitti Reynolds, Rebecca Forkner, Rebecca Klaper, Matt Wood, Kim Holton, Joe McHugh, Darald Batzer, Don Champagne, Hefin Jones, Angela Russon, Dick Southwood, Arthur Chung, Judith Marshall and Katy Watt.

At Blackwell Science, we are enormously grateful to Ian Sherman our editor, for his patience and understanding as well as having to read the entire book, to Simon Rallison who started the whole thing off, and to Jonathan Rowley who had to put the book together. The project would have been impossible without the excellent services of the Radcliffe Science Library, the Oxford Forestry Institute library, St Anne's College library, and the Zoology Department libraries, all in Oxford, and the Science Library at the University of Georgia.

We were all in our own ways inspired to study ecological entomology in the early days by our mentors, Tony Dixon, John Lawton, Dick Southwood, and the late George Varley. Without these people, none of us would be in a position to attempt this book.

Martin Speight
Mark Hunter
Allan Watt
June 1998

Chapter 1: An Overview of Insect Ecology

In this chapter we provide a brief overview of the major concepts in insect ecology, and attempt to present a taste of what is to come in the nine detailed chapters that follow. Unavoidably, a little repetition may therefore occur, in that topics briefly discussed in this chapter will appear again in more detail elsewhere in the book. This overlap is entirely intentional on our part, and we would encourage the random reader to explore this introductory chapter first, and then turn to the details in whichever subject and later chapter takes his or her fancy. Readers who know what they are looking for can of course proceed directly to the relevant chapter(s).

We have assumed that most readers will have some knowledge of insect taxonomy, and we have not attempted to provide an in-depth coverage of this topic. Some information is, however, provided to set the scene. Readers who wish to know more are recommended to obtain one of various excellent entomology textbooks such as those by Gillott (1995) and Gullan and Cranston (1996), the magnificent two-volume work on insects of Australia (CSIRO 1991), or the introduction to insect biology and diversity by Daly, Doyen and Purcell (1998).

1.1 History of ecology and entomology

The science of ecology is broad ranging and difficult to define. Most of us think we know what it means, and indeed it can imply different things to different people. It is best considered as a description of interactions between organisms and their environment, and its basic philosophy is to account for the abundance and distribution of these organisms. In fact, ecology encompasses a whole variety of disciplines, both qualitative and quantitative, whole organism, cellular and molecular, from behaviour and physiology, to evolution and interactions within and between populations. In 1933, Elton suggested that ecology represented, partly at least, the application of scientific method to natural history. As a science, Elton felt that ecology depended on three methods of approach: field observations, systematic techniques, and experimental work both in the laboratory and in the field. These three basic systems still form the framework of ecology today, and it is the appropriate integration of these systems, which provides our best estimates of the associations of living organisms with themselves and their environment, which we call ecology.

Insects have dominated the interests of zoologists for centuries. Those who could drag themselves away from the charismatic but species-poor vertebrates soon found the ecology of insects to be a complex and rich discipline, which has fascinated researchers for at least 200 years. At first, the early texts were mainly descriptive of the wonders of the insect world. Starting in 1822, Kirby and Spence published their four volumes of *An Introduction to Entomology*. This was a copious account of the lives of insects, written over three decades before Darwin first produced *The Origin of Species* in 1859. The fact that Kirby and Spence did not have the benefit of explaining their observations on the myriad interactions of insects and their environment as results of evolution did not detract from the clear fascination that the world of insects provided. This fascination has withstood the test of time. Half a century or more after Kirby and Spence, Fabre published insightful and exciting accounts of the lives of insects, starting before the turn of the century (Fabre 1882), wherein he presented in great detail observations made over many years of the ecologies of dung beetles, spider-hunting wasps, cicadas and praying mantids. These books epitomized the first of Elton's three approaches to ecology, that of field observations. Later work began to enhance the third approach, that of experimental insect ecology, summarized in *Insect Population Ecology* by Varley *et al.* (1973). This book was probably the first and certainly most influential, concise account of insect ecology derived from quantitative, scientific research. A year later, Price's *Insect Ecology* appeared, which is now in its third edition (Price 1997).

Table 1.1 Summary of associations in ecology.

Type	Interaction level	Insect example
Competition		
Intraspecific	Individuals within a population of one species	Lepidoptera larvae on trees
Interspecific	Individuals within populations of two different species	Bark beetles in tree bark
Herbivory, or phytophagy	Between autotroph and primary consumer	Aphids on roses
Symbiosis		
Mutualism	Between individuals of two species	Ants and fungi
Sociality	Between individuals of one species	Termites
Predation	Between primary and secondary consumer trophic levels	Mantids and flies
Parasitism	Between primary and secondary consumer trophic levels	Ichneumonid wasps and sawflies

1.2 Ecological associations

The term 'association' used by Elton covers a great many types of interaction, including those between individuals of the same species, between species in the same trophic level (a trophic level is a position in a food web occupied by organisms having the same functional way of gaining energy), and between different trophic levels. These associations may be mutually beneficial, or, alternatively, involve the advancement or enhancement of fitness of certain individuals at the expense of others. Of course, insects can live in association with other organisms without having any influence on them, or being influenced by them; such associations are 'neutral'.

Ecological associations among insects operate at one of three levels, i.e. at the level of individual organisms, at a population level, or over an entire community. Table 1.1 summarizes some of these basic associations. With the exception of primary production, it should be possible to discover insect examples for all other kinds of ecological interactions. The fact that insects are so widespread and diverse in their ecological associations is not surprising when it is considered just how many of them there are, at least on land and in fresh water, how adaptable they are to changing and novel environmental conditions, and how long they have existed in geological time.

1.3 The Insecta

1.3.1 Structure

This is not a book about insect morphology or physiology in the main, but both are certainly worthy of some consideration, as the ability of an organism to succeed in its environment is dictated by form and function. Here, we merely provide a few basic details, and the reader should refer to the textbooks cited in section 1.1 for more information.

The basic structure of a typical adult insect is shown in Fig. 1.1 (CSIRO 1979). Comparing the figure with the photograph of a grasshopper (Fig. 1.2) will enable any new student of entomology to quickly realize how insects evolved a highly technologically efficient set of specialized body parts and appendages. The three basic sections (called tagmata) of an insect's body are admirably adapted for different purposes. The head specializes in sensory reception and food gathering, the thorax in locomotion, and the abdomen in digestion and reproduction. All but a minimum number of appendages have been lost when compared with ancestors, leaving a set of highly adapted mouthparts and a pair of immensely stable tripods, the legs. Throughout the book, we shall refer to these and other body structures in terms of their evolution and ecology. This basic plan is of course highly variable, and the most specialized insects, for example, blow fly larvae (the fisherman's maggot), bear little or no superficial similarity to this plan.

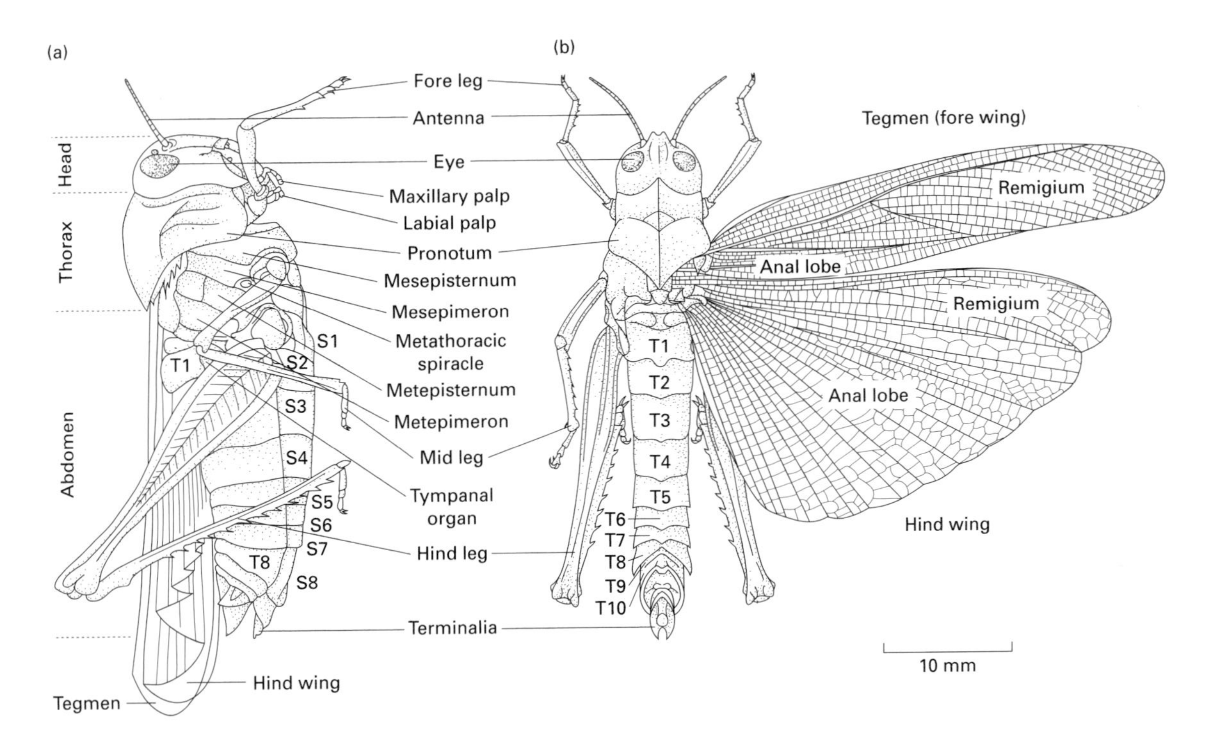

Fig. 1.1 The external structure of an adult insect, illustrating general features of a non-specialized species. (From CSIRO 1979.)

Fig. 1.2 Side view of grasshopper showing general structure.

1.3.2 Taxonomy

We hope that the reader does not get too bogged down with taxonomic nomenclature or classification. We are concerned with what insects do rather than what their names are, and although it is essential that detailed taxonomy is eventually carried out as part of an ecological investigation, in this book we have provided scientific names more as a reference system for comparison with other published work than as something people should automatically learn. However, some introduction to the taxonomic relationships of the insects will help to set the scene.

The class Insecta belongs within the superclass Hexapoda of the subphylum Uniramia in the phylum Arthropoda. The placement of the hexapods within the Uniramia indicates some affinity with the myriapods, and a lack of any recent associations with the other two subphyla of the arthropods, the Crustacea and the Chelicerata. In fact, according to Willmer (1990), the latter two groups arose separately a very long time ago, and indeed the notion of a phylum called the Arthropoda is merely based on a 'representation of a grade of organization, arising whenever soft-bodied worms develop toughened cuticles'. The Uniramia, consisting of myriapods and hexapods, however, do seem to have a common ancestor in the so-called proto-annelids, but Willmer again has suggested that there are no recent links between the two groups despite superficial similarities.

Less controversial is the organization and relatedness of groups within the hexapods. Though the Insecta are by far the most numerous, they share the superclass with three other classes, each of which comprises just one order each, the Diplura, Protura and Collembola (springtails) (Gullan & Cranston 1996). These latter three were until fairly recently included amongst the apterygote (wingless) insects, but their morphological and physiological features are more likely to mimic those of insects by virtue of convergent evolution than by true relatedness.

The taxonomic classification of the Insecta used in this book follows that described by Gullan and Cranston (1996). There are two orders within the Apterygota (wingless insects): the Thysanura (silverfish) (Fig. 1.3) and the Archaeognatha (bristletails). All other insects belong to the pterygote group (winged insects) and this is in turn divided into the Exopterygota (also known as the Hemimetabola), where wings develop gradually through several nymphal instars, and the Endopterygota (also known as the Holometabola), where there is usually a distinct larval stage separated from the adult by a pupa.

Figure 1.4 summarizes the classification of insects, and indicates roughly the number of species so far described from each order. This great species richness within the Insecta is thought to have resulted from low extinction rates throughout their history (see below) (Labandeira & Sepkoski 1993), and it is important to realize that nearly 90% of insect species belong to the endopterygotes, indicating the overwhelming advantages for speciation provided by the specialized larval and pupal stages. As will be seen throughout the book, lay-people and entomologists alike tend to concentrate on adult insects, but the ecology of larval or nymphal stages may well have much more relevance to the success or otherwise of the species or order, as well as having much more direct impact on human life. After all, it is the larvae of peacock butterflies that devour their host plant, nettle, not the adults.

Fig. 1.3 Silverfish (Thysanura).

Success of a group of organisms is not just measured in terms of the number of species accrued in the group, but can also be discussed as the range of habitats or food types dealt with, the extremes of environments in which they are able to live, how long the group has been extant, and the relative abundance of individuals. Ecologically, it may be more useful to break down the insect orders into functional groups according to life style or feeding strategies, rather than to merely count the number of species.

1.4 Fossil history and insect evolution

The nature of insect bodies, at least the adults of both Exo- and Endopterygota and to some extent the nymphs of the Exopterygota, makes them very suitable for fossilization in a variety of preserving media from sediments to amber (Plate 1.1, opposite p. 158). The tough insect cuticle is composed of flexible chitin, and/or more rigid sclerotin. Chitin is certainly an ancient widespread compound, known in animals from at least the Cambrian period, more than 550 Myr ago (Miller 1991). The fossil record of insects is therefore relatively complete when compared with that of most other animal groups. Table 1.2 shows the earliest known fossils from the major insect orders that are still extant today. Though conventional wisdom might suggest that insects such as Odonata (dragonflies and damselflies) are primitive or early when compared with, for example, Hymenoptera (bees, ants and wasps), there is little evidence from the fossil record of a clear progression or development from one group to the next. Even with modern molecular techniques such as nucleotide sequencing to investigate the relatedness of organisms, it is difficult to explain the lineages of most modern insect orders. It is likely that the divergence of these orders is very ancient, and may have occurred too rapidly for easy resolution (Liu & Beckenbach 1992).

Most major orders were already distinguishable by 250 Myr ago, and only a few are known to be much more ancient. The most ancient winged insects probably included primitive cockroaches, the Palaeodictyoptera, whose fossils date back about 370 Myr to late Devonian time (Kambhampati 1995; Martinez-Delclos 1996), illustrating the significance of a scavenging lifestyle in terms of early adaptability. Certain ways of life, however, do appear to be more recent

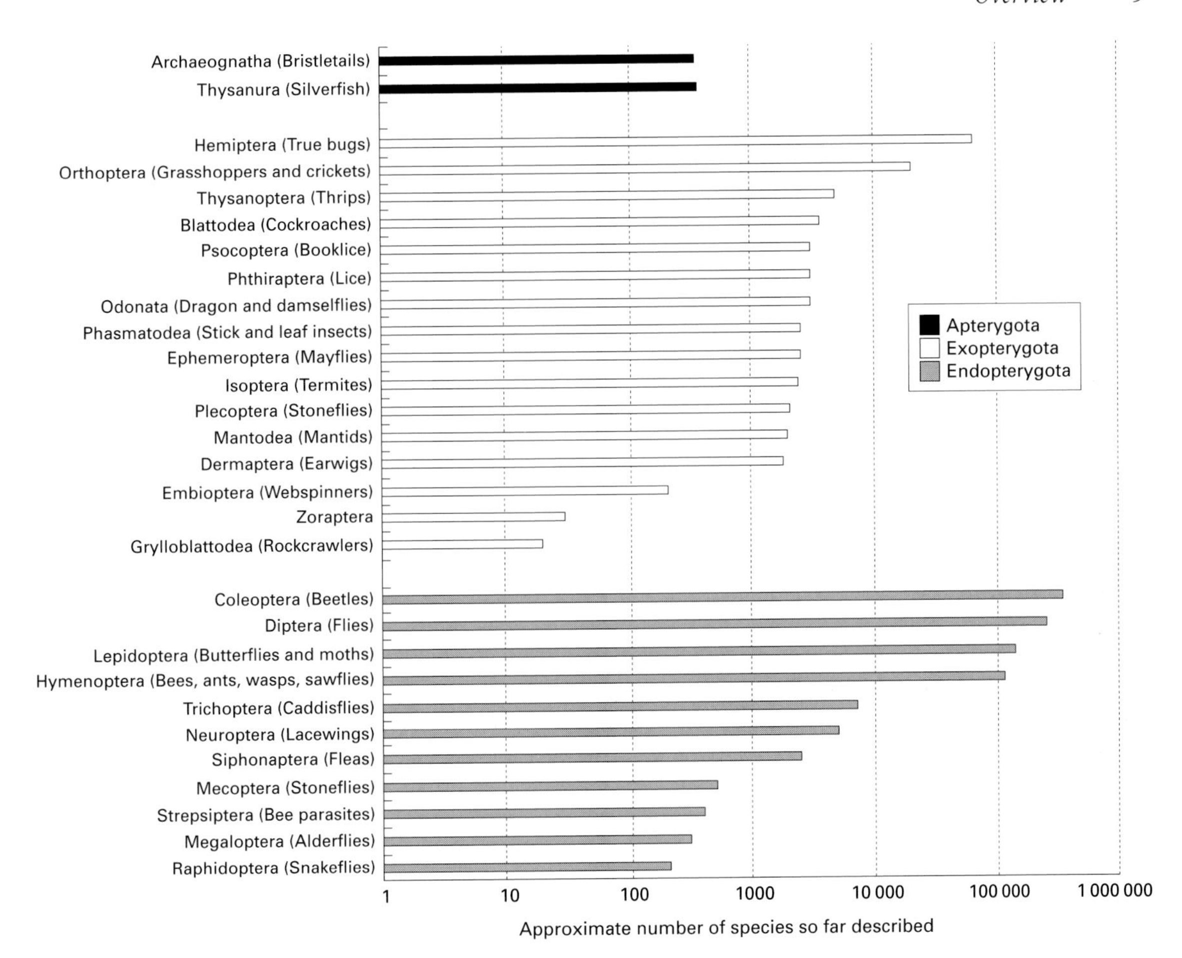

Fig. 1.4 The orders of insects, and the approximate number of species in each. (Data from Gullan & Cranston 1996.)

than others. Specialist predators such as the praying mantids (Mantodea) are not found as fossils until the Eocene period, around 50 Myr ago, and sociality might be expected to be an advanced feature of insect ecologies; termites, for example, do not appear in the fossil record until the Cretaceous period, around 130 Myr ago (Gullan & Cranston 1996).

Various major orders or 'cohorts' have become extinct in the last 250 Myr, but relatively few when compared with many other major animal phyla, such as the molluscs and the tetrapods (Fig. 1.5). The average time in which an insect species is in existence is conjectured to be an order of magnitude greater than that for, say, a bivalve or tetrapod (May *et al.* 1995), possibly well in excess of 10 Myr. At the family level, it would appear that no insect families have become extinct over the past 100 Myr or so. In some cases, it is thought that species we find alive today may extend much further back in time even than that. Dietrich and Vega (1995), by examining fossil leaf-hoppers (Hemiptera: Cicadellidae) from the Tertiary period, suggested that modern genera of these sap-feeders existed as early as 55 Myr ago.

It would seem therefore that evolutionary pressures have not dramatically altered the fundamental interactions between insects and their environment for a very long time indeed, despite some major changes to terrestrial ecosystems. How the insects have managed to sustain such taxonomic constancy through large-scale climatic fluctuations is puzzling (Coope 1994), but this puzzle forms the very basis for investigations of insect ecology. Insects do show vari-

Table 1.2 Fossil history of major insect orders alive today. (Data from Boudreaux 1987; Gullan & Cranston 1996.)

Order	Earliest fossils	Myr ago
Archaeognatha	Devonian	390
Thysanura	Carboniferous	300
Odonata	Permian	260
Ephemeroptera	Carboniferous	300
Plecoptera	Permian	280
Phasmatodea	Triassic	240
Dermaptera	Jurassic	160
Isoptera	Cretaceous	140
Mantodea	Eocene	50
Blattodea	Carboniferous	295
Thysanoptera	Permian	260
Hemiptera	Permian	275
Orthoptera	Carboniferous	300
Coleoptera	Permian	275
Strepsiptera	Cretaceous	125
Hymenoptera	Triassic	240
Neuroptera	Permian	270
Siphonaptera	Cretaceous	130
Diptera	Permian	260
Trichoptera	Triassic	240
Lepidoptera	Jurassic	200

ations in the patterns of dominance of different species and families as climates have changed. In Chile, for example, as the ice retreated from the south central lowlands about 18 000 years ago, the beetle fauna was characterized by species of moorland habitat. However, by 12 500 years ago, fossil beetle assemblages from the same region consisted entirely of rainforest species (Hoganson & Ashworth 1992). Thus, insects may not be prone to global extinctions, but are able to move from one region or another as conditions become more or less favourable for their ecologies. They can be thought of as ecologically 'malleable' rather than 'brittle'.

Undoubtedly, the single most important development in the evolution of insects has been the development of wings. From fossil evidence, it is clear that the appearance of the Pterygota (winged insects) coincided with the tremendous diversification of insects that began in the Palaeozoic era (Kingsolver & Koehl 1994). Equally clear are the obvious advantages of flight to an animal, including avoiding predators, finding mates, and locating food, breeding sites and new habitats. Any organism that has to walk or hop everywhere is bound to be at a disadvantage

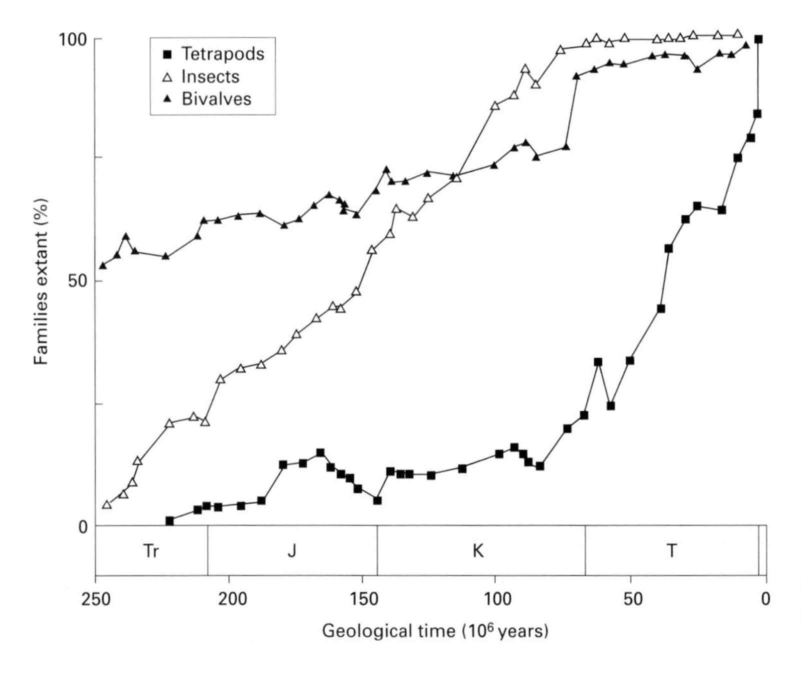

Fig. 1.5 The number of families extant at various time periods as a percentage of those alive today. (From May *et al.* 1995.)

when compared with one that can fly! It seems to be the case that insects only ever possessed wings that operated as flight organs on the second and third thoracic segments (the meso- and metathorax; see Fig. 1.1) (Wootton 1992), and the fundamental problem concerns the evolutionary steps that led to this final, fully bi-winged state. Various theories exist to explain the origin of wings in insects, and their intermediate functions. One of the most popular suggests that so-called proto-winglets originated from unspecialized appendages on the basal segment of the legs (the pleural hypothesis, according to Kingsolver and Koehl (1994)). These winglets were primitively articulated and hence movable, right from the start. Fossil mayfly nymphs (Ephemeroptera) from Lower Permian strata (of around 280 Ma) show both thoracic and even abdominal proto-wings, though this does not imply that that such appendages evolved initially in fresh water. However, other theories that utilize genetic examination of both crustacean and insect species have suggested that wings arose from gill-like appendages called epipodites present on aquatic ancestors of the pterygote insects (Averof & Cohen 1997). Whatever the origin of proto-wings, there is still confusion as to what purpose or purposes they might have served before they became large enough to assist with aerodynamics. Sexual display and courtship, thermoregulation, camouflage and aquatic respiration have all been suggested, but perhaps the most traditional idea involves some assistance in gliding after jumping. Whichever early route, once the winglets had increased in size sufficiently to be aerodynamically active, they would have been able to prolong airborne periods and, with the assistance of long tails (cerci), stable gliding would have become possible. If we assume that the proto-winglets were articulated from their earliest forms, then fairly easy evolutionary steps to fully powered and controlled flight can be imagined after the gliding habit was perfected. This ability to fly is likely to have been a great advantage in the colonisation of new ecospace provided by early tree-like plants such as pteridophytes.

1.5 Habits of insects

Not only have many species of insects been on Earth for many millions of years, but their various modern ecological habits also appeared at an early stage. Table 1.3 shows some major events in the development of insect ecology over the last 400 Myr. The great radiation of insect species is thought to have begun about 245 Myr ago, in early Triassic time (Labandeira & Sepkoski 1993), but judging by various insect fossils, it is clear that insects were exploiting terrestrial habitats maybe 100 Myr earlier. For example, evidence of

Table 1.3 A summary of evolutionary events in the ecology of insects, derived from fossil evidence (time scales very approximate).

Event	Period	Approx. time (Ma)	Reference
Most recent aphid families present	Early Tertiary	45	Heie (1996)
Radiation of higher (cyclorrhaphan) flies (Diptera)	Tertiary (Paleocene)	55–65	Grimaldi (1997)
Evidence of complex insect damage to angiosperm leaves	Cretaceous	65–145	Scott *et al.* (1992)
Midges feeding on the blood of dinosaurs	Cretaceous	88–93.5	Borkent (1996)
Differentiation of various weevil families	Middle Cretaceous	100	Labandeira & Sepkoski (1993)
Earliest fossil ant	Early Cretaceous	130	Brandao *et al.* (1989)
Radiation of major lepidopteran lineages on gymnosperms	Late Jurassic	150	Labandeira *et al.* (1994)
Establishment of intracellular symbionts in aphids	Permian–Jurassic	160–280	Fukatsu (1994)
Insect grazing damage on fern pinnules and gymnosperm leaves	Late Triassic	220	Ash (1996)
First evidence of pollenivory	Early Permian	275	Krassilov & Rasnitsyn (1996)
First evidence of leaf mines and galls	Late Carboniferous	300	Scott *et al.* (1992)
Evidence from tree-ferns of insect feeding by piercing and sucking	Carboniferous	302	Labandeira & Phillips (1996)
First evidence of wood boring by insects	Early Carboniferous	330	Scott *et al.* (1992)
Earliest fossil insect (bristletail) with significant structural data	Early Devonian	400	Labandeira *et al.* (1988)

their activities is detectable in wood and leaf remains from the Carboniferous and Permian periods. Conventional wisdom suggests that insect species richness has increased predominantly because of the appearance and subsequent radiation of angiosperm plants, but examination of the numbers of new species in the fossil record suggests that the radiation of modern insects was not accelerated particularly by the expansion of the angiosperms in Cretaceous time. Instead, the basic 'machinery' of insect trophic interactions was in place very much earlier (Labandeira & Sepkoski 1993). One example from Nishida and Hayashi (1996) describes how fossil beetle larvae have been found in the fruiting bodies of a now-extinct gymnosperm from the Late Cretaceous period in Japan.

In 1973, Southwood described the habits of the major insect orders in terms of their main food supplies, thus defining their general trophic roles (Fig. 1.6). If it is assumed that each order has evolved but once, and that all species within the order are to some extent related so that they represent a common ancestral habit, then it can be seen that the major 'trophic roles' at order level are scavenging (detritivory) followed by carnivory. Herbivory (or phytophagy), feeding on living plants (or the dead parts of living plants such as heart wood), is represented by only eight major orders, suggestive of what Southwood called an 'evolutionary hurdle'. It is very significant, however, to realize that over 50% of all insect species occur within the orders that contain herbivores, suggesting that once the 'hurdle' was overcome, rapid and expansive species radiation was able to take place.

1.6 The numbers of insects: species richness

One problem that has at the same time excited and bewildered entomologists since the very earliest days has been the sheer number and variety of insects. Linnaeus named a large number of species of insect in the 18th century in the 10th edition of his *Systema Naturae*, probably thinking that he had found most of them. At that time, only a small percentage of the world had been explored by entomologists (or indeed, anyone else), so it is not surprising that the earliest workers had no concept of the diversity and richness of insect communities, even less the complexities of their interactions. Two hundred years later we are not very much closer to estimating the total number of insect species on the planet with any degree of accuracy, though we do have a better feel for the subject than did Linnaeus. We have now described more than 1.5×10^6 organisms in total, of which more than 50% are insects. Figure 1.7 illustrates the dominance of the Insecta in the list of described species (May 1992). It is revealing to

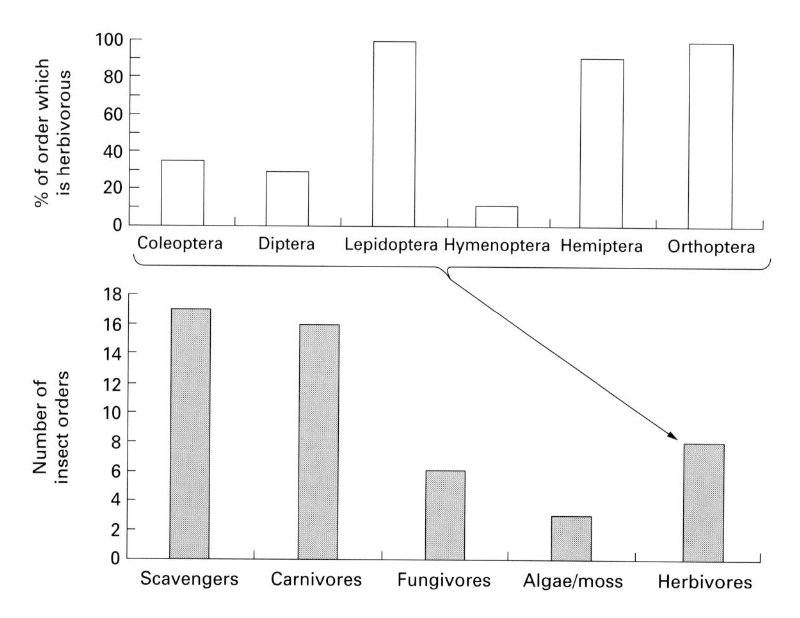

Fig. 1.6 Major feeding guilds in the Insecta, showing the importance of herbivory. (From Southwood 1973.)

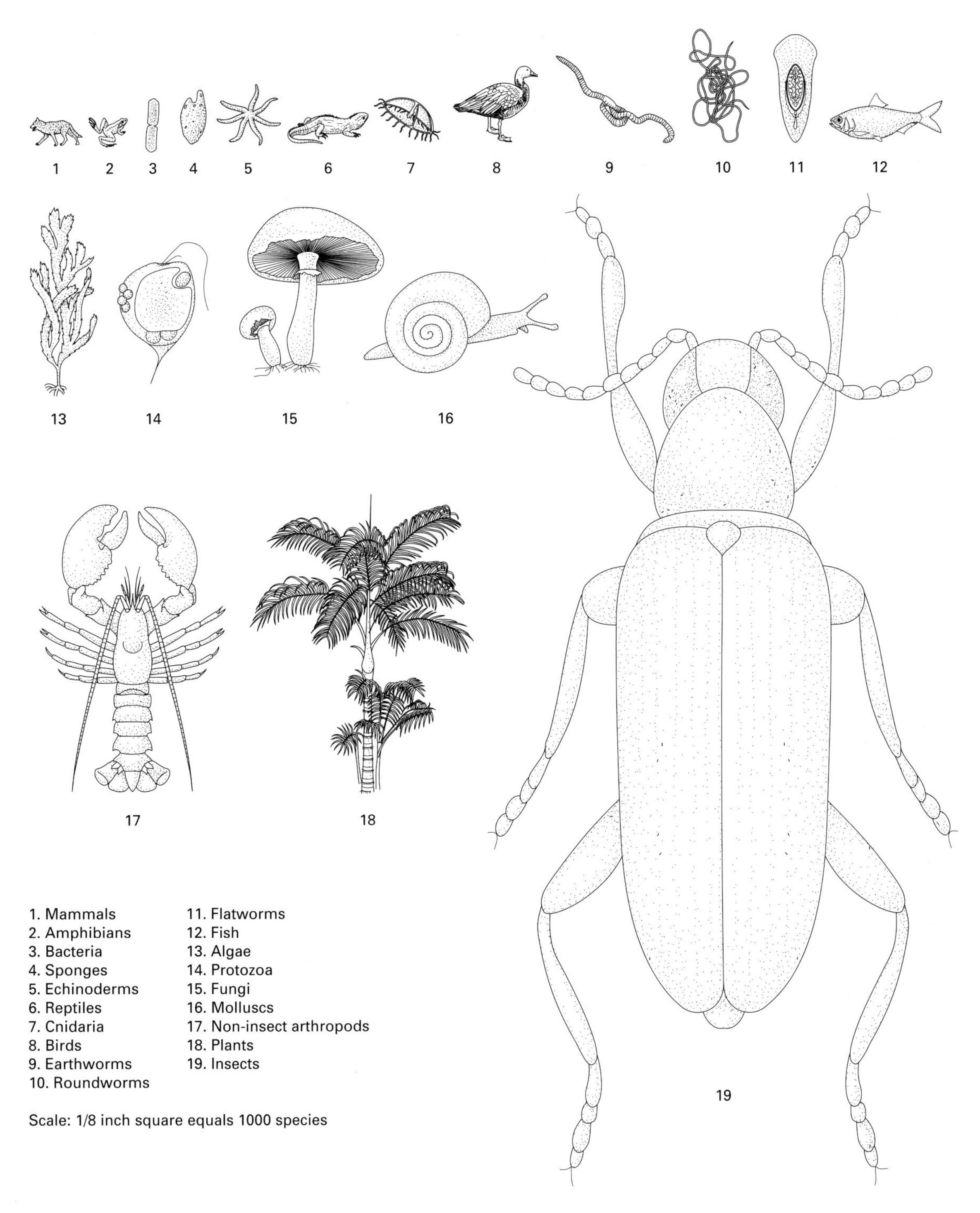

Fig. 1.7 Relative species richness of various animal and plant groups shown as proportional to size of picture. (From May 1992.)

compare the number of species of insects with those of vertebrates (merely a subphylum of the Chordata, when all said and done): around 47 000 insect species have been described, compared with a trivial 4000 or so for mammals. Furthermore, it is likely that the vast majority of vertebrate species have now been discovered. Insect species meanwhile continue to roll in apace. One problem with this is the distribution and abundance of taxonomists, rather than that of the species of other organisms they are dedicated to naming. Gaston and May (1992) have pointed out that there are gross discrepancies between both the number of taxonomists and the groups that they study, and also the regions of the world wherein they collect their animals. For every taxonomist devoted to tetrapods (amphibians, reptiles, birds and mammals), there is only 0.3 for fish, and around 0.02–0.04 for invertebrate species. To make matters worse, if published ecological papers are anything to go by, the study by Gaston and May suggested that 75% or so of authors come from North America, Europe and Siberia together, whereas the areas of supposed highest species richness, the humid tropics, are seriously understudied.

As Gaston and Hudson (1994) have discussed, the total number of insect species in the world is an important but elusive figure. Exactly why it is important is also rather elusive, and different scientists would argue differently as to why we need to know just how many species there are in the world. First, by invoking Gause's Axiom, which states that no two species may coexist in the same habitat if their niche requirements completely overlap, then by assessing the number of species in a given place, we can comment on the complexity of niche separation and thus the implied heterogeneity of the particular habitat. (Niche ecology is considered in Chapter 4.) Second, a great deal of concern is being expressed world-wide about the decline of natural habitats through anthropogenic activities such as the logging of tropical rainforests and the urbanization of temperate regions (see Chapter 8). One consequence of habitat decline is species extinction, and with insects, as indeed with all other organisms, it is impossible to assess extinction rates effectively until we know how many species exist in the first place. Certainly, we cannot give a percentage extinction figure unless we know how many species were there originally. Third, in applied ecology, it is important to predict stabilities of agro- or forest ecosystem, in terms of pest outbreak potential (see Chapter 10). If we are to rely more heavily in the future on population regulation by insect natural enemies, we need to inventory beneficial species in order to select targets for manipulation. One of the aims of evolutionary biologists is to explain the relationships between organisms, both extinct and extant, and the Insecta provide by far the most diverse group for this type of study. As all species have arisen through natural selection, a knowledge of the number of species must be an important base line for discussing their origins. Finally, the intrinsic wonder of insect life wherever we turn in the non-marine world must beg the question: 'Just how many are there?'

Different researchers have used different methods to estimate the number of species in the world. Table 1.4 summarizes some of these conclusions. There is a surprising variation in opinion, spanning a whole order of magnitude from 2 or 3 million to a staggering 30 million or even more, though as we shall see, these higher figures seem relatively implausible (Gaston *et al.* 1996).

The methods for arriving at these estimates vary considerably (May 1992), and are considered in detail in Chapter 8. Only a brief résumé is presented here. In the case of well-known animal groups such as birds and mammals, the tropical species so far described (probably most of them) are twice as numerous as temperate ones. If it is assumed that this ratio also holds true for insects (though this has not been tested), and that we have described the majority of temperate species of insect, then we reach an estimate of around 1.5–2 million species in total. As for all such estimates, some of the basic assumptions may be suspect. This theory assumes that insects with an average size orders of magnitude smaller than those of birds and mammals would show a proportional increase in species number as their tropical habitats diversify curvilinearly when compared with their temperate ones. Work in Borneo by Stork (1991) showed that the mean number of insect species in a rainforest canopy was 617 per tree, with one tree sample containing over 1000 species. This compares with the species numbers recorded from native British tree species, with an average of around 200 (Southwood 1961) (see below), showing that tropical species might outnumber their temperate counterparts by a good deal more than 2 : 1.

Taxonomic biases are also to be found within the Insecta. Some orders of insects are better known than

Table 1.4 Estimates of the numbers of insect species in the world.

Estimated number of insect species ($\times 10^6$)	Method of estimation based on (also see text)	Reference
2–3	Hemiptera in Sulawesi	Hodkinson & Casson (1991)
3	Nos of insects on each plant species	Gaston (1991)
3–5	Tropical vs. temperate species	May (1992)
6	Butterflies in UK	Hammond (1992)
≤10	Biogeographic diversity patterns	Gaston & Hudson (1994)
30	Beetle–tree associations	Erwin (1982)

others, and have been collected heavily from only certain regions of the world. Butterflies, for example, have fascinated amateur entomologists far longer than almost any other group of insects, and so we should by now know most of the temperate species within this order, and also have a fairly good notion of the tropical ones too. There are 67 species of butterfly in the UK, and between 18000 and 20000 world-wide. In total, we have 22000 or so British insect species encompassing all the orders, so by using the ratio of 67:22000 and applying it to the 20000 global butterfly species, we reach a grand total of around 6 million insect species world-wide. This estimate again, of course, relies on linear relationships of habitat diversity between temperate and tropical communities.

Perhaps the most extravagant estimate of insect species richness comes from Erwin (1982), who predicted a figure of 30 million species for tropical forests on their own. Amongst other crude assumptions, this estimate relies heavily on the concept that tropical insect herbivores are plant-species specialists, so that a 'one beetle one tree' system has to operate. However, it is now suspected that host-generalist rainforest insects are more common than might have been thought (Basset 1992; Williams & Adam 1994), so that rainforest trees may sustain herbivore faunas ranging from highly specialized to highly polyphagous.

1.7 Variations in species number

Merely attempting to estimate the total number of insect species in the world, though a worthwhile exercise, conceals a myriad of ecological interactions that influence the number of species found in any particular habitat. Many of these processes will be considered in detail in later chapters, but they are summarized here as an introduction.

1.7.1 Habitat heterogeneity

As stated above, if we invoke Gause's Axiom, or the Competitive Exclusion Principle, it is clear that homogeneous habitats with relatively few 'available niches' should support fewer species than heterogeneous ones. The latter are likely to allow for more coexistence of species by enabling them to partition resources within the habitat and hence avoid interspecific competition (see Chapter 4). From this, we may conclude that habitats that are more diverse in terms of greater plant species richness, for example, should exhibit greater insect species richness as a consequence. Undoubtedly, once Southwood's evolutionary hurdle of herbivory (see above) has been overcome, plants provide an enormous variety of new habitats and niches for insects (Fernandes 1994). Hutchinson's 'environmental mosaic' (1959) concept describes the system well, and mosaics of mixed natural habitats such as small-sized crops, fields and natural habitats in the same landscape maximize insect species richness and diversity (Duelli *et al.* 1990). Such habitat mosaics are also thought to decrease the probability of the extinction of rare species (see Chapter 8).

Early seral stages of succession, for example, tend to possess higher plant species richness than those close to climaxes, so that it might be expected that early succession is also typified by higher insect species richness as well. A study on bees in set-aside fields in Western Europe illustrates this phenomenon well (Gathmann *et al.* 1994). Various types of crops and fallow fields in an agricultural landscape were assessed for bee species richness, and it was found that habitats with greater floral diversity offered better and richer food resources for flower visitors. Set-aside fields that were mown, and hence reduced to early successional stages, showed a greatly increased plant species richness, coupled with double the species richness of bees when compared with unmown, late successional stage fields.

Finally, a complex derivation of plant species richness involves hybrid zones, areas of habitats where two or more plant species hybridize to produce a third, F1, phenotype. These areas seem to provide a great diversity of resources for herbivorous insects. In Tasmania, Australia, for example, where two species of eucalyptus hybridize, this area was found to be a centre for insect (and fungal) species richness (Whitham *et al.* 1994). In this study, out of 40 insect and fungal taxa, 53% more species were supported by hybrid trees than in pure (parent plant) zones. In a different study, Floate *et al.* (1997) found that not only were populations of leaf-galling aphids in the genus *Pemphigus* (Hemiptera: Aphididae) 28 times as abundant in hybrid zones of the host tree cottonwood (*Populus* spp.), but that, in the author's opinion at least, preserving such small hybrid zones could have a disproportionate beneficial role in maintaining insect biodiversity. However, the role of hybrid zones in determining insect numbers remains controversial (see Chapter 3).

It is not only herbivore communities that might be expected to show an increased species richness as their plant-derived habitat also becomes more species rich. Consumer trophic levels such as specialist predators or parasites can also be influenced in this way. Parasitoid guilds constitute an important part of the biodiversity of terrestrial ecosystems (Mills 1994), and because many of them are host specialists, high species richness of their insect hosts results in a similarly higher parasitoid richness. This system parallels that of plant–herbivore interactions well. Insect herbivores feed on many different parts of the plant (see below), so that one plant species can often support many insect species. Similarly, it is possible to recognize a series of guilds of parasitoids, each composed of various species, depending on the life stage of the host insect (Fig. 1.8).

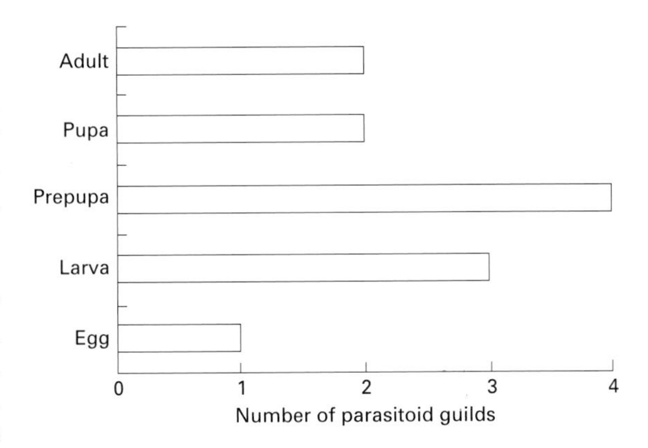

Fig. 1.8 Number of guilds of parasitoid insect attacking different life stages of insects. (From Mills 1994.)

1.7.2 Plant architecture

The term 'architecture' when applied to plants was coined by Lawton and Schroder (1977). It describes both the size, or spread, of plant tissues in space, and the variety of plant structures (Strong *et al.* 1984), from leaves to shoots to wood to roots. Figure 1.9 shows the great diversity of niches thought to be available for just one, albeit the largest, order of insects, the Coleoptera (Evans 1977). All imaginable parts of the tree are utilized by one species of herbivore or another. Clearly, different plant types vary widely in their structures, and thus architectures. Ferns are very much less complex than trees, and hence in terms of habitat heterogeneity, a tree would be expected to support more insect species than would a fern, everything else being equal (Fig. 1.10). If the ratios of species richness of herbivorous insects between more complex plant types or species and less complex ones are considered (Table 1.5) (Strong *et al.* 1984), it can be seen that, in all cases, the architecturally more complex plant has considerably more species associated with it. Summing over the ratios in the table, we can 'guesstimate' that a tree should have something like 50 times as many species associated with it than a monocotyledon, although this is probably somewhat of an overestimate. A most striking case compares Lepidoptera on trees and shrubs with those on herbs and grasses (Niemala *et al.* 1982), where there are over 10 times as many species on the former plant types than on the latter.

1.7.3 Plant chemistry

Chemicals in plants that may have an influence on the ecology of insects fall into two basic categories:

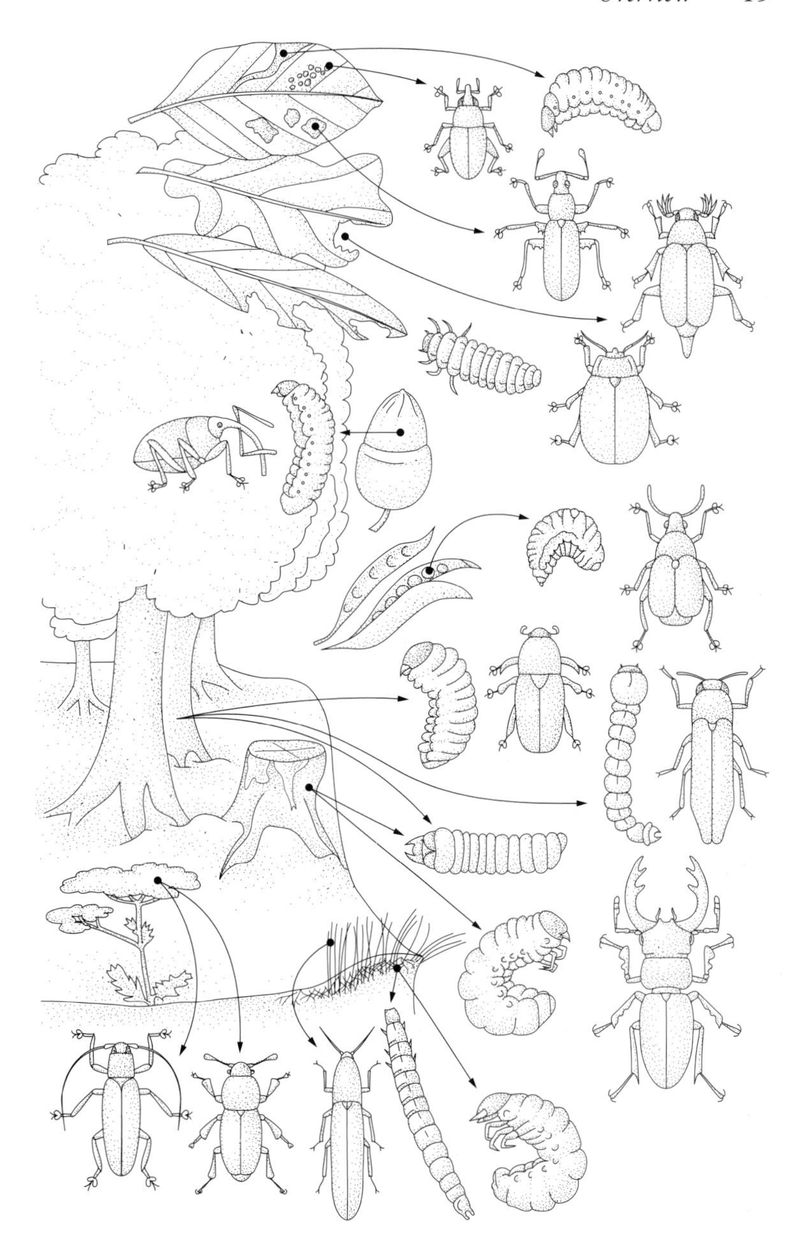

Fig. 1.9 An illustration of the diversity of beetles found living on a single tree. (From Evans 1977.)

food and defences. Insect herbivores are particularly limited by the suboptimal levels of organic nitrogen provided by plants as food (White 1978), and plant defences reduce the efficiencies of feeding processes even further. However, although plant chemistry influences the abundance of insects, it is difficult to detect a major influence on insect species richness. Jones and Lawton (1991) looked at the effects of plant chemistry in British umbellifers, and though there was thought to be some influence on species richness via changes in natural enemy responses to hosts or prey on biochemically diverse plants, they concluded that there was no evidence that plant species with complex or unusual biochemistries supported less species-rich assemblages of insects. Essentially then, though a 'niche' on a toxic or low-nutrient plant may be harder to utilize, the number of niches (i.e. habitat heterogeneity) remains the

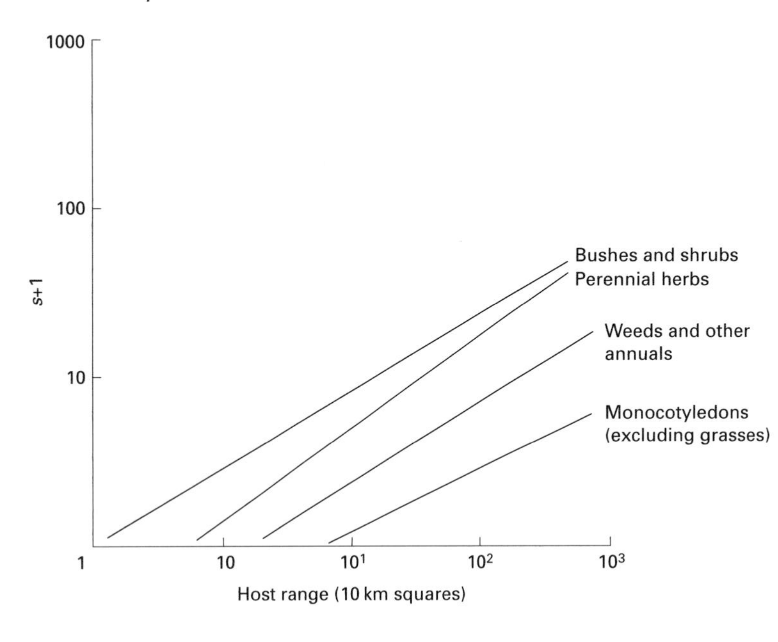

Fig. 1.10 Number of insect species on plants with different architectures and varying commonness. (From Strong *et al.* 1984.)

Table 1.5 Ratio of numbers of herbivore insect species on plants of different structures. (From Strong *et al.* 1984.)

More complex plant	Ratio	Less complex plant
Trees	27 : 1	Shrubs
Shrubs	21 : 1	Herbs
Bushes	1.3 : 1–25 : 1	Perennial herbs
Perennial herbs	1.5 : 1–27 : 1	Weeds
Weeds	1.3 : 1–29 : 1	Monocotyledons

same. This result is difficult to reconcile with the fact that most insect herbivores (about 80%) are specialists, unable to feed on plants with different chemistries. This is considered in detail in Chapter 3.

1.7.4 Habitat abundance (time and space)

For insects to adapt to a new habitat or resource of any sort, there must be sufficient opportunity for natural selection to do its job, and for selection pressure to 'push' a species into a new and stable form. This type of opportunity consists of either the regional abundance of the resource and/or how long in evolutionary time the resource has been available. The best illustrations of this principle again come from insects feeding on plants.

In the UK, both the abundance through time, measured as the number of Quaternary (last million years) remains, and the commonness, measured as the number of kilometre squares of the UK wherein the species of tree was recorded, have a significant positive influence on the numbers of insects on the trees, as noted by Southwood (1961) and Claridge and Wilson (1978). Both studies showed that the *Quercus* genus (oaks), which is both the most common and has been in the landscape for the longest time, shows the highest number of insects associated with it. Newly introduced tree species such as sycamore and horse chestnut have relatively few (see Chapter 8). In summary, there appears to be tremendous variation in numbers of insect species on British trees, all of which may be assumed to possess roughly the same architecture.

1.7.5 Habitat size and isolation

A fundamental concept in ecology is the equilibrium theory of island biogeography, first proposed by MacArthur and Wilson (1967). Simply put, the theory makes predictions about how island area and distance from a source of colonists affect immigration and extinction rates (Schoener 1988) (see also Chapter 8). The term 'island' can indicate an island in the true, geographical sense, but it can also mean any patch of relatively homogeneous habitat that is surrounded by a different one. The latter broad category could include woodlands in farmland, ponds in fields, or even one plant species surrounded by different ones. Theoretically, there should be a balance between the migration of species into an island or habitat patch and extinction rate, such that for a given size of island there is an equilibrium number of species present (Durrett & Levin 1996). The species–area relationship is firmly established in ecology (Hanski & Gyllenberg 1997), and it describes the rate at which the number of animal and/or plant species increases with the area of island available. The model is represented by the simple equation

$$s = a^z$$

where s is the number of species, a is the island area, and z is a constant.

There are a host of assumptions to the theory (see Chapter 8), including those that insist that habitat type, age and degree of isolation of 'islands' being compared are the same. There is some uncertainty about the existence of a maximum value for variability or heterogeneity of physical environments (Bell *et al.* 1993). It is possible that as islands become larger, their heterogeneity increases apace. However, field data fitted to the species–area model suggest that the curve will asymptote; in other words, the exponent value, z, is less than one. In fact, there are surprisingly few examples where z has been estimated with confidence for insects. Values range from 0.30 and 0.34 for ants and beetles, respectively, on oceanic islands (Begon *et al.* 1986) to 0.36 for ground beetles (Coleoptera: Carabidae) in Swedish wooded islands (Nilsson & Bengtsson 1988).

As mentioned above, the species–area model has some shortcomings (Williams 1995). There are no limits to the function that describes it, and the model is unable to handle zero values, so that if, by some quirk of fate, a particular island has no species representing the particular animal or plant group under investigation, then that island cannot be included in the model fitting. Despite these drawbacks, the island biogeography theory does fit observed data on occasion, but particularly few insect examples exist. One experimental illustration was carried out by Grez (1992), who set up patches of cabbages containing different numbers of plants, ranging from four to 225. Later sampling of the insect herbivores on and around the cabbages showed that species richness of insects was highest in the large host plant patches, which also showed an enhanced presence of rare or infrequent species within the general area, when compared with the smaller patches. In a field study, Compton *et al.* (1989) were able to detect a significant but rather weak tendency for larger patches of bracken (*Pteridium aquilinum*) to support more species of insect herbivores than smaller ones, in both Britain and South Africa. On the other side of the Atlantic Ocean, the Florida Keys have gained their fauna of longhorn beetles (Coleoptera: Cerambycidae) from both the islands of the West Indies, including the Bahamas and Cuba, and the mainland, represented by the peninsula of south Florida. Both species–area and species–distance relationships for cerambycids were found to conform to the island biogeographic theory (Browne & Peck 1996), even though the islands making up the Keys have been fragmented by rising sea levels for only the last 10 000 years or so.

The fragmentation of habitat islands may be a natural event over geological time, as evidenced by the Florida Keys above, but many aspects of human activity, such as logging and road building, over a very much shorter time scale result in fragmentation as well (Plate 1.2, opposite p. 158) (see Chapter 8). This produces a further decrease in 'island' size and an increase in isolation. Animal and plant populations remaining in these fragments constitute metapopulations, that is, local populations undergoing constant migration, extinction and colonization on a regional rather than global scale (Husband & Barrett 1996). A British example of a metapopulation study is that by Hill *et al.* (1996) on the silver spotted skipper butterfly, *Hesperia comma* (Lepidoptera: Hesperiidae). The larvae of this butterfly are grass feeders, preferring short swards where the conditions are warm. Patches of this type of habitat occur on chalk downland in the south of England, and studies of the metapopulation dynamics of the species showed that habitat patches were more likely to be colonized if they were relatively large and close to other large, occupied patches. Furthermore, adult butterflies were more likely to move between large patches close together, whereas local populations in small isolated patches were more likely to go extinct. The consequences of metapopulation dynamics for wildlife conservation are discussed in Chapter 8, but the importance of habitat size and location relative to other patches in the maintenance of species richness can already be seen.

1.7.6 Longitude and altitude

On a local scale, insect species richness may still vary even when habitat heterogeneity and abundance, or patch size are constant. These variations would seem to be influenced by environmental factors such as latitude, longitude and altitude, presumably via climatic interactions. Several examples describe how insects in a particular locale are most species rich at a certain height above sea level. Libert (1994) found that many species of butterfly observed in Cameroon, West Africa, exhibited a preference for the tops of hills, whereas in Sulawesi, Indonesia, hemipteran communities showed highest species richness at elevations between 600 and 1000 m (Casson & Hodkinson 1991). The explanations of these observations are no doubt complex, and detailed microclimatic and vegetational data would be required to investigate causality. Weather conditions can influence insect species richness. In tropical Australia, leaf beetle (Coleoptera: Chrysomelidae) communities were found to be most species rich during the hottest and wettest times of the year (December to March) (Hawkeswood 1988). It can often be difficult in ecology to assign cause and effect with confidence, but various effects of weather and climate on insect ecology are considered in detail in Chapter 2. Suffice it to say for now that, in this example, leaf beetle richness was not correlated with plant richness, though it might be expected that these climatic influences would operate at least in part via the host plants of these totally herbivorous species.

On a larger geographical scale, it is usual to expect that insect species richness will decrease towards the poles. In North America, of 3550 species studied, 71% occurred south of a line along state boundaries from the Arizona–California to Georgia–South Carolina borders (Danks 1994), whereas in Scandinavia, out of a wide range of insect families examined, the number of species was generally highest in the southern provinces, declining to the north and north-west (Vaisanen & Heliovaara 1994). As always in insect ecology, exceptions to neat rules crop up. Also in Finland, Kouki *et al.* (1994) showed that the species richness of sawflies (Hymenoptera: Symphyta) showed an opposite latitudinal trend, so that species richness is highest in the north, not the south. Kouki *et al.* explained this by the fact that the principal host plant group, willows (*Salix*), are also most species rich further north.

1.8 The number of insects: abundance

Species richness is only one measure of the success of insects; their abundance is another equally important one. No-one could deny that a plague of locusts, although very species poor, is the epitome of success, at least as judged by the number of individuals and their rapacious ability to devour plant material of many kinds. Unfortunately, it is rather more difficult to obtain reliable estimates of the number of individuals (abundance) of a species than it is merely to score the species present in or absent from a habitat (species richness). Common but immobile or concealed species such as aphids or soil and wood borers may be underestimated or even completely overlooked by most ecological sampling systems.

None the less, numerous attempts have been made to estimate the numbers of individual insects either globally or locally, and the literature is full of statistics describing how many insects are to be found in certain habitats. Table 1.6 shows some of these. Reviewing the table does not, in fact, reveal any constancy, nor does it tell much of a story in most cases. The numbers of army worm, *Spodoptera frugiperda*, in Texas appear to be enormous (Pair *et al.* 1991), until the size of the area in which they were estimated is brought into the equation. This huge abundance represents an average of only four adults emerging from pupae in the soil per square metre of corn crop, which seems to be a much less impressive figure. However, adult *Spodoptera* may be carried over 1000 km on the wind (see Chapter 2) to be deposited on distant crops where resulting larvae of the next generation are likely to cause huge economic losses if not controlled. Surprisingly perhaps, it is only when we look very closely at tiny and, in most people's perceptions, inconsequential, insects in novel habitats that we find just how common some groups are. An example of booklice (Psocoptera) (Fig. 1.11) from Norway spruce canopies in lowland England shows how immensely abundant they are in a seemingly sterile habitat (Ozanne *et al.* 1997). Using insecticidal mist-blowing of tree canopies, an astonishing 6500 per m^2 were collected. As a side issue, this example illustrates the importance and magnitude of detritivore pathways in some ecosystems. Booklice feed mainly on fungi, lichens and fragments of leaf and bark material in dark tree canopies, and are likely to form the staple diet of a myriad of predatory insects and other arboreal arthropods.

Table 1.6 The scale of insect population size; examples from the literature (all insects adult unless otherwise stated).

Insect	Country or region	Site	Estimated number (max.)	Size of plot	Per unit area (per m²)	Reference
Aedes triseriatus (Diptera: Culicidae)	USA	Scrap tyreyard	4 492	0.03 ha	14	Pumpuni & Walker (1989)
Archanara geminipuncta (Lep.: Noctuidae)	—	Reed beds	180 000			Tscharntke (1992a)
Ceutorhynchus napi (Col.: Curculionidae)	—	Rape field	750 000	5 ha	15	Debouzie & Ballanger (1993)
Collembola	India	Forest soil	14 160	1 m²	14 160	Vats & Narula (1990)
Culex tarsalis (Diptera: Culicidae)	California	Marsh	914 000			Reisen & Lothrop (1995)
Culex tritaeniorhynchus (Diptera: Culicidae) larvae	S. Korea	Rice field	14 900	1 m²	14 900	Baik & Joo (1991)
Deraecoris nebulosos (Heteroptera: Mridae)	Mississippi	Cotton	137 000	1 ha	14	Snodgrass (1991)
Hylesinus varius (Col.: Scolytidae)	Spain	Olive grove	147 700	100 ha	0.15	Lozano & Campos (1993)
Operophtera brumata (male) (Lep.: Geometridae)	Hungary	Oak woodland	7 200	1 ha	1	Ambrus & Csoka (1992)
Plutella xylostella (Lepidoptera: Plutellidae)	S. Africa	Cabbages	36	1 plant		Dennill & Pretorius (1995)
Semenotus japonicus (Col.: Cerambycidae)	Japan	Cedar plantation	34 000	1 ha	3	Ito & Kobayashi (1991)
Soil litter arthropods	India	Broadleaf forest	51 484	1 m²	51 484	Vats & Handa (1988)
Spenonomus spp. (Col.: Bathsciinae)	Belgium	Cave system	44 000			Tercafs & Brouwir (1991)
Spodoptera frugiperda (Lep.: Noctuidae)	Texas	Corn	7 940 000 000	200 000 ha	4	Pair *et al.* (1991)
Spodoptera frugiperda (Lep.: Noctuidae) larvae	Louisiana	Corn	300 000	1 ha	30	Fuxa (1989)

Fig. 1.11 Booklice (Pscoptera).

The densities of insects that commonly act as disease vectors, such as mosquitoes, tsetse flies and aphids, are especially significant when it is considered that these can be low-density pests, where only a very few (as few as one) individuals are required to pass on diseases such as malaria, sleeping sickness or potato leaf curl (see Chapter 9). Although it is one thing to wonder at the sheer numbers of insects in a locust swarm, it is quite another to appreciate the potential for harm inherent in one small individual.

1.8.1 Variations in insect numbers

Of much more fundamental interest to ecologists is the manner in which insect population densities vary within a population through time. Some of the most basic ecological processes have been explored in response to such variations, and the explanation of the patterns and processes in population density changes has taxed ecologists the world over, and still does. Large variations in the population densities of insects occur under two different headings: population cycles and population eruptions (Speight & Wainhouse 1989). Cycles are periodic, with some degree of predictability about the time between peak numbers; high densities are rapidly followed by large declines. Eruptive populations, on the other hand, often remain at low densities for long periods before outbreaks occur suddenly and often unexpectedly. Once such outbreaks have developed, they may be sustained for some time.

1.8.2 Cycles

Most cycles of insect populations have been observed in species that inhabit perennial, non-disturbed habitats such as forests. Figure 1.12 shows a typical population cycle of a forest insect, the larch budmoth, *Zeiraphera diniana* (Lepidoptera: Tortricidae) (Baltensweiler 1984). This species shows remarkably regular cycles of abundance in the Engadine Valley in Switzerland, where outbreaks have a periodicity of 8 or 9 years (Plate 1.3, opposite p. 158). Mean population density may vary 20000-fold within five generations, though 100000-fold increases have been observed locally. Many other forest insects show similarly predictable cycles; the Douglas fir tussock moth, *Orgyia pseudotsugata* (Lepidoptera: Lymantriidae), has exhibited regular outbreaks at 7–10 year intervals in British Columbia since the first recorded observations in 1916 (Vezina & Peterman 1985). Unless climatic patterns are themselves cyclic (Hunter & Price 1998), most insect population cycles can usually be attributed to biotic interactions, such as competition and predation, that have a delayed action on population growth rates. Delayed density dependence will be considered in detail in Chapter 5, and is introduced below. Population cycles are rarely as neat as in the case of *Zeiraphera*, and long-term examinations of abundance data for insect species, though revealing some degree of cycling, suggest that the levels of peaks and troughs are less regular or predictable.

1.8.3 Regulation

In ecological terms, we recognize that population cycles in insects are at least partially under the influence of a process known as regulation. Regulation describes the way in which a population's abundance varies through time as a decrease in population growth rate as population density increases (Fig. 1.13). Declines in population growth rates with density can be manifested by (a) increases in the rate (proportion) of mortality that the population suffers, (b) decreases in birth rate, (c) increases in emigration rates, or (d) decreases in immigration rates. When

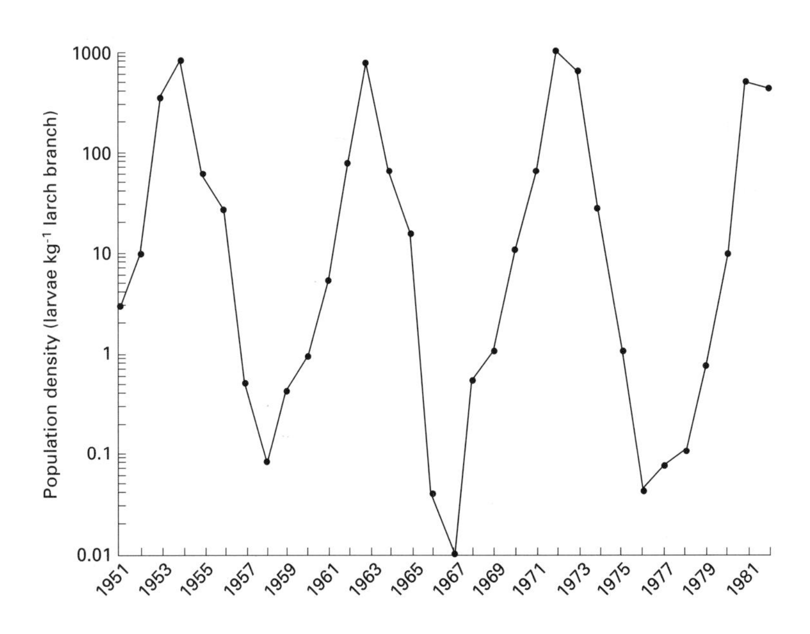

Fig. 1.12 Population cycles for *Zeiraphera diniana* on larch in Switzerland. (From Baltensweiler 1984.)

birth, death or movement rates vary with density (i.e. they are density dependent), they have the potential to maintain an insect population around some equilibrium density. If the density dependence occurs on a time delay, the population can overshoot this equilibrium, and exhibit cyclic behaviour. In general, the higher the insect's fecundity, or the longer the time lag, the more dramatic the oscillation will be.

Various biotic factors are known to be potentially regulatory (i.e. may act in a density dependent fashion), and each is considered in detail in later chapters. However, it is useful to separate such factors into those that act within a trophic level, such as competition (see Chapter 4), and those that act between trophic levels, either from below (so-called 'bottom-up'), via food supply, or from above (so-called 'top-down'), via the action of natural enemies such as predators, pathogens or parasitoids (see Chapter 5).

In reality, these factors interact, so that, for example, competition often acts via the amount of food available for individual insects. Experimental studies frequently demonstrate this sort of competition. Cycles in populations of Indian meal moth, *Plodia interpunctella* (Lepidoptera: Pyralidae), for example, in laboratory containers appeared to be caused by density dependent competition for food amongst the larvae (Sait *et al.* 1994), but field conditions are, of course, likely to be more complex. Though the cinnabar moth, *Tyria jacobaea* (Lepidoptera: Arctiidae), suffers periodic crashes in abundance because of competition for larval food (Van der Meijden *et al.* 1991), recovery after a crash is still delayed even with food available, presumably because of the activities of natural enemies and reduced food quality. Clearly, competition for food might be something to avoid if possible. Various behavioural mechanisms have evolved that minimize competition, from eating siblings in tropical damselflies (Fincke 1994), to avoiding oviposition on host plants already bearing eggs in butterflies (Schoonhoven *et al.* 1990), or by selecting different types or size of food when competition becomes intense, as in stoneflies (Malmqvist *et al.* 1991).

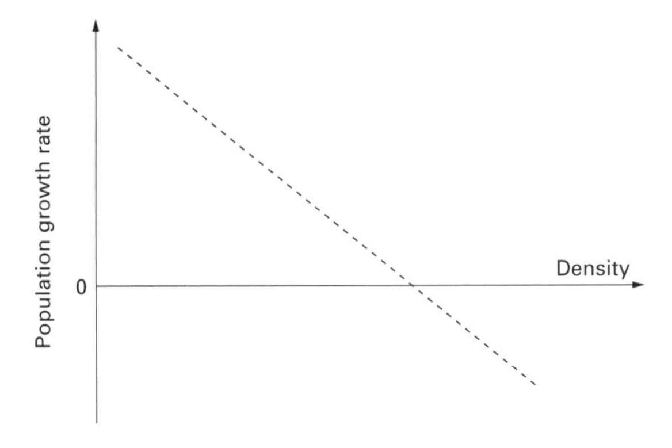

Fig. 1.13 Theoretical relationship between population density and growth rate under the influence of density dependent factors.

The role of natural enemies in the population regulation of insects is discussed in full in Chapter 5. Classic work by Varley and Gradwell (1971) has shown that, as herbivore populations vary with time, so do the abundances of various predators and parasitoids that feed on them (Fig. 1.14). *Philonthus decorus* (Coleoptera: Staphylinidae), is a predatory rove beetle which eats the pupae of winter moth, *Operophtera brumata* (Lepidoptera: Geometridae), in the soil, whereas *Cratichneumon culex* (Hymenoptera: Ichneumonidae) is a parasitic wasp, and *Lypha dubia* and *Cyzenis albicans* are both parasitic flies (Diptera: Tachinidae), all of which attack the larval stages whilst feeding in the canopy. Two points are worthy of note. First, the peaks of all the enemies seem to be 1 year (or in the case of *Cyzenis*, 2 years) later than that of the winter moth, illustrating the phenomenon of delayed density dependence described above. Second, it may be tempting to attribute the changes in winter moth abundance to regulation from its enemies. As moth larvae population increases from year to year, so do the levels of predation and parasitism, thus knocking the herbivore numbers back again. In this instance, however, only pupal predation by *Philonthus* was shown to be density dependent and hence regulatory (Varley & Gradwell 1971). The other enemies, in fact, were merely tracking the variations in the host, with no regulatory impact. These host variations relate to nutrient and defence quality and quantity, and will be discussed in Chapter 3.

Many pathogens are now known to act as regulators of insect populations, with fungi, bacteria and especially viruses having a great impact on occasion. Forest insects known to be at least partially regulated by naturally occurring viruses include nun moth, *Lymantria monacha* (Lepidoptera: Lymantriidae) (Bakhvalov & Bakhvalova 1990), rhinoceros palm beetle, *Oryctes rhinocerus* (Coleoptera: Scarabaeidae) (Hochberg & Waage 1991), and browntail moth, *Euproctis chrysorrhoea* (Lepidoptera: Lymantriidae)

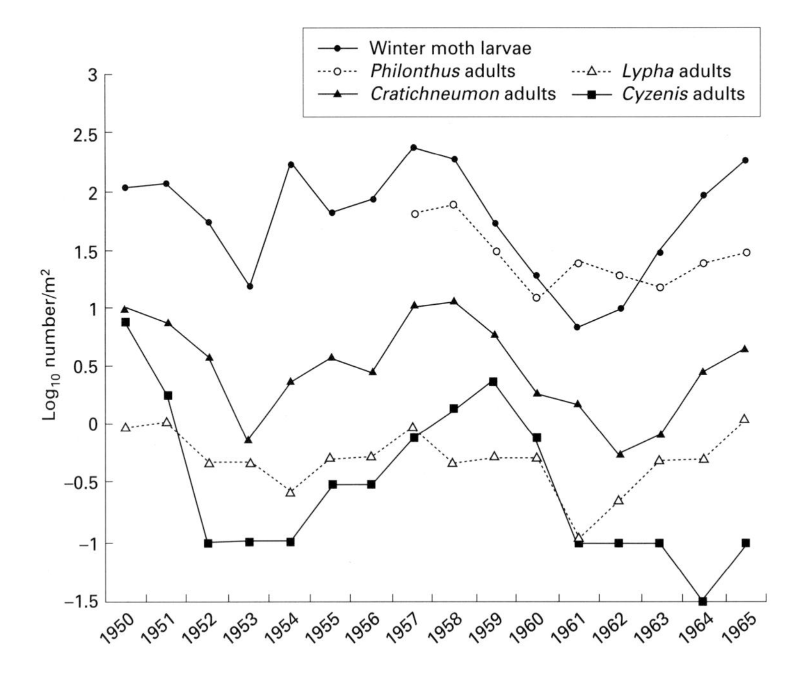

Fig. 1.14 Densities of winter moth larva and its natural enemies in Wytham Wood, Oxfordshire. (From Varley & Gradwell 1974.)

(Speight *et al.* 1992). Chapter 10 provides more details.

The host plant may also influence the density of herbivorous insects via a factor or factors that cause feedbacks with time lags of suitable duration (Haukioja 1991). One example involves the autumnal moth, *Epirrita autumnata* (Lepidoptera: Geometridae) in northern Fennoscandia. Large-scale defoliation of the host tree, mountain birch, induces changes in the food quality of foliage produced subsequently, such that the reproductive potential of the moth is significantly reduced (see also Chapter 3). So for several years following outbreaks, insect populations are suppressed. However, *Epirrita* larvae also feed on apical buds of birch, which causes a change in plant hormone balance resulting in luxuriant growth of new leaves. This new foliage is particularly suitable for herbivores. Thus the insect population begins to build up again (Haukioja 1991). These complex reactions to herbivore density result in statistically significant cycles of the autumnal moth of 9–10 years duration.

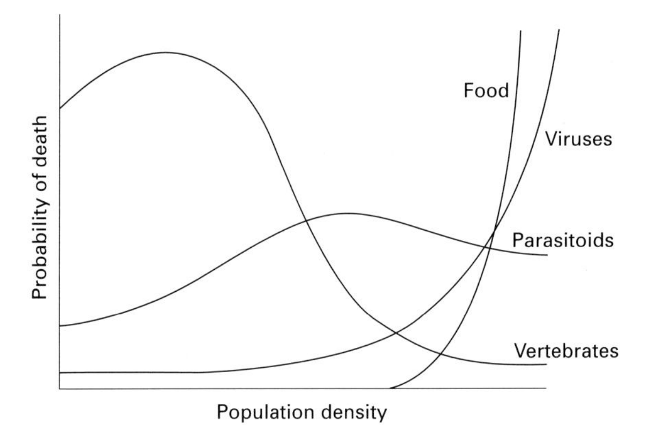

Fig. 1.15 Relative importance of various mortality factors on insect populations at different densities. (From Berryman *et al.* 1987, after Campbell 1975.)

Controversy has raged for years concerning the relative importance of natural enemies vs. resource (e.g. food) limitation in the population ecology of insects, and other animals. For now, it is clear that there are no easy answers. The science and practice of biological pest control relies on the ability of predators, parasitoids and pathogens to regulate pest populations, whereas the ecology of pest outbreaks, and its links to crop husbandry of all types, emphasizes resource limitation as the driving force in insect epidemiology. Figure 1.15 summarizes the potential relationships between various potential mortality factors, and suggests that their relative importances vary as the population of the insect on which they act varies (Berryman 1987). In general terms, the figure suggests that regulation from natural enemies such as predators and especially parasitoids might be expected to be effective up to a medium host or prey density, but at high, epidemic levels, regulation via pathogens such as viruses, if present, is more important. Food limitation is often the most important regulatory factor in high-density insect populations. These controversies are of fundamental relevance to insect pest management (see Chapter 10).

1.8.4 Eruptions

Eruptive outbreaks can develop when environmental changes, such as consecutive seasons of favourable weather acting directly on the insect or indirectly via its food supply, permit the rapid growth, dispersal and/or reproduction of the insect population in question. In essence, regulation at low density is lost because of environmental conditions, and unpredictable outbreaks result.

Locusts are a classic example of this phenomenon. Globally, many species of locust are known to erupt into plagues from time to time, the most common being the desert locust, *Schistocerca gregaria*, the migratory locust, *Locusta migratoria*, the tree locust, *Anacridium melanorhodon*, and the Australian plague locust, *Chortoicetes terminifera* (all Orthoptera: Acrididae) (Wright *et al.* 1988; Showler 1995). Various species of *Schistocerca* are also serious pests in South America (Hunter & Cosenzo 1990).

Locust plagues (Fig. 1.16) have been ravaging crops for many centuries. Swarms of locusts invading southern Europe from Africa were described in Roman times by Pliny the Elder, and countries from Hungary to Spain were particularly badly attacked in the 14th, 16th and 17th centuries (Camuffo & Enzi 1991). In 1693, for example, a swarm of *L. migratoria* built up on the north-western shore of the Black Sea and between the rivers Danube and Theiss. Some of the swarm invaded the Tyrol, but most entered Austria via Budapest. Some ended up in Czechoslovakia and Poland, whereas others headed west into Germany. A few individuals actually reached the British Isles (Weidner 1986). Locust eruptions are, of course, still occurring. In the 1980s, for example, devastating plagues occurred in Algeria (Kellou *et al.* 1990), Argentina (Hunter & Cosenzo 1990), Australia (Bryceson 1989), Peru (Beingolea 1985), Tchad (Ackonor & Vajime 1995), China (Kang *et al.* 1989), the Arabian Peninsula (Showler & Potter 1991), and Sudan (Skaf *et al.* 1990). In the 1990s, successive generations of locusts gave rise to localized eruptions for 18 months as far west as Mauritania in West Africa and as far east as India (Showler 1995). The damage caused by these eruptions is enormous, as is the cost of control. During one desert locust plague in Africa, 1.5×10^7 L of insecticide were used, at a cost of around US$200 million (Symmons 1992). Not only that, but locust plagues can adversely affect other animal populations. In central Saudi Arabia, many species of birds adapted to feeding on grasses and seeds were recorded inhabiting grassland and savanna areas (Newton & Newton 1997). A plague of desert locusts in spring and summer 1993 combined with poor spring rains to reduce the bird populations to the lowest numbers and species diversity recorded during the 28 month study.

Fig. 1.16 Swarm of immature locusts in Mauritania, W. Africa. (Courtesy of M. de Montaigne, FAO.)

The origin of locust swarms and their subsequent migrations are basically controlled by climatic factors (Camuffo & Enzi 1991) (see Chapter 2). Locust eggs may stay dormant in soil for months, waiting for random events to bring rain to their area. Depending on the type of soil, eggs survive best during rains (Showler 1995), and rain also promotes the growth of vegetation on which the hatching nymphs, or hoppers, can feed (Hunter 1989; Phelps & Gregg 1991). The 'El Niño' phenomenon is a southward-flowing ocean current off the coasts of Peru and Ecuador. Apparently cyclical changes in the pattern of its flow are the cause of environmental and climatic disturbances that cause widespread damage every few years. In Peru in 1983, El Niño caused up to 3000 mm of rain in an otherwise arid region, resulting in enormous locust swarms (Beingolea 1985). In fact, the detection by satellite imagery of new areas of green vegetation caused by rain is one of the most important tools in international locust plague monitoring programmes (Bryceson 1990; Hielkema 1990). Once the nymphs have depleted local resources, the migratory phase of a locust plague ensues, whose scale and direction is mainly dependent on winds (Symmons 1986; Camuffo & Enzi 1991). Under the right weather conditions, plagues can last for months or even years. Eventual declines occur because of extremely dry conditions (Wright *et al.* 1988), or drops in temperature (Camuffo & Enzi 1991). It can be seen therefore that population eruptions as typified by locust plagues are rarely if ever under tight regulation by density dependent factors. Rather, density independent factors such as climate are the most influential (see also Chapter 2).

1.9 Insects and humans

So far, we have briefly considered the success of insects as a function of their enormous species richness and their huge number of individuals. Finally, we need to introduce the all-important associations that humans have with insects, to see how the most successful group of animals interact, for better or for worse, with ourselves. In this way, we move into the realm of applied ecology.

1.9.1 Pests

The term 'pest' is entirely anthropocentric and subjective. It tries to attach a label to an organism that, via its ecological activities, causes some sort of detriment to humans, their crops or livestock. Ecologically, an insect pest is merely a competitor with humans for another limited resource, such as a crop. A crop, after all, is merely a rather special type of host plant community from the viewpoint of a herbivorous insect. It is only because humans planted the crop for their own uses that this herbivore then assumes pest status.

Crop losses on a global scale are rather difficult to assess, but clearly, insects can have an enormous detrimental impact on humans and their activities. Crop losses averaged from many published studies reach almost 45% of total yield annually, a colossal loss in food and other products. Let us take cotton as an example. This is one of the most important cash crops, with the capacity to earn international money for many countries, both developed and developing. It is grown on a colossal scale; the world total cotton production for the 1992–93 period exceeded 20×10^6 tonnes, harvested from an area approaching 34×10^6 ha of farmland (Luttrell *et al.* 1994) (Fig. 1.17). Despite long and bitter experiences of the development of high levels of resistance in insect pests to insecticides over the years, cotton growers still rely predominantly on this method of pest management (see Chapter 10). In the USA, for example, despite an average of six treatments with insecticide per growing season, nearly 7% of the entire crop is still lost to insect damage at one stage of growth or another (Luttrell 1994). Australia seems even worse. Here, an average of 10 or even 12 applications of insecticide are carried out per year (Luttrell *et al.* 1994), with an estimated cost in 1992 of A$90 million (£45 million) (Fitt 1994).

The previous examples involved developed countries. Even more significant perhaps is the fact that developing countries may not have the infrastructure, knowledge or technology to manage their pests effectively or safely. In Rwanda, as just one example, a country noted for its extreme socio-economic and political difficulties, insect pests of various types attack common beans, a vital crop for subsistence farmers. Trutmann and Graf (1993) reported losses of between 158 and 233 kg/ha, which is equivalent to a national loss to Rwanda of dry beans worth somewhere in the region of an amazing US$32.7 million (£22 million) per year!

In fact, as the 20th century comes to a close, we are

Fig. 1.17 Cotton crop. (Courtesy of C. Hauxwell.)

no closer to removing these threats. Instead, we are constantly attempting to develop new and often highly sophisticated techniques to combat insect pests, based these days on a sound knowledge of the target insect's ecology, how it interacts with its environment, and the influence of other organisms, natural enemies in particular. This is the subject of Chapter 10.

Insect pests also interact directly with humans via stings, rashes and more serious medical conditions. The tiny hairs on the larvae, pupae and egg masses of browntail moth, *Euproctis chrysorrhoea* (Lepidoptera: Lymantriidae) cause very serious reactions in people in the UK, from serious rashes (urticaria), to temporary blindness and even death via anaphylactic shock (Sterling & Speight 1989) (Plate 1.4, opposite p. 158). The pine processionary caterpillar, *Thaumetopoea pityocampa* (Lepidoptera: Lymantriidae) causes similar dermatitic and conjunctivitic reactions in Continental Europe (Lamy 1990), and contact with *Lonomia achelous* (Lepidoptera: Saturniidae) larvae is known to bring about haemorrhagic diathesis, clinical bleeding with reduced blood clotting, which, on rare occasions, can prove fatal (Arocha-Pinango *et al.* 1992). Bee and wasp stings are, of course, a regular occurrence, but with the spread of Africanized bees to the warmer parts of the USA, for example, worries are increasing about the potential medical effects of multiple stings (Schumacher & Egen 1990).

1.9.2 Vectors

A particular type of insect pest is one that is able to carry diseases from one mammalian host to another, or from one plant to another. These hosts may both be human, or one may be another mammal such as another primate or a rodent. They may be wild plants that provide reservoirs of diseases for infection of crops, or they may both be crops. Both local and global epidemics are vectored in this way by insects; the rising importance or malaria, sleeping sickness, plague, encephalitis and so on illustrate the vital need to explore the intimate ecology of insect–disease associations, in attempts to reduce the colossal and direct impact on human lives. Malaria, for example, is estimated to infect around 300 million people in the world (Collins & Paskewitz 1995), with over 80% or so of cases occurring in sub-Saharan Africa (Torre *et al.* 1997). Such problems may appear to those living in developed nations as irrelevant or remote. However, human suffering on such a large scale must impinge on all our lives. A steadily increasing number of Europeans, for instance, are treated in hospital after returning from trips to tropical countries where the disease is rife (see Chapter 9).

So important are tropical diseases becoming, that in early 1996, the US Congress approved over US$50 million for research on disease transmission in tropical countries. Though the Diptera dominate in terms of the number of human and livestock diseases with which they are associated, other orders such as the Hemiptera (true bugs) and the Siphonaptera (fleas) also have an enormous impact. Bubonic plague, the 'Black Death', which is carried by fleas, is thought to have killed around a third of the entire population of Britain in a pandemic that first appeared in England in 1348 (Kettle 1984).

Insects also vector numerous pathogenic organisms that cause extremely serious diseases in annual and perennial plant species. The list of major problems includes Dutch elm disease, barley yellows virus and potato leaf curl, and insects range from Hemiptera such as aphids and hoppers, to Coleoptera (beetles). The details of the ecology of insect–pathogen–host associations are considered in Chapter 9.

1.9.3 Beneficials

Insects that in some way damage ourselves or our livelihoods tend to be uppermost in peoples' emotions, and though we are also familiar with bees and their activities, a very large number of insect species that are important to us are largely ignored. Chapter 10 describes the vital importance of predatory and parasitic insects in the ever-growing field of biological pest control, and the revenue from diverse systems such as pollination of crops and the silk industry is staggering. In Italy, for instance, the value of crop pollination by insects is estimated at around 2000 billion lire (US$1.2 billion). The profit from honey, wax and other beekeeping products is about 30 billion lire (US$1.8 million) (Longo 1994). In the USA in recent years, two parasitic mites have caused drastic declines in feral honeybee populations, and crop losses associated with a consequent reduction in pollination run into billions of dollars. Silk has been produced from silkworms (Lepidoptera: Saturnidae) commercially for many hundreds of years, and is now estimated to

Table 1.7 Nutritional value of insects used as food in rural Mexican communities. (From Ramos-Elorduy *et al.* 1997.)

Dietary component	Content or range (%)	Insect with highest dietary content
Dry protein	15–81	Wasp larvae
Fat content	4.2–77.2	Butterfly larvae
Carbohydrate	77.70	Ants
Essential amino acids	46–96*	—
Protein digestibility	76–98	—

* Percentage of total requirements.

be worth US$1200 million globally. On a much broader scale, the roles played by a myriad of insects in food webs are of supreme importance to the functioning of a very large number of terrestrial and non-marine aquatic ecosystems. The lives of many bird species, such as blue and great tits in an English oakwood, for example, are utterly dependent on a plentiful supply of lepidopteran larvae when the chicks are in the nest, and partridges in farmland have a similar reliance on insect food for their young.

1.9.4 Aesthetics

Most humans, if they have any interest at all in wildlife, will tend to think about higher vertebrates such as birds and especially mammals when it comes to considering the aesthetics of animals around them. The majority of conservation systems in the world are still heavily biased towards this minor subphylum, but the value of insects for amenity and as natural and important components of habitats is increasing. Conservation projects are now directed on occasion at targets such as butterflies or dragonflies, which have really no economic value; they are encouraged for their own sake. The ecology of insect conservation and augmentation is considered further in Chapter 8. Insects can also take part in education, and they are considered to be particularly appropriate to interest and excite pre-college students (Matthews *et al.* 1997). The handling, rearing and simply admiration of moths and butterflies, beetles, stick insects and mantids, cockroaches and locusts can play a very significant role within an educational framework, even beginning at primary school level.

1.9.5 Food

To a surprisingly large number of people in the world, insects provide a vital source of protein and other nutrients. Insects are undoubtedly very useful nutritional sources. Table 1.7 provides some details of the nutritional value of a variety of insects used by rural people in the state of Oaxaca in Mexico (Ramos-Elordy *et al.* 1997). In some regions, insects are eaten every day as part of the staple diet. They may be roasted, fried or stewed, usually as larvae. Ants, bees and wasps are apparently the most popular, though as the table suggests, butterfly caterpillars (*Phasus triangularis*) are full of fat, and also high in calories.

Deep-fried grasshoppers can be bought on street corners in Bangkok, Thailand (Plate 1.5, opposite p. 158), and the Tukanoan Indians of the north-west Amazon eat over 20 species of insect, the most important being beetle larvae, ants, termites and caterpillars (Dufour 1987). Insects provide up to 12% of the crude protein derived from animal foods in men's diets, and 26% in those of women. In Irian Jaya (Indonesia), the Ekagi people regularly eat large species of cicada (Hemiptera: Cicadidae) (Duffels & van Mastrigt 1991), and the mopane worm, the larva of the mopane emperor moth, *Imbrasia belina* (Lepidoptera: Saturniidae) in southern Africa has become a cash 'crop', with an annual production of nearly 2000 tonnes. Ironically, it is possible that conservation projects attempting to boost populations of large grazing mammals in game reserves in the region may deprive locals of this food source in that, in Botswana at least, local absences of mopane worm may be caused by extensive herbivory on their host plant (Styles & Skinner 1996).

1.10 Conclusion

As we noted at the beginning, all of the concepts introduced in this chapter will appear again later in the book, where they are considered in detail.

Although each chapter is necessarily separate, it is important to try to consider insect ecology as a series of interlocking systems. For instance, to fully understand how an insect population can be of great economic significance in a crop, we need to look at its relationships with its host plant, its enemies, its competitors, and with the climatic conditions in its environment, although these aspects are presented in separate chapters.

Chapter 2: Insects and Climate

2.1 Introduction

The New Shorter Oxford English Dictionary defines climate as 'the prevailing atmospheric phenomena and conditions of temperature, humidity, wind, etc. (of a country or region),' whereas weather is 'the condition of the atmosphere at a given place and time with respect to heat, cold, sunshine, rain, cloud, wind, etc.' We shall use climate both to describe major patterns such as wind, rain and heat (or lack of them), and local ones that operate on a smaller scale. The latter category can be labelled microclimate, which is 'the climate of a very small or restricted area, or of the immediate surroundings of an object, especially where this differs from the climate generally'. Whether large scale such as typhoons or droughts, or small scale such as relative humidity, these abiotic factors can have fundamental influences on the ecology of insects, ranging from reproductive success and dispersal to growth and both intra- and inter-specific interactions. This chapter will look at a series of climatic factors, one by one, and illustrate the various ways in which they may impinge on the ecologies of insects.

2.2 Temperature

All insects can be considered to be poikilotherms, that is, their body temperature varies with that of the surroundings. Although some, such as bees or moths, can elevate the temperature of their flight muscles by rapid contractions before take off, their basic metabolism is a function of the temperature of their surroundings, such that within a certain range, the higher the temperature, the faster metabolic reactions are able to proceed. This means that any processes in the insect such as growth, development or activity are all dependent on temperature. In Fig. 2.1 we see a simple illustration of how temperature affects the development time, measured in days, of larvae and pupae of the pea weevil, *Bruchus pisorum* (Coleoptera: Bruchidae) (Smith & Ward 1995). As the temperature rises, the time-spans of both the second instar and the pupal stage decrease markedly, although not linearly. In fact, the development time reaches a minimum, in this case at around 30°C, whereupon it begins to increase again.

Development time is essentially the reciprocal of development rate. Figure 2.2(a) shows the mean development time of the pine false webworm, *Acantholyda erythrocephala* (Hymenoptera: Pamphiliidae), a sawfly whose ecology is basically the same as that of a defoliating lepidopteran caterpillar. Development rate increases as development time decreases, although in this case, there is no second increase in development time at high temperatures. Instead, at least for the eggs of the webworm, development stops altogether at around 30°C, and the eggs fail to hatch (Fig. 2.2b) (Lyons 1994). It is also important to note that no eggs hatch at temperatures below 6°C either—clearly there is a fairly restricted temperature range over which insect eggs can survive and mature.

Of course, carnivorous insects such as predators and parasitoids will also be affected by temperature in similar ways to the herbivorous bruchid described above. Another function of metabolic rate is activity, and in Fig. 2.3 it can be seen that there is a significantly higher probability of a parasitic wasp, *Trichogramma minutum* (Hymenoptera: Chalcidae) attacking the eggs of the spruce budworm moth, *Choristoneura fumiferana* (Lepidoptera: Tortricidae) when it is able to spend longer periods at higher temperatures (Bourchier & Smith 1996). Simply put, the wasp is able to move more rapidly at higher temperatures, and hence to search out and lay eggs in more host eggs per unit time. We notice also that as relative humidity (RH) increases, the odds of parasitism occurring decline. Humid conditions, especially over 85% or so, are usually associated with rain, and it is well known that small parasitoids have difficulty moving around in such wet conditions.

However, there are prices to pay for the advantages conferred on insects by living at relative high temperatures. One of the drawbacks is reduced longevity;

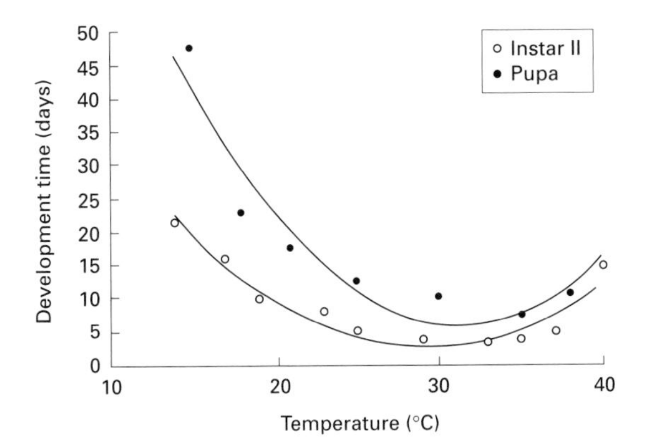

Fig. 2.1 Development time in days of second instar larvae and pupae of pea weevil, *Bruchus pisorum*, under differing conditions of constant temperature. Data points from published work; curves from polynomial fits to indicate trends. (From Smith & Ward 1995.)

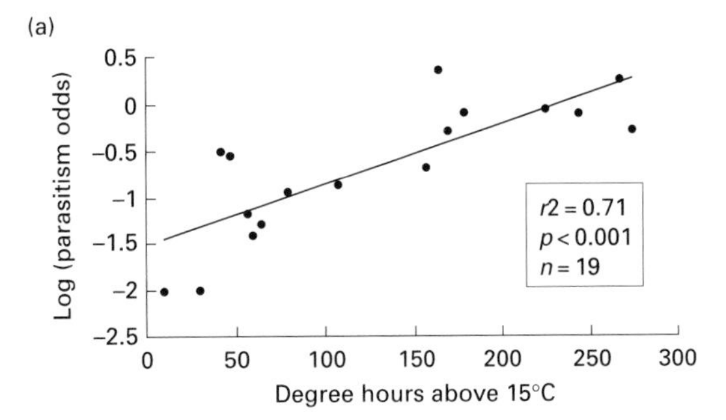

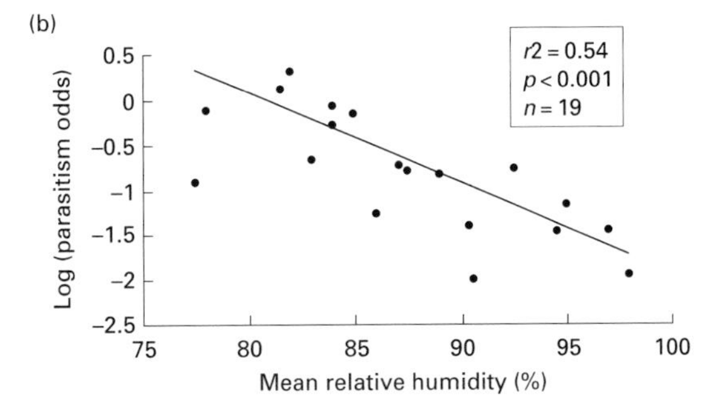

Fig. 2.3 Relationship between (a) the number of hours above 15°C, and (b) the mean relative humidity, in a 3-day period following a release of the parasitoid *Trichogramma minutum*, and the probability of parasitism of spruce budworm egg masses. (From Bourchier & Smith 1996.)

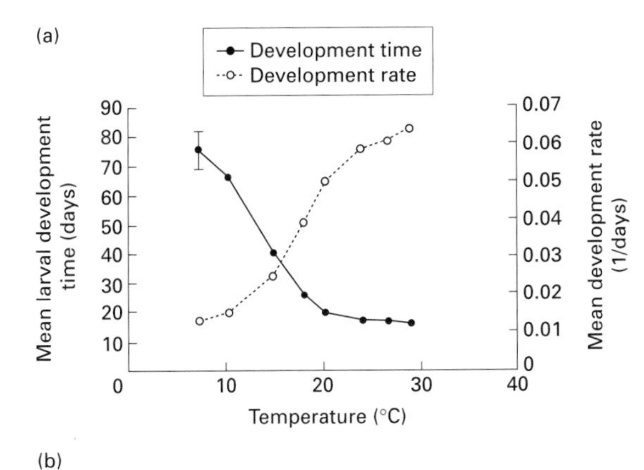

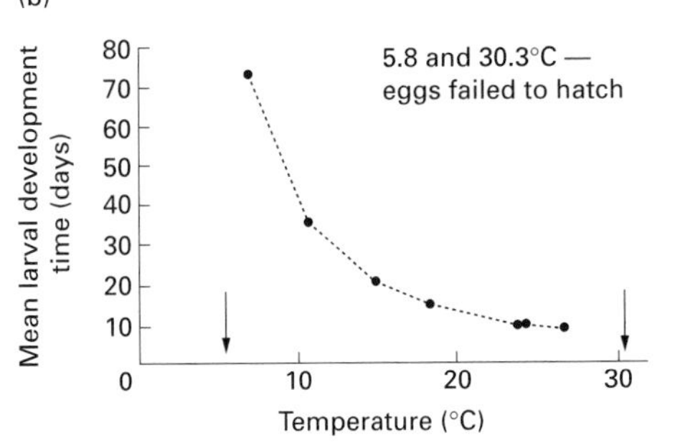

Fig. 2.2 Mean development time for (a) larvae and (b) eggs of the pine false webworm reared at constant temperatures. (From Lyons 1994.)

insects live for shorter times at higher temperatures, everything else being equal. In the case of the parasitic wasp *Meteorus trachynotus* (Hymenoptera: Braconidae), which attacks larvae of the spruce budworm in North America, adult females lived for a much shorter time when exposed to higher temperatures (Fig. 2.4) (Thireau & Regniere 1995). A 40 day adult lifespan at 15°C was reduced to a mere 10 days or so at 30°C.

Living in cold conditions may not always be deleterious either. One advantage is that the energy demands made by metabolism are low, so that if resources such as food are in short supply, insects can survive longer without starving. The gypsy moth, *Lymantria dispar* (Lepidoptera: Lymantriidae) is an extremely serious defoliator of broad-leaved trees such as oak in North America, and its is essential for its young larvae to find nutritious and palatable leaf material on which to feed as soon as possible after they hatch. In Fig. 2.5 it can be seen that the starva-

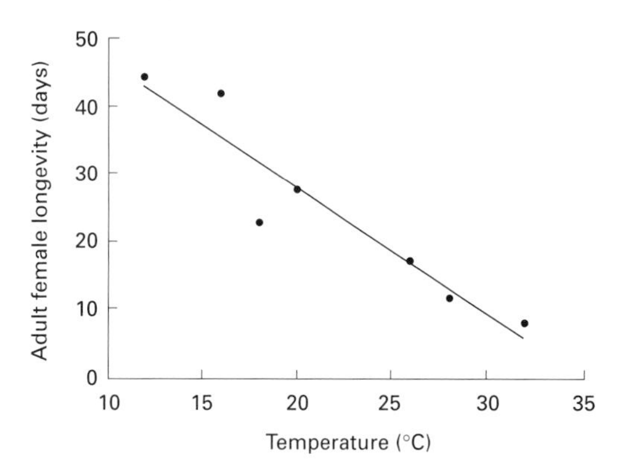

Fig. 2.4 Longevity of adult female parasitoids, *Meteorus trachynotus*, at various temperatures. (From Thireau & Regniere 1995.)

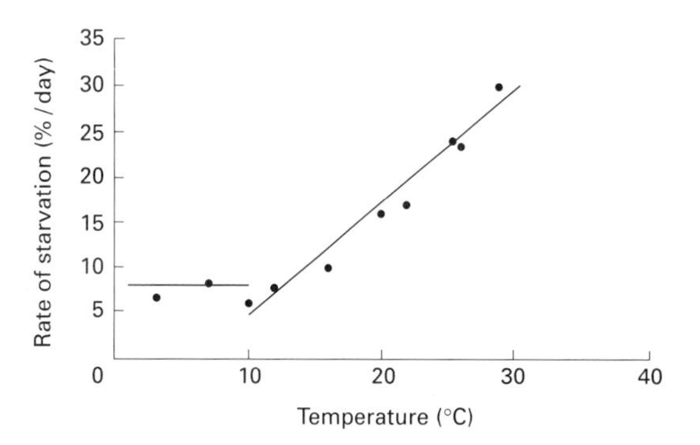

Fig. 2.5 Median starvation rate of newly hatched gypsy moth larvae as related to temperature. (From Hunter 1993.)

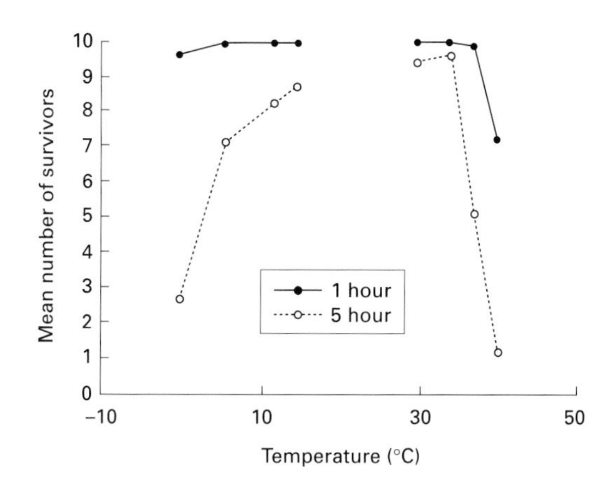

Fig. 2.6 Mean survival of pecan aphids, *Monelliopus pecanis*, exposed to different temperature regimes for 1 and 5 h. (From Kaakeh & Dutcher 1993.)

tion rate of these insects increases fairly linearly with temperature after about 10°C, so that a warm spring with little new leaf on the trees can rapidly cause the larvae to starve to death. Below 10°C, however, there is a suggestion that no change occurs; the insects remain in a kind of suspended animation until warmer weather returns (Hunter 1993). Development then proceeds by the insects 'accumulating' day-degrees. Put very simply, if an insect requires 100 day-degrees to develop from egg to adult, then in theory this could consist of 10 days at 10°C or 5 days at 20°C. Normally, however, there is a base temperature below which the accumulation does not take place, akin to the horizontal line below 10°C in Fig. 2.5 (Hunter 1993). For example, the scale insect *Hemiberlesia rapax* (Hemiptera: Diaspididae), a pest of kiwifruit in New Zealand, has a total development time of 1056 day-degrees above a base temperature of 9.3°C (Blank *et al.* 1995). One advantage of possessing such data for insect pests is that it becomes possible to predict fairly accurately when a certain life stage will be attained after a reference point such as egg laying, as long as temperature records are kept for the habitat in which the insect is active (see Chapter 10).

Simple life and death (survival) may also be influenced by temperature, particularly at both extremes of warmth and cold. The pecan aphid, *Monelliopus pecanis* (Hemiptera: Aphididae), was exposed in experiments to both low and high temperatures for two periods of time, 1 and 5 h. Figure 2.6 shows that the aphids could survive fairly well even at temperatures near freezing if exposed for only 1 h, but many deaths occurred if they were forced to spend 5 h at these low temperatures. At very high temperatures, survival was reduced even after a 1 h exposure (Kaakeh & Dutcher 1993).

So what does all this mean for insect ecology? Obviously, the chemistry that drives metabolism is more properly considered to be a physiological process rather than an ecological one, but at this level, it is unrealistic to separate one from another. Let us take, for example, a situation where seed-boring beetles such as the bruchid mentioned above are growing in cold conditions: they are forced to remain in the juvenile stages for a relatively long time, where they are exposed to potentially fatal problems such as predation or disease. A faster development rate enables the insects to reach adulthood and reproduce with fewer

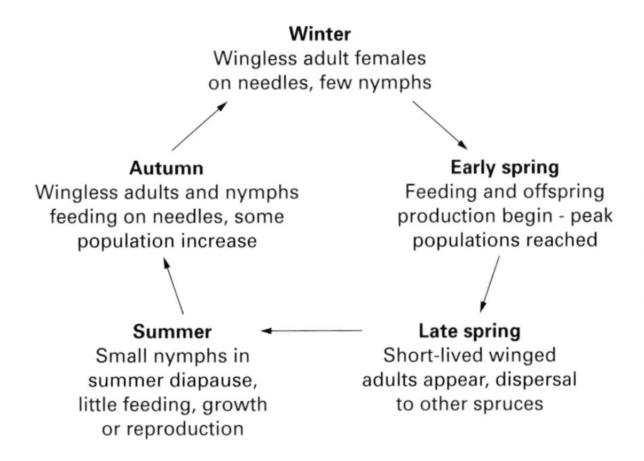

Fig. 2.7 Diagram of the anholocyclic lifestyle of green spruce aphid *Elatobium abietinum* in the UK. (Details from Bevan 1987.)

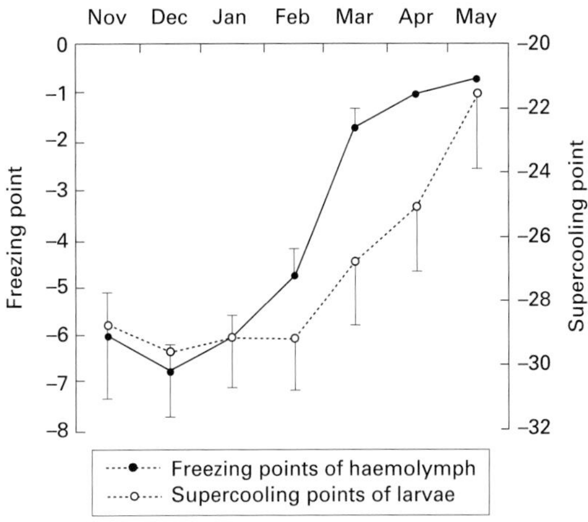

Fig. 2.8 Mean freezing and supercooling points of overwintering larvae of *Scolytus laevis*. (From Hansen & Somme 1994.)

risks. Herbivores and carnivores alike are able to function faster and more efficiently at higher temperatures, and though they may live for a shorter time, on balance they can feed, develop, reproduce and disperse when the climate is warm. However, mechanisms exist which enable insects to survive through severe winters as well, and their ecologies are keenly adapted to the changing seasons. A complete example of this adaptation is shown by the green spruce aphid, *Elatobium abietinum* (Hemiptera: Aphididae). Its life cycle is shown in Fig. 2.7. In the UK, the life cycle of this aphid is known as anholocyclic, as compared with holocyclic, which occurs in continental Europe. The former system is devoid of sexual reproduction in the main, so that offspring are produced entirely parthenogenically, and are all females. Most aphids living in temperate climates are adapted to surviving winter freezing by passing the cold period as resistant eggs, but *Elatobium* risks freezing to death as an adult. The advantage of this system is that this species is able to make a very early start in the spring in terms of feeding and reproduction, as soon as temperatures allow. In fact, the only really serious mortality factor that impinges on the spruce aphid is freezing during cold snaps in early spring when feeding has recommenced.

Insects such as *E. abietinum* inhabiting temperate or even arctic regions must be able to survive extremes of cold, frequently far below those experienced by the pecan aphid in the example described above. They accomplish this by a process known as supercooling, where tissues are able to withstand the freezing of their fluids for extended periods without damage. The supercooling point (SCP) is considered to be the absolute lower limit for survival. Another example of an insect able to perform this feat is the elm bark beetle *Scolytus laevis* (Coleoptera: Scolytidae). This European species is a potential vector of Dutch elm disease (see Chapter 9), and exhibits supercooling to withstand winter temperatures in Scandinavia (Hansen & Somme 1994). In Fig. 2.8, it can be seen that the mean supercooling point for larvae of this species reached as low as –29°C in midwinter, and increased to a 'mere' –21°C in May. The haemolymph of the insects actually froze at around –7°C in December, rising to around 0°C in the spring. When frozen to temperatures corresponding to their minimum SCPs, larvae were able to recover muscular contractions for a while, but later died. In reality, larvae were able to survive temperatures as low as –19°C in winter, but were killed at that temperature in spring. So, any situation where temperatures under the bark of elm trees do not fall below –19°C or so will allow *S. laevis* to survive quite happily, and hence, in this case, such regions cannot be considered to be safe from the risk of Dutch elm disease (Hansen & Somme 1994).

The basis for this supercooling ability seems to be enhanced concentrations of glycerol in the body fluids, which acts as an antifreeze, and the reduction of supercooling ability in spring results from declining levels of this chemical. Many other insects are capable of synthesizing glycerol. The spruce budworm, *Choristoneura fumiferana* (Lepidoptera: Tortricidae) produces higher levels of glycerol when exposed to lower temperatures, so that as winter sets in, the insect is able to respond by setting up its supercooling abilities in advance of serious cold (Han & Bauce 1995). However, if these cold spells arrive too early, and the larvae are not in the appropriate physiological state to synthesize glycerol, they may still be frozen to death. The biotic environment also has a role to play in freeze protection for insects. Just as with the bark beetle described above, where an insulating layer of bark will to some extent protect from, or at least delay, extremes of cold, so monarch butterflies, *Danaus plexippus* (Lepidoptera: Nymphalidae), need to have their overwintering sites covered by forest canopy. This canopy layer insulates them from freezing winter storms at high altitude in Mexico (Anderson & Brower 1996). Another problem for these butterflies is wetting. If it rains heavily before freezing, then their supercooling abilities are unable to prevent them from dying. The influences of rainfall on insect ecologies are discussed in section 2.4.

2.3 Daylength (photoperiod)

The length of daylight during a 24 h day–night cycle, known as photoperiod, may not exactly qualify as a climatic factor *per se*, but it does have a fundamental influence on the development and ecology of insects that live in seasonal climates. The daylength experienced by an insect larva provides information about the progression of the seasons, and the ability to vary growth and development rates enables the insect to achieve efficient timing relative to favourable conditions (Leimar 1996). *Kytorhinus sharpianus* (Coleoptera: Bruchidae) is a wild bean weevil from Japan that oviposits and develops as a larva inside both immature fresh and mature legume seeds (Ishihara & Shimada 1995). In Fig. 2.9 it can be seen that the durations of the various stages in the life cycle from egg to adult vary according to the photoperiod at constant temperature. The whole cycle can be accomplished between 75 and 80 days when the insect receives 15 or 16 h of daylight per day–night cycle, but this period increases dramatically as the hours of daylight diminish to 14 and then 12 h. With only 12 h of daylight the pupal stage is never reached until longer hours of light return.

As in the bruchid example above, daylength can be

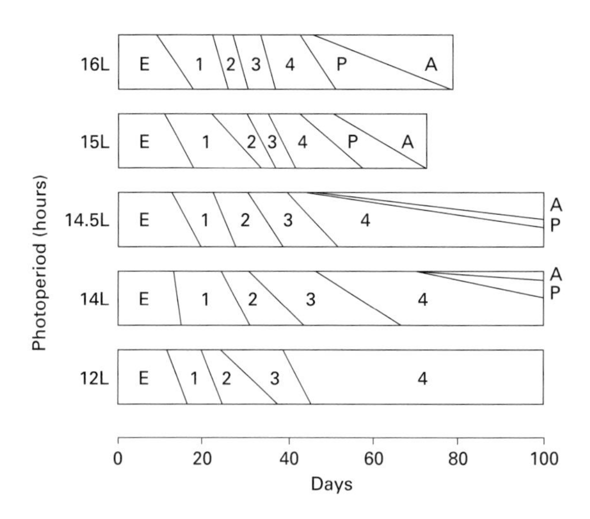

Fig. 2.9 Development schedules for a bruchid beetle under five photoperiod cycles at 24°C. E, egg; 1–4, larvae instars; P, pupa; A, adult. (From Ishihara & Shimada 1995.)

used as a signal or trigger by insects to enter a quiescent period, diapause, within which to wait out potentially deleterious conditions such as summer heat or drought, or winter cold. The summer diapause may also be used by adult insects such as plant feeding chrysomelid beetles (Coleoptera: Chrysomelidae) to lie quietly and become sexually mature before reproduction in the late summer or autumn (Schops *et al.* 1996). Indeed, some insects show both types of diapause. The burnet moth, *Zygaena trifolii* (Lepidoptera: Zygaenidae), occurs over much of Europe, but in Mediterranean regions it may undergo two types of diapause in 1 year (Wipking 1995). First, a facultative diapause may take place, lasting between 3 and 10 weeks in summer depending on weather conditions, followed by an obligate winter diapause, which may last several months if low temperatures persist. As with most climatic effects, it is difficult to separate the two influences of temperature and daylength in both the onset and the cessation of diapause, but daylength has been shown to have a fundamental effect in some cases. Two examples are shown in Figs 2.10 and 2.11. In the first, when larvae of the parasitic wasp, *Cotesia melanoscela* (Hymenoptera: Braconidae), were exposed to long daylengths (greater than 18 h), they developed virtually continuously all the way to adulthood, whereas when exposed to short daylengths (less than 16 h), they entered a diapause, which halted development in the cocooned prepupal stage

(Nealis *et al.* 1996). In ecological terms, this system makes sure that adult parasitoids appear only during the long days of summer when their host insects, in this case gypsy moth larvae, are most likely to be abundant. In the second example, which concerns the sapfeeder *Empoasca fabae* (Hemiptera: Cicadellidae), the life cycle proceeds all the way to the adult stage, but under short day conditions, reproduction (measured in terms of oviposition rate and preoviposition period) is reduced (Taylor *et al.* 1995). In this way, as the shorter days of late summer ensue, the insect slows down its reproductive efforts and in fact begins a diapause-mediated migration to overwintering areas.

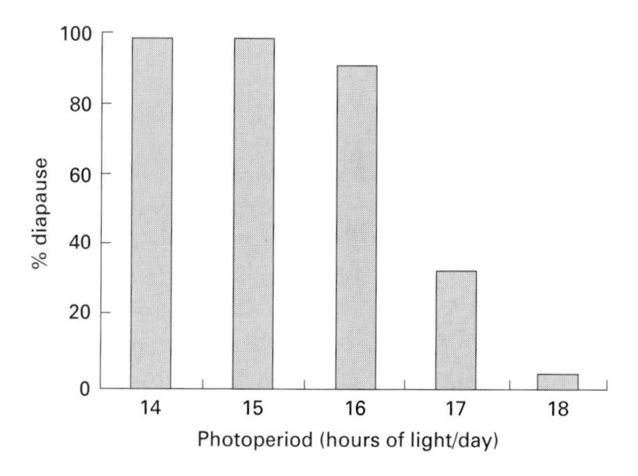

Fig. 2.10 Percentage diapause in cohorts of *Cotesia melanoscela* reared at 21°C in relation to photoperiod. (From Nealis *et al.* 1996.)

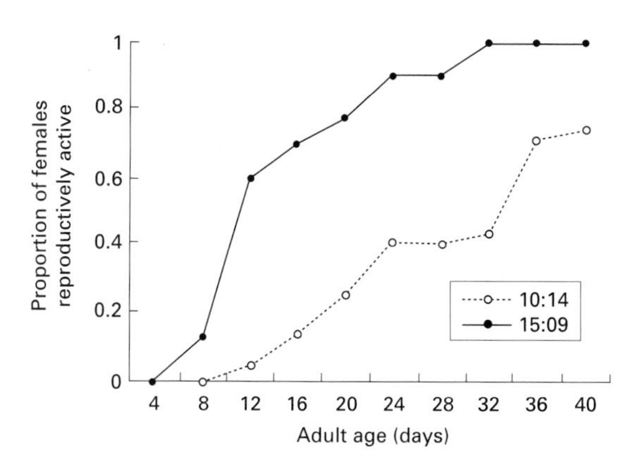

Fig. 2.11 Proportion of female *Empoasca fabae* ovipositing by a given adult age at constant temperature (24°C) and two different daylengths. (From Taylor *et al.* 1995.)

2.4 Rainfall

The effects of rainfall on insects can be direct or indirect. Heavy rain can knock aphids off their host plant and both beetles and bugs may be killed by violent thunderstorms (T.R.E. Southwood personal communication). Lack of rain can cause desiccation and death. Seasonal rains influence the ways in which host plants flush and grow and provide food for herbivores (and hence, of course, all those in higher trophic levels above them), whereas droughts can stress trees, rendering them susceptible to insect attacks. Rainfall also affects humidity, which combines with temperature and wind to dictate local microclimatic conditions. Finally, of course, water itself is a vital habitat for around 3% of all insect species, the aquatic ones.

The immature stages of orders such as dragonflies (Odonata), mayflies (Ephemeroptera), stoneflies (Plecoptera) and caddisflies (Trichoptera), are dependent on ponds and streams, as are myriads of species of flies (Diptera), beetles (Coleoptera) and bugs (Hemiptera) for at least part of their life cycles. Rainfall will affect the water levels and current strengths of these habitats, and also creates a number of more specialized and ephemeral aquatic habitats in some rather unlikely places. These include bromeliad 'pools' high in rainforest canopies, and also bamboo

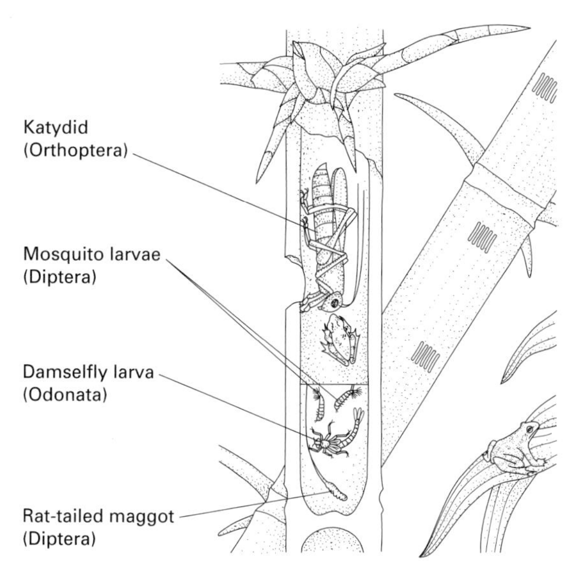

Fig. 2.12 Inhabitants of a bamboo internode community from Amazonia. (From Louton *et al.* 1996.)

internode communities illustrated in Fig. 2.12. In this example from Amazonia, young bamboo shoots have slits cut in them by ovipositing katydid grasshoppers (Orthoptera: Tettigoniidae). As the shoot grows, reaching on average 45 mm in diameter, the oviposition scars deteriorate into a series of open slots and then to a single opening. Rain fills the cavity up to the level of the slot, and a succession of insects and other taxa colonize the little pool and its environs. These include mosquitoes (Diptera: Culicidae), rat-tailed maggots (Diptera: Syrphidae), and damselflies (Odonata: Zygoptera). Even amphibians take up residence to prey on the emerging adult insects (Louton *et al.* 1996).

Rainfall patterns can influence the long-term abundance of insect populations. Figure 2.13 shows tree-ring chronologies from 24 mixed conifer stands in northern New Mexico, from which large-scale outbreaks of the western spruce budworm, *Choristoneura occidentalis* (Lepidoptera: Tortricidae), have been detected over the last 300 years (Swetnam & Lynch 1993). The simple link is that heavy defoliation by the moth's larvae causes significant reductions in that period's ring widths. Superimposed on these are rainfall patterns, of which the older ones have been reconstructed by examining the growth rings of an insect-free tree species, limber pine. Although much variation in this type of data is to be expected, the figure shows fairly clearly that periods of increased and decreased budworm activity coincide with wetter and drier periods, respectively. Analysis of variance was carried out on these data, and Swetnam and Lynch felt that there was 'compelling evidence' for a climate (rainfall)–budworm association that has existed for at least three centuries. The underlying mechanisms are rather harder to determine, however. The plant-stress hypothesis (see Chapter 3), which suggests that drought-struck trees will provide better nutrients for and possibly fewer defences against herbivores, does not seem to apply here; wetter weather promotes the insects, not the reverse. Complex interactions may exist between precipitation, insect natural enemies and diseases, and much more basic research is needed to elucidate such mysteries.

Insect abundance is certainly linked to seasonal variations in rainfall, with some species being more abundant in the dry season, whereas others proliferate only during the rains. In the eucalyptus forests of northern Australia, Fensham (1994) found conflicting distributions of various insect taxa according to season (Fig. 2.14). Figure 2.14 shows but two examples of herbivorous taxa from that work, one a sapfeeder (Hemiptera: Psylloidea) and the other a defoliator (Orthoptera). The psyllids were very much more common in the late dry season than at any other time of year, whereas the grasshoppers, although present in the early and late dry seasons, were most abundant during the wet (rainy) season. It

Fig. 2.13 Tree ring chronologies of Douglas fir in New Mexico over nearly 300 years depicting spruce budworm outbreaks, with reconstructed March to June rainfall patterns for the same period. (From Swetnam & Lynch 1993.)

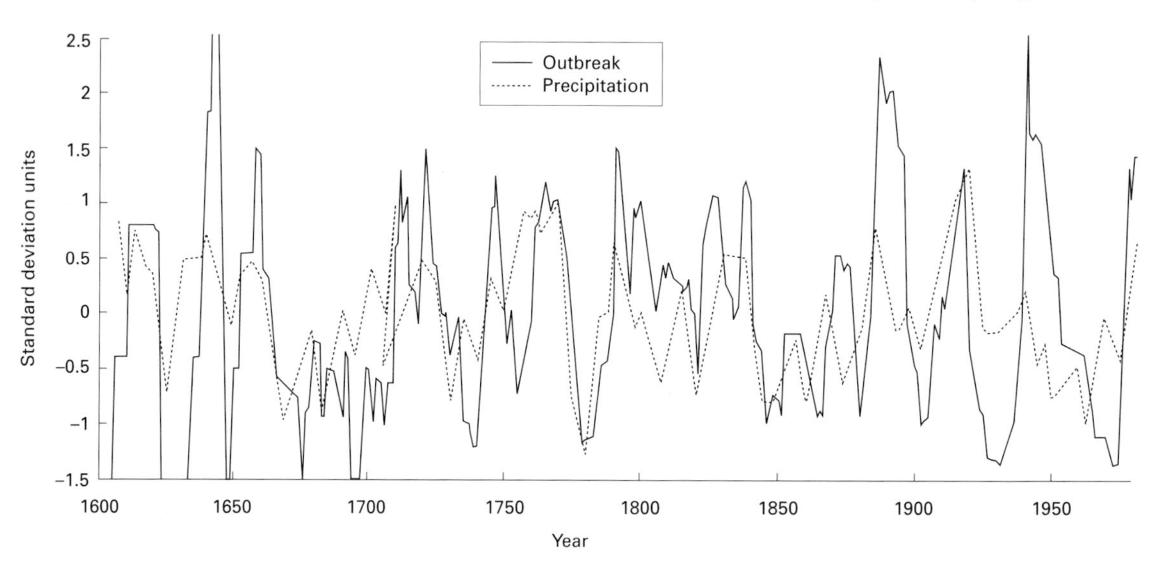

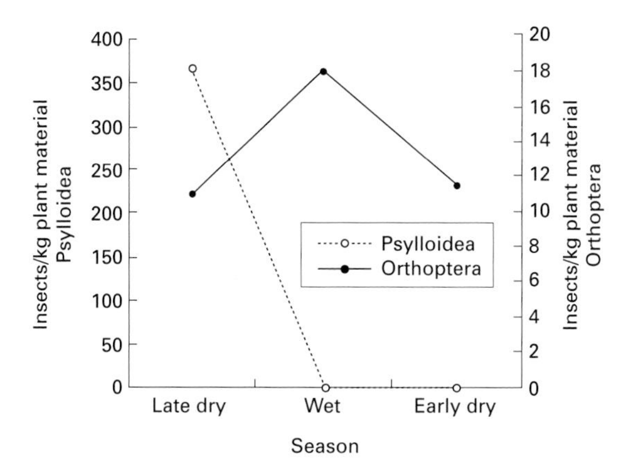

Fig. 2.14 Mean seasonal abundance of two groups of phytophagous insects in low eucalypt forest in northern Australia. (From Fensham 1994.)

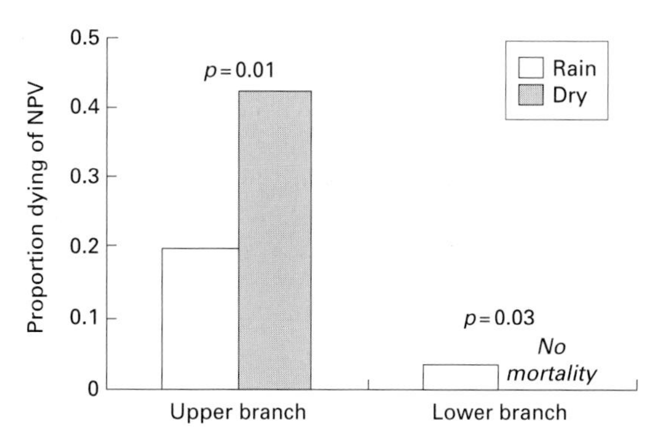

Fig. 2.15 Mean proportion of third instar gypsy moth larvae killed by nuclear polyhedrosis virus (NPV) on upper and lower tree branches in wet and dry weather. (From D'Amico & Elkinton 1995.)

is suggested that the sapfeeders are receiving nutrients from sap produced by the regrowth of trees in response to fires that sweep through these forests in the late dry season. The defoliating Orthoptera, on the other hand, are benefiting from the relatively luxuriant production of new leaves during the rains.

It has been suggested above that the actions of natural enemies of insects may be affected by rainfall, and insect-specific viruses provide an example. Nuclear polyhedrosis viruses (NPVs) are discussed in detail in Chapter 10. Suffice it to say for now that these mainly species-specific pathogens are able to remain infective outside the host-insect body, attached to foliage and bark, for considerable lengths of time. In the case of the gypsy moth, *Lymantria dispar* (Lepidoptera: Lymantriidae), in the USA, leaves of the host tree, red oak, were inoculated with gypsy moth NPV by using infected first instar caterpillars caged onto foliage in the upper canopy (D'Amico & Elkinton 1995). When these insects died, they released NPV onto the leaves, on which fresh, healthy third instar larvae were subsequently placed. Similar larvae were also placed on leaves on lower branches, situated below the NPV-inoculated upper ones but not covered with NPV. Some trial branches were then protected from rain, and the mortalities of the larvae measured (Fig. 2.15). Significantly less larvae died from NPV infection on the upper branches in wet conditions: the rain was able to wash some of the virus off the upper leaves and hence reduce the pathogen concentration. However, more larvae died of NPV on the lower branches exposed to rain, as NPV washed off the upper branches reached the lower ones, where some larvae picked up the infection.

Clearly, the effects of rainfall on insects are complex and not always immediately easy to explain. It is also difficult to separate them from other allied climate patterns, wind in particular, as combinations of the two can result in areas of low pressure and weather convergences that redistribute insect populations and at the same time provide fresh plant food for them to eat (see section 2.5). Rain may not even have an influence at the time, but instead may promote insect performance some months later. This phenomenon is well illustrated by the example of the seasonal outbreaks of African armyworm, *Spodoptera exempta* (Lepidoptera: Noctuidae). Armyworms can reach enormous numbers (Plate 2.1, opposite p. 158), and have been serious pests of cereals and pasture in sub-Saharan Africa for over 100 years (Haggis 1996). In Kenya, long-term rainfall records were used to predict the number and size of armyworm outbreaks, and it was found that the number of outbreaks were negatively correlated with rainfall in the 6–8 months preceding the start of the armyworm season (Fig. 2.16). In other words, severe outbreaks are very often preceded by periods of drought. The explanation lies again with NPVs from which epidemic armyworm populations suffer. In wet, humid conditions, the larvae become stressed and succumb to viral infections, whereas in dry weather, the sun rapidly kills viruses. In these dry spells, as long as there is still enough rain for the host plants to grow, then low-density populations of armyworm can persist for some time, in fact until the next outbreak. So dependable are the links between rain and later armyworm outbreaks that Haggis (1996) has been able to construct a prediction system for Kenya, which is illustrated in Fig. 2.17. In almost all cases, the forecasts produced by this system concerning the likelihood of armyworm outbreaks were correct, although it should be noticed that the system provides opportunities to modify decisions based on updated rainfall conditions.

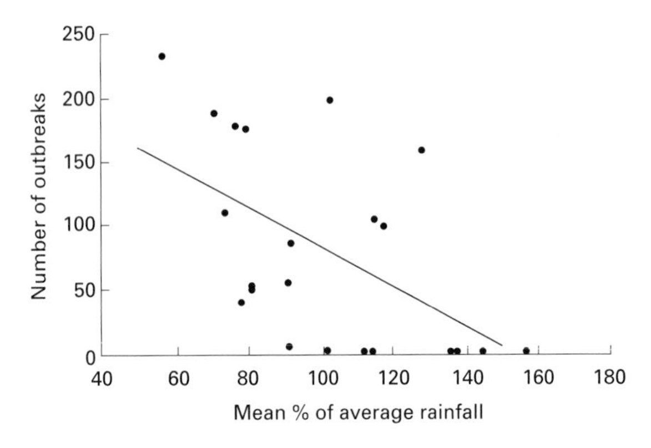

Fig. 2.16 Mean percentage of average rainfall in April to May in Kenya over 23 years related to the number of armyworm outbreaks in the subsequent season. (From Haggis 1996.)

Fig. 2.17 Stages in forecasting local or widespread armyworm outbreaks next season from current season's rainfall in SE Kenya (based on 22 years data). Numbers in parentheses denote number of years when forecast was fulfilled/number of years with rainfall sequence. Provisional forecast with thin border; final forecast with bold border. (From Haggis 1996.)

2.5 Wind

As with rain, wind can have a variety of influences on the ecology of insects. It can carry them many miles to new habitats and regions, it can bring rain to produce new food, it can destroy trees to provide new breeding sites, and it can transmit chemicals from one insect to another, or from a host plant to an insect.

Many insects appear to undertake enormous migrations covering hundreds if not thousands of miles on occasion. Many species regularly move from continental Europe across the English Channel to southern and eastern England, and a few, such as the monarch butterfly, *Danaus plexippus* (Lepidoptera: Nymphalidae) for example, have been known to make it all the way across the Atlantic Ocean to Europe. Undoubtedly, they do not perform this feat unaided, and in fact it is air currents or wind that assists in both distance and direction travelled. Another butterfly related to the monarch, *Danaus chrysippus*, is a tropical species that occurs in a number of genetic races across Africa. Smith and Owen (1997) studied the seasonal cycles of abundance of several of these races in and around Dar es Salaam on the Tanzanian coast, and found that these oscillations were related to migratory activity of the butterflies. This activity in turn was related to the cyclical north–south movement of wind, coupled

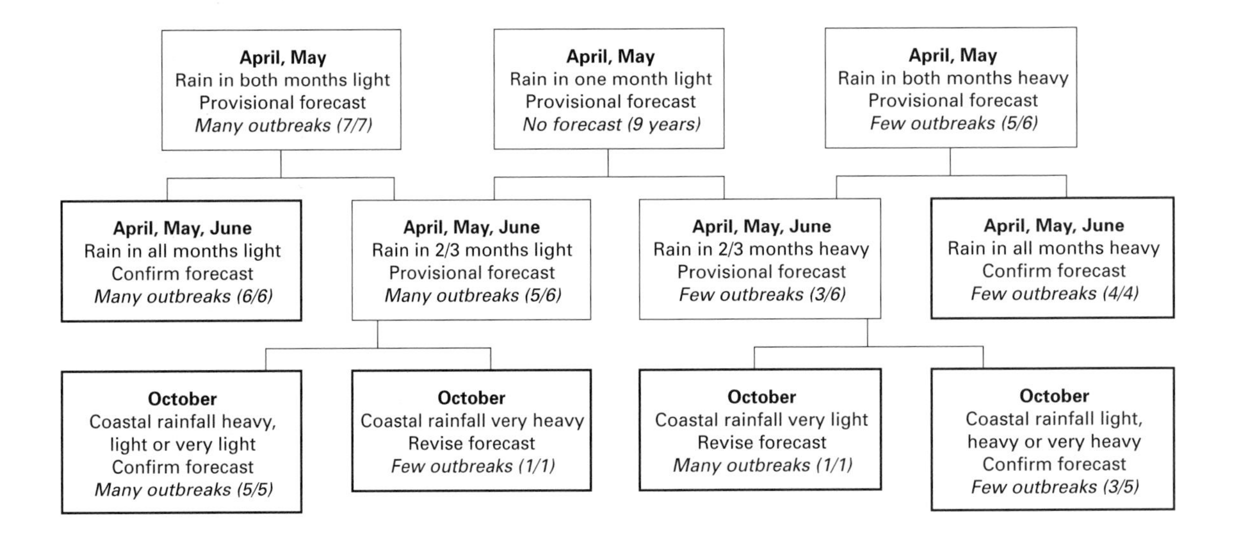

with the allied rainfall and temperature regimes imposed by the monsoons (Fig. 2.18). The problem, however, with the sorts of data shown in the figure is attributing cause and effect. Smith and Owen cited the Australian common name for *D. chrysippus* as 'wanderer'—clearly an insect known to move about the place! Large butterflies such as this one can certainly fly against the wind when they have to, and so it is not yet clear in this example whether or not the monsoon winds are carrying the insects on their seasonal migrations, or whether it is simply the wind that creates the right environmental conditions such as temperature and food plant to which the butterflies return at certain times of the season.

Wind is certainly a vital component of more general weather patterns, giving rise to fronts and convergence zones. Atmospheric convergence occurs on a number of scales (Drake & Farrow 1989). In the tropics and middle latitudes, belts of convergent and ascending air occur mainly within the Intertropical Convergence Zone (ITCZ), which produce heavy rain and storms. In Fig. 2.19 it can be seen how two converging winds are able to concentrate airborne insects into a relatively localized region. The insects are assumed to rise to their flight ceiling and subsequently land to feed and reproduce. As the same convergence zone produces rain, semiarid tropical regions soon bloom with luxuriant vegetation on which the large densities of new generation insect larvae feed voraciously. Serious African pests such as the desert locust, *Schistocerca gregaria* (Orthoptera: Acrididae), and the armyworm, *Spodoptera exempta* (Lepidoptera: Noctuidae), are two classic examples of this phenomenon. Smaller-scale wind convergences regularly occur as well, and have been attributed to the unpredictable and sudden concentrations of insect pests in small patches of forest, such as happens in southern India with the teak defoliator moth, *Hyblaea puera* (Lepidoptera: Hyblaeidae) (Fig. 2.20) (Nair & Sudheendrakumar 1986).

One of the most complex systems so far studied in which wind patterns influence the movement of insects is in the Far East (Fig. 2.21), where trajectory

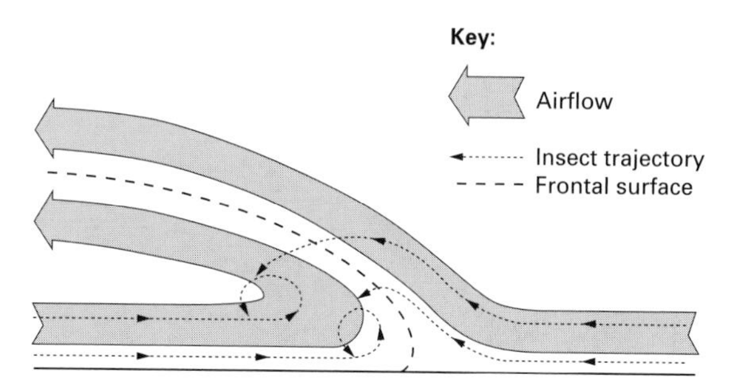

Fig. 2.19 Vertical section through a frontal convergence, showing a likely mechanism of entrapment, and hence concentration, by the front. The insects are assumed to rise until they reach their flight ceiling, which will be higher in the warm air at the right than in the cooler undercutting air that is advancing from the left. (From Drake & Farrow 1989.)

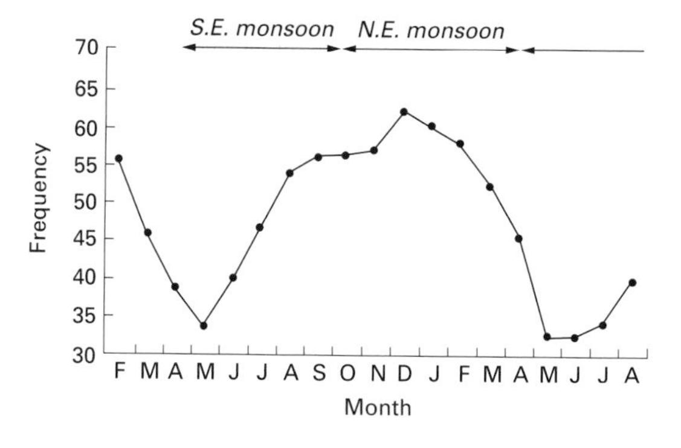

Fig. 2.18 Frequencies (3 month moving average) of the orange phenotype of the butterfly *Danaus chrysippus* in Dar es Salaam, Tanzania, relative to monsoon patterns. (From Smith & Owen 1997.)

Fig. 2.20 Teak (*Tectona grandis*) defoliated by teak defoliator moth larvae (*Hyblaea puera*) in Kerala, India.

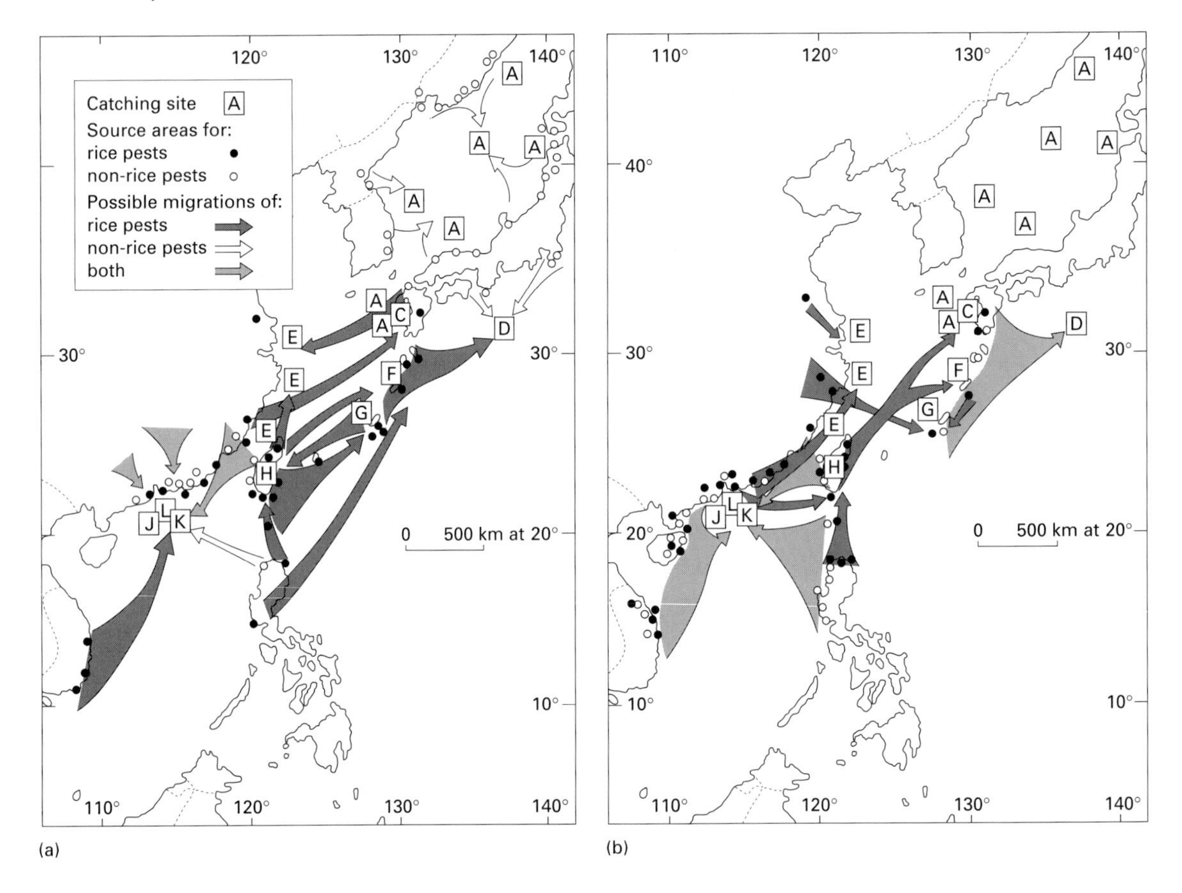

Fig. 2.21 Schematic diagram of movements of rice pests and other insects and their likely source area at (a) 10 m and (b) 1.5 km above the ground in spring in south-eastern and eastern Asia. (From Mills *et al.* 1996. © CABI Publishing.)

analysis of winds at both 10 m and 1.5 km height was used to determine the direction and extent of wind-borne movements of insect pests of rice (Mills *et al.* 1996). Aerial traps caught two species predominantly, the brown rice plant-hopper, *Nilaparvata lugens*, and the white-backed plant-hopper, *Sogatella furcifera* (Hemiptera: Delphacidae). Northward migrations of these and other species were detected along a broad front in prevailing summer monsoon and trade winds (Fig. 2.21). However, because most wind trajectories lasted no more than 40 h in any one direction, coupled with the fact that many of the associated weather systems were mobile and hence rather unpredictable in extent and direction, insect catches were very variable. This indicates that although general insect movement occurs between the tropics and temperate areas during the spring and summer, it is difficult to predict where the highest populations will finally be. Undoubtedly, some movement within the tropical zone also occurs on a regular basis.

Winds can also aid in the movement of insects over much smaller distances than discussed so far. *Bemisia tabaci* (Hemiptera: Aleyrodidae) is a world-wide aphid pest of various agricultural crops, including sweet potato, tomato, tobacco and cotton, and to achieve integrated pest management (IPM; see Chapter 10), it is important to understand how far the insect is able to migrate. In this way, it should be possible to predict the likelihood of new crops becoming infested, especially if wind directions are known. Figure 2.22 and Table 2.1 illustrate the movement of whitefly, within fields of melons in the USA (Byrne *et al.* 1996). The figure shows that insects were caught up to 2.7 km from the source field, but, because in this trial there were no traps further away, it is highly likely that the insects could travel much further. The figure also indicates that two distinct peaks of insect abundance were in evidence. This can be explained by the fact that two types of whitefly exist within the population; one group are called 'trivial' flying morphs, which move only in the immediate vicinity, whereas the other group contains strong migratory individuals who take off under the influence of cues from daylight. If the wind then picks up individuals from this latter group, they can be carried considerable distances. As the table shows, most whiteflies were caught approximately downwind, and in fact the published data from this trial suggest that that the higher the wind speed in any given direction, the greater the percentage of insects caught in traps in the corresponding downwind quadrant.

It is not only adult insects that can be dispersed by the wind. Sometimes, larvae can also be carried considerable distances, especially if they are attempting to escape from intraspecific competition when they occur in large densities on a limited food supply. One example of larval dispersal by wind involves the gypsy moth, *Lymantria dispar* (Lepidoptera: Lymantriidae). In North America, as in Europe, female adult gypsy moths are flightless (except in the Asian strain), and in fact the pest population disperses mainly as first instar larvae (Plate 2.2, opposite p. 158), which use 'ballooning' as a means of colonizing new areas (Diss *et al.* 1996). Recently hatched larvae produce fine silk threads that enable them to catch the wind and sail away from their overcrowded host tree. Most of them settle again within 120 m or so of the take-off point, but it is thought that some may be transported for many kilometres. Again, the strength and direction of the wind will be of paramount importance, although the final key to the success of the operation will be whether or not the little larvae can by chance reach a new, suitable host plant.

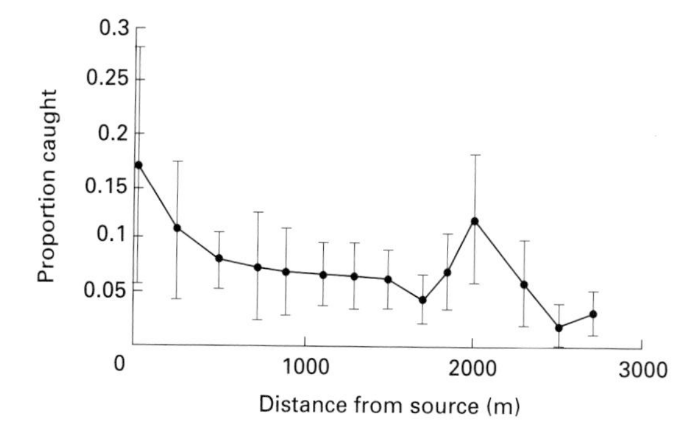

Fig. 2.22 Proportion of whiteflies trapped in 200 m distance classes from source field; average over eight trapping dates (From Byrne *et al.* 1996.)

Another risk for insects who rely on wind to help them disperse is that of going too far. Strong winds may well carry small insects out of their habitat range altogether, and so some species will fly only in winds of a certain velocity, and refuse to take off if it becomes too strong. A small amount of wind helps movement and host finding; too much could kill by carrying individuals to unsuitable areas. *Diachasmimorpha longicaudata* (Hymenoptera: Braconidae) parasitizes the larvae of fruit flies, and hence has biological control potential for these extremely serious pests (see Chapter 10). Messing *et al.* (1997) studied the flight behaviour of this species in a cage, and found that wind speeds of only 0.3 m/s elicited a clear response (Fig. 2.23). Adult parasitoids averaged 5–8 flights/min during the calm period, which later rose to ten or more, whereas the number of flights decreased abruptly to less than 2 flights/min during wind pulses. Later wind-tunnel studies found that wind speeds up to 4 m/s stimulated flight, whereas when the speed approached 0.8 m/s, flight was suppressed (Messing *et*

Table 2.1 Prevailing wind direction, wind speed, and percentage of whiteflies captured downwind in 1992. (From Byrne *et al.* 1996.)

Date	Wind direction	Whiteflies captured in downwind quadrant (%)	Wind speed (m/s)
11 Aug.	SE	NW (55%) SW (17%)	0.3–0.6
16 Sep.	NE	SW (43%) NW (51%)	0.6–1.1
25 Sep.	NW	NE (33%) SE (45%)	0.8–1.6
9 Oct.	NE	SW (68%) NW (30%)	0.6–1.3

al. 1997). Clearly, the latter experiments produced somewhat different results from the former, but both suggest that a little wind may be beneficial to the insects; too much is avoided. In species such as this one, which have to find specific hosts in heterogeneous habitats, some air movement is advantageous.

Though no host odours were used in the trials described in the previous example, one vital property of wind is its ability to convey chemical messages to insects from point sources. These messages can include information about the distance and location of a specialized food plant, or of a suitable and receptive mate. The beauty of a wind-driven system like this is that as long as the wind flows in a laminar (non-turbulent) manner the chemical 'plume' should provide an unbroken guide to the location of the source. Insects can then fly up a concentration gradient of chemical signal to locate their prize precisely!

Most herbivorous insects are host plant specialists to one degree or another (see Chapter 3), and it is therefore of prime importance that they are able to locate suitable hosts without wasting too much energy. This may be a difficult task in a habitat that contains many species of plant, only one or two of which are in fact suitable, and it is common for insects to locate their host from a distance by olfaction, or odours carried downwind. This system works particularly well if the host plant itself has a distinctive chemical 'signature'. Ragwort (*Senecio jacobaea*) is one such plant, whose chemical constituents include toxins such as cyanides. None the less, some insects have evolved to feed happily on this poisonous plant, including the ragwort flea beetle, *Longitarsus jacobaea* (Coleoptera: Chrysomelidae). Using wind tunnels, Zhang and McEvoy (1995) investigated the response by adult beetles to host plants placed 60 cm upwind (Fig. 2.24). Both male and female adults orientated themselves in the upwind direction if they had been starved, though satiated insects showed no particular response to wind direction. Obviously, insects are able to turn off their food-finding systems if they are not hungry. Under field conditions, it is easy to imagine that specialist herbivores such as the flea beetles would be much more efficient at finding their host plant during a mildly windy day than on one where the air was completely still. However, a strong wind, even if did not break up the chemical plume, might well prevent insects flying upwind to the plant—they might be able to smell it, but would be unable to reach it!

A large number of insect species use chemical signals, pheromones, to locate mates. One sex, more often than not the female, sits in one place and releases complex chemical messages, which disperse around her. In still air, these volatile compounds remain in close proximity to the emitter, and can diffuse over only short distances. However, in a light wind, just as with host plant odours, the pheromone forms a plume or concentration gradient up which males can fly once their immensely sensitive chemoreceptors have detected the scent. A very large body of expertise has built up concerning the uses of sex-attractant pheromone systems for the monitor-

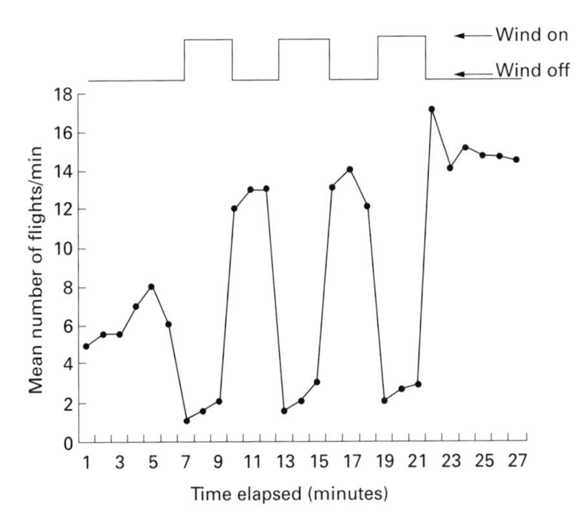

Fig. 2.23 Flight response of 10 caged female parasitoids to 0.3 m/s pulses of wind. (From Messing *et al.* 1997.)

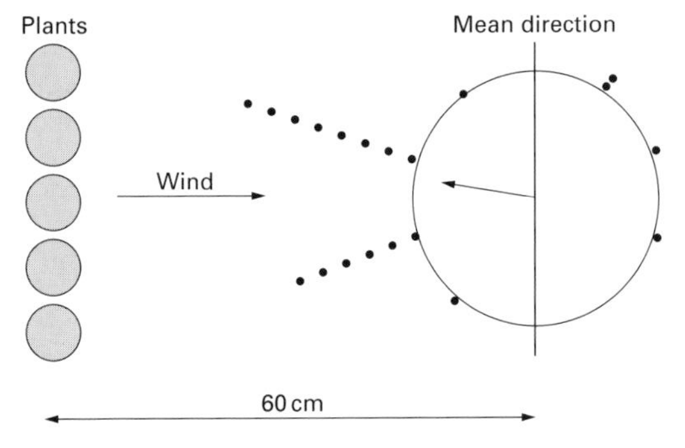

Fig. 2.24 Movement directions of a group of starved male flea beetles with their host plant, ragwort, placed upwind of their release point. (From Zhang & McEvoy 1995.)

ing and management of many farm and forest pests (see Chapter 10), but the interactions of insects, pheromones, wind and habitat can be exceedingly complex and need to be understood fully before they can be harnessed successfully. The codling moth, *Cydia pomonella* (Lepidoptera: Torticidae), is a very widespread pest of pome fruit such as apples and pears, and because its larvae (the grub in the apple) are concealed they are hard to control chemically or, to some extent, biologically too. The knowledge of timing and extent of adult flight and egg-laying activity is a key to the pest's management, for which pheromones are used in a mating disruption strategy (see Chapter 10). However, as field trials have shown, the nature of the habitat (orchard) and the wind speed and direction can have a fundamental influence on how moths respond to artificial pheromones (Milli *et al.* 1997). In field trials in Germany, it was found that codling moth pheromone could be detected up to 60 m downwind of a release point, and that stronger winds lifted the compound higher into the air than more moderate ones. However, as Fig. 2.25 shows, winds blowing into an orchard from untreated (no pheromone) grassland actually produced a zone near the edge of the crop which became depleted in pheromone, where pests might be able to proliferate. In addition, a wind blowing along the edge of the orchard rather than into it tended to concentrate pheromone along the edge and in fact into the grassland, where there was less wind resistance (no trees), thus wasting the treatment (Milli *et al.* 1997).

2.6 Climate change

One of the most interesting and controversial topics in applied ecology today is that of climate change and global warming. In this chapter, we have shown how climate can influence the ecologies of insects in many varied ways, and it is of great significance to speculate about how the actual and potential changes in the world's weather patterns have affected and will affect insects. Some work has already been carried out and given definite results, whereas other systems involve modelling and speculation. The essential assumptions about global warming are centred on the increase in the concentration of atmospheric carbon dioxide. General circulation models (GCMs) of the equilibrium response to the doubling of current levels of CO_2 indicate increases in global mean temperature in the range of 4 ± 2°C (Jeffree & Jeffree 1996). Though the CO_2 levels themselves may influ-

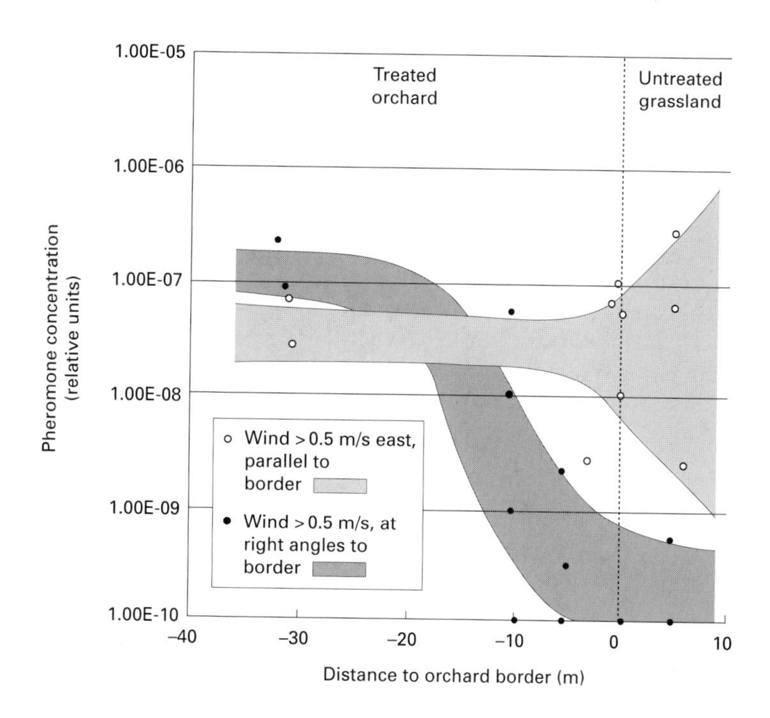

Fig. 2.25 Codling moth pheromone density profiles according to wind direction in an apple orchard. (From Milli *et al.* 1997.)

ence phytophagous insects either directly or, more likely, via the changed physiology of their host plants, there has been most speculation about the effects of this temperature increase on insect development.

Rather a large number of experiments have been carried out on a variety of insects to investigate the effects of temperature increases of around 2 or 3°C. One example concerns the aphid *Acyrthosiphon svalbardicum* (Hemiptera: Aphididae) on Spitsbergen in northern Norway, where cloches were used to cover experimental host plants and the reproductive ecology of the aphids feeding on these plants was investigated over a summer season (Strathdee *et al.* 1993). Figure 2.26 gives one set of results from these manipulations, and shows that reproductive output increased dramatically when the temperature was raised by 2.8°C. This resulted in an 11-fold increase in the number of overwintering eggs produced at the successful completion of the aphid's life cycle by the end of the season. In other words, if global warming is able to raise temperatures in this arctic region, *A. svalbardicum* will be able to sustain much higher population densities. This will depend, of course, on concurrent responses of its hosts and natural enemies. For pest species, elevated temperature may have very serious consequences. Another aphid, the green spruce aphid, *Elatobium abietinum* (Hemiptera: Aphididae), discussed in section 2.2, is probably the most serious pest of Sitka spruce in the UK, and its only limiting factor is mortality through freezing in late winter, as the adults begin to feed again. If winters in the west of the UK become on average warmer, and spring weather comes earlier in the year, then this already occasionally serious pest could become much more critical, and even jeopardize the use of Sitka as the UK's number one softwood species.

Other pests may also extend their geographical ranges as the climate warms. Simulation models abound that describe the likely expansion of insects through Europe as temperatures rise, and Fig. 2.27 shows one example. The Colorado potato beetle, *Leptinotarsa decemlineata* (Coleoptera: Chrysomelidae), is a notorious defoliator of potato crops, in both the larval and adult stage. At the moment, it is mainly restricted to southern parts of Europe, as shown in the figure, and its occasional forays into the UK are unlikely to result in establishment. However, using GCMs, Jeffree and Jeffree (1996) have produced predictions based on increases in regional temperature equivalent to a doubling of CO_2 levels in the atmosphere, and as shown in the figure, suggest that the potential geographical range of summer and winter temperatures suitable for the Colorado beetle could extend into Britain as far as the north of England. The consequences for agricultural production could be extremely serious.

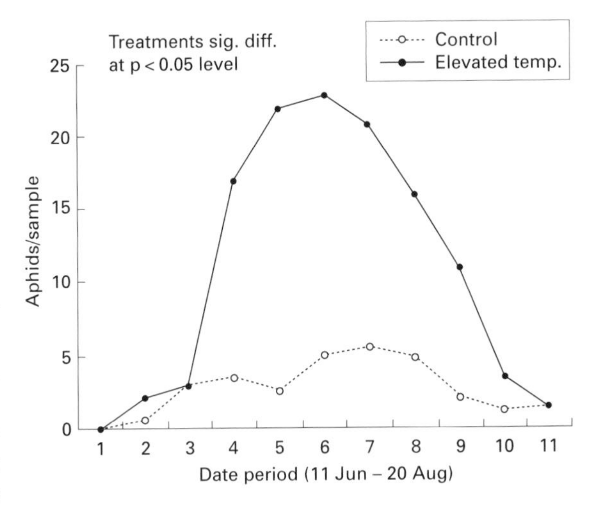

Fig. 2.26 Population densities of aphids comparing control plants with plants placed under cloches and experiencing mean temperature elevations of 2.8°C at the leaf surface. (From Strathdee *et al.* 1993.)

It is not just pests that may be able to extend their ranges as global warming proceeds. The speckled wood butterfly, *Pararge aegeria* (Lepidoptera: Satyridae), has already expanded its populations into new areas over the last decade or so, as shown in Fig. 2.28 (Pollard *et al.* 1996). This species feeds as a larva on various grass species, and is normally found at the edges of woodland rides or hedges where there is a mixture of dappled sunshine and shade. The new sites that have been colonized in recent years tend to be towards the east of England. It is not clear how significant the role of weather has been in these expansions; habitat change must also play a part, but it is clear that, ecologically, these new sites are now much more suitable for the butterfly than they once were.

Not all effects of warming are as predictable or clear as the foregoing examples. As has already been mentioned, rises in temperature will affect not just the insect; if it is a herbivore, the host plant may also be influenced. One of the best known insect–plant interactions in ecology must be that of the winter moth, *Operophtera brumata* (Lepidoptera: Geometridae), and its host plant, oak. It has long been known that the

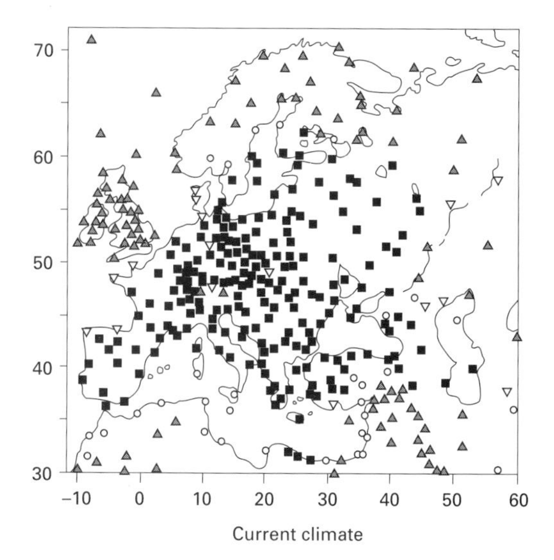

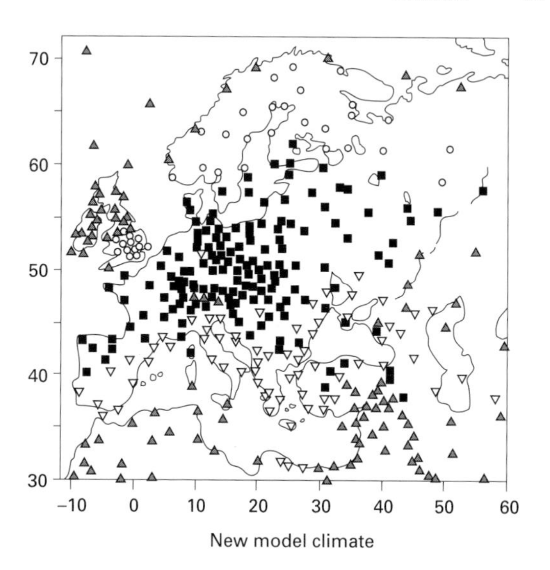

Fig. 2.27 Changes in potential geographical distribution of Colorado beetle in Europe, resulting from a doubling of CO_2 levels and hence an increase in global mean temperature of 4 ± 2°C. Squares, locations within the present distribution that fall into the new model's prediction; circles, locations currently unoccupied but that are potentially suitable under the new model; inverted triangle, locations currently occupied but becoming marginal under new model; triangle, locations currently unoccupied and likely to remain so under new model. (From Jeffree & Jeffree 1996.)

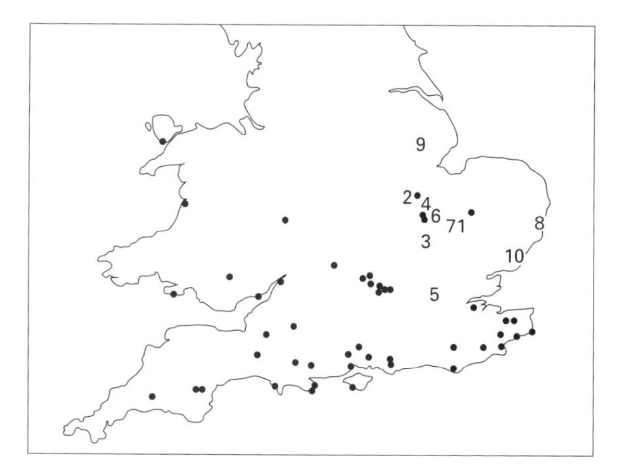

Fig. 2.28 Location of sites with populations of *Pararge aegeria*. Filled circle, populations present at the start of monitoring and monitored for 6 years or longer. Numbered sites are those colonized during the period of monitoring in order of assumed colonization: 1, Chippenham Fen; 2, Barnack Hills and Holes; 3, Potton Wood; 4, Holme Fen; 5, Hampstead Heath; 6, Woodwalton Fen; 7, Wicken Fen; 8, Walberswick; 9, Moor Farm; 10, Stour Wood. (From Pollard *et al.* 1996.)

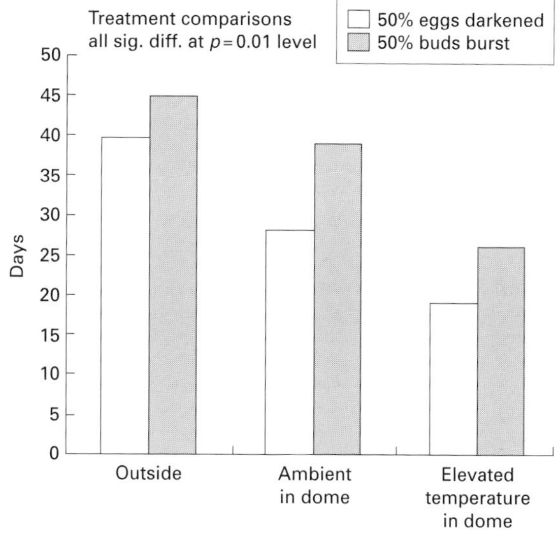

Fig. 2.29 Time taken (from 1 March 1994) for 50% egg darkening and 50% oak bud burst in ambient temperatures inside and outside a solar dome, and with a +3°C elevated temperature. (From Buse & Good 1996.)

synchrony between hatching winter moth larvae and the budburst of oaks in the spring is critical for the moth's success (Wint 1983), and hence any climate-induced changes in this synchrony would have fundamental consequences for the insect's population dynamics. Buse and Good (1996) looked at the effect of a 3°C rise on this synchrony using experimental manipulations in solar domes, and found that both eggs and trees were similarly advanced under higher temperature regimes (Fig. 2.29). As the eggs hatched more rapidly, and the buds burst earlier, the synchrony was not upset. The 4–5 day delay between egg hatch and maximum leaf flush measured for larvae in field trials was also maintained at the higher temperatures.

Finally, let us not forget that aquatic insects might also be influenced by a rise in water temperature, in similar ways to their terrestrial counterparts. Hogg and Williams (1996) manipulated the temperature of a Canadian stream to investigate the effects of a 2–3.5°C rise in water temperature. Various stream arthropods were studied, and Fig. 2.30 presents the results for one species of insect, the stonefly, *Nemoura trispinosa* (Plecoptera). Stonefly nymphs are commonly found on the gravelly bottoms of flowing streams, where they mainly feed on algae, though some can also be predators. The warmer stream system saw earlier emergences of adult stoneflies, with new imagos appearing on the wing about 2 weeks in advance of those from the cooler water.

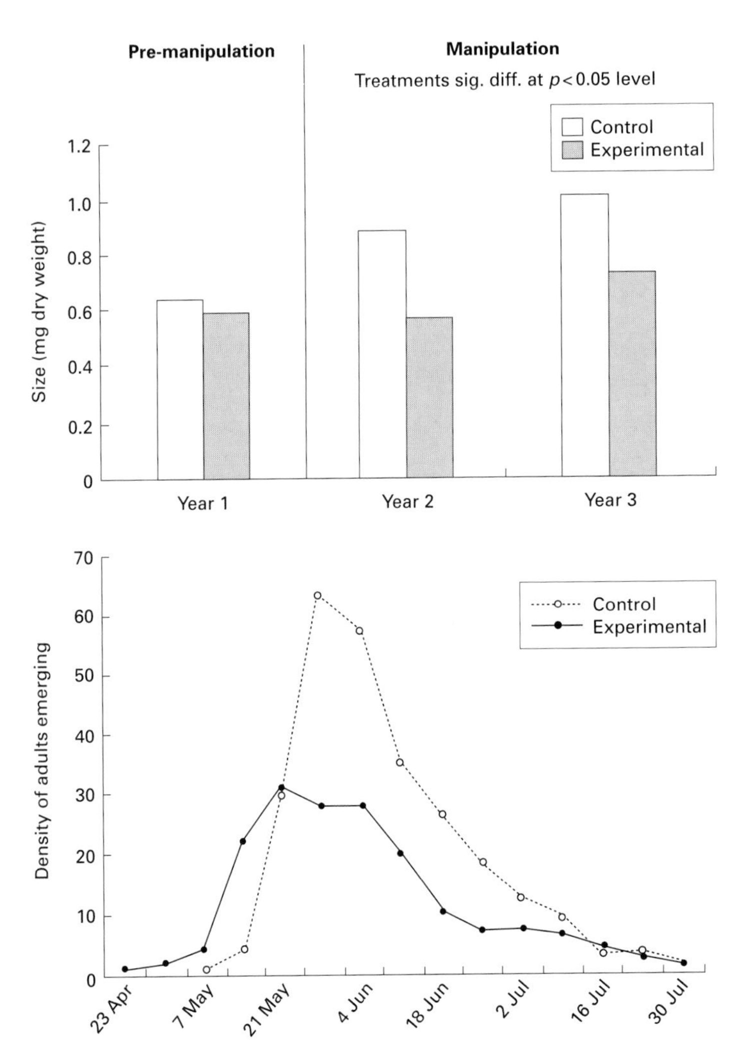

Fig. 2.30 Effect of higher stream temperature on mean size and emergence dates for female stoneflies. (From Hogg & Williams 1996.)

However, the density of new adults was somewhat reduced by warmer conditions, and size of adult females was significantly lower in warm than in cool water. There are obvious consequences for the ecology of this species. Smaller and fewer adults would tend to indicate that the population would be less reproductively successful, simply as a function of lowered fecundities. There may also be genetic implications. Hogg and Williams suggested that in areas where different streams are at varying temperatures, the earlier emergence of adult insects from the warmer ones might provide sufficient time for dispersal to the colder regions and hence a promotion of interbreeding between different habitats.

2.7 Conclusion

It is clear that climatic variables such as temperature, wind and rain can have enormous consequences for the ecology of insects. Many aspects of host exploitation, growth, reproduction and dispersal are fundamentally influenced by these abiotic factors, but it must be remembered that, in general terms, these interactions cannot act in a density-dependent fashion (see Chapters 1 and 4). Instead, most climatic effects are density independent, and, in isolation, cannot regulate insect populations around equilibrium densities. However, abiotic factors can cause large fluctuations in insect abundance, even if they do not regulate populations. Moreover, they can interact markedly with biotic factors such as competition and predation. The relative impacts of climate, competition and predation on insect population change will be considered in some detail at the end of Chapter 5.

Chapter 3: Insect Herbivores

3.1 The trouble with plants as food

As was pointed out in Chapter 1, over 50% of extant insect species feed on plants. The fact that these insect species belong to only eight major orders suggests that, while the evolution of phytophagy may involve fundamental hurdles for insects, the radiation that results from 'jumping those hurdles' is dramatic (Southwood 1961). What are the hurdles associated with eating plants? Difficulties include adequate nutrition, problems of attachment, maintaining water balance, and battling potential 'evolutionary arms races' during which plants are hypothesized to protect themselves from herbivory by periodic development of novel physical and chemical defences (Chapter 6). Plant tissue contains lower concentrations of nitrogen than insect tissue, and gaining sufficient nitrogen appears to be a fundamental problem for insect herbivores (McNeill & Southwood 1978; Mattson 1980; Fig. 3.1). Moreover, the proportions of various amino acids differ between animal and plant proteins, and insect tissues have a higher energy content than those of land plants (Strong *et al.* 1984). Overall, these differences are reflected in the low assimilation and growth efficiencies of insect phytophages compared with those of predatory insects (2–38% for phytophages vs. 38–51% for predators, Southwood 1973).

Of course, before an insect can take a bite of plant tissue, it has to be able to attach itself to the plant. Insects exhibit a multitude of adaptations for holding on to plants, including modified tarsi, the ability to roll leaves, and the capacity to live within plant tissue (leaf miners, gall formers, stem borers) (Juniper & Southwood 1986). For insects that live on the exterior surfaces of plants, such as Orthoptera, many Lepidoptera, and Thysanoptera, there is the added problem of minimizing water loss when exposed to the drying influence of air currents. Adaptations to avoid water loss include a waterproof exoskeleton (morphological), rolling leaves or producing webbing (behavioural, e.g. Hunter and Willmer 1989), and the reabsorption of water in the hindgut (physiological). Of course, maintaining water balance is necessary for flight, and therefore adult insects might be considered in part preadapted for phytophagy.

Finally, insect herbivores have to overcome the myriad of defences exhibited by plants. Physical defences include tough leaves, hairs and spines (trichomes), sticky secretions, and shiny surfaces. Chemical defences include a dazzling array of both toxic substances (e.g. nicotene) and compounds that may reduce the digestibility of food (e.g. some tannins). Plant defences will be discussed in detail later in this chapter.

3.2 Feeding strategies of herbivorous insects

Just as there are many ways to skin a cat, there are many different ways that phytophagous insects eat plant tissue. Most easily observed are those that live freely on the surface of plants, for example leaf-chewing insects, that consume portions of leaves. Grasshoppers and locusts, stick insects, many Lepidoptera and Coleoptera, and some Hymenoptera are considered to be leaf-chewers, and they share similar cutting mandibles (Plates 3.1 & 3.2, opposite p. 158). When leaf-chewing insects become very abundant, they can defoliate their host plants. The gypsy moth, *Lymantria dispar* (Lepidoptera: Lymantriidae), and the winter moth, *Operophtera brumata* (Lepidoptera: Geometridae), are examples of important defoliators of deciduous forest in North America and Europe, respectively.

Not all free-living insects are leaf-chewers. Some (but far from all) insects that suck the phloem of plants live exposed on plant surfaces. Phloem-sucking insects are dominated by the Hemiptera-Homoptera. These include well-known insects such as aphids, plant hoppers, leaf hoppers, and shield bugs (Plate 3.3, opposite p. 158). The mouthparts of sucking insects are modified to form stylets to penetrate plant tissue. Many sap-sucking insects are impor-

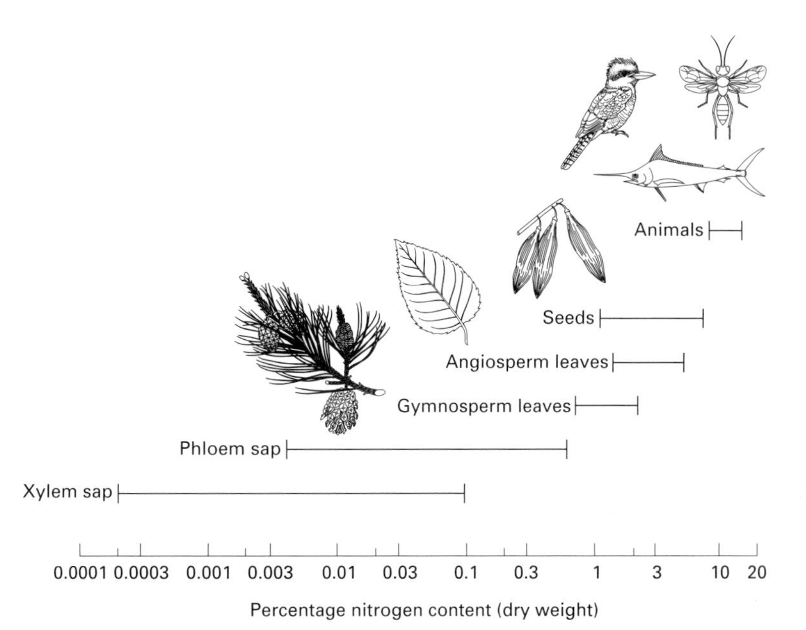

Fig. 3.1 A comparison of the nitrogen content of plant and animal tissue. Low nitrogen levels may have represented a fundamental hurdle to the evolution of phytophagy in insects. (From Strong *et al.* 1984.)

tant agricultural pests, not only because they reduce the growth of their host plants, but also because they can spread plant pathogens that can decimate crop production. Diseases transmitted by insects will be considered in detail in Chapter 9.

Other insect species with sucking mouthparts feed on xylem rather than phloem. They consume one of the most nutrient-poor parts of the plant (especially due to low nitrogen concentration) and yet are both diverse and abundant. Spittlebugs (Homoptera: Cercopidae) are the most polyphagous herbivores known. The spittle that surrounds these insects as nymphs is derived from a combination of fluid voided from the anus and a mucilagenous substance excreted from epidermal glands on the seventh and eighth abdominal segments. While we might question the personal hygeine habits of spittle bugs, anyone who has walked through a meadow in summer can testify to their abundance. Cicadas (Homoptera: Cicadidae) are also xylem feeders and can be so abundant that their biomass over one hectare is greater than the biomass of cows over one hectare of rangeland. Some cicadas are 'periodical' and take 13 or 17 years to develop.

Yet other insects feed between the upper and lower surfaces of leaves, and such species are called leaf-miners. Leaf-miners occur in four orders of insects (Coleoptera, Hymenoptera, Diptera, and Lepidoptera) and the larvae leave characteristic trails or blotches on leaves as they feed (Plate 3.4, opposite p. 158). High densities of leaf-miners, such as often occur on white oak in eastern North America, can significantly reduce the photosynthetic area of plants. Other leaf-miners, particularly in the Dipteran family Agromyzidae, are important pests of agricultural crops. A fourth group of phytophagous insects form galls on the leaves, shoots, stems, and reproductive parts of their host plants. Galls are made when the insects subvert the natural development of growing tissue of the plant so that it forms a chamber in which the nymphs or larvae live and feed (Plate 3.5, opposite p. 158). Most galling insects are Diptera, Hymenoptera, or Hemiptera-Homoptera, although some mites (also in the phylum Arthropoda, but not related to insects) form similar gall-like structures on plants.

Another important strategy for feeding on plants is stem- or shoot-boring. Many of the most serious timber pests are wood-boring species such as the bark beetles (Coleoptera: Scolytidae), some longhorn beetles (Coleoptera: Cerambycidae), and jewel beetles (Coleoptera: Buprestidae). Bark beetles can form extensive galleries in the phloem of trees, and cut off nutrient supplies to and from the crown. Like sucking insects, wood-boring species are often associated with pathogens, particularly fungi, that can infect living trees and cause extensive mortality. Although many wood-boring species can only attack dead, dying, or stressed trees, the most serious timber pests will, if numbers are sufficiently high, attack and kill healthy trees (e.g. *Dendroctonus micans* in Europe and *D. pseudotsugae* in the United States) (Plate 3.6, opposite p. 158). Other strategies for feeding on plants include root feeding, seed predation, flower predation, and nectar consumption (either during pollination, or as a 'nectar robber'). Many feeding strategies employed by insects, particularly those relying on the digestion of cellulose, are enhanced by interactions with mutualistic organisms such as bateria and fungi. Insect–mutualist interactions are considered in detail in Chapters 6 and 7.

3.2.1 Classifying insects into feeding guilds

One assumption that many insect ecologists have made is that insects that feed in similar ways should exhibit similar ecologies. This is the basis of the 'feeding guild' as described by Root (1973). Insect herbivores can be divided among a series of guilds based on their feeding strategy, such as gall-former, sap-sucker, leaf-chewer, or leaf-miner. These guilds are not taxonomically based; leaf-chewing insects, for example, occur in at least five disparate orders. However, many ecologists have found it useful to group insects into guilds in order to study the ecological interactions between insects, their hosts, their natural enemies, and climate. Guilds are useful if, and only if, they provide us with generalities that would be missed by taxonomic studies alone. For example, that wood-boring species appear to suffer more from the effects of intraspecific competition than from the effects of predation is a generality based upon their way of life, not their taxonomy (Denno *et al.* 1995, see Chapter 4 for details). Likewise, that sap-sucking insects may be less vulnerable to phenolic-based defences (e.g. tannins) in plant tissue than are leaf-chewers is another generality based on the study of guilds (Hunter 1997). In addition, sap-suckers may belong to the only guild that consistently responds in a positive way to the addition of nitrogen to its host-plant (Kyto *et al.* 1996). However, some authors have found little evidence for common ecological factors operating on members of the same feeding guild (Fritz *et al.* 1994). They argue that different species within the same feeding guild respond in idiosyncratic ways to ecological pressures, and that generalities made at the guild level are suspect. The utility of grouping insects by feeding guild remains a matter of debate and further experimental work will be needed to determine the value of grouping insect herbivores by feeding strategy for ecological analysis.

3.3 Plant defences

Unless a species of insect herbivore benefits its host by pollination, seed dispersal, or confers some other advantage to the plant (see section 3.4), the impact of insects on plants should generally be deleterious. The removal of photosynthetic area, the loss of pollen, nectar or seeds, the introduction of pathogenic organisms, or the interuption of nutrient, water, or hormone transport can all be associated with attack by phytophagous insects. If these plant parasites reduce plant fitness, it should be expected that plants will evolve mechanisms by which they can reduce the impact of insect herbivores. The wide variety of traits expressed by plants that are associated with the avoidance, resistance, or tolerance of herbivory are generally called 'plant defences'. All plant defences are based on interference with one of the following steps of herbivory: (i) locating a host; (ii) accepting that host as suitable; (iii) attaching physically in some way to the host; (iv) avoiding natural enemies while on the host; (v) tolerating the microclimate during feeding; and (vi) gaining sufficient nutrition from the host during a period of time defined by the availability of suitable plant tissue and the insect's own life history. Any plant trait that interferes with one or more of the above steps can loosely be defined as a 'defence.'

Before describing the major classes of plant defences, however, it is worth pointing out that not all traits expressed by plants that appear to ameliorate the effects of insect herbivores are necessarily adaptations resulting from depredation by specific insect herbivores (Janzen 1980; Hartley & Lawton, 1990). Similarly, defensive traits that appear tightly linked to particular herbivore species may confer additional advantages to plants beyond those of protection (Schmitt *et al.* 1995). The origin and maintenance of defensive traits will be considered later in this chapter.

3.3.1 Avoidance: escape in space and time

One obvious mechanism by which plants can reduce

levels of herbivory is to be separated in space or time from their major herbivores. A plant might avoid phytophages in space if it were rare, or if its occurrence in a particular area was unpredictable. Rare or ephemeral plant species are, by definition, difficult to find. There is certainly some evidence that widely distributed plants support more species of insect herbivore than do rarer plants. In Chapter 1, data were presented that suggested that the number of insect species associated with British trees increased with the abundance of the tree species (Claridge & Wilson 1977). This pattern was firmly established during studies of agricultural systems where the number of species attacking a crop increased with the area planted. Examples include pests of cacao (Strong 1974), sugarcane (Strong *et al.* 1977), and Agromyzid leaf-miners associated with crops in Kenya (Chemengich 1993). Rare plants are likely to accumulate lower levels of herbivory for both ecological and evolutionary reasons. In ecological time, rare plants might be difficult to find, even by herbivores that are able to feed on the plant should they manage to locate it. The probability of finding a suitable individual will decrease with decreasing plant density. In addition, rare or ephemeral plants may not provide a consistent resource on which insect populations can build over time: if the same plant species occurs in the same place at the same time each year, there is a greater probability that large populations of insect herbivores will develop.

It seems unlikely, however, that 'rarity' is an adaptive strategy exhibited by plants to avoid herbivory. Natural selection should maximize an organism's lifetime reproductive output (or, strictly, its inclusive fitness (Hamilton 1964)). Sacrificing reproductive output to avoid insect herbivores is an unlikely evolutionary outcome. It is more likely that rare and ephemeral plants are responding to other limitations of the environment, such as lack of suitable habitat or competition with other species for limited resources. Most modern theories of plant succession, for example, emphasize limitations imposed by the abiotic environment or competitors to explain the distribution and abundance of plant species (Connell & Slatyer 1977; Tilman 1987). Low levels of herbivory are likely to be a consequence, rather than a cause, of rarity.

There is much better evidence that escaping herbivory in time rather than space is an adaptive strategy exhibited by plants. The phenological development of particular plant tissues (leaves, fruits, seeds, etc.) appears to have been modified in some plant species by the pressures of insect herbivores. Most oak trees in Britain, for example, support lower insect densities and suffer lower levels of defoliation if they burst bud late in the season rather than earlier (Hunter *et al.* 1997; Fig. 3.2). Because leaves generally decline in quality as food with age (Mattson 1980), insect larvae that feed on leaves that are even a few days older than average exhibit slower growth rates and lower pupal weights than larvae feeding on younger leaves (Wint 1983). The selection pressure on plants to burst bud late must be balanced, however, by the needs of a long growing season to accumulate photosynthate (sugars accumulated by plants from carbon dioxide during the process of

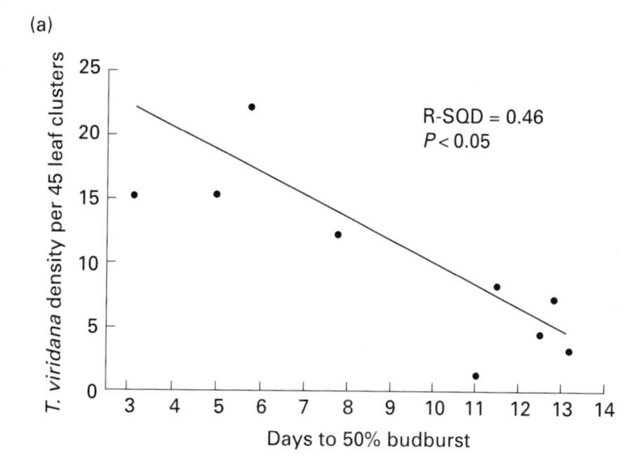

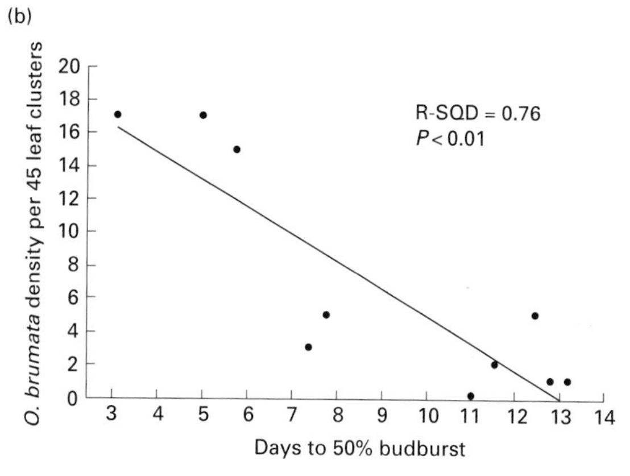

Fig. 3.2 Densities of both (a) *Tortrix viridana* (Lepidoptera: Tortricidae) and (b) *Operophtera brumata* (Lepidoptera: Geometridae) are higher on trees that burst bud early than on those that burst bud late. (From Hunter *et al.* 1997.)

photosynthesis). None the less oaks, which support a species-rich insect fauna because of their abundance and long evolutionary history in Britain, are among the last woodland trees in Britain to burst bud. Similar phenological constraints appear to operate for many tree species in tropical dry forests that burst bud before the start of the wet season. Since trees require relatively large amounts of water during leaf expansion, bursting bud during the dry season would appear counter-productive. However, insect densities are at their greatest during the wetter months, and several authors have suggested that early bud burst reduces the pressures of herbivory on valuable young, expanding leaves (Aide 1992; Borchert 1994). Overall, we can conclude that the degree of synchrony between tissue production by plants and the life-history of insect herbivores has a major impact on the distribution and abundance of insects on plants.

3.3.2 Physical defences

Plants exhibit a variety of physical traits that appear to reduce their susceptibility to insect herbivores. Hairs on the surfaces of leaves, called trichomes, have been associated with reduced levels of herbivory. Trichomes can be either purely structural, presenting simple barriers to attachment or consumption, or glandular, in which case the trichomes secrete sticky or noxious compounds (Fig. 3.3). Variation among different genotypes of soybean and cowpea in the densities of non-glandular trichomes have been correlated with variation in damage by several species of phytophagous insects including pyralid moths and coreid bugs (Turnipseed 1977; Jackai & Oghiakhe 1989). Likewise, whitefly density on 14 genotypes of soybean appears to be related to the erectness of the trichomes on the surface of leaves (Lambert *et al.* 1995). In a manipulative experiment, leaves of the biennial plant mullein, *Verbascum thapsis*, were more readily colonized by the aphid *Aphis verbascae* after the trichomes had been removed with an electric razor (Keenlyside 1989).

At least one species of aphid, *Myzocallis schreiberi* (Homoptera: Calliphididae), appears to have modified both morphology and behaviour to ameliorate the effects of non-glandular trichomes: rather than placing its tarsi flat on the surface of its oak host, *M. schreiberi* literally 'tiptoes through the trichomes' on the bottom surface of *Quercus ilex* leaves by standing on the tips of its tarsi (Kennedy 1986; Fig. 3.4). Other aphid species have stylets that are much longer than is typical for their body size, apparently to reach the surface of leaves through the dense covering of trichomes on the leaves of their hosts (Keenlyside 1989). Despite adaptations by certain insect species to overcome non-glandular trichomes, they are still effective barriers to insect colonization on some plants: differences in aphid density between two sympatric species of alder (*Alnus incana* and *Alnus glutinosa*) appear to be directly related to differences in trichome density (Gange 1995).

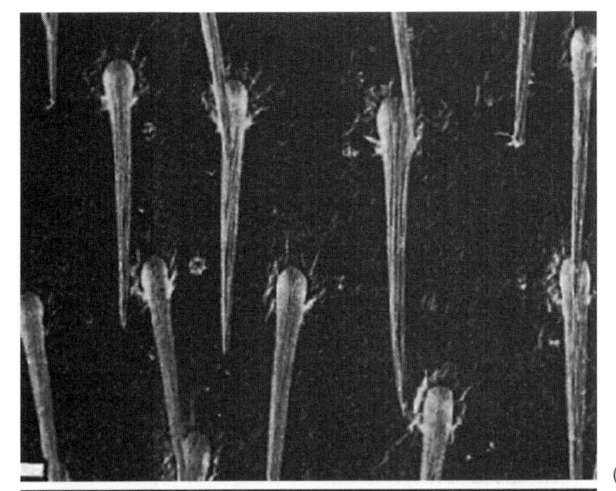
(a)

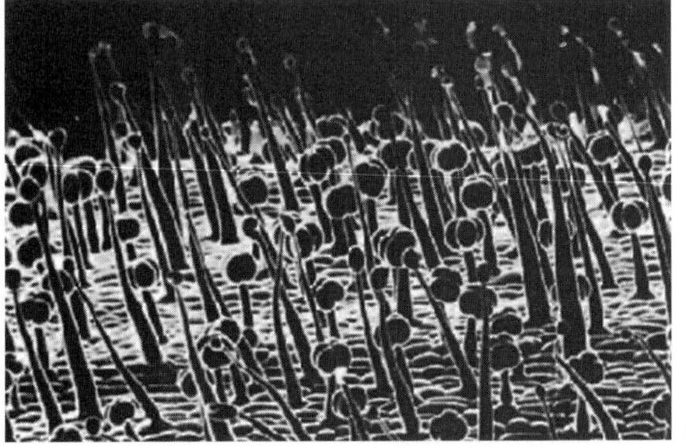
(b)

Fig. 3.3 Trichomes on plant surfaces. (a) Non-glandular trichomes in the pitcher of *Sarracenia purpurea* (From Jeffree 1986). (b) Glandular trichomes on the foliage of *Solanum berthaultii*. (From Ryan *et al.* 1982.)

Glandular trichomes produce a variety of secretions that are either sticky enough to trap potential insect herbivores, or sufficiently noxious to deter herbivory. In a fascinating series of papers (Walters *et al.* 1989a, b; 1990), Ralph Muma and coworkers have demonstrated that there is genetic variation for resis-

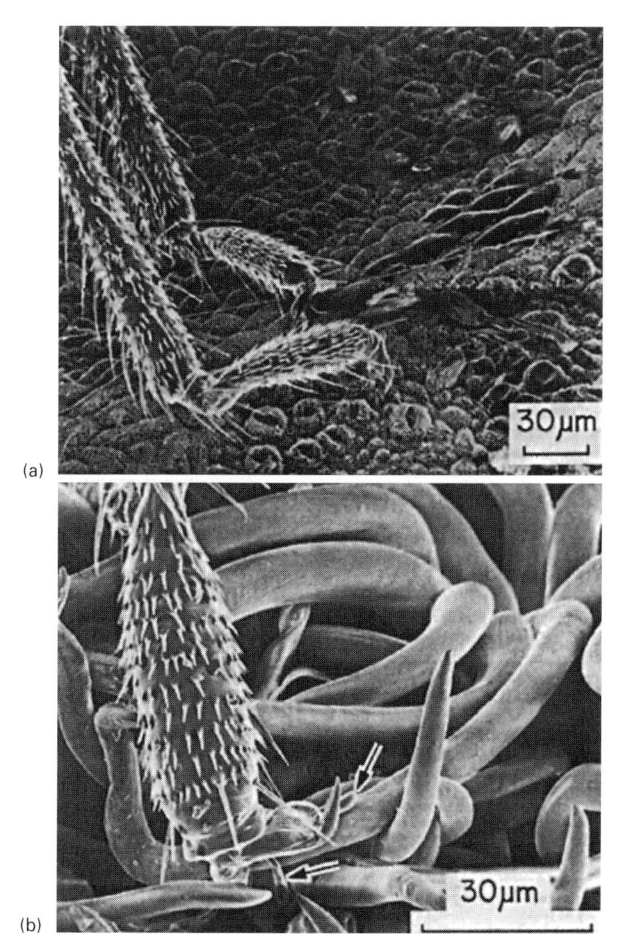

Fig. 3.4 (a) Most aphids place their tarsi on the leaf surface while walking. (b) *Myzocallis schreiberi*, by contrast, walks on tip-toe through the trichomes on the under-side of *Quercus ilex* leaves. (From Kennedy 1986.)

tance to herbivores on *Geranium* plants based on glandular trichomes. *Geranium* trichomes produce sticky secretions that can trap insects, particularly aphids, that attempt to feed on the plants. The 'stickiness' of the secretions varies with genotype, and is responsible for variation in resistance. However, the relative resistance of the genotypes varies with the temperature of the environment. Apparently, the stickiness of the secretions, based on the melting point of the anacardic acid-based compounds produced by the trichomes, varies with temperature. One genotype may be the most resistant at high temperatures and the least resistant at low temperatures. This is an example of a 'genotype-by-environment'

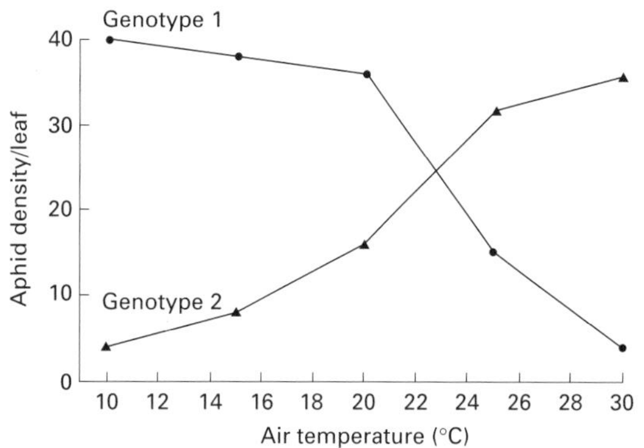

Fig. 3.5 Hypothetical example of a gene-by-environment interaction, where the relative susceptibility of two plant genotypes to an insect herbivore varies with temperature.

interaction, whereby the relative expression of a particular trait such as resistance among different genotypes varies with the environment (Fig. 3.5). Genotype-by-environment interactions are probably a common feature of insect–plant interactions (Fritz & Simms 1992; Weis & Campbell 1992).

Recent work on the glandular trichomes of wild potato, *Solanum berthaultii* (see Fig. 3.3), suggests that the secretions produced by the trichomes contain feeding deterrents that influence host-choice by the Colorado potato beetle, *Leptinotarsa decemlineata* (Yencho & Tingey 1994). Feeding deterrents produced by glandular trichomes combine a physical defence (the hairs) with a chemical defence (the feeding deterrent). Several other physical defences will be familiar to many readers, such as the stinging hairs of nettles and the thorns of Acacia trees. In both these cases, the physical defence is 'inducible', so that the density or length of the defensive structure increases following attack. Inducible defences are considered in more detail in section 3.3.4.

At the other extreme from trichomes, some authors have suggested that waxy, shiny leaves act as physical defences against insect herbivores. The assumption is that shiny leaves are difficult for insects to hold onto or colonize (Juniper & Southwood 1986; but see Jayanthi *et al.* 1994). For example, leaf-mining by larvae of *Plutella xylostella* (Lepidoptera: Plutellidae) is apparently inhibited by the shiny surface of varieties of glossy-waxed cabbage, *Brassica oleracea* (Eigenbrode *et al.* 1995). However, the important consequence of reduced mining activity is that larvae are more susceptible to a group of generalist natural enemies that are ultimately responsible for lower survival on shiny plants. This is an example of a 'tri-trophic interaction' whereby variation in plant quality influences herbivorous insects indirectly by changing their susceptibility to natural enemies (Price *et al.* 1980). In the study by Eigenbrode *et al.* shiny leaves were shown to improve foraging efficiency of the natural enemies of *P. xylostella*. This illustrates the importance of experiments that examine the *mechanisms* underlying patterns observed by insect ecologists: without the experimental work, the natural conclusion might have been that glossy leaf surfaces acted directly to reduce colonization by the insect herbivore.

A second factor that can confound studies of the effects of shiny leaves on insect herbivores is that glossy leaves are generally the result of surface waxes, and waxes can contain a variety of chemical compounds that either stimulate or deter feeding by insects. For example, the surface waxes of the tomato cultivar *Lycopersicon hirsutum* f. *typicum* contain three sesquiterpenes, carbon-based compounds that are thought to act as defences. Removal of the sesquiterpenes by washing the leaves with methanol increases the 10-day survival of beet armyworm, *Spodoptera exigua*, larvae from 0% to 65% (Eigenbrode *et al.* 1994). Indeed, surface extracts of plants such as wild tobacco, *Nicotiana gossei*, have been shown to exhibit antiherbivore activity against a number of commercially important insect pests, and are under investigation for the development of 'biorational' pesticides (Neal *et al.* 1994). Moreover, surface waxes can contain compounds that stimulate feeding or oviposition by insect herbivores. The wild crucifer *Erysium cheiranthoides*, for example, contains both glucosinolates that stimulate oviposition by the cabbage butterfly, *Pieris rapae*, and cardenolides that deter oviposition (Hugentobler & Renwick 1995). One important conclusion from the study of plant surface characteristics is that different defensive traits (trichomes, waxes, defensive chemicals, etc.) interact in complex ways to determine the ultimate susceptibility of plants to their insect herbivores.

Most insect ecologists agree, however, that tough plant tissue generally acts as an antiherbivore defence (Feeny 1970; Mattson 1980, but see Feller (1995)). Growing tissue is generally softer than mature tissue because growing tissue requires the flexibility to expand. After expansion, the leaves of most plants become tougher. The factors that cause leaves to become tough are still a source of debate (Casher 1996; Choong 1996). However, the thickness of the leaf per unit area, the number of structural bundle sheaths, the amount of lignin, and the volume of leaf occupied by cell wall appear to be the primary factors dictating toughness. Lignin and cellulose, neither of which are easily digested by insects (Chapters 6 and 7), can be considered as very general plant 'defences' in that they are consistently associated with tough or indigestible tissues.

Variation in leaf toughness, within and among species, is often associated with variation in levels of herbivory. For example, authors have reported negative correlations between leaf toughness and insect development (Stevenson *et al.* 1993), insect preference (Bergivinson *et al.* 1995a,b), and levels of defoliation (Sagers & Coley 1995). In both temperate and tropical systems (Hunter 1987; Coley & Aide 1991), the vast majority of defoliation that an individual leaf receives occurs during the first few weeks of growth. The leaves of some tropical plants appear to be almost invulnerable to herbivores after leaf expansion is complete (Coley & Aide 1991).

There are several potential mechanisms by which tough leaves could reduce insect densities and subsequent levels of defoliation on plants. Tough leaves may take longer to consume, and therefore result in slower growth rates of insect herbivores (Stevenson *et al.* 1993). Slow growth rates are thought to increase the probability of mortality from natural enemies or adverse climatic conditions. Alternatively, larval and adult insects may avoid individuals or species with tough leaves, thereby leading to lower levels of herbivory (Roces & Hoelldobler 1994). Finally, one study has demonstrated that leaf toughness is associated with mandibular wear in leaf-feeding beetles on willow (Raupp 1985). Mandibular wear causes yet further declines in feeding rate and extends beetle development.

As was noted for other defensive traits, leaf toughness is often correlated with other putative defences. For example, within the leaves of several species of deciduous oak, increases in leaf toughness are correlated with declines in water content, declines in nitrogen content, increases in condensed tannin concentrations, and declines in hydrolysable tannin concentrations (Feeny 1970; Hunter & Schultz 1995; Fig. 3.6) (tannins are a major class of carbon-based chemical 'defences' in some plant species, and will be dis-

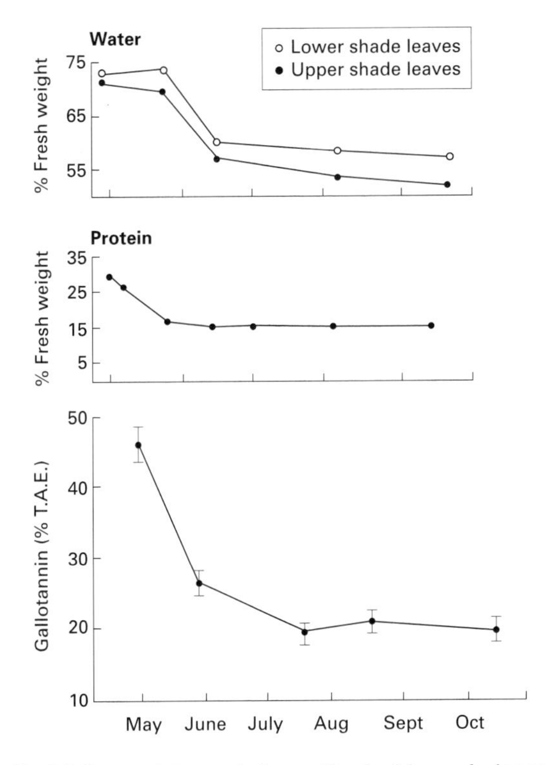

Fig. 3.6 Seasonal changes in the quality of oak leaves for insect herbivores. When several measures of foliage quality change simultaneously, it becomes difficult to assess which has the greatest impact on insect populations. Data collated from Strong *et al.* (1984) and the authors' unpublished data.

cussed in some detail below). In a similar example, lower levels of herbivory on the neotropical shrub *Psychotria horizontalis* are associated with both increased foliar toughness and higher levels of leaf tannins (Sagers & Coley 1995). When two or more plant traits correlate simultaneously with levels of herbivory, it is extremely difficult to determine which factor (or factors) are responsible for the observed trend. Even experimental work can be inconclusive: treatments that cause increases or decreases in leaf toughness often cause increases or decreases in other plant traits at the same time (Foggo *et al.* 1994; Sagers & Coley 1995). However, the frequency with which leaf toughness is correlated with low levels of herbivory in studies of plant–insect interactions makes it a strong candidate as an adaptive response by plants to reduce the pressures of herbivory.

One further physical defence of plants is worthy of note because of its importance in forest pest management. Trees that are attacked by wood-boring beetles often respond with copious flows of resin at the site of attack. The resin, released under pressure, is thought to physically exclude beetles from the tree by a process called 'pitching out'. Resin flow is considered by most forest entomologists to be an important component of the resistance of trees to attack by bark beetles. 'Oleoresin pressure' is a measure of the flow rate of resin from wounds to the boles of trees, and is affected by levels of water deficit of trees. The reduction in resin pressure that can follow sustained drought conditions is thought to be responsible, in part, for the increased susceptibility of some trees to bark beetle attack following periods of low rainfall (Schwenke 1994; Lorio *et al.* 1995). However, even an apparently simple defence such as 'pitching-out' may be more complicated than it first appears. A variety of volatiles in resin are used by bark beetles to locate their hosts (Hobson *et al.* 1993) and are components of the pheromones used by beetles to attract conspecifics (Birch 1978). At the same time, other resin volatiles appear to act as deterrents (Nebeker *et al.* 1995). It seems likely that the resin flow caused by beetle attack has both physical and chemical components that influence the success of colonizing beetles in complex ways.

3.3.3 Chemical defences

The field of insect–plant interactions has been dominated by studies of the effects of plant chemistry on the preference (choices made by individuals) and performance (growth, survival, and reproduction) of insect herbivores. Since the classic paper by Ehrlich and Raven (1964) that considered apparent associations between host-choice by butterflies and the occurrence of specific chemical compounds in the tissues of their hosts, there has been an explosion of studies that consider the effects of plant compounds on the distribution, abundance, and behaviour of phytophagous insects. Thousands of different chemical compounds have been isolated from plant tissue and there are undoubtedly many more yet to be identified. Some common classes of plant-derived compounds are given in Table 3.1. Many of the chemical compounds isolated from plant tissue appear to have no primary metabolic function for either growth or reproduction. Consequently, these compounds are called 'secondary plant metabolites' (or secondary compounds) and are often considered to serve defensive functions (Harborne 1982; Rosenthal & Beren-

Table 3.1 Common classes of secondary metabolite in plant tissue; the numbers of structures of each are probably underestimates and new plant compounds are discovered daily by phytochemists. (After Harborne 1982.)

Class	Approx. number of structure	Distribution	Physiological activity
Nitrogen compounds			
Alkaloids	5500	Widely in angiosperms, especially in root, leaf and fruit	Many toxic and bitter tasting
Amines	100	Widely in angiosperms, often in flowers	Many repellent smelling; some hallucinogenic
Amino acids (Non-proteins)	400	Especially in seeds of legumes but relatively widespread	Many toxic
Cyanogenic glycoides	30	Sporadic, especially in fruit and leaf	Poisonous (as HCN)
Glucosinolates	75	Cruciferae and ten other families	Acrid and bitter (as isothiocynates)
Terpenoids			
Monoterpenes	1000	Widely, in essential oils	Pleasant smells
Sesquiterpene lactones	600	Mainly in Compositae, but increasingly in other angiosperms	Some bitter and toxic, also allergenic
Diterpenoids	1000	Widely, especially in latex and plant resins	Some toxic
Saponins	500	In over 70 plant families	Haemolyse blood cells
Limonoids	100	Mainly in Rutaceae, Meliaceae and Simaroubaceae	Bitter tasting
Cucurbitacins	50	Mainly in Cucuribitacae	Bitter tasting and toxic
Cardenolides	150	Especially common in Apocynaceae, Asclepiadaceae and Scrophulariaceae	Toxic and bitter
Carotenoids	350	Universal in leaf, often in flower and fruit	Coloured
Phenolics			
Simple phenols	200	Universal in leaf, often in other tissues as well	Anti-microbial
Flavonoids	1000	Universal in angiosperms, gymnosperms and ferns	Often coloured
Quinones	500	Widely, especially Rhamnaceae	Coloured
Other			
Polyacetylenes	650	Mainly in Compositae and Umbelliferae	Some toxic

baum 1991). Berenbaum (1995) has pointed out, however, that primary metabolites also influence the preference and performance of insects on plants. In other words, the ratios of certain amino acids and proteins might be considered to be 'defensive'. And certain enzymes, such as proteinase inhibitors (themselves protein-based), influence the ability of insects to extract nutrition from plant material (Green & Ryan 1972) and are therefore 'defensive'. Clearly, it is difficult to make generalities about what kinds of compounds are defensive and Duffey and Stout (1996) have suggested that terms such as 'toxin', 'digestability-reducer', and 'nutrient' signify ecological outcomes rather than specific properties of molecules. In addition, it would be naïve to assume that all secondary plant metabolites represent evolutionary adaptations directed against insect herbivores (Schultz 1992). Indeed, one single 'secondary compound', such as the dihydrochalcone phloridzin from apple leaves, has been considered to be a metabolic end product (Challice & Williams 1970), an auxin repressor (Growchowska & Ciurzynska 1979), a rooting cofactor (Bassuk *et al.* 1981), an antifungal defence (Raa 1968; Tschen & Fuchs 1969), an antibacterial defence (MacDonald & Bishop 1952), and an antiherbivore defence (Montgomery & Arn 1974; Hunter *et al.* 1994), depending upon the focus of the investigator. Moreover, insects are just one of many types of natural enemy that might influence the evolution of plant chemical defences (Denno & McClure 1983; Hunter *et al.* 1992). Clearly, plant compounds can play a variety of primary and secondary roles, and searching for a single dominant selective agent for the production of specific chemicals in plant tissue may often be unrealistic.

None the less, there is overwhelming evidence that intra- and interspecific variation in the production of plant chemicals can influence the preference and performance of insect herbivores (Rosenthal & Berenbaum 1991). Until recently, about 80% of insect herbivore species were thought to be specialists (monophagous), feeding on plant species within a single family. Conversely, 20% were thought to be oligophagous or polyphagous (feeding on several, or many plant families, respectively). These figures come from early studies of the British insect fauna such as aphids (Homoptera: Aphididae) (Eastop 1973) and British Agromyzidae (Spencer 1972) that show high levels of specialization. Recent work in the tropics (Chapter 8) has questioned whether monophagy world-wide is really as prevalent at 80% of the fauna, and realistic figures may be difficult to obtain.

However, it is clear that many insect species will feed on just a few plant species, and host chemistry is thought to be the dominant factor responsible for the tight linkage between most insect herbivores and specific plant taxa (Ehrlich & Raven 1964; Becerra 1997; but see Bernays and Graham 1988). Effects of plant chemistry on insects can occur over evolutionary time and ecological time. The evolution of insect–plant associations is considered in detail in Chapter 6, and we concentrate on interactions in ecological time in this chapter. Suffice it to say that there is good evidence to support a dominant role for plant chemistry in the evolution of insect–plant associations (Ehrlich & Raven 1964; Becerra 1997).

Effects of secondary metabolites on insects in ecological time can be either negative (apparently defensive) or positive (acting as oviposition or feeding stimulants). The glucosinolates present in many crucifers, for example, appear to deter most generalist herbivores, but act as oviposition stimulants for the cabbage butterfly, *Pieris rapae* (Hugentobler & Renwick 1995). It seems logical that specialist insect herbivores would use a variety of cues, including plant chemistry, to locate and initiate consumption of their given hosts, and there are dozens of examples in the literature of oviposition stimulants (e.g. Huang *et al.* 1994; Nishida 1994; Honda 1995; Tebayashi *et al.* 1995) and feeding stimulants (Mehta & Sandhu 1992; Bartlet *et al.* 1994; Nakajima *et al.* 1995) isolated from plant tissue.

3.3.4 Broad classes of chemical defence

In an endeavour to find some commonality among the many thousands of secondary metabolites present in the tissues of plants, there have been several attempts to categorize chemical defences based on their expression or mode of action. One such attempt was based on dividing chemical defences between two broad groups; digestibility reducers and toxins (Feeny 1976; Rhoades & Cates 1976). These authors argued that the type of defence that a plant would exhibit should depend upon its probability of colonization by insect herbivores. Long-lived and abundant plant species would, over evolutionary time, be colonized by a large diversity of herbivore species. Because the reproductive rate of insects is so much higher than the reproductive rate of long-lived plants, insects would be likely to develop resistance to specific toxins in plant tissue in much the same way that pests develop resistance to insecticides (Chapter 10). It was argued therefore that long-lived, or 'apparent' plants, should invest in compounds that reduced the digestibility of their tissues rather than in narrow-action toxins (Feeny 1976; Rhoades & Cates 1976). Tannins, large polyphenolic molecules (Fig. 3.7), were considered to be examples of secondary plant metabolites that reduced the digestibility of plant tissue. Specifically, tannins have the property of precipitating protein from solution by hydrogen bonding (and, under certain conditions, by covalent bonding), and consequently can render plant proteins unavailable for digestion by insect herbivores. Conversely, rare or ephemeral plant species ('unapparent' plants) should be more likely to invest in toxins because their probability of frequent encounter by herbivores, and therefore the selection pressure for resistance, should be lower (Feeny 1976; Rhoades & Cates 1976). Broadly

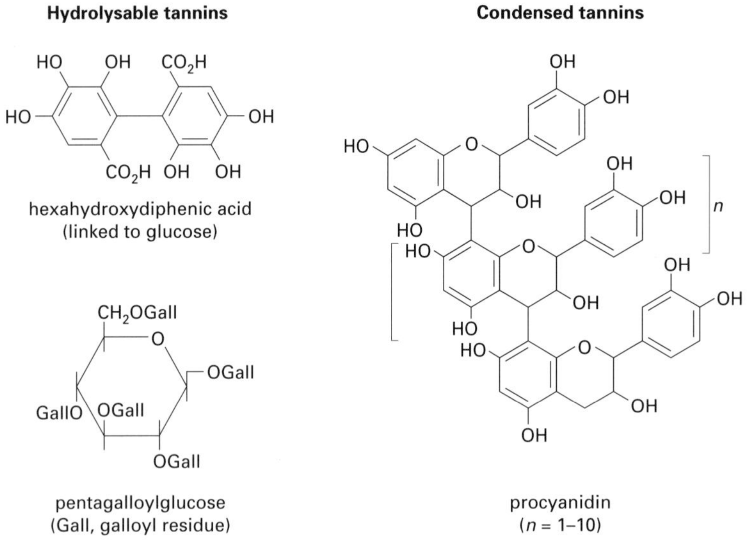

Hydrolysable tannins

hexahydroxydiphenic acid
(linked to glucose)

pentagalloylglucose
(Gall, galloyl residue)

Condensed tannins

procyanidin
(n = 1–10)

Fig. 3.7 Two major classes of tannin in plant tissue. Hydrolysable tannins are derivatives of simple phenolic acids such as gallic acid and its dimeric form, hexahydroxydiphenic acid, combined with the sugar glucose. Condensed tannins are oligomers formed by condensation of two or more hydroxyflavanol units. Gall, galloyl residue. (From Harborne 1982.)

speaking, tannins do appear to be more common in the tissues of apparent plants than unapparent plants. However, there have been several criticisms levelled at 'apparency theory'.

Firstly, tannins are a very diverse group of secondary metabolites, with a great variety of chemical forms (Zucker 1983; Waterman & Mole 1994). Based on an analysis of structure, it seems likely that some tannins act more like toxins than digestibility reducers (Zucker 1983). Secondly, not all apparent plants invest heavily in compounds that affect digestibility. Maples and aspens, for example, are common tree species that invest as much in toxin-like molecules as in tannins (Schultz *et al.* 1982; Lindroth & Hwang 1996). Thirdly, apparency theory requires that there are significant barriers to the evolution of adaptations by insects to combat digestibility-reducing compounds. Yet many insect species maintain conditions in their mid-gut, such as high pH (Berenbaum 1980), the presence of surfactants (Martin & Martin 1984) or a specific redox potential (Appel 1993; Appel & Maines 1995) that ameliorate or negate the precipitation of tannin–protein complexes. None the less, there are strong indications that tannins can reduce the performance of a wide variety of herbivores, including insects (Rossiter *et al.* 1988; Bryant *et al.* 1993; Hunter & Schultz 1995; Fig. 3.8).

A second way to classify chemical defences is to divide them between 'constitutive defences' which are present continually in plant tissue, and 'induced defences' which are produced, or increase in concentration, following attack by herbivores (Rhoades 1985). The recognition that some plants respond to defoliation by changing their concentrations of secondary metabolites is relatively recent. Green and Ryan (1972) reported that potato and tomato plants respond to herbivory by Colorado potato beetles (Coleoptera: Chrysomelidae) by producing proteinase inhibitors, compounds that can inhibit certain digestive enzymes in the mid-guts of insects. Subsequent work has demonstrated the existence of

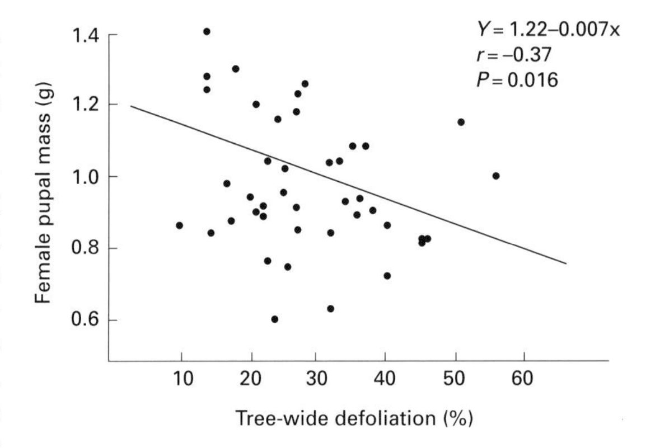

Fig. 3.8 Defoliation of oak trees causes increases in foliar tannin concentrations, which, in turn, reduce female pupal mass of the gypsy moth *Lymantria dispar* (Lepidoptera: Lymantriidae). Because pupal mass is related to fecundity, oak tannin can reduce gypsy moth birth rates. (From Rossiter *et al.* 1988.)

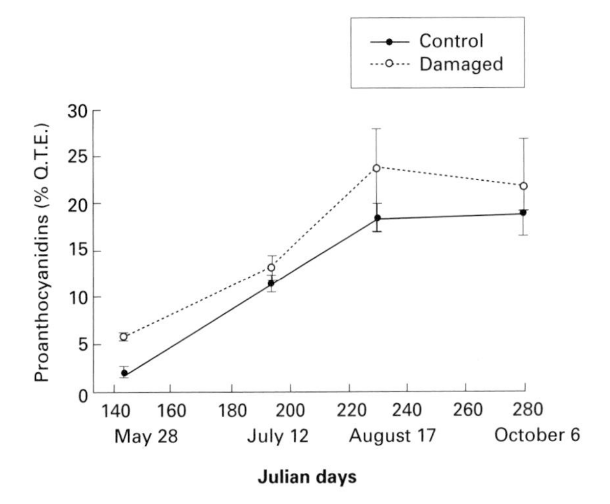

Fig. 3.9 Wound-induced increases in the condensed tannin (proanthocyanidin) content of red oak, *Quercus rubra*, foliage following damage by the gypsy moth *Lymantria dispar* (Lepidoptera: Lymantriidae). Damaged trees suffered about 55% leaf area removed (LAR) compared with 6% on controls. (From Hunter & Schultz 1995.)

induced defences in a wide variety of plant species including birch, oaks, sedges, and alder, to name but a few. Authors have disagreed about whether induced defences are the result of evolutionary pressures on plants by insects, and even whether such defences are powerful enough to influence insect populations (Haukioja & Niemela 1977; Baldwin & Schultz 1983; Edwards & Wratten 1985; Hartley & Lawton 1987). However, this is one area where disagreement has slowly faded over time. There is now little doubt that many, if not most, plant species undergo chemical changes following insect attack, and some of those changes can deter subsequent herbivory. Interested readers should consult a recent text on the subject (Karban & Baldwin 1997). An example of damage-induced increases in the condensed tannins of oak foliage is shown in Fig. 3.9.

A third way to categorize the types of chemical defences that plants exhibit is by referring to the resources generally available to the plants to produce them. The resource-availability hypothesis (Coley *et al.* 1985) asserts that the levels of light, nutrients, and other resources available to plants over evolutionary time influence the type and magnitude of defensive compounds present in different plant species (readers are warned not to confuse this hypothesis with the carbon–nutrient balance hypothesis described in section 3.3.5, which focuses on differences among individual plants in ecological time). According to the resource-availability hypothesis, species native to resource-rich environments usually have rapid growth rates and short leaf lifetimes. Such species should employ low concentrations of mobile defensive substances, such as alkaloids and cyanogenic glycosides (Table 3.1), that can be recovered from leaves prior to senescence. Conversely, plant species typical of low-resource environments have slow growth rates and long leaf lifetimes. These species should protect themselves with high concentrations of immobile compounds such as tannins. In long-lived leaves, a one-time investment in immobile defences should be less expensive than the task of continually replacing mobile defences as they are rapidly turned over (Coley *et al.* 1985).

In the authors' view, the resource-availability hypothesis has not received sufficient empirical investigation as yet. Problems of phylogeny (relatedness among the plant species under comparison) make interpretations of data difficult because related species are likely to express more similar defensive strategies, even in different environments, than are unrelated species (Baldwin & Schultz 1988). We have yet to find any empirical studies that have controlled adequately for phylogeny while replicating sufficiently to test rigorously the hypothesis. Theoretical models of defence allocation suggest that defensive strategy should vary with herbivore pressure (Lundberg & Astrom 1990) and that reallocation of mobile defences may actually be more advantageous in slow-growing species (Van Dam *et al.* 1996). As with all modelling studies, however, the conclusions reached are only as valid as the assumptions built into the models, and we feel that there is much room for further empirical work on resource-based theories of defence allocation among plant species.

An exciting development in the study of plant chemical defences is the recognition that some compounds with antiherbivore activity originate not from the plant itself, but from endophytic fungi that 'infect' leaf tissue. Most of the work on fungal-derived toxins has been done on grasses and sedges, many of which act as hosts to the fungi *Acremonium lolii* and *Balansia cyperi* (Ahmad *et al.* 1985; Clay *et al.* 1985). The fungi produce ergot alkaloids (potent antiherbivore compounds) and Lolitrem A and B (mammalian neurotoxins). The plants appear to benefit from the presence of the endophytes without significant costs to either growth or reproduction (Keogh & Lawrence 1987). Infection by fungi confers resistance in grasses to a wide variety of insect herbivores including two species of armyworm in the genus *Spodoptera* (Lepidoptera: Noctuidae) (Ahmad *et al.* 1987; Clay & Cheplick 1989), the Argentine stem weevil *Listronotus bonariensis* (Coleoptera: Curculionidae) (Prestidge & Gallagher 1988), house crickets *Acheta domesticus* (Ahmad *et al.* 1985), and the hairy chinch bug *Blissus leucopterus* (Hemiptera: Lygaeidae) (Mathias *et al.* 1990). The production of ergot alkaloids by fungi in grasses may even be 'induced' by herbivore damage (Bultman & Ganey 1995). Although not all insects are equally susceptible to the fungal toxins, they are under consideration as novel pesticides (Findlay *et al.* 1995). Recently, studies of endophyte–insect interactions have moved beyond grasses to include oaks, firs, and cottonwood trees (Wilson 1995; Findlay *et al.* 1995; Gaylor *et al.* 1996). It is too early to say whether endophytes have the same general effects on the herbivores of trees that they have on the herbivores of grasses, but it promises to be an important area for future research.

3.3.5 Spatial and temporal variation in chemical defences

Insect herbivores foraging for food in the environment face a complex mosaic of chemical defence in space and time. As previously described, plant species vary enormously in the types and concentrations of chemical defence in their tissues. Even within one plant species, however, there can be significant variation in chemical defence among different tissues of the same individual, among different individuals in the same population, and among different populations of the same species (Denno & McClure 1983; Hunter *et al.* 1992). For example, concentrations of the phenolic aglycone phloretin, the hydrolysis product of the dihydrochalcone phloridzin, are two and a half times higher in terminal shoot leaves than in the fruit leaves of apple trees (cv Delicious) (Hunter & Hull 1993; Fig. 3.10). Spatial variation in apple leaf phenolics is associated with the distribution of apple herbivores within individual trees (Hunter *et al.* 1994). Recently, spatial variation in the distribution of organisms has become increasingly important to the study of their population dynamics (Hassell *et al.* 1987; Pulliam 1988; Gilpin & Hanski 1991, see also Chapter 4). For insect herbivores, spatial variation in secondary metabolites is an important component of their distribution in space and, consequently, a potential force influencing population change (Hunter *et al.* 1996).

Temporal changes in the concentrations of plant secondary metabolites are also well documented. The concentrations of tannins in oak foliage, for example, change as the leaves age. Gypsy moth (*Lymantria dispar*) larvae that feed on oak leaves in spring in the north-eastern United States feed during the period in which leaf chemistry is changing most rapidly (Fig. 3.11). Oak leaf tannins influence the growth rates and fecundity of gypsy moth (Rossiter *et al.* 1988; Fig.

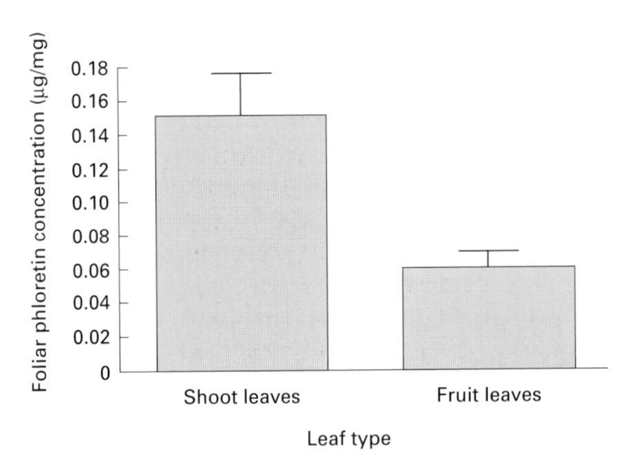

Fig. 3.10 Within-tree variation in the secondary chemistry of apple foliage. Phloretin concentrations are 2.5 times higher in shoot leaves than in fruit leaves. (From Hunter & Hull 1993.)

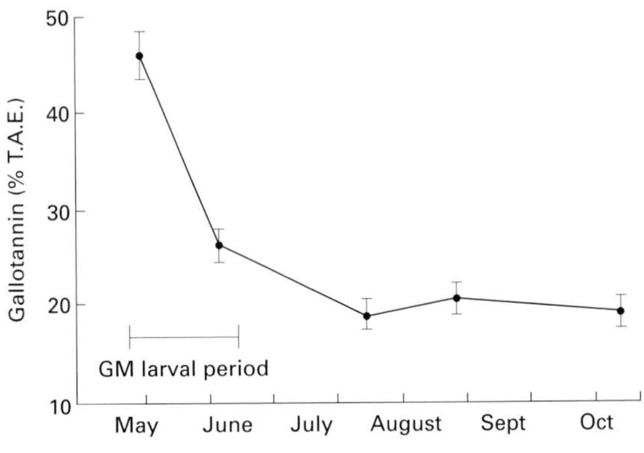

Fig. 3.11 Temporal changes in gallotannin concentrations in the foliage of red oak. Gypsy moth larvae (GM) feed during the period when foliage quality changes most rapidly. (M.D. Hunter, unpublished data.)

3.9). Likewise, the concentration of phenolic resin on the exterior leaf surfaces of the creosote bush, *Larrea tridentata*, declines as the leaves increase in size. Age-specific concentrations of phenolic resin have been shown to influence the feeding behaviour of creosote bush insects such as the Argentinian grasshoppers in the genus *Astroma* (Schultz 1992).

What are the origins of intraspecific variation in chemical defence? Like most traits exhibited by organisms, phytochemical variation arises from genetic variation, the influence of the environment, and genotype-by-environment interactions (Fritz & Simms 1992). There is good evidence that different plant genotypes produce different concentrations of secondary metabolites. For example, the production of furanocoumarins, toxins activated by sunlight that are found in wild parsnip plants, varies with plant genotype (Zangerl & Berenbaum 1990). In turn, spatial variation in furanocoumarin production determines foraging patterns of the parsnip webworm, *Depressaria pastinacella* (Lepidoptera: Oecophoridae) (Zangerl & Berenbaum 1993). In fact, genetically based variation among individuals within a plant species is one prerequisite for the evolutionary and coevolutionary theories that have dominated the literature of plant–insect interactions (Ehrlich & Raven 1964; Feeny 1976; Rhoades & Cates 1976; Edmunds & Alstad 1978) and continues to be a fundamental component of entomological research.

Additionally, environmental variation can have a direct effect of the secondary metabolism of plants. Svata Louda and colleagues have shown, for example, that gradients in water availability of only 30 m in length in prairie systems cause predictable changes in the concentrations of methylglucosinolates in *Cleome serrulata* (Capparacaeae) (Fig. 3.12). In turn, methylglucosinolate concentrations influence the levels of leaf damage and seed predation that plants suffer from insects (Louda *et al.* 1987a). Likewise, experimentally induced increases in light availability result in decreases in certain glucosinolate compounds in *Cardamine latifolia* (Brassicaceae), and cause four-fold increases in subsequent herbivory (Louda & Rodman 1983).

One theory that has attempted to explain environmentally induced variation in certain plant defences is the carbon: nutrient balance hypothesis (Chapin 1980; Bryant *et al.* 1983; 1993). According to this theory, carbon-rich defensive compounds such as tannins and terpenes should occur in greater concentrations in low-nutrient or high-light environments. Briefly, individual plants growing under low-nutrient conditions have high carbon/nutrient ratios. Carbon-rich secondary metabolites act as sinks for 'excess' carbon, and pathways that generate carbon-rich defences are favoured. Under high-nutrient conditions, however, carbon/nutrient ratios are lower, and pathways associated with growth and reproduction are favoured over defence. The carbon/nutrient balance hypothesis has provided a framework for much valuable research on environmentally based variation in plant defence, yet it remains controversial (Gershenzon 1994). The results of some experiments in which light or nutrient conditions are varied provide equivocal results. None the less, a significant number of studies support, at least partially, the carbon : nutrient balance hypothesis (Muzika & Pregitzer 1992; Bryant *et al.* 1993; Holopainen *et al.* 1995; Hunter & Schultz 1995).

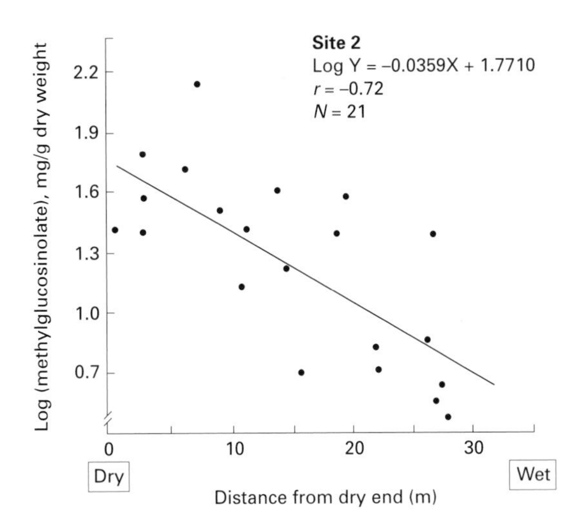

Fig. 3.12 A decline in the concentration of methylglucosinolates in the foliage of *Cleome serrulata* from the dry to the wet end of a 30 m soil moisture gradient. Insect damage to both leaves and seed capsules increases as methylglucosinolate concentrations fall. (From Louda *et al.* 1987a.)

An additional hypothesis, based on the availability of resources for plants, is the 'plant stress hypothesis' of insect outbreak (White 1974; 1984). According to White, plants that are under physiological stress represent higher quality of food for insect herbivores than plants growing under 'optimal' conditions. The mechanism underlying the original hypothesis was that some plants respond to stress with increases in soluble nitrogen and free amino acids in their tissues. Most support for the plant stress hypothesis comes from studies of forest or range insects, particularly those of economic importance. For example, drought stress increases the susceptibility of Norway spruce to the spruce needle miner, *Epinotia tedella* (Lepidoptera: Olethreutidae) (Muenster-Swendsen 1987). Susceptibility of balsam fir stands in Quebec to attack by spruce budworm, *Choristoneura fumiferana* (Lepidoptera: Tortricidae) is also associated with water availability (Dupont *et al.* 1991). Likewise, several studies have demonstrated that *Eucalyptus* species under moisture stress become more susceptible to wood-boring (Paine *et al.* 1990; Hanks *et al.* 1995) and leaf-chewing (Stone & Bacon 1994) insects. Where the mechanism of increased susceptibility under stress has been investigated, the conclusions have been varied. Certainly, there is some evidence that higher nitrogen availability increases the quality of stressed plants for insect herbivores (Gange & Brown 1989; Thomas & Hodkinson 1991). However, not all plants respond to stress in this way (Louda & Collinge 1992), and other mechanisms have been suggested. For example, stress-induced changes in leaf size (Stone & Bacon 1994), leaf toughness (Foggo *et al.* 1994), plant architecture (Waring & Price 1990), resin production (Schwenke 1994; Lorio *et al.* 1995), and plant physiology (Louda & Collenge 1992) have also been associated with increased susceptibility to insect attack. Table 3.2 lists some insect species that have been shown to increase in density following drought stress.

It should be no surprise that the plant stress hypothesis, like most hypotheses in insect ecology, has been the subject of considerable controversy (Larsson 1989). At least some of the controversy arises from the fact that there are many different kinds of stress, many idiosyncratic responses by

Table 3.2 Some outbreaks of forest and range insects associated with drought. (From Mattson & Haack 1987.)

Species	Family	Genus of host	Location	Reference
Coleoptera				
Agrilus bilineatus	Buprestidae	*Quercus*	USA	Haack & Benjamin (1982)
Corthylus colambianus	Scolytidae	*Acer*	USA	McManus & Giese (1968)
Dendroctonus brevicomis	Scolytidae	*Pinus*	USA	Vite (1961)
D. frontalis	Scolytidae	*Pinus*	USA	Craighead (1925); St George (1930)
D. ponderosae	Scolytidae	*Pinus*	Canada	Thompson & Shrimptom (1984)
Ips calligraphus	Scolytidae	*Pinus*	USA	St George (1930)
I. grandicollis	Scolytidae	*Pinus*	USA	St George (1930)
		Pinus	Australia	Witanachchi & Morgan (1981)
Ips spp.	Scolytidae	*Pinus, Picea*	Europe, Africa	Chararas (1979)
Scolytus quadrispinosa	Scolytidae	*Carya*	USA	Blackman (1924); St George (1930)
S. ventralis	Scolytidae	*Abies*	USA	Berryman (1973); Ferrell & Hall (1975)
Tetropium abietis	Cerambycidae	*Abies*	USA	Ferrell & Hall (1975)
Homoptera				
Aphis pomi	Aphididae	*Crataegus*	Switzerland	Braun & Fluckiger (1984)
Cardiaspina densitexta	Psyllidae	*Eucalyptus*	Australia	White (1969)
Hymenoptera				
Neodiprion sertifer	Diprionidae	*Pinus*	Sweden	Larsson & Tenow (1984)
Lepidoptera				
Bupalus piniarius	Geometridae	*Pinus*	Europe	Schwenke (1968)
Choristoneura fumiferana	Tortricidae	*Abies, Picea*	Canada	Wellington (1950); Ives (1974)
Lambdina fiscellaria	Geometridae	*Abies, Picea*	Canada	Carroll (1956)
Lymantria dispar	Lymantriidae	*Picea*	Denmark	Bejer-Peterson (1972)
Orthoptera				
Several species	Acrididae	Grasses	World-wide	White (1976)

plants, and equally diverse responses by insect herbivores. Although water deficit is a common cause of stress in plants, other factors such as mammalian browsing and sun exposure (Linfield *et al.* 1993), root disturbance (Gange & Brown 1989; Foggo *et al.* 1994), and nutrient availability (Mopper & Whitham 1992) can alter susceptibility to insect herbivores. The timing, duration, location, and intensity of stress have also been shown to affect the responses of colonizing insects (Louda *et al.* 1987b; English-Loeb 1989; Mopper & Whitham 1992; McMillin & Wagner 1995). Even the genotype of individual plants within a population determines the magnitude of its stress responses (Mopper *et al.* 1991; McMillin & Wagner 1995). Perhaps most importantly, not all insects respond to plant stress in the same way. For example, although both leaf-chewing and leaf-mining insects occur at higher densities on stressed individuals of the crucifer *Cardimine cordifolia*, aphid densities are unaffected by stress (Louda & Collinge 1992). In a striking example, Waring and Price (1990) described the responses of eight species of leaf-galler in the genus *Asphondylia* (Diptera: Cecidomyiidae) to water stress of their host plant, creosote bush (*Larrea tridentata*). Five species were more abundant on stressed plants, two on unstressed plants, and one showed no preference (Table 3.3). Even within one genus, then, insects vary in their responses to plant stress.

Direct contradictions to the plant stress hypothesis are also common. Corn seedlings, for example, respond to water stress with increased concentrations of DIMBOA and DIBOA, two cyclic hydroxamic acids that deter both insect pests and pathogens (Richardson & Bacon 1993). Leaf-miners in the genus *Cameraria* occur in greater numbers on irrigated oak trees than on oaks under water stress (Bultman & Faeth 1987). The assumption that outbreaks of the pine beauty moth, *Panolis flammea* (Lepidoptera: Noctuidae) on lodgepole pine were caused by site-induced stress on deep peat soils was refuted by experimental manipulation: survival of larvae did not differ between 'stressed' and 'unstressed' trees (Watt 1988).

Table 3.3 Mean densities (per 10 g foliage) of eight galling insects in the genus *Asphondylia* (Diptera: Cecidomyiidae) on water-stressed and unstressed creosote bushes, *Larrea tridentata*; data from Waring & Price (1990) and species identification numbers follow their designations.

Species	Density (unstressed)	Density (stressed)	Site favoured
1	7.0	27.0	Stressed
2	6.6	6.6	None
3	10.4	18.9	Stressed
6	2.7	6.2	Stressed
7	30.7	144.1	Stressed
8	4.4	2.2	Unstressed
10	3.0	0.1	Unstressed
11	7.1	30.5	Stressed

The most direct assault on the plant stress hypothesis has been the 'plant vigour hypothesis' which contends that insect herbivores prefer, and perform better on, vigorous, not stressed, plants (Price 1991). Detailed studies by Peter Price and colleagues on gall-forming sawflies in the genus *Euura* and their willow hosts has provided a large body of evidence that more vigorous plants, and the most vigorous parts within plants, are preferred by some herbivores (Price 1989; 1990; 1991; Fig. 3.13). The plant vigour hypothesis has been supported by Price and others for an increasing number of insect–plant associations such as 13 of 15 phytophages on mountain birch trees (Hanhimaeki *et al.* 1995), five species of tropical insect herbivore on five different host plants (Price *et al.* 1995), the rosette galls of *Rhopalomyia solidaginis* (Diptera: Cecidomyiidae) on *Solidago altissima* (Raman & Abrahamson 1995), and leaf-galling grape phylloxera on clones of wild grape (Kimberling *et al.* 1990).

Even insect herbivory on eucalypts, considered to be a classic case of the plant stress hypothesis (Paine *et al.* 1990; Stone & Bacon 1994; Hanks *et al.* 1995), appears to be controversial: Landsberg and Gillieson (1995), after a landscape-level analysis of eucalypt production, soil factors, and climate, found that insect herbivory was positively correlated with eucalypt production and unrelated to stress. It is difficult to reconcile the plant stress hypothesis and the plant vigour hypothesis into a single theory to explain patterns of insect attack. Price (1991) points out that the vigour hypothesis may be better suited to galling and shoot-boring insects that rely on rapidly growing tissue, yet examples from other insect guilds exist as well. The key point, however, is that variation in the resources available to plants (environmental variation) can influence the susceptibility of plants to attack by insect herbivores.

3.3.6 Plant hybrid zones

A subset of the studies on plant genotype and insect

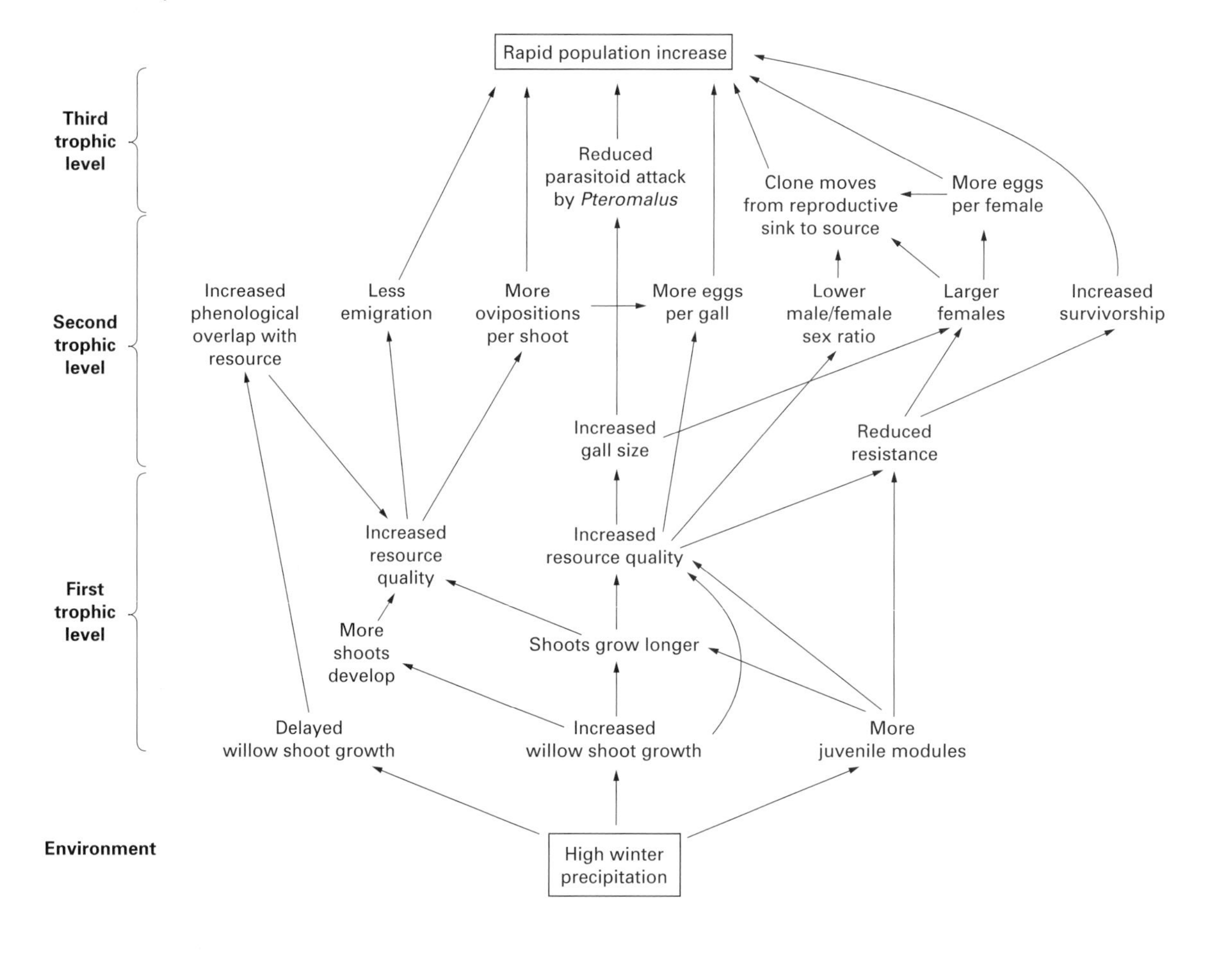

Fig. 3.13 Price and coworkers have argued that more vigorous plants are more likely to be attacked by insect herbivores. In this diagram, populations of the gall-forming sawfly, *Euura lasiolepis*, increase on the vigorous willow plants that result from high levels of winter precipitation in Arizona. The effects of precipitation on plant vigour are felt throughout the trophic web. (From Price *et al.* 1998.)

herbivores has focused on plant hybrid zones, regions where two or more congeneric species of plant produce hybrid offspring. Insect ecologists became interested in plant hybrid zones when Whitham (1989) demonstrated that hybrid zones of *Populus* species acted as 'sinks' for insect herbivores. In other words, hybrid poplars supported higher densities of insect herbivores and suffered higher levels of herbivory than either of the parental genotypes. The assumption was that the defences of the parental species (chemistry, morphology, phenology, etc.) were somehow weakened or disrupted by hybridization. The concordance between insect densities and parental vs. hybrid genotypes can be so strong that the distributions of insects can be used to segregate closely related plant taxa, their hybrids, and even complex backcrosses (Floate & Whitham 1995).

The 'hybrid susceptibility' hypothesis has been controversial. Boecklen and Spellenberg (1990) reported that densities of leaf-mining Lepidoptera and gall-forming Hymenoptera were both lower in two different hybrid zones of oak (*Quercus* species), a direct contradiction of the hypothesis (Fig. 3.14). In subsequent work, Aguilar and Boecklen (1992) found that leaf-miners and gall-formers in a third oak hybrid zone occurred at intermediate densities when compared with the parental oak genotypes. They argued that the responses of insect herbivores to hybridization were idiosyncratic, depending both upon the insect and hybrid zone in question. The efficacy of natural enemies of herbivores is also influenced by plant hybridization (Preszler & Boecklen 1994; Gange 1995), a so-called tri-trophic interaction (Price *et al.* 1980). Although several authors feel that there is no general response by insect herbivores to plant hybrid zones (Fritz *et al.* 1994; Gange 1995), it appears that at least some hybrid zones act as sinks for herbivores (Floate *et al.* 1993; Siemens *et al.* 1994; Whitham *et al.* 1994; Christensen *et al.* 1995). Perhaps the best conclusion at present is that hybrid zones represent regions of strong selection pressure on insects and plants, and are worthy of further study.

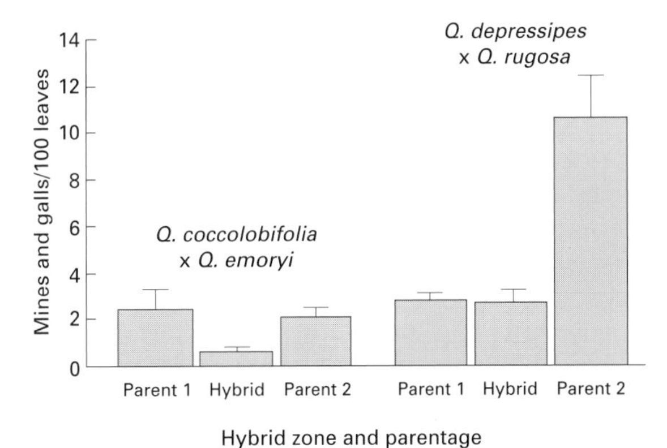

Fig. 3.14 Combined densities of galling and mining insects in two oak (*Quercus*) hybrid zones. In neither hybrid zone are insects more common on hybrid trees than on parental trees. Data are the means of 100 trees per genotypic class. (From Boecklen & Spellenburg 1990.)

3.3.7 Natural enemies of herbivores as components of plant defence

One final type of plant defence, often described as an 'indirect defence', is the attraction of natural enemies of insect herbivores (predators or parasitoids) by plants. Plants can provide a variety of resources for predators and parasitoids, including shelter, food, and information on the location of their herbivorous prey. One illustration of an indirect defence is the extra-floral nectaries produced by some plant species (Fig. 3.15). Extra-floral nectaries usually play no role in pollination because they are distant from the sites of pollen transfer. Rather, they act as sources of food (sugars and amino acids) used by natural enemies. The assumption is that by attracting natural enemies such as ants, predatory wasps, or parasitoids, plants will gain an advantage by increased levels of predation on their insect herbivores. Extra-floral nectaries and ants have been collected from 35 million-year-old fossils of *Populus*, suggesting that this kind of

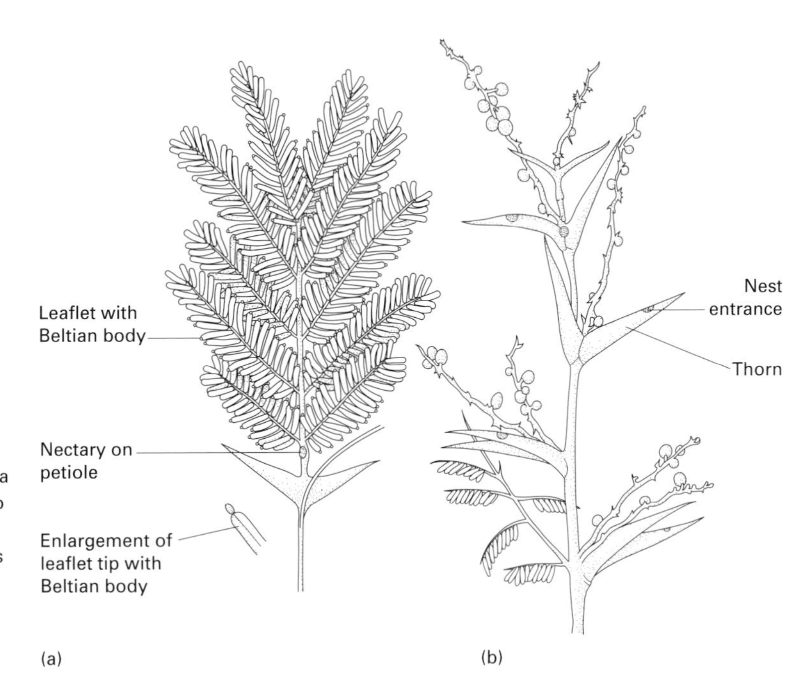

Fig. 3.15 Features of a bull's horn acacia that assist in its mutualistic relationship with ants. (a) Extra-floral nectaries and Beltian bodies provide ants with sugars and proteins, respectively. (b) Hollow thorns provide shelter for *Pseudomyrmex* ants. (From Wheeler 1910; Price 1996.)

ant–plant mutualism (where two organisms both benefit from an association, Chapter 6) may be very ancient (Pemberton 1992). There is evidence, for example, that the extra-floral nectaries of bracken, *Pteridium aquilinum*, attract wood ants, *Formica lugubris*, in Britain. Ants have been shown to lower the population densities of some native insect herbivores on bracken, although their greatest impact may be to deter colonization by naïve, or maladapted, herbivores (Strong *et al.* 1984). Extra-floral nectaries and ant–plant mutualisms are discussed in some detail in Chapter 6.

Plants can also provide shelter for natural enemies. Some ant–plant associations are obligate, called myrmecophytism, whereby a single species of trees is always inhabited by a single species of ant (Fiala & Maschwitz 1991, 1992). Ant–*Acacia* associations are classic examples of this phenomenon. Some *Acacia* trees have hollow thorns which are colonized by ants. In return for shelter, the ants are supposed to reduce levels of herbivory on the *Acacias*. For example, ants living within *Acacia farnesiana* in Santa Rosa National Park in Costa Rica remove about 50% of eggs laid by bruchid beetles (seed predators) on seed pods (Traveset 1990). In Kenya, the symbiotic ants associated with *Acacia drepanolobium* are even thought to deter giraffe browsing (Madden & Young 1992). Similarly, the hollow stems (internodes) of *Cecropia* and *Macaranga* trees provide nest sites for ants which are then supposed to reduce levels of defoliation on these rapidly growing tropical trees. The evolution of ant–plant mutualisms will be considered in more detail in Chapter 6.

One of the most intriguing forms of indirect defence exhibited by plants is the production of volatile chemicals called 'synomones' that are meant to attract predators and parasitoids to the site of attack by herbivores. It has long been known that parasitoids can use chemical cues from their insect hosts, or volatiles in insect frass, to locate their prey. However, several studies have demonstrated that volatiles released from damaged plant tissue are also attractive to parasitoids. One early example was the demonstration that *Cyzenis albicans* (Diptera: Tachinidae) is attracted to volatiles from oak leaves that have been damaged by the winter moth, *Operophtera brumata* (Hassell 1976). Recently, the volatile responsible has been identified as borneal. Because borneal is not produced by damaged apple leaves, it may explain the lack of aggregation responses of *C. albicans* to apple tree defoliation that is readily observed on oaks (Roland *et al.* 1995).

A natural extension of the attraction of natural enemies to damaged plant tissue has been the suggestion that, over evolutionary time, plants have developed chemical signals specifically to attract predators and parasitoids. As Turlings *et al.* (1995) have pointed out, to function effectively as signals for natural enemies, the emitted volatiles should be clearly distinguishable from background odours, and specific to the herbivore concerned. There is little value in attracting a parasitoid that cannot attack the herbivore currently feeding on the plant. In addition, the volatiles must be emitted during times when natural enemies forage. An increasing number of studies suggest that synomones meet these criteria. Marcel Dicke and colleagues have done much of the pioneering work on synomones. They have argued that synomones solve the 'reliability-detectability' problem for natural enemies (Vet *et al.* 1991). Insect host cues are very reliable for searching parasitoids, but are hard to detect. Normal plant volatiles are easy to detect, but unreliable because the host may not be present. Synomones are both detectable (plant-derived) and reliable (the product of insect attack). For example, Lima beans, *Phaseolus lunatus*, that are infested with two-spotted spider mites, *Tetranychus urticae*, produce two terpenoids (E)-beta-ocimene and (E)-4,8-dimethyl-1,3,7-nonatriene, that are not produced by uninfested plants. These two terpenoids attract the predatory mite *Phytoseiulus persimilis* which feeds on the herbivorous species (Takabayashi *et al.* 1991; Dicke *et al.* 1993). The terpenoids are not just volatiles released from damaged tissue: Dicke *et al.* (1993) have shown that there is a plant-wide (systemic) response generated by an elicitor that travels from the petiole of infested leaves to other leaves throughout the plant.

Other plant–herbivore–enemy complexes appear to behave in similar ways. Maize plants attacked by the stem-borer *Chilo partellus* produce a plant-wide volatile attractive to the parasitoid *Cotesia flavipes* (Potting *et al.* 1995). Similarly, cabbage plants attacked by the large white butterfly *Pieris brassicae* emit a mixture of volatiles attractive to parasitic wasps (*Cotesia glomerata*) that attack the herbivores. In this case, beta-glucosidase in the regurgitate of *P. brassicae* larvae is thought to act as the elicitor that stimulates synomone production (Mattiacci *et al.* 1995). In other words, the saliva of the insect herbivore provides the cue to the plant to produce the synomones that will attract natural enemies! It may be that plants have evolved the capacity to respond to insect herbivore saliva to manage synomone release. A further example comes from the beet armyworm. A compound named 'volicitin' (*N*-(17-hydroxylinolenoyl)-L-glutamine) from larval saliva was painted onto corn seedlings by Alborn *et al.* (1997). The corn seedlings began to release synomones attractive to the parasitoids of beet armyworm larvae following application of volicitin. There now seems little doubt that some plants produce volatile compounds *de novo* when attacked by insect herbivores, and that herbivore derived cues can be responsible. Judging by the burgeoning literature on the topic, synomone-induced attraction of natural enemies may be much more common than previously thought (Sabelis & Afman 1994; Drukker *et al.* 1995; Turlings *et al.* 1995; Finidori *et al.* 1996).

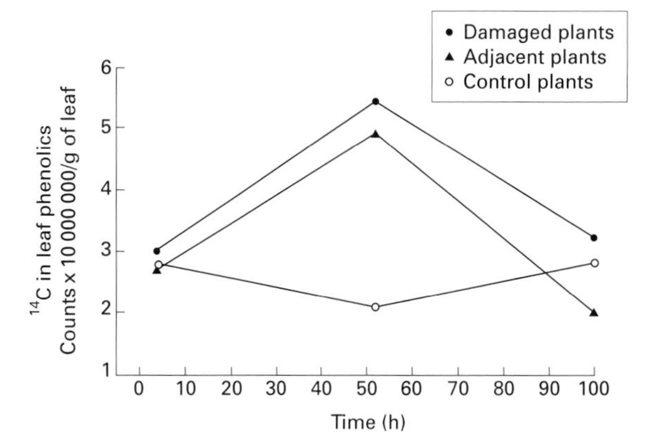

Fig. 3.16 ^{14}C in foliar phenolics of poplar ramets that either received experimental damage, were adjacent to damaged ramets, or were controls. Increased phenolic synthesis in plants adjacent to those damaged experimentally provides evidence for inter-plant communication. Data redrawn from Baldwin and Schultz (1983).

3.3.8 Interplant communication and 'talking trees'

In the course of their studies on synomones, Dicke and colleagues have reawakened interest in an intriguing and controversial subject, communication among individual plants. In 1983, Baldwin and Schultz published a paper in the journal *Science* which appeared to describe interplant communication. They were studying the 'induced defences' of poplars and sugar maples, in which leaf damage causes an increase in polyphenolic concentrations (carbon-rich defences) in their foliage. As well as recording increases in polyphenolics in damaged plants, Baldwin and Schultz (1983) observed increases in the same compounds in adjacent plants that had not been damaged (Fig. 3.16). They argued that undamaged plants must be able to detect volatile cues from damaged plants, and so 'raise their own defences' before attack. Quickly labelled the 'talking tree' hypothesis, this study caught the public imagination, and was reported in the tabloids as well as in scientific journals. In the same year, an independent research study (Rhoades 1983) found a similar phenomenon on willow trees. The idea that plants might be warning each other of impending defoliation was met with disbelief by many. In one dismissive article, Fowler and Lawton (1985) described the results as statistically flawed, evolutionarily unreasonable, and explainable by other, more likely, factors. It is important to realize, however, that Baldwin and Schultz envisaged plants as 'listening' for evidence of herbivory, rather than co-operating with one another by direct signals. None the less, their results proved difficult to reproduce, and interest in the subject declined.

In recent years, the evidence for interplant communication has increased, supporting the original conclusions of Baldwin and Schultz. The synomones produced by damaged Lima bean plants, for example, are apparently detected by undamaged plants, which respond by producing synomones of their own. Specifically, when uninfested leaves of Lima bean are placed on wet cotton wool that has had infested leaves on it previously, the uninfested leaves start to produce synomones that attract predatory mites (Takabayashi *et al.* 1991). In a similar experiment, uninfested cotton seedlings that were exposed to volatiles from infested seedlings became both deterrent to herbivorous mites and attractive to predatory mites (Bruin *et al.* 1992). These studies have caused a reassessment of interplant communication (Bruin *et al.* 1995, Shonle & Bergelson 1995) and should stimulate further research in this fascinating area.

3.3.9 When is a defence not a defence?

It is rarely possible to say unequivocally that a particular trait exhibited by a plant is a direct result of natural selection imposed by an insect herbivore.

What we call 'defences' may have arisen for a number of different reasons and, serendipitously, may influence the preference or performance of insects on their hosts. It is all too easy for entomologists, fixated as they are with insects, to forget that plants are exposed to many other selection pressures from the abiotic (non-living) environment and from other members of the biotic environment (competitors, decomposers, pathogens, and non-insect herbivores). It is likely that natural selection has moulded most traits expressed by plants in response to multiple selection pressures. Trichomes, for example, have properties in addition to deterring herbivores. In Indian mustard, *Brassica juncea*, trichomes act as sites in which toxins from the soil such as cadmium are accumulated (Salt *et al.* 1995). Whether or not environmental toxin accumulation in trichomes led to the later development of trichomes as a defence against insects is unknown. Trichomes can also improve the water status of leaves by reducing water loss and increasing water uptake. The hairs can trap and retain surface water, and assist in its final absorption into the mesophyll (Grammatikopoulous & Manetas 1994). There seems little doubt that trichomes can play a critical role in avoiding drought stress for some plant species.

Likewise, the constitutive and induced chemical defences in most leaf tissue may play more than one role. Induced resistance is thought by some to be an active defence against herbivores (Rhoades 1979, 1985) and by others to be an inevitable result of changes in carbon–nutrient ratios (Myers & Williams 1984; Bryant *et al.* 1983, 1993). Yet chemical changes following the damage of leaf tissue are also associated with resistance to plant pathogens and wound repair (Cadman 1960; Friend 1979; Harborne 1988; Hartley & Lawton 1990; Schultz 1992). Multiple functions can be attributed to most plant 'defences' against insects. Is one function any more important than any other? What ecological factors were most important in their evolutionary development? It is unlikely that we will be able to answer these questions, and it may be pointless to try. It is much more important to understand the current role of putative defences in the ecology of plants than their precise evolutionary origins. The controversy surrounding the tight evolutionary linkages between insects and plant defences will be considered in the discussion of coevolution in Chapter 6.

3.3.10 Costs of resistance and tolerance to herbivores

Why are all plants not maximally and successfully defended against attack by insect herbivores? One possibility is that there is a dynamic, evolutionary 'arms race' between insect herbivores and their hosts, with the development of novel plant defences being followed by adaptations of herbivores to resist these defences. This evolutionary arms race will be considered along with pollination in the chapter on evolutionary ecology (Chapter 6). A second (not mututally exclusive) possibility is that the production of defences is costly to plants. Indeed, this has been a common assumption by insect-plant ecologists for many years, especially those working on induced defences. Presumably the advantage of an induced defence is that it is activated (and the costs paid) only when herbivores are known to be present. No costs are paid when the herbivore is absent (Rhoades 1985). If defences truly are costly for plants, there ought to be trade-offs between the production of defences and other needs such as growth or reproduction. Ultimately, of course, costs should be reflected in reductions in lifetime reproductive success. However, since size and reproductive output are commonly linked in plants, growth is often substituted for reproduction in studies of defence costs.

For example, Coley (1986) studied the costs and benefits of tannin production in a neotropical tree, *Cecropia peltata*. She grew seedlings of equal age under uniform conditions, and measured their tannin concentrations, leaf production, and palatability to insect herbivores. The results demonstrated that plants with higher tannin content suffered lower levels of herbivory (benefit of defence) but produced fewer leaves (cost of defence) (Fig. 3.17). Since *Cecropia* is a pioneer tree species that colonizes disturbed areas, rapid growth is presumably advantageous so that reproduction occurs before competitively superior, but slower-growing trees, dominate the area. *Cecropia* appears to face a trade-off between growth and defence (Coley 1986). Similarly, there is a negative correlation between growth and tannin production in the neotropical shrub *Psychotria horizontalis*. The tannins in this shrub confer some protection to the plants against insect herbivores (Sagers & Coley 1995). Tradeoffs between growth and defence presumably help to explain natural variation in defence

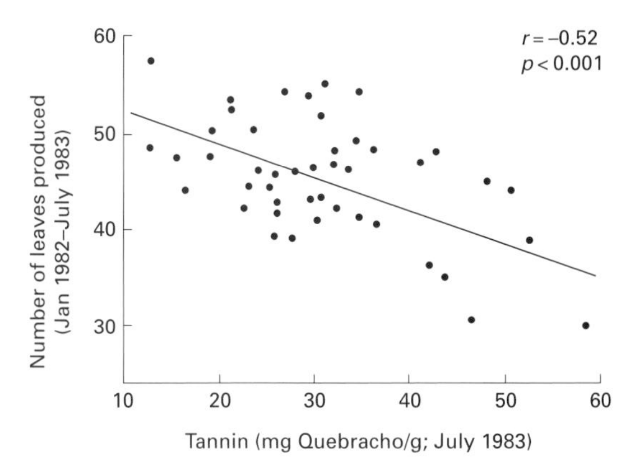

Fig. 3.17 Relationship between tannin concentration in the foliage of *Cecropia* and the number of leaves produced per plant. Investment in tannin appears to occur at the expense of leaf production, and may represent the cost of defence. (From Coley 1986, with permission from Springer-Verlag New York, Inc.)

production among individual plants. Simply put, there is no perfect way to 'be a plant' in the presence of herbivores. Some individuals are heavily defended but grow slowly. Others are less well defended, suffer insect attack, but grow more rapidly. Both strategies can work, both can be maintained in populations over time, and result in the natural variation we observe in plant populations.

However, substituting growth for reproduction when measuring the costs of defence may be misleading. This is illustrated by the perennial plant *Senecio jacobaea*. Although vegetative (growth) costs associated with the production of chemical defences have been measured in *S. jacobaea*, there are no apparent reproductive costs (flowers and seeds) to pyrrolizidine alkaloid production (chemical defence) or the production of tough leaves (Vrieling 1991). Indeed, costs of defence have proven difficult to detect in many different studies, generating some anxiety as to the relevance of evolutionary theories based on optimal allocation to defence (Simms & Rausher 1989; Adler & Karban 1994). Defence costs may not be particularly high if defences last a reasonable length of time without renewal. For example, there may not be as rapid a turnover of some chemical defences, such as the monoterpenes of peppermint, as was previously thought (Gershenzon 1994; Fig. 3.18). Similarly, there are no measurable costs associated with the

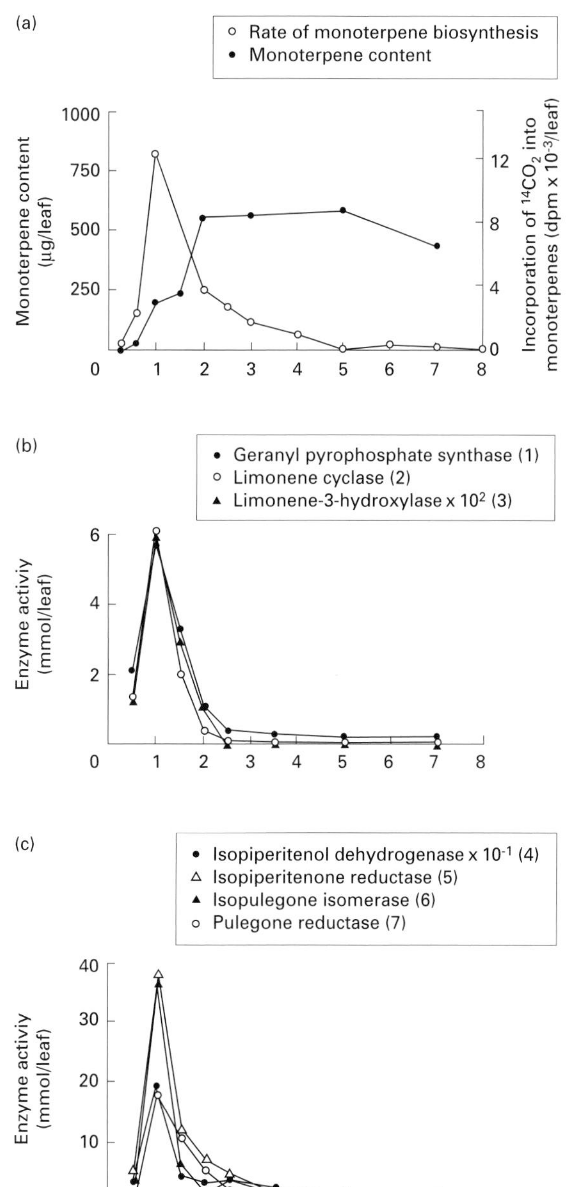

Fig. 3.18 (a) Monoterpene content vs. monoterpene synthesis in peppermint. Content is stable after synthesis declines, showing that monoterpenes are not being turned over. Enzymes involved in monoterpene biosynthesis (b,c) are virtually inactive after leaves are 2 weeks old. (From Gershenzon 1994.)

production of the thorns on bramble plants that deter herbivores (Bazely *et al.* 1991). Measurable costs of defence have proven illusive in the desert shrub, *Gossypium thurberi* (Karban 1993), in *Plantago lanceolata* (Adler *et al.* 1995), and in the annual morning glory, *Ipomoea purpurea* (Simms & Rausher 1989). Indeed, only about 33% of studies have shown detectable costs of resistance. In a recent review of 88 published studies, Bergelson and Purrington (1996) found that costs were most commonly detected in plants resistant to herbicides, followed by pathogens. Costs of resistance to herbivores were detected least frequently. Moreover, costs were detected more often in crop plants than in natural plant communities.

The difficulties associated with measuring resistance costs have led some insect ecologists to explore alternative costs imposed by herbivores. One of these is 'tolerance'. Tolerance is defined as the ability of plants to compensate, in part, for the effects of defoliation on plant fitness. In other words, a plant with high tolerance maintains higher levels of growth or reproduction for a given level of defoliation than a plant with low tolerance. As originally formulated (Van Der Meijden *et al.* 1988), the hypothesis suggests that there is a tradeoff between tolerance and resistance, and that both strategies act to reduce the impact of insect herbivores on plant fitness. Certainly, some plants appear to be able to tolerate herbivory more than others, losing little in overall growth or reproduction despite defoliation (Rosenthal & Kotanen 1994). Given that costs of resistance are difficult to measure, are there measurable costs of tolerance instead? If tolerance is expensive to plants, we might expect to see lower reproductive rates from tolerant plants than less tolerant plants *in the absence of herbivory* (Simms & Triplett 1994) (Fig. 3.19). To date, there are few published studies that have explored the costs of tolerance, but several researchers are actively exploring the issue. For example, Fineblum and Rausher (1995) have studied the effects of damage to the apical meristems of the annual morning glory, *Ipomoea purpurea*, by insects, and have shown that genotypes of *I. purpurea* with relatively high levels of tolerance to this type of damage pay the 'cost' of lower constitutive resistance. On the same plant species, Simms and Triplett (1994) demonstrated that individuals that were more tolerant to disease had lower fitness in the absence of disease, suggesting a significant cost of tolerance. Studies of the ecological and evolutionary implications of tolerance to herbivory are in their infancy, and it may prove a fruitful area for future investigation.

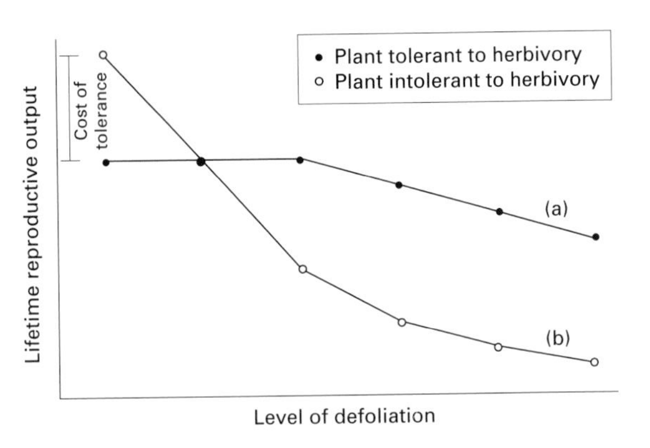

Fig. 3.19 Hypothetical costs associated with a plant's tolerance to insect herbivory. A tolerant plant (a) exhibits a slower decline in fitness than an intolerant plant (b) as defoliation increases. There may be a cost for this tolerance in reduced fitness when herbivores are absent.

3.4 Overcompensation

Some studies of plant tolerance to herbivory have revealed that certain plant species appear to overcompensate for defoliation. In other words, the growth or reproduction of the plant is actually higher in the presence of herbivores than it is in the absence of herbivores. This has led some workers to speculate that some levels of herbivory may be beneficial to plants, and increase their fitness (Dyer & Bokhari 1976; Owen & Wiegert 1976). It has further been suggested that tight coevolution between plants and herbivores may have resulted in 'plant-herbivore mutualisms' by which some degree of defoliation is 'expected' by plants, and rates of plant production rise following damage by herbivores (Paige & Whitham 1987). Given the limited resource budget of plants and the rigours of natural selection, it is difficult to understand why plants that are capable of increasing growth or reproduction following damage would not grow or reproduce maximally in the absence of herbivores. This argument was forcibly made by Belsky *et al.* (1993) who concluded that, 'There is no evolutionary justification and little evidence to support the idea that plant-herbivore mutualisms are likely to evolve. Neither life-history theory nor recent theoretical models provide plausible explanations for the benefits of herbivory.'

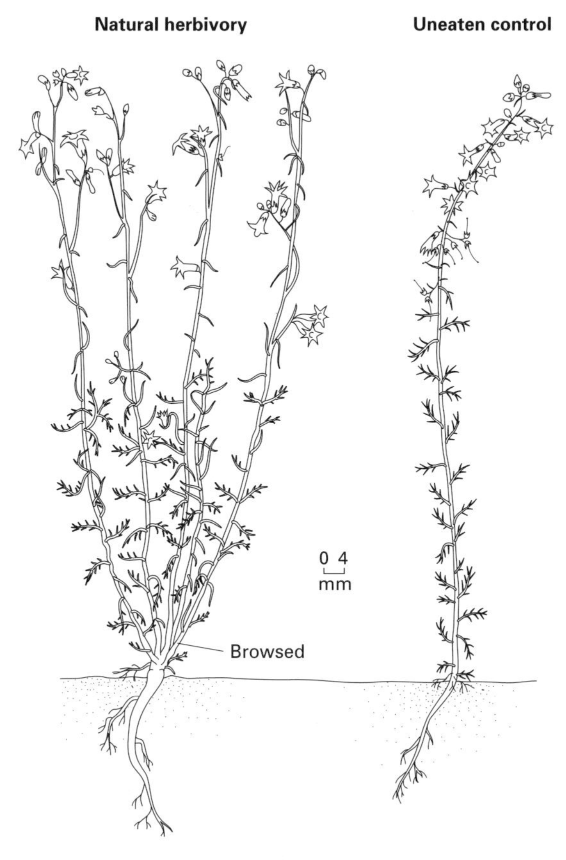

Fig. 3.20 Illustration showing the positive effects of herbivory on scarlet gilia, *Ipomopsis aggregata*. Plants damaged by herbivores have multiple stems and multiple inflorescences. Control plants have a single stem and inflorescence. (Adapted from Paige and Whitham 1987, with permission from The University of Chicago Press.)

Although convincing evolutionary explanations for overcompensation may be illusive, there is now little doubt that overcompensation occurs in some circumstances. For example, flower production by Scarlet Gilia, *Ipomopsis aggregata*, increases following herbivory: undefoliated plants produce only one inflorescence whereas defoliated plants produce multiple inflorescences (Fig. 3.20). The result is a 2.4-fold increase in relative fitness following defoliation (Paige & Whitham 1987; Paige 1992). Similarly, growth of blue grama grass increases following herbivory by the big-headed grasshopper, *Aulocara elliotti* (Williamson *et al.* 1989), and the production of tall-grass prairie increases following defoliation (Turner *et al.* 1993).

Recent work on grasshopper crop and midgut secretions suggests that there may be animal biochemical messengers that indirectly increase the growth rates of defoliated plants by up to 295% (Moon *et al.* 1994; Dyer *et al.* 1995). Termed 'reward feedback' by Mel Dyer and colleagues, these biochemical messengers may exert major control on ecosystem productivity (Dyer *et al.* 1995). In section 3.3.7 above, we described the intriguing notion that plants might recognize chemical messengers in insect herbivore saliva and, in response, produce volatiles to attract parasitoids. We should at least consider the possibility that insect herbivores can also produce salival messengers to manipulate the growth patterns of their host plants. However, several studies have found little or no evidence of overcompensation following defoliation (Bergelson & Crawley 1992; Edenius *et al.* 1993; Oba 1994) and it appears that many factors, including plant species (Alward & Joern 1993; Escarre *et al.* 1996), resource availability (Alward & Joern 1993), the type of defoliation (Hjalten *et al.* 1993), and defoliation history (Turner *et al.* 1993) all influence overcompensation responses. It has been suggested that overcompensation is at one end of a continuum of plant responses to herbivory, from negative to positive, and that we should expect to observe considerable variation in plant responses to defoliation in natural systems (Maschinski & Whitham 1989).

Chapter 4: Resource Limitation

4.1 Sources of limitation on insect populations

Most insects have enormous potential for population growth. Imagine if there was no limitation on the growth of insect populations: what would happen? We can explore the importance of population limitation by imagining a population of insects, growing without any biological constraints (Table 4.1). As an example, we will use data from the winter moth, *Operophtera brumata* (Lepidoptera: Geometridae), in Europe to show that the population becomes very large, very quickly. In fact, if we assume that our *O. brumata* population has a constant rate of growth (or constant per capita rate of increase, λ) we can estimate the number of *O. brumata* in the population at any time in the future from the simple expression

$$N_t = \lambda N_0 \quad (4.1)$$

where N_t is the number of *O. brumata* at time t and N_0 is starting population size. Varley *et al.* (1973) have published long-term data on the winter moth that allow us to perform these calculations. First, we know that in any given year, about 89% of all winter moth die from abiotic (non-living) sources of mortality. In Chapter 2 we learned that abiotic factors (floods, freezing temperatures, fire, etc.) can be very powerful sources of mortality on insects. We can calculate that, if 89% of larvae die from abiotic sources of mortality and females lay approximately 150 eggs each, then the per capita rate of increase (λ) of an unconstrained winter moth population would be about eight (Table 4.1). If we start with a population size of 20 moths, and we know that λ of an unconstrained *O. brumata* population is approximately equal to eight, we could calculate the number of moths in 35 years time as follows:

$$\text{population size in 35 years} = 8^{35} \times 20 = 8.11 \times 10^{32} \text{ moths.} \quad (4.2)$$

If an average winter moth weighs 40 mg and the Earth weighs 5.976×10^{24} kg, then after only 35 years of population growth, our population of winter moth would have a combined weight of over 5000 times the mass of the Earth!

It should be clear from this example that the growth of real insect populations must be limited in some way. Specifically, rates of mortality must increase with density, or rates of natality (birth) must decrease with density (Fig. 4.1). When rates of birth and death vary with density, we call them density dependent. Density-dependent mortality and density-dependent natality can limit the growth of insect populations. Competition for limited resources and the effects of natural enemies are the factors most commonly thought to introduce density-dependent mortality or natality into insect populations. The circumstances in which natural enemies can limit insect populations will be considered in detail in Chapter 5. Here, we focus on competition.

4.2 Competition for limited resources

Begon *et al.* (1990) defined competition in the following way: 'Competition is an interaction between individuals, brought about by a shared requirement for a resource in limited supply, and leading to a reduction in the survivorship, growth, and/or reproduction of the competing individuals concerned.' There are several important consequences that arise from defining competition in this way. First, organisms must overlap in their use of some limited resource; if the resource is unlimited, then competition does not occur. As an example, the readers of this book overlap completely in their use of oxygen with the authors of this book, but we do not compete with each other for the resource because it is not limiting. Similarly, if two species of herbivorous insect feed exclusively on leaves from the same plant species, they do not compete for the resource unless leaves, or at least leaves of high nutritional quality (Chapter 3), are in limited supply.

A second consequence of this definition is that competition has the potential to limit the growth of

Table 4.1 Unconstrained population growth of the winter moth, *Operophtera brumata*.

Initial population size	= 20
Number of females	= 10
Fecundity (average)	= 150
Next generation	= 10 × 150
	= 1500
Abiotic mortality	= 89%
Survival	= 11%
	= 1500 × 0.11
	= 165
Per capita rate of increase, λ	= N_{t+1}/N_t
	= 165/20
	= 8.25
Population size in 35 years	= $\lambda^{35} \times N_0$
	= $8^{35} \times 20$
	= 811 000 000 000 000 000 000 000 000 000 000 moths

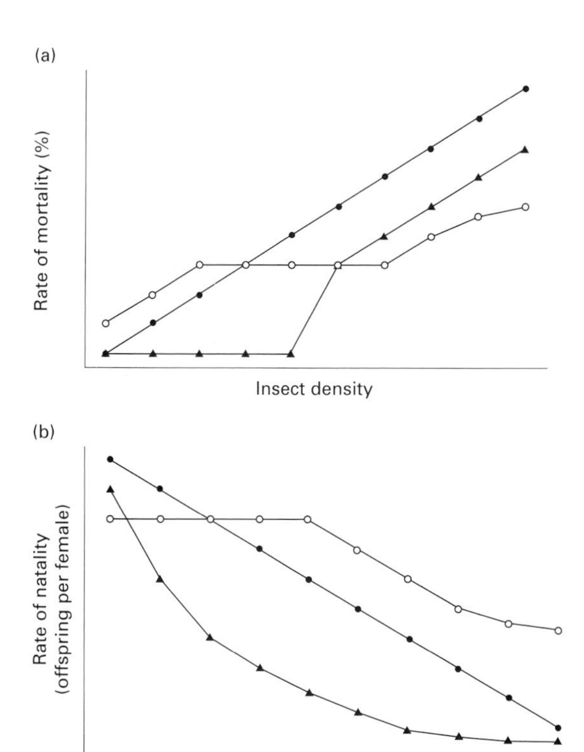

Fig. 4.1 Hypothetical examples of density-dependent rates of (a) death and (b) birth acting on insect populations. Although the exact form of the relationships can vary (i.e. the slopes of the lines can be different), death rates rise and birth rates fall with increasing insect density.

insect populations by its influence on birth rate or death rate. If the limited supply of some essential resource (space, food, shelter, etc.) causes a reduction in birth rates or an increase in death rates as population density increases, then competition for that resource is density dependent. We stress here that, just because competition has the potential to limit the growth of an insect population, it does not follow that competition regularly acts to limit natural insect populations; other factors such as predation can maintain populations below levels at which competition ever occurs (Chapter 5). Indeed, the regularity with which competition influences insect populations has been the subject of considerable controversy. We will describe this controversy later in the chapter (see section 4.5.1).

One final consequence of the definition of competition is that the process can usually be envisaged as a negative–negative interaction (Fig. 4.2). Unlike predation or mutualism, both parties engaged in competition suffer as a result of the interaction (we discuss asymmetric, or unequal, competition later in the chapter). We might therefore expect that insects will attempt to avoid competition wherever possible. Avoidance can take several forms, including strategies that are expressed in ecological time and strategies expressed over evolutionary time. In other words, some are day-to-day events, whereas others have become characteristic ecological traits of populations, developed over the course of evolution. An example

Effect on species A	Effect on species B	Interaction
+	+	Mutualism
+	–	Predation
–	–	Competition
–	0	Amensalism
+	0	Comensalism

Fig. 4.2 Potential types of interaction between insect species. Amensalism is a special form of asymmetric competition, where only one species suffers as a result of the interaction (section 4.5.3).

of a strategy to avoid competition expressed in ecological time might be for insects to leave a site already occupied by a potential competitor. Females of the codling moth, *Cydia pomonella* (Lepidoptera: Tortricidae), for example, avoid laying eggs beside other codling moth eggs in response to a chemical cue in egg tissue (Thiery *et al.* 1995). Similarly, larvae of the ladybird beetle, *Cryptolaemus montrouzieri* (Coleoptera: Coccinellidae), produce a chemical cue that deters oviposition by females of the same species (Merlin *et al.* 1996). Such 'oviposition-deterring pheromones' are probably a common way for insects to avoid sites already occupied by potential competitors (McNeil & Quiring 1983; Quiring & McNeil 1984).

Strategies expressed over evolutionary time to avoid competition might include fixed shifts in resource utilization that minimize similarity among potential competitors. One example is the late-season feeding habit of oak leaf-miners in the genus *Phyllonorycter* (Lepidoptera: Gracillariidae) (West 1985). The miners begin development after early season leaf chewers have pupated, so avoiding the consumption of their mines by defoliating Lepidoptera. The late-season feeding of *Phyllonorycter* has become a fixed population trait over evolutionary time. In general, both behavioural and morphological changes can reduce overlap in resource use, and these will be discussed in the section on niche breadth (see section 4.3).

4.2.1 Types of competition

There are several more precise definitions of competition that are useful for understanding the struggle for limited resources. Intraspecific competition describes competition for limited resources by members of the same species, whereas interspecific competition refers to competition for limited resources among members of different species. There is considerable appeal to the notion that individuals of the same species will use resources in very similar ways and are therefore more likely to compete with one another than with individuals of different species. Indeed, there has been much support in recent years for the view that intraspecific competition plays a more dominant role in the population and community ecology of insects than does interspecific competition (Connell 1983; Strong *et al.* 1984). However, the relative dominance of competitive interactions within and among species is a complex issue that remains topical (Faeth 1987; Hunter 1990; Denno *et al.* 1995; see section 4.5.1).

We can also identify the form of competition among individuals. Scramble competition refers to an interaction whereby limited resources are utilized evenly (or approximately so) by individuals. For example, if dragonfly nymphs (Odonata) in a pond are competing for the limited resource of prey items, and each nymph removes approximately the same biomass of prey from the stream, we might say that dragonfly larvae are 'scrambling' for resources: when their food supply is depleted, all nymphs suffer equally. This is also sometimes referred to as exploitation competition. In contrast, if some individuals obtain an unequal share of resources by limiting the access of weaker individuals to that resource, or injure those individuals as they forage, we refer to this as contest or interference competition. Dragonfly adults that maintain territories around the edges of ponds, and exclude weaker individuals from their territories, may obtain a disproportionate share of limited resources (e.g. food, access to mates, etc.) by contest competition (Tsubaki *et al.* 1994). Contest competition is thought to lead to more stable population dynamics than scramble competition because, as resources decline, some individuals will still have access to sufficient resources for growth, survival, and reproduction (i.e. fewer individuals, but with all necessary resources). Scramble competition, in contrast, can lead to severe over-exploitation of the habitat whereby few if any individuals receive sufficient resources. In such circumstances, reproduction may decline swiftly, emigration will be favoured, and population crashes can occur (Fig. 4.3). In reality, most insect populations that compete for limited resources probably exhibit a blend of contest and scramble competition, with some individuals obtaining a greater proportion of resources than others, yet

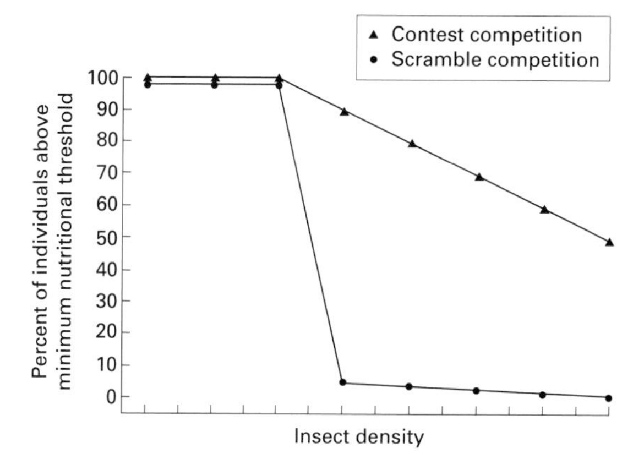

Fig. 4.3 A comparison of scramble and contest competition. Because resources are divided approximately evenly under scramble competition, most individuals fall below the minimum requirements for survival at the same point. Contest competition is more stable because some individuals maintain sufficient resources whereas others have none.

suffering some consequences of declining resource availability. We should remember, however, that neither contest nor exploitation competition will occur unless resources are limiting.

4.3 The concept of the niche

The niche concept dates back to the early 1900s. Grinnell (1904) equated an organism's niche with its distribution in space. Elton (1927) expanded the concept to incorporate the behaviour of the organism within the definition of the niche: he envisaged the niche as the 'functional role' of the organism in the community. This was the first time that the niche was considered to be interaction-based, and this view has dominated niche theory ever since. For our purposes, an insect's niche is the pattern of resource utilization that results from its interactions with the abiotic environment and other organisms (biota) in that environment. Inherent in this definition of the niche is the concept of constraint. Insects are constrained in their use of resources by the abiotic environment (suitable temperature, humidity, etc. to support life) and the biotic environment (interactions with food, mutualists, and natural enemies). For example, the niche of the monarch butterfly, *Danaeus plexippus* (Lepidoptera: Danaeidae) is constrained by the abiotic conditions to which it is adapted, the availability of its host plants in the genus *Asclepias*, and its interactions with natural enemies such as orioles and grosbeaks.

Niche theory and interspecific competition have been inextricably linked over the years by the simple principle that two species should not be able to coexist if they have identical niches (Gause's Axiom): any slight variation in competitive advantage would allow the superior competitor to exclude the weaker competitor. This assertion spawned the study of 'maximum tolerable niche overlap'. Just how similar can two insect species be to one another before competitive exclusion will occur? Unfortunately, there seems to be no easy answer to this question. Although some ecologists have suggested on theoretical and empirical grounds that niche overlap has some maximum value (Hutchinson 1957; MacArthur & Levins 1967; May 1973), there has been considerable controversy surrounding both the validity of the concept of maximum tolerable overlap and appropriate ways to measure overlap in nature (Abrams 1983). Although the study of maximum tolerable niche overlap is still prevalent in some other fields of ecology, we detect a loss of interest in this topic among insect ecologists, and refer interested readers to broader ecological textbooks (Begon *et al.* 1990; Putnam 1994) for further discussion.

4.4 Theoretical approaches to the study of competition

We begin our description of theoretical approaches to the study of competition with the following caveat: we do not believe that most competition models, particularly the simple models presented here, accurately reflect competition among insects in natural populations. Algebraic models, by their very nature, behave according to the assumptions built into them, and many of the assumptions that underlie competition models do not hold true in natural populations. Yet there are two main reasons for presenting some simple models of competition here. First, the structure of the simple models can be used to develop more complex models with a greater degree of realism. Although complex models are beyond the scope of this text, students who are interested in theoretical approaches to studying insect populations should be familiar with the basic structure upon which more realistic models can be developed. Second, models often have their greatest value when they fail to reflect nature. Models are really hypotheses about the way that we believe natural populations operate. When we test these hypotheses experimentally (i.e. try to make them match our observations of real insects), and find them lacking, we learn that there is some basic assumption in our model that is incorrect. In other words, the failure of a model to match up to empirical and experimental observation forces us to reconsider our views of how competition operates, and to investigate the phenomenon further. We begin with a description of some simple competition models, then describe the predictions that we can derive from them, and conclude with experimental evidence that supports or refutes the models' predictions.

4.4.1 Competition models

Competition models for insects have developed in parallel with competition models for other taxa. We can distinguish two fundamental types of population model. The first type, continuous time models, is suitable for insect species that have overlapping generations (i.e. daughters begin to reproduce before mothers stop reproducing) and that exhibit continuous reproduction. The second type, discrete time models, is more suitable for insect populations without overlapping generations and with seasonal reproduction. Neither type of model matches the real world perfectly. For example, many species of aphid (Hemiptera: Homoptera) exhibit overlapping generations. Indeed, aphids have 'telescoping generations' in which the aphid's granddaughters are already formed in the body of the daughters before the mother gives birth! Yet these same species of aphid often pass the winter in the egg stage. Clearly, many aphids fulfil both the criteria of overlapping generations and seasonal reproduction. Should we use continuous time or discrete time models to study their population dynamics? The answer tends to be a compromise, depending on the insect species in question and the purpose of the model. We present only continuous time models for competition here because they illustrate the same general principles as discrete time models and provide a simple introduction. We encourage interested readers to explore discrete time models of competition in other texts (e.g. May 1978).

4.4.2 A continuous time model for competition

The most simple continuous time model of competition is the logistic growth equation of Verhulst (1838):

$$\frac{dN}{dt} = rN\left[\frac{(K-N)}{K}\right] \qquad (4.3)$$

where N is population size, r is the per capita exponential growth rate of the population (the natural log of λ in eqn 4.1), and K is the 'carrying capacity' of the environment for species N. The carrying capacity represents the maximum average population size that the environment can support.

In English as opposed to 'math-speak', the above model states that 'the rate of population change depends upon population growth rate, the number of individuals in the population, and how far away current population size is from carrying capacity'. The term $(K-N)/K$ is designed to modify population growth rate, r, such that it is positive when N is below K, negative when N is above K, and zero when $N = K$ (choose some numbers for N and K, and try this yourself).

This model has a very simple equilibrium solution: independent of initial density, N will eventually reach K and stay there, unless some external force moves N from K (Fig. 4.4). If some environmental perturbation causes an increase or decrease in N, it will again return to K. From where does this solution arise? By definition, the population will reach equilibrium when $dN/dt = 0$ (the rate of population change is zero at equilibrium). Readers should try rearranging eqn 4.3 with the left-hand side (dN/dt) equal to zero. There are three possible solutions: $N = 0$, $r = 0$, or $[(K-N)/K] = 0$. If $N = 0$, then the population does not exist, so of course it cannot change size! When $r = 0$, we already know that the population is not changing: the per capita rate of change is zero. If $[(K-N)/K] = 0$, then $N = K$. This is the solution of interest. Simply, when the population reaches equilibrium (stops changing), it does so at carrying capacity (Fig. 4.4).

This is the simplest model that describes intraspecific competition—individuals within one species are competing for the limited resources that determine carrying capacity. We can expand this model to consider competition among different species. To do that, we need an additional equation for the second species and some way to influence the population growth rate of each species according to the density of the other. The simplest form of the two-species competition equations, developed by Lotka (1932) and Volterra (1926), are

$$\frac{dN_1}{dt} = r_1N_1\left[\frac{(K_1 - N_1 - \alpha N_2)}{K_1}\right] \quad (4.4)$$

and

$$\frac{dN_2}{dt} = r_2N_2\left[\frac{(K_2 - N_2 - \beta N_1)}{K_2}\right]. \quad (4.5)$$

We now have two equations, one for each of the competing species, and each species has its own specific r, N and K (denoted by the subscripts 1 and 2). We have also modified eqn 4.3 by adding the term αN_2 (or βN_1). The terms α and β are called the competition coefficients and are designed to transform individuals of the competing species into 'equivalent units' of the original species: α in eqn 4.4 reflects the effect of each individual of species 2 on species 1, whereas β in eqn 4.5 reflects the effect of each individual of species 1 on species 2. In other words, if α in eqn 4.4 is 0.75, it means that, as far as species 1 is concerned, each individual of species 2 uses about 0.75 of the resources of each individual of species 1. More precisely, if the population size of species 2 increases by 100, the population size of species 1 will drop by only 75. This suggests that individuals of species 1 suffer more from competition with each other than they do from competition with species 2: for species 1, intraspecific competition is more intense than interspecific competition.

In contrast, if β in eqn 4.5 is 1.25, an increase of 100 individuals of species 1 would cause the population size of species 2 to fall by 125. If this were the case, we would consider interspecific competition stronger than intraspecific competition for species 2.

We can determine equilibrium densities for both competing species as before by setting dN/dt equal to zero in eqns 4.4 and 4.5. By the magic of mathematical manipulation, the equations reduce to

$$N_1 = K_1 - \alpha N_2 \quad (4.6)$$

and

$$N_2 = K_2 - \beta N_1. \quad (4.7)$$

In other words, the equilibrium density for species 1 is no longer a single density, but varies depending on the density of species 2 and its competition coefficient. The same is true for species 2. We notice that eqns 4.6 and 4.7 represent equations of straight lines of the general form $y = mx + c$. The competition coefficients represent the slopes of these lines. Ecologists have come up with the expression 'zero-growth isocline' to describe the range of equilibrium densities (the straight lines) for each of the species. They are called zero-growth isoclines because they describe the range of densities for each species in which population rate of change (dN/dt) equals zero. The zero-growth isocline for species 1 is shown in Fig. 4.5. The line must cross the x-axis where $N_2 = 0$. When we set N_2 to zero in eqn 4.6, we obtain the value $N_1 = K_1$. Similarly, the line must cross the y-axis when $N_1 = 0$. Substituting $N_1 = 0$ into eqn 4.6, the line must cross the y-axis when $N_2 = K_1/\alpha$. We can then calculate the zero-growth isocline for species 2 in the same way using eqn 4.7. The zero-growth isocline for species 2 must cross the y-axis at K_2, and the x-axis at K_2/β. Here's the fun part: although we have established the equilibrium conditions (zero growth) for the two

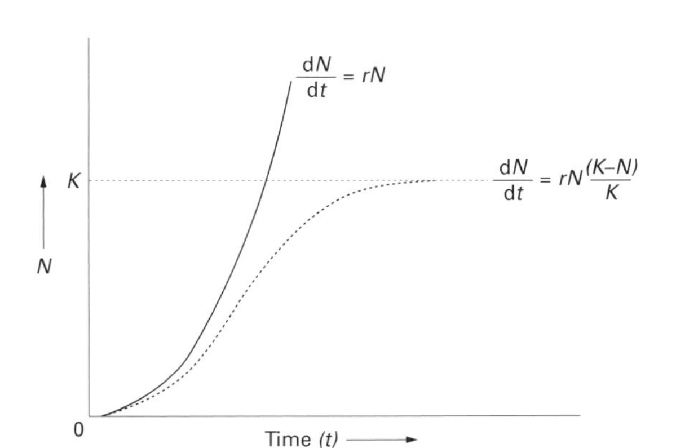

Fig. 4.4 Unlike exponential growth, logistic population growth brings a population to carrying capacity.

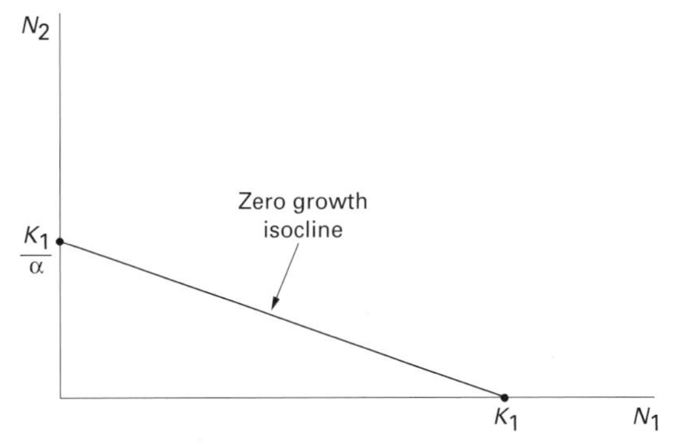

Fig. 4.5 The zero-growth isocline for species 1. Each point along the line represents an equilibrium density for species 1. It should be noted that the equilibrium density declines for increasing densities of species 2.

species, we have yet to establish whether or not they have a combined equilibrium where both species can coexist together. We determine this by drawing the isoclines for both species on the same graph, and there are four possible combinations (Fig. 4.6). A common equilibrium exists only if the lines cross each other (Fig. 4.6c,d) and that equilibrium is stable only if the lines cross as shown in Fig. 4.6(d).

Spend a few moments looking at Fig. 4.6. Starting with Fig. 4.6(a), whatever the initial conditions, species 1 will eventually reach carrying capacity (K_1) and species 2 will become extinct. This occurs because the isocline for species 2 falls completely within that of species 1. At any combination of densities between the two isoclines, the population trajectory leads towards K_1 so that the density of species 1 is increasing while the density of species 2 declines. The reverse is true in Fig. 4.6(b): trajectories will lead the population of species 2 to K_2 and the population of species 1 to extinction because the species 2 isocline encloses the species 1 isocline. In Fig. 4.6(c), the winner of the competitive interaction depends on the starting densities of the two species and their relative growth rates. If conditions are such that population growth results in the population trajectory entering the stippled area of the graph (where the species 2 isocline covers the species 1 isocline) then species 2 will drive species 1 to extinction. Conversely, if the starting conditions cause the population trajectory to enter the striped area of the graph, species 1 will drive species 2 to extinction. Where the lines exactly overlap in Fig. 4.6(c), both species are at equilibrium. However, the equilibrium is unstable, and any perturbation that causes a change in density of either species will result in the extinction of one or the other.

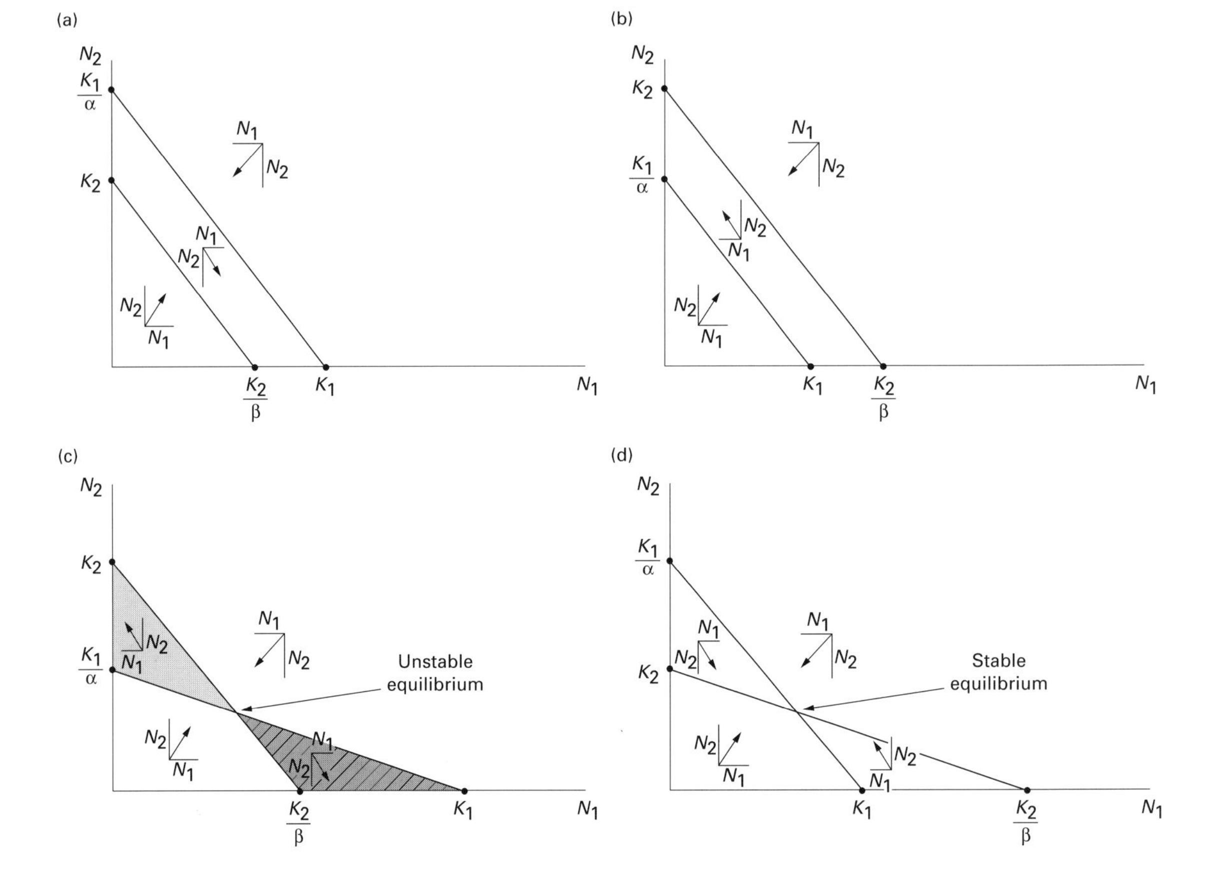

Fig. 4.6 Zero-growth isoclines for two insect species in competition. Stable coexistence is possible only in the conditions shown in (d), where the isoclines cross such that the carrying capacities of each species fall within the isocline of the other species.

Only Fig. 4.6(d) represents a stable equilibrium with coexistence of both species. The carrying capacities of both species (K_1 and K_2) are enclosed below the isoclines of the other species. Simply put, this means that species 1 will inhibit its own population growth before it could ever drive species 2 to extinction and vice versa. This is another way of saying that, overall, intraspecific competition is greater than interspecific competition. From the x-axis, we can see directly that coexistence is possible if $K_1 < K_2/\beta$. From the y-axis, we see that coexistence is possible if $K_2 < K_1/\alpha$. By substituting and rearranging these equations, we obtain an additional condition for coexistence: $\alpha\beta < 1$. Earlier in this section we saw that, if a competition coefficient is greater than one, interspecific competition is greater than intraspecific competition. Here, we have refined the concept to suggest that, as long as the product of the two competition coefficients is less than one, coexistence of competing species is possible.

4.4.3 Assumptions and predictions of Lotka–Volterra competition models

There are some important assumptions that underlie the simple competition models described above. We recall that the assumptions that are built into models can be considered hypotheses about the way the natural world operates. Assumptions of the Lotka–Volterra models include the following.

1 Competition coefficients are constant. The coefficients α and β in eqns 4.4 and 4.5 are not permitted to vary with any changes in the abiotic or biotic environment.

2 The only density-dependent factors operating on populations N_1 and N_2 are limitations imposed by their own density (intraspecific competition) and limitations imposed by the density of the second species (interspecific competition). No other potentially density-dependent factors, such as predation or disease, act to limit either population.

3 There are no abiotic sources of mortality acting on the populations, and there is no stochastic (random) variation in the power of biotic factors.

In addition, the major prediction of the Lotka–Volterra competition equations is that coexistence among competing species is possible only if intraspecific competition is stronger than interspecific competition (or, more precisely, if $K_1 < K_2/\beta$, $K_2 < K_1/\alpha$, and $\alpha\beta < 1$). Let us see how well these assumptions and predictions reflect the real world.

4.5 Competition among insects in experimental and natural populations

Early laboratory studies with insects demonstrated that several assumptions of Lotka–Volterra theory were unrealistic. Chief among these is the assumption that competition coefficients are constant. For example, using laboratory populations of stored-product beetles, Birch (1953) demonstrated that competitive dominance among two species, *Rhizopertha dominica* (Coleoptera: Bostrichidae) and *Calandra oryzae* (Coleoptera: Curculionidae), varied with temperature. At 29.1°C, *C. oryzae* is the dominant competitor, and its abundance increases at the expense of that of *R. dominica*. This reverses at 32.3°C, where *R. dominica* becomes dominant (Fig. 4.7). This simple study suggests that competition coefficients are temperature dependent. In a similar study, Ayala

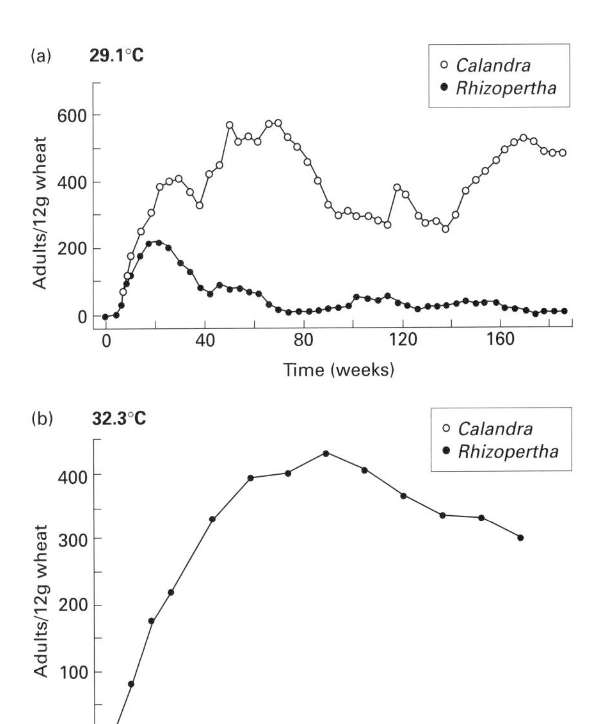

Fig. 4.7 The competitive dominance of two beetle species switches in response to changing temperature. It seems probable that fluctuations in abiotic forces in nature help to maintain coexistence of competing species. (After Birch 1953.)

(1970) found that the coexistence of two fruit fly species (*Drosophila pseudoobscura* and *D. serrata*) (Diptera: Drosophilidae) was temperature dependent. Below 23°C, *D. pseudoobscura* excluded *D. serrata* from laboratory populations, whereas, above 23°C, *D. serrata* excluded *D. pseudoobscura*. When temperatures were maintained at 23°C, both species could coexist indefinitely.

We can draw two very important conclusions from Ayala's study. First, fluctuating environmental conditions are likely to mediate coexistence among species of insect through reversals of competitive advantage. We know that temperatures in nature fluctuate daily and seasonally, and this suggests that neither *D. pseudoobscura* nor *D. serrata* would maintain a long-term advantage over the other species. Without maintaining a long-term advantage, neither species should exclude the other, and therefore coexistence should occur. The role of fluctuating environmental conditions in mediating coexistence of species has been formalized mathematically by several researchers (Chesson & Warner 1981; Commins & Noble 1985) and environmentally mediated reversals in competitive advantage that promote coexistence have been shown to occur in natural insect populations. For example, coexistence of the winter moth, *Operophtera brumata* (Lepidoptera: Geometridae), and the green oak tortrix, *Tortrix viridana* (Lepidoptera: Tortricidae), on the English oak depends upon climatic variation. The winter moth is competitively superior when spring temperatures are relatively high, whereas the green oak tortrix is competitively superior when spring temperatures are low (Hunter 1990).

A second important conclusion that we can draw from Ayala's study is illustrated in Table 4.2 and Fig. 4.8. We notice that, at 23°C, where both species could coexist, their population densities were much lower than their densities when reared alone. We can use the data in Table 4.2 to calculate the competition coefficients between the species at 23°C. Simply, the effect of *D. pseudoobscura* on *D. serrata* can be calculated as

$$\frac{(1251-278)}{252}=3.86 \tag{4.8}$$

and the effect of *D. serrata* on *D. pseudoobscura* can be calculated as

$$\frac{(664-252)}{278}=1.48. \tag{4.9}$$

We notice that both competition coefficients are greater than one, and that their product is necessarily also greater than one. According to Lotka–Volterra theory (section 4.4.3), coexistence between species is not possible when the product of their competition coefficients exceeds unity. Clearly, the theory does not reflect competition among insects realistically, even in simple laboratory populations. Gilpin and Ayala's (1973) solution to this disparity was to curve the zero-growth isoclines of the two species (Fig. 4.8) so that the lines intersected at a lower combined density. It should be recalled that the slopes of the zero-growth isoclines represent the competition coefficients, so by curving the lines, Ayala is suggesting that competition coefficients are frequency (or density) dependent (Ayala 1971). In essence, the

Table 4.2 Densities of *Drosophila serrata* and *D. pseudoobscura* in laboratory experiments of competition. (Data from Ayala 1970.)

	Population size
Species raised separately	
D. pseudoobscura	664
D. serrata	1251
Species raised together	
D. pseudoobscura	252
D. serrata	278

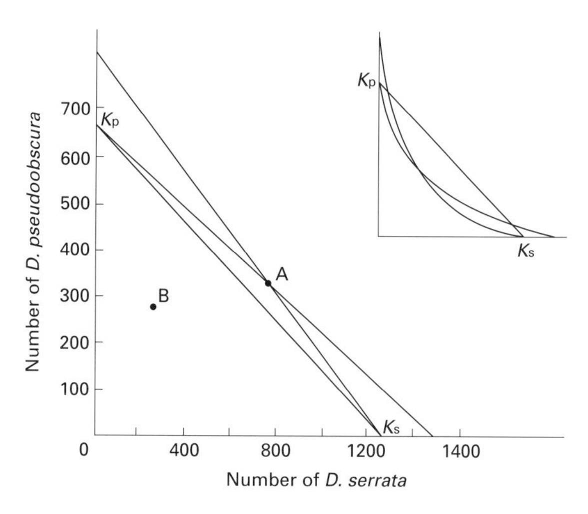

Fig. 4.8 Competition between *Drosophila pseudoobscura* and *D. serrata*. Calculations based on Lotka–Volterra theory suggest that coexistence should occur at the densities marked at point A. Experimental work by Ayala (1970) demonstrated, however, that coexistence occurs at point B, and the zero-growth isoclines must curve.

curved isoclines suggest that the relative importance of intraspecific competition over interspecific competition increases with population density. This might occur if, as populations grow, scramble competition is replaced by interference (contest) competition and, in concert, species are more likely to interfere with members of their own species than with members of another species. The message to take away from these two simple laboratory experiments is that competition coefficients, that is, the strengths of interactions between competing species, are likely to vary in response to biotic and abiotic ecological factors.

4.5.1 The relative strengths of intra- and interspecific competition

If intraspecific competition is generally a more powerful force than interspecific competition, then coexistence among competing species will occur simply because populations will limit their own population growth before reaching levels at which they exclude their competitors. In literature reviews of manipulative experiments on competition among animals, both Connell (1983) and Schoener (1983) concluded that intraspecific competition was indeed often stronger than interspecific competition. However, the debate on the relative importance of interspecific competition for insect populations in general, and phytophagous insects in particular, is far from resolved. We will describe studies of insect herbivores first, then follow with a discussion of interactions among insects at other trophic levels.

In their classic paper, Lawton and Strong (1981) suggested that 'interspecific competition is too rare or impuissant to regularly structure communities of insects on plants'. They were in general agreement with the views of Hairston *et al.* (1960), who regarded the overall abundance of green plants as evidence that insect herbivores rarely deplete their food resources, and should not be limited by competition. Rather, those workers considered natural enemies the most likely factors regulating insect herbivore populations. Although most insect ecologists agree that competition among herbivorous insects can occur during periodic outbreaks (Varley 1949; Bylund & Tenow 1994; Carroll & Quiring 1994), the general view during the 1980s was that competition among insects was probably less important than previously conceived. For example, in a review of 31 insect life table studies (tables that categorize mortality factors operating during each generation), Strong *et al.* (1984) found convincing evidence of intraspecific competition in only six. It should be noted that, in the early to mid-1980s, there were not many rigorous experimental studies of competition among insects from which to draw generalizations, and that the reviews of Lawton and Strong (1981) and Strong *et al.* (1984) did much to stimulate a wave of manipulative experiments to study competition among insects.

4.5.2 Intraspecific competition among insect herbivores

Some studies have recorded significant intraspecific competition among phytophagous insects. In a detailed series of experiments, Ohgushi has shown that populations of the herbivorous lady beetle *Epilachna japonica* (Coleoptera: Coccinellidae) compete strongly for their thistle host plants (Ohgushi & Sawada 1985; Ohgushi 1992). In this system, female beetles resorb eggs rather than lay them on defoliated thistle hosts (Table 4.3) and density-dependent birth rates regulate populations with remarkable stability over time. Density-dependent colonization of hosts may also be common in bark beetles, some of which switch from producing aggregation pheromones to producing dispersion pheromones when the density of adults on a tree surpasses some threshold (Borden 1984).

In section 4.5.1, we described the views of Hairston *et al.* (1960) and Lawton and Strong (1981), who suggested that insect herbivores should not be limited by competition because they rarely deplete their host plants. However, we learned in Chapter 3 that there is significant variation within and among individual plants in their quality as food for insects, and that not all green tissue is necessarily edible. As a consequence, some insect herbivores may be 'trapped' in fierce competition with conspecifics for the limited high-quality plants (or plant parts) within a sea of unpalatable plant tissue. For example, Whitham (1978, 1986) has demonstrated that galling aphids in the genus *Pemphigus* (Homoptera: Aphididae) are clumped on high-quality individual leaves within their cottonwood (*Populus*) host plants. Most leaves on cottonwood trees cannot support high rates of aphid population growth, and intense intraspecific competition for suitable leaves results. Even within leaves, some positions are better than others. When 'stem mothers' are experimentally removed from the

Cage	No. of adults		% Leaf damage	% Females with egg resorption
	Male	Female		
A	1	1	20	0
B	1	1	20	0
C	1	1	20	0
D	1	2	50	100
E	4	5	80	100
F	4	7	90	100

Table 4.3 Females of the herbivorous lady beetle *Epilachna japonica* resorb eggs if defoliation of their host plant is too high; by resorbing eggs, females are more likely to survive the winter to reproduce the following year. (Data from Ohgushi 1992.)

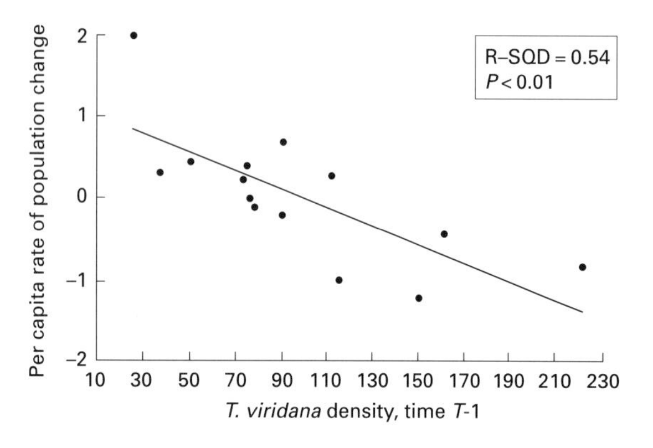

Fig. 4.9 The per capita rate of change of *Tortrix viridana* is negatively associated with its population density in the previous year (time $T-1$). Years of high density are followed by low population growth rates, and vice versa. (From Hunter *et al.* 1997. © National Academy of Sciences, USA.)

most suitable positions on individual leaves, there is an increase in the reproductive output of other aphids on those leaves.

In some cases, statistical analyses of long-term population data can be used to infer the strength of ecological forces such as competition. For example, the analysis of long-term population data of the green oak tortrix, *Tortrix viridana* (Lepidoptera: Tortricidae), has indicated that densities may be regulated by intraspecific competition (Hunter *et al.* 1997). By studying 16 years of sampling data, it was found that the per capita rate of increase of a European *T. viridana* population was strongly and negatively related to its density in the previous year (Fig. 4.9). Simply put, if densities are high in one year, the population declines in the following year. Conversely, years of low population density are followed by years of increased population growth. In this system, parasitoids and predators appear to have little impact on *T. viridana* populations and, although there is occasional interspecific competition with the winter moth (above), long-term dynamics are best explained by intraspecific competition among foraging larvae (Hunter 1998).

4.5.3 Interspecific competition among insect herbivores

As we mentioned previously, there has been much debate on the frequency and strength of competitive interactions among different species of insect herbivore. Certainly, there is compelling evidence that interspecific competition is rare in some systems. In a comparison of the insect faunas feeding on bracken fern, *Pteridium aquilinum*, in Britain, Papua New Guinea, and New Mexico, Lawton (1982, 1984) observed very different numbers of herbivore species using the same resources in different regions, and idiosyncratic gaps in resource utilization (Fig. 4.10). Lawton concluded that there was evidence of 'empty niches' on bracken in some regions and no corresponding increase in the densities of resident species. In other words, the lack of certain taxa in species-poor bracken communities did not provide any apparent increase in resources for the remaining species. These results are consistent with the view that the pool of colonizing species is exhausted, and that the community is not 'saturated' with competing species of herbivore.

In a study of leaf-miners on oak, Bultman and Faeth (1985) found that leaf-miner mortality was in general higher when individuals shared an oak leaf with members of the same species than with members of a different species, suggesting that intraspecific effects were more important than interspecific effects (Table 4.4). Likewise, Rathcke (1976) demonstrated that members of a stem-boring guild of insects in tall

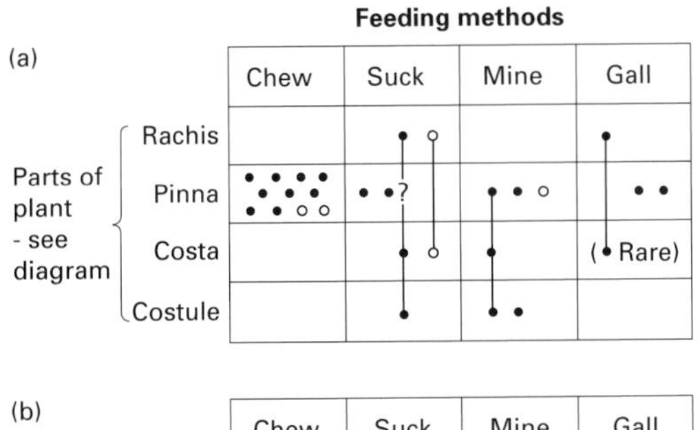

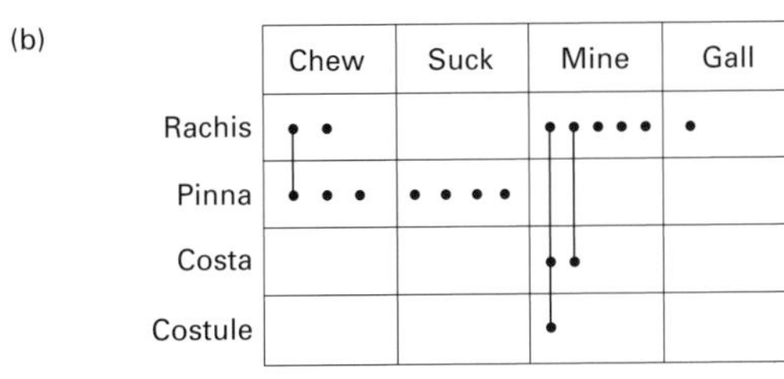

Fig. 4.10 'Gaps' in the use of bracken resources by insect herbivores. These data suggest that the insect herbivore communities are not 'saturated' by competing species. Open circles refer to species recorded in less than 50% of years studied. (Data from Lawton 1982, 1984, redrawn from Strong *et al.* 1984.)

Table 4.4 The mortality of leaf-miners on three species of oak tree; in general, rates of mortality are higher in the presence of conspecifics than in the presence of heterospecifics. (Data from Bultman and Faeth 1985.)

	Cohabitants				
Species	Alone	Conspecifics	Heterospecifics	d.f.	χ^2
Bucculatrix cerina					
Survival	926	326	116[a]	2	20.75†
Mortality	306	101	8[b]		
Cameraria sp.nov.					
Survival	53	0	8	2	4.00
Mortality	531	12	40		
Tischeria sp.nov.					
Survival	87	18[b]	19	2	8.47*
Mortality	347	165[a]	74		
Stigmella sp.					
Survival	422	21[b]	47[a]	2	13.11†
Mortality	630	48	34[b]		
Stilbosis quadricustatella					
Survival	60	3[b]	5	2	5.80*
Mortality	168	33	10		

* $P < 0.05$, † $P < 0.01$.
Numbers followed by superscript letter make major contributions to overall χ^2.
[a] Observed > expected. [b] Observed < expected.

grass prairie avoid conspecifics (members of the same species) more often than heterospecifics (members of different species). Finally, Strong (1982) studied eight species of hispine beetle (Coleoptera: Chrysomelidae) that live in leaf rolls on *Heliconia* at one site in Costa Rica. Despite considerable dietary overlap, including many different species found living in the same leaf roll, there was no evidence of significant interspecific interactions among the beetle species.

However, the concept that phytophagous insects can become trapped into intense competition by variation in host-plant quality, described above for *Pemphigus* aphids (section 4.5.2), can be applied to interspecific competitive interactions also. For example, Gibson (1980) observed that two species of grassbug (*Notostira elongata* and *Megalocerea recticornis*) (Hemiptera: Miridae) that feed in limestone grasslands in England are forced to shift between host-plant species during the season to track nitrogen availability. Because nitrogen limitation affects both herbivore species, however, they are forced into interspecific competition on the same plant species (see Chapter 3 for a discussion of insect herbivores and nitrogen limitation). Similarly, Fritz *et al.* (1986) suggested that a stem-galling sawfly on willow had negative effects on three other willow sawfly species because they were constrained to feed on a subset of available willow genotypes.

Some of the confusion that exists over the relative importance of interspecific competition for insect herbivores may have arisen because of just such forces that constrain the foraging of phytophagous insects. As originally described, by Lawton and Strong (1981) and Strong *et al.* (1984), interspecific competition did not regularly structure insect communities on plants. Unfortunately, a subset of the studies that those researchers used to illustrate their point was population-level studies, and this was taken by many ecologists to mean that competition did not influence the population dynamics of insect herbivores. There is a profound difference between population dynamics and community structure. Using Gibson's (1980) study of grassbugs (above), it is clear that nitrogen limitation determines host utilization patterns by grassbug species: it is the major determinant of community structure. None the less, the herbivore species are forced to compete with one another, and this influences their population dynamics. It is ironic that competition probably has its greatest impact on insect herbivores when other forces structure the community. It should be remembered that interspecific competition is a negative–negative interaction where both species suffer, and we should expect insects to develop strategies, over ecological or evolutionary time, to avoid competition. Only where some more powerful force constrains the foraging of insects should we expect to see the effects of competition still operating. In other words, when competition structures communities, interactions should decline over time to trivial levels, leaving only the 'ghost of competition past' as a reminder of once fierce interactions (Connell 1980). Ghosts of competition past might include morphological or behavioural adaptations that reduce overlap in resource use. An extreme ghost of competition past might be the competitive exclusion of one insect species by a second species. Exclusion is likely to be a transient phenomenon and therefore difficult to observe. In contrast, communities structured by forces other than competition may retain significant competitive interactions that influence population dynamics.

In any case, the tide has turned once more, and a significant number of insect ecologists are re-evaluating the role of interspecific competition in the population ecology of insect herbivores. This has arisen in part because techniques such as life table analysis, previously used to infer the strength of competitive effects (Strong *et al.* 1984), can miss critical aspects of life history, particularly oviposition preference, during which resource limitation can be expressed (Price 1990). In addition, life tables do not reliably detect density dependence when insects are patchily distributed (Hassell 1985; Hassell *et al.* 1987). Patchy distributions can greatly enhance the regulatory power of competitive interactions (De Jong 1981; Chesson 1985). In addition, a plethora of experimental studies during the late 1980s and 1990s have suggested that interspecific competitive interactions can occur through subtle and unexpected pathways.

For example, the recognition that many plants respond to damage by herbivores with changes in their chemistry (Chapter 3) led several ecologists to search for temporally separated interactions among phytophagous insect species. The possibility existed that damage by one insect species could influence populations of a second insect species through wound-induced changes in plant quality, even if the insects lived at different times of the year.

Across-season competitive effects were demonstrated by West (1985) in studies of the English oak,

Quercus robur. He demonstrated that spring defoliation by two caterpillars, *Operophtera brumata* (Lepidoptera: Geometridae) and *Tortrix viridana* (Lepidoptera: Tortricidae), on oak leaves resulted in reduced leaf nitrogen contents. Low nitrogen levels adversely affected the survival of *Phyllonorycter* (Lepidoptera: Gracillariidae) species of leaf-miner that attacked oak leaves late in the season (Fig. 4.11). In the same study system, Silva-Bohorquez (1987) showed that late-season aphids were also negatively affected by spring defoliation. Similarly, Faeth (1985, 1986) demonstrated that spring defoliation on *Quercus emoryi* caused declines in late-season leaf-miner success although, in this case, effects seemed to be caused by the attraction of parasitoids to previously damaged foliage. In a more recent reanalysis, Faeth (1992) has concluded that such cross-season effects may be weaker than intraspecific competition among leaf-miners in this system. None the less, evidence had accumulated that temporally separated competitive effects may be common among insect herbivore populations. For example, gall-forming and phloem-feeding insects that feed during autumn avoid leaves damaged by the gypsy moth, *Lymantria dispar* (Lepidoptera: Lymantriidae) during spring (Hunter & Schultz 1995; Fig. 4.12).

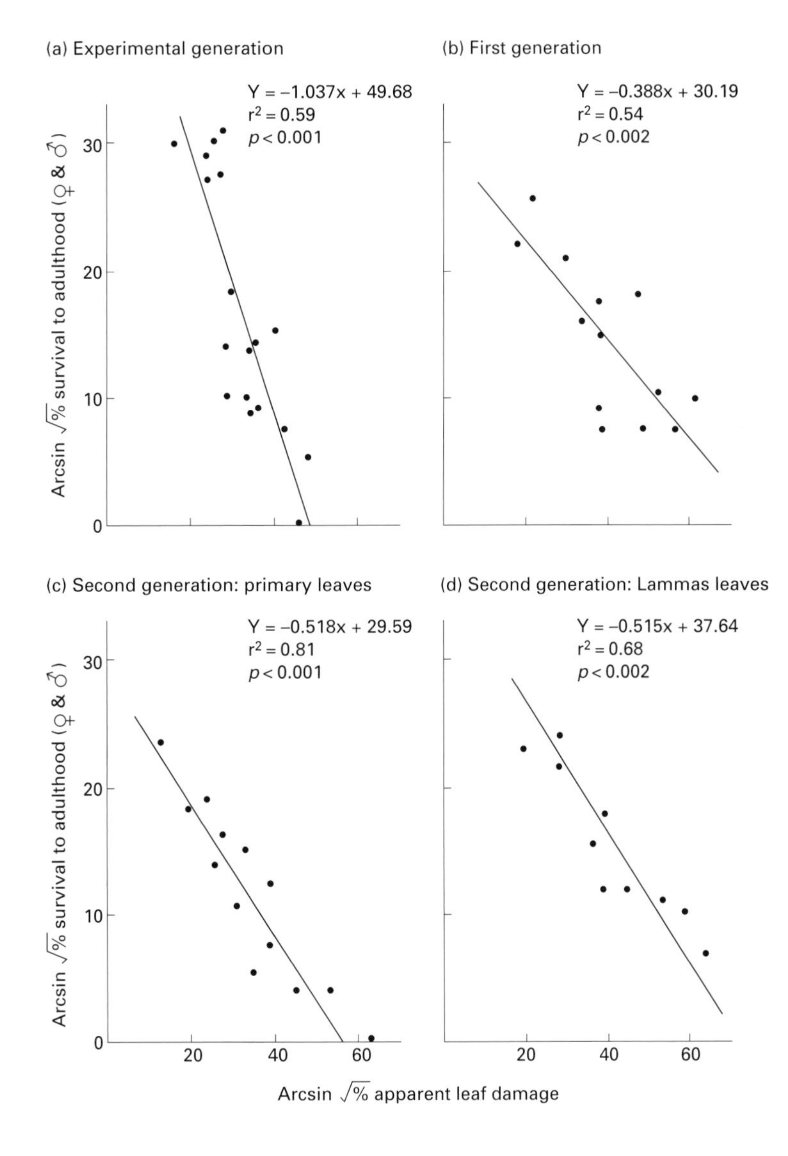

Fig. 4.11 Spring defoliation by oak caterpillars reduces the survival of late-season leaf-miners. This across-season interaction results from defoliation-induced declines in foliage quality. (From West 1985.)

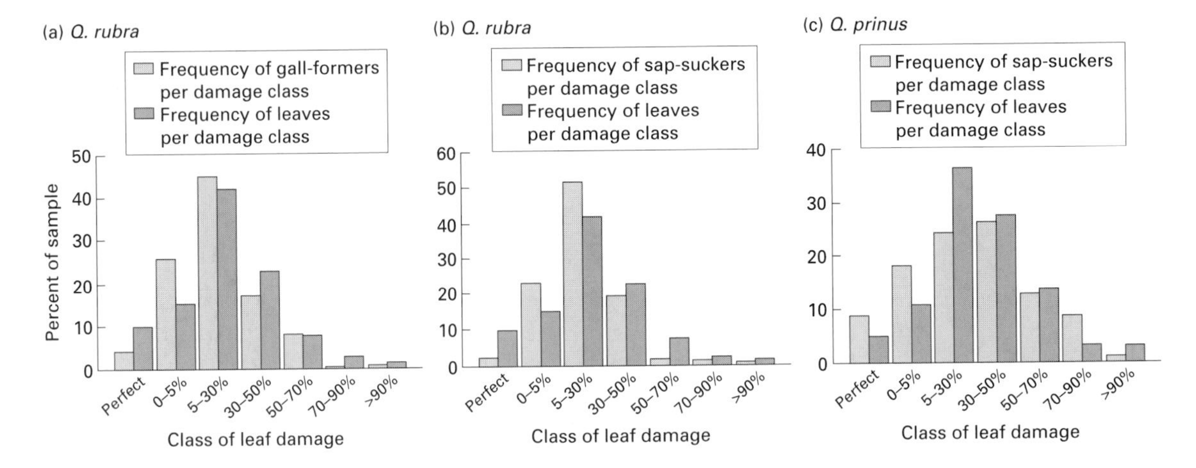

Fig. 4.12 Frequency distributions of the damage level to individual oak leaves, and the distributions among those leaves of two oak insect guilds. Both gall-formers (a) and sap-feeders (b) on *Quercus rubra*, and sap-feeders on *Q. prinus* (c), are found less frequently on leaves high in spring damage than expected by chance.

Plant-mediated competitive effects can occur on much shorter time-scales too. High densities of phloem-feeding insects can cause rapid declines in the nutritional quality of their hosts for other insect species (McClure 1980; Denno & Roderick 1992). Aphids, in particular, have been shown to exhibit reduced survival (Itô 1960) and increased emigration (Edson 1985; Lamb & Mackay 1987) following competition-induced declines in foliage quality. We should point out, however, that some insects appear to respond positively to previous infestation of their host plants by heterospecifics (Williams & Myers 1984; Kidd *et al.* 1985; Hunter 1987).

Is consensus possible in the debate on the importance of competition for insect herbivore populations? A recent review of 193 pair-wise interactions among insect herbivore species (Denno *et al.* 1995) has provided the most detailed analysis to date. We will review some of the conclusions of Denno *et al.* (1995) and we encourage interested readers to refer to the original paper. Of the 193 pair-wise interactions analysed, 76% provided evidence for interspecific competition, 6% demonstrated facilitation, and 18% recorded no interactions among insect herbivore species. The frequency of competition was therefore much greater than that described in previous reviews (Lawton & Strong 1981; Strong *et al.* 1984; but see Damman 1993).

Insect species with haustellate (piercing and sucking) mouthparts provided more evidence of competition than species with mandibulate (chewing) mouthparts (93% vs. 78% of species tested, respectively). In short, interspecific competition appeared particularly prevalent among phytophagous Hemiptera. Within the mandibulate herbivores, interactions were more common among enclosed feeders such as stem-borers, wood-borers, and seed feeders than among external feeders such as folivorous Orthoptera, Coleoptera, Hymenoptera, and Lepidoptera (89% vs. 59% of species tested). Moreover, competitive interactions among pairs of enclosed feeders occurred with greater frequency than interactions among free-living folivores. In fact, folivores appear to exhibit the lowest levels of interspecific competition of all herbivore guilds. As many reviews that have discounted interspecific competition among insects have been biased towards folivores (Lawton & Strong 1981), this may explain in part the continuing debate on the relative importance of competition. What is clear is that some enclosed feeders, such as nut feeders (Harris *et al.* 1996), galling insects (Craig *et al.* 1990; Akimoto & Yamaguchi 1994) and leaf-miners (Bultman & Faeth 1985), compete frequently.

Most competitive interactions reviewed by Denno *et al.* (1995) were asymmetric. Asymmetric competition occurs when one species suffers to a greater extent from the interaction than does a second species. The extreme example of asymmetric compe-

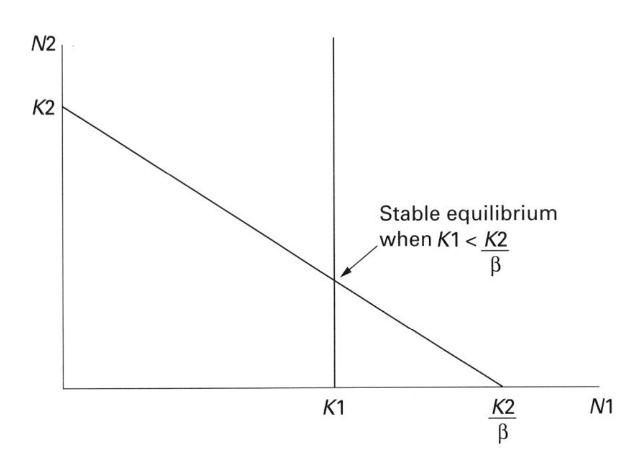

Fig. 4.13 Asymmetric competition (amensalism) between two insect species. Species 1 is unaffected by species 2, whereas densities of species 2 decline with increases in species 1.

tition, where one of the species is entirely unaffected by the presence of the other, is known as amensalism. A theoretical depiction of amensalism is shown in Fig. 4.13. It should be noted that the zero-growth isocline for species 1 is vertical, suggesting that the density of species 1 remains constant over any range of density of species 2. Asymmetric competitive interactions may be more stable than symmetric interactions because a decline in species 2 does not result in any increase in species 1: there is no positive feedback on the population of species 1 that would tend towards exclusion of species 2.

The prevalence of asymmetric competition (84% of cases analysed by Denno *et al.* (1995)) agrees with the landmark paper of Lawton and Hassell (1981), which suggested that amensalism was a common feature of interactions among insects. This, at least, appears to be a point of consensus. Strong asymmetry has been noted in insect phytophages, carrion feeders, aquatic detritivores, predators, and nectarivores (Lawton & Hassell 1981). For example, the carrion flies (Diptera: Calliphoridae) *Calliphora auger* and *Chrysomyia rufifacies* both dominate *Lucilla cuprina* on sheep carrion, and are essentially unaffected by *L. cuprina* (Andrewartha & Birch 1954). Similarly, *Danaus plexippus* (Lepidoptera: Danaidae) is a dominant competitor of *Oncopeltus* (Heteroptera: Lygaeidae) species on milkweed (Blakely & Dingle 1978). None the less, symmetric competition appears relatively common among sap-feeding insects, and may also be common among bark beetle species that respond negatively and reciprocally to the pheromones of other species (Birch *et al.* 1980; Coulson *et al.* 1980; Flamm *et al.* 1987). For example, the two bark beetle species, *Ips typographus* and *Pityogenes chalcographus* (Coleoptera: Scolytidae), respond negatively to each other's aggregation pheromone during colonization of Norway spruce (Byers 1993).

4.5.4 Competition among different guilds of insect herbivore

In section 4.3, we described the underlying assumption of niche theory that insects should compete most strongly with individuals that are very similar, occupying similar niches. However, insect species can still compete with one another if they occupy rather different niches. In 1973, Janzen pointed out that species of root-feeder, leaf-chewer, stem-borer, etc., might still compete for shared host plants because they are linked by the common resource budget of the plant (Janzen 1973). With the additional caveat that plant resources must be limiting for insect herbivores if competition is to occur, we might expect to see competition among insects in different feeding guilds (Chapter 3) if they share a host-plant species.

Examples of interspecific competition among insect herbivores in different feeding guilds are growing in number. For example, flower-feeding thrips can deter pollinators from plants (Karban & Strauss 1993), and shoot-feeding aphids on the crucifer *Cardamine pratensis* reduce the densities of root-feeding aphids on the same host plant (Salt *et al.* 1996). Similarly, leaf-miners on the foliage of the annual composite *Sonchus oleraceus* reduce the density of root herbivores (Masters & Brown 1992). Indeed, root-feeders seem generally to suffer more than folivores in across-guild competition (Jones *et al.* 1988; Denno *et al.* 1995). For example, Moran and Whitham (1990) studied both root-feeding and above-ground gall-forming aphids that co-occur on plants in the genus *Chenopodium*. Although the gall-formers and root-feeders never encountered one another, they were linked by their common exploitation of phloem sap. On susceptible *Chenopodium* plants, the gall-former reduced densities of the root-feeder by an average of 91%, often eliminating the root-feeder altogether. In contrast, there was no measurable effect of the root-feeders on the gall-formers. Likewise, leaf-miners are often the 'losers' in competitive interactions with leaf-chewers (Faeth 1985; West 1985), although some chewers may avoid mined leaves (Hartley & Lawton 1987).

Insects may also compete with totally unrelated taxa: herbivorous insects may compete with taxa as divergent as rust fungi (Hatcher *et al.* 1994) and rodents (Brown & Davidson 1977; Davidson *et al.* 1984).

4.6 Competition among insects other than herbivores

If the study of competition among insect herbivores has been controversial, the study of competition among other groups of insects can best be characterized as under-developed. Thanks to reviews by Lawton and Strong (1981) and Strong *et al.* (1984), there has been considerable interest in experimental investigations of competition among insect herbivores. Experiments designed to test the prevalence of competition among other groups of insects have not been as common in recent years. Some general patterns appear to be emerging, but we advocate caution until a substantial body of literature is available from which to draw firm conclusions.

One fairly reliable conclusion is that competition among Hymenoptera, particularly ants, may be widespread (Lawton & Hassell 1981). For example, ants in the genus *Camponotus* frequently kill, maim, and displace *Aphaenogaster* species in confrontations at feeding sites. Likewise, interspecific competition between *Pogonomyrmex rugosus* and *P. desertorum* can be intense. The introduction of fire ants (*Solenopsis invicta*) into the southern USA has been characterized by intense intra- and interspecific competition. Intraspecific competition among fire ants occurs in several stages, including competition among queens during the founding of colonies (Balas & Adams 1996), density-dependent brood raids between young colonies that result in 'stealing' of broods and workers (Adams & Tschinkel 1995), and group fighting of individuals from older colonies at feeding sites. Overall, intraspecific competition among fire ants results in the even spacing of colonies on the landscape (Adams & Tschinkel 1995). Aggression by fire ants is not limited to conspecifics: laboratory studies have shown that *S. invicta* will kill species of termite and may be responsible for declines in native *Solenopsis* species. We can see from these studies of ants that, as competitive interactions become increasingly asymmetric and contest-based, there is a very fine line between competition and predation.

Recent work has suggested that hymenopteran parasitoids may also suffer regularly from intra- and interspecific competition. Parasitoids lay eggs in or on other insects, and their larvae then kill the host (they are described in some detail in Chapter 5). Superparasitism occurs when individual parasitoids lay eggs in hosts that have been previously attacked by other individuals of the same or different species. Competition may then occur for the limited resource of the host insect (Kaneko 1995; Marris & Caspard 1996). The 'winner' of competition that results from superparasitism is generally thought to depend on which individual or species attacks first (Bokononganta *et al.* 1996), although the first attacker is not always competitively dominant.

One striking example of intraspecific competition comes from the gregarious ectoparasitoid *Goniozus nephantidid* (Hymenoptera: Bethylidae) (Hardy *et al.* 1992). Female wasps lay clutches of up to 20 eggs on the caterpillars of microlepidopterans. Clutch size increases as larval host size increases, suggesting that host body size is limiting (Fig. 4.14a). In addition, when the number of wasp larvae per host is increased experimentally, there is a concomitant decrease in female body size at maturation (Fig. 4.14b). Larger female wasps both live longer and lay more eggs (Hardy *et al.* 1992).

Interspecific competition among hymenopteran parasitoids also appears to be common (Kato 1994; Monge *et al.* 1995; Reitz 1996). One recent study has demonstrated the facultative production of an 'extraserosal envelope' around the developing embryo of the parasitoid wasp *Praon pequodorum* (Hymenoptera: Aphidiidae). The envelope, which separates the chorion and trophamnion of the developing embryo, is produced only when eggs are laid in hosts previously attacked by the heterospecific parasitoids *Aphidius ervi* and *A. smithi*. The envelope is not produced in singly or conspecifically superparasitized aphid hosts (Danyk & Mackauer 1996). This suggests that the extraserosal envelope has a defensive function and protects developing embryos from physical attack by mandibulate larvae of potential (interspecific) competitors. It is not yet clear whether limitations of host availability and competitive interactions support the suggestion of Hawkins (1992) that the dynamics of parasitoids mirror donor control (available food influences parasitoid densities, but parasitoids do not influence food availability).

Available data do suggest, however, qualitative similarities between competition among parasitoids and competition among insect herbivores. For

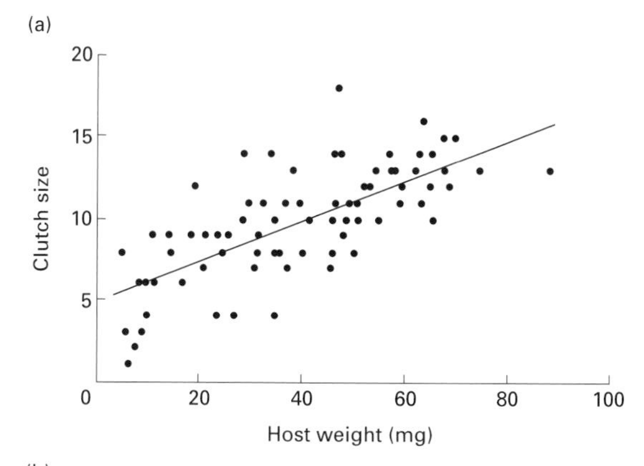

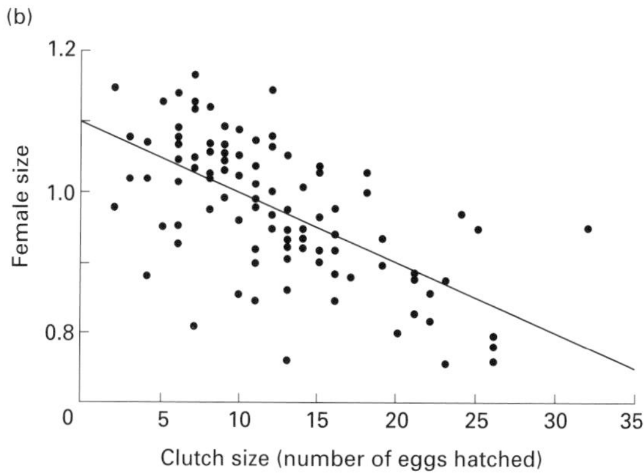

Fig. 4.14 Evidence of intraspecific competition among individuals of the hymenopteran parasitoid *Goniozus nephantidid*. The size of clutch (a) that a female parasitoid will lay on her caterpillar host increases with host size, suggesting that host body size is limiting. The larger the clutch laid on a host of a given size (b), the lower the body size of females that emerge. (From Hardy *et al.* 1992.)

example, interspecific competition among parasitoids is often asymmetric (Kato 1994; Monge *et al.* 1995; Reitz 1996), and can occur with unrelated taxa. Chilcutt and Tabashnik (1997) have documented competition between a parasitoid wasp, *Cotesia plutellae*, and a bacterial pathogen for the lepidopteran host, *Plutella xylostella*. The outcome of the interaction depended upon the degree of susceptibility of *P. xylostella* to the pathogen. In susceptible hosts, the parasitoid did not affect performance of the pathogen, but the pathogen had a significant negative effect on the parasitoid. In moderately resistant hosts, the interaction between the parasitoid and pathogen was symmetrical and competitive. Highly resistant hosts were not susceptible to infection by the pathogen, and this created a refugium from competition for the parasitoid. In a similar study, Nakai and Kunimi (1997) demonstrated that larvae of the endoparasitoid *Ascogster reticulatus* were negatively affected by infection of their lepidopteran host with a granulosis virus. Virus capsules were shown to accumulate in the guts of developing parasitoids, and this lowered rates of pupation and eclosion. Interactions between pathogens, insect hosts, and parasitoids provide fascinating opportunities for future research as well as potential difficulties for integrated pest management (Chapter 10). Parasitoids have also been shown to compete with beetles (Evans & England 1996; Heinz & Nelson 1996) and ants (Itioka & Inoue 1996).

Competition among insects is not restricted to terrestrial habitats. Intraspecific competition among individual larvae of the predatory mosquito *Topomyia tipuliformis* has been demonstrated within tiny pools of water (phytotelmata) held in plant tissue. In this case, the interaction is an extreme form of interference (contest) competition based on cannibalism (Mogi & Sembel 1996). Similarly, individual midge larvae in the species *Cricotopus bicintus* (Diptera: Chironomidae) compete for available periphyton (Wiley & Warren 1992). Although it is not yet clear how general the phenomenon is, competition among aquatic insect larvae for algal biomass has certainly been documented (Dudgeon & Chan 1992). We will return, in Chapter 7, to the role of aquatic insects in the regulation of primary production.

We do not wish to present an exhaustive list of all the various insect feeding groups, and the prevalence of competition in each. Indeed, we doubt that there are sufficient data to accomplish the task with any rigour. Suffice it to say that competition has been observed among most groups studied, including insect pollinators (Inoue & Kato 1992; Rathke 1992; Roubik 1992), predators (Vanbuskirk 1993; Moran *et al.* 1996), and decomposers (Trumbo & Fernandez 1995). In contrast', competition has not been observed in other studies of insects in these groups (Rosemond *et al.* 1992; Minckley *et al.* 1994). Like most ecological phenomena, the strength of competition is likely to vary in space, in time, and with community composition. Specifically, abiotic forces (Chapter 2), natural enemies (Chapter 5), and mutualism (Chapter 6) will interact with competition to determine the variation in insect populations and communities that we observe in space and time.

Chapter 5: Natural Enemies and Insect Population Dynamics

5.1 Introduction

Ecologists have long sought explanations for fluctuations in the abundance of insects and other animals. Attention usually tends to focus on the role of natural enemies—predators, parasites, parasitoids and pathogens. The reason for the emphasis placed on natural enemies, particularly predators, is largely because, as Price (1975) wrote: 'predation . . . is certainly one of the most visible aspects of mortality'. This visibility stems first from the variety of natural enemies (particularly the many insect species of predator and parasitoid) and their obvious roles. It is therefore hardly surprising that the action of parasitic wasps, predatory ladybirds and other natural enemies has been observed by naturalists for over a hundred years: '. . . very frequently it is not the obtaining of food, but the serving as prey to other animals, which determines the average numbers of a species' (Darwin 1866). In addition, the many adaptations of insects against attack by natural enemies, the successful cases of biological control, and population models have all given weight to the view that natural enemies have a dominant role in the population dynamics of insect herbivores.

In this chapter, we aim first to briefly describe the variety of insect and other natural enemies. We will then examine the role of natural enemies in insect population dynamics, and finally we will discuss the ways in which the effects of natural enemies combine with other factors, such as climate and host plant chemistry, to determine insect abundance.

5.2 The variety of natural enemies

5.2.1 Insect predators

A predator may be defined as an insect or other animal which kills its prey immediately, or perhaps soon after attacking it. Predators also tend to kill many prey individuals. There are exceptions, of course, and many insect species may be both predatory and obtain their food resources by other means. For example, many predatory ant species also consume the honeydew produced by aphids, psyllids or coccids, and some earwigs are both phytophagous (i.e. plant-feeding) and predatory. Other insects have different feeding strategies at different stages in their life cycles e.g. some staphylinid beetles are parasitoids when juvenile and predatory as adults.

Many different taxonomic groups of insects contain predatory species, or species that are partly predatory (Figs 5.1–5.5). The major wholly or largely predatory groups are the dragonflies and damselflies (Odonata), mantids (Mantodea), ant-lions, lacewings, scorpion flies and other Neuroptera. Within the Hymenoptera, the ants (Formicidae), although they adopt a diversity of lifestyles, are the major predatory family. Within the Coleoptera there are several predatory families such as the tiger beetles and ground beetles (Carabidae), the soldier beetles (Cantharidae) and the diving beetles (Dytiscidae). The Staphylinidae includes many predators, parasitoids, scavengers and species that show more than one type of feeding strategy. Perhaps the best known predatory beetles are the ladybirds (Coccinellidae), but there are some notable plant-feeding coccinellids. The 'true bugs' (Hemiptera, suborder Heteroptera) contain both predators and plant-feeding species. Some families (such as the largest family, the mirid or capsid bugs (Miridae)) contain examples of both lifestyles but the assassin bugs (Reduviidae), damsel bugs (Nabidae), pond skaters (Gerridae), water scorpions (Nepidae) and other families contain mostly predatory species. Some Diptera, notably the robber flies (Asilidae) and the larvae of many hoverflies (Syrphidae), are predators. Although thrips (Thysanoptera), grasshoppers and crickets (Orthoptera) are largely plant-feeding, and include many serious crop pests, some are predatory, including some bush-cricket species (Tettigoniidae). Clearly, there is no such thing as a typical insect predator.

Fig. 5.1 Praying mantis (Mantodea). (Courtesy of P. Embden.)

Fig. 5.2 Diving beetle larva (Coleoptera: Dytiscidae). (Courtesy of P. Embden.)

Fig. 5.3 Rove beetle (Coleoptera: Staphylinidae).

Fig. 5.4 Ladybird larva and eggs (Coleoptera: Coccinellidae).

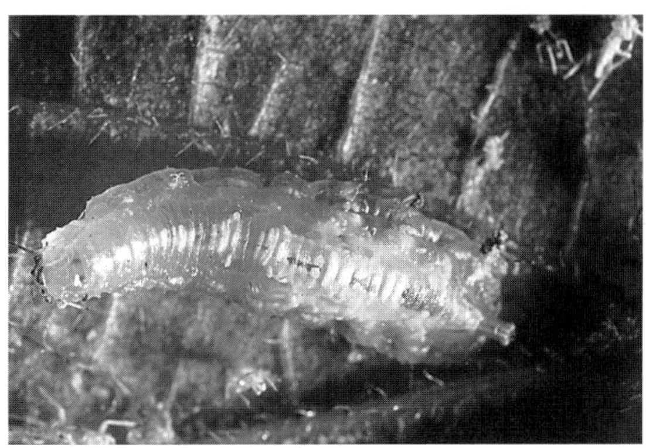

Fig. 5.5 Hover fly larva (Diptera: Syrphidae).

5.2.2 Other predators

Insects are subjected to predation from a wide range of insect and other species. The latter include other arthropods, such as centipedes (Chilopoda), scorpions, false scorpions (Pseudoscorpiones), harvestmen (Opiliones) and spiders (Araneae). In addition, there are many predatory, as well as plant-feeding and scavenging, mite (Acari) species.

Many vertebrates are partly or wholly insectivorous and, as we will see below, bird and small mammal predators can play a significant role in the population dynamics of many insects such as several important forest pests.

5.2.3 Parasitic insects

A parasite may be defined as a species that, like a predator, obtains its nutritional requirements from another species, but, unlike a predator, usually does not kill its prey (Askew 1971). Parasites may live on or in the body of their hosts (ectoparasites and endoparasites, respectively). Ectoparasites include lice (Phthiraptera—Mallophaga and Anoplura) and fleas (Siphonaptera). These parasites, particularly lice, spend much of their lives on their hosts. Other ectoparasitic insects, in contrast, feed only briefly on their hosts. This type of parasite, which includes many vectors of serious diseases of humans and their livestock, is the subject of Chapter 9.

Most parasitic insects are parasitic as larvae but not as adults, feed within or sometimes on their hosts, and eventually kill them. They are known as parasitoids (Askew 1971). Approximately 10% of all insect species are parasitoids (Eggleton & Belshaw 1992). Most, about 75%, of these are Hymenoptera; the remaining 25% are Diptera or Coleoptera. Within the Hymenoptera, the so-called Parasitica are almost all parasitic—the Ichneumonidae, Chalcidae and Braconidae are the major parasitoid families in the Parasitica (Figs 5.6–5.8).

In most species of parasitoids, the adult female locates hosts but sometimes these hosts are located by first instar larvae of the parasitoid (Eggleton & Belshaw 1992). The least common method of host location is for the host to ingest the parasitoid's egg. One example of the latter is *Cyzenis albicans* (Diptera: Tachinidae), a fly parasitoid that can develop only in the winter moth, *Operophtera brumata* (see section 5.3.4).

Some ecologists classify many plant-feeding insects as parasites (e.g. Price 1975). This chapter considers parasites only as natural enemies of other insects.

5.2.4 Pathogens

Insects are attacked by a range of pathogenic bacteria, viruses, fungi, protozoans and nematodes. Examples

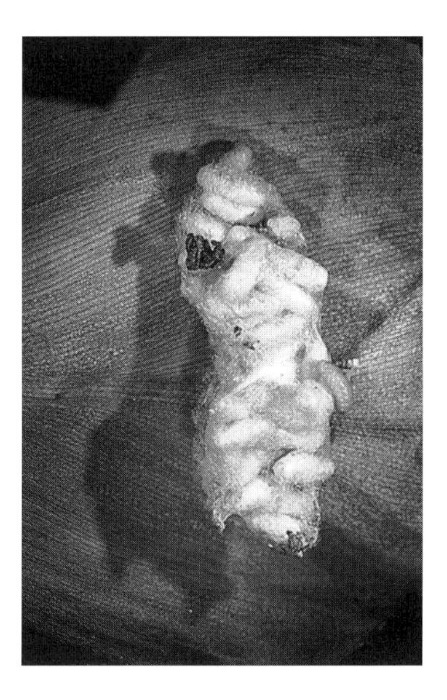

Fig. 5.7 Hymenopteran parasitoid larva and pupae emerged from the cadaver of lepidopteran larval host.

Fig. 5.6 Adult hymenopteran parasitoid ovipositing in lepidopteran larval host.

Fig. 5.8 Adult tachinid parasitoid (Diptera).

include the bacterium *Bacillus thuringiensis*, the fungi *Entomophthora* and *Beauvaria* species, and the nuclear polyhedrosis viruses (NPVs). The increasing role of pathogens in biological control of insect pests is discussed in Chapter 10.

5.3 The impact of natural enemies on insect populations

5.3.1 Observations, experiments or models?

Ecologists have used different approaches to study the impact of natural enemies: observations, experiments and models. In this context, we define observations as studies of insects and their natural enemies in their natural environment without manipulation of either, experiments as studies where either or both are manipulated in some way (e.g. through the artificial introduction or exclusion of natural enemies), and models as mathematical representations of natural enemy–insect interactions.

Each of these approaches has its strengths and weaknesses, and each has contributed in many ways to our understanding of the impact of natural enemies. Although we will present them as separate approaches below, each owes much to the other two. For example, observations and experiments are used to test the predictions of models. It should be noted that we avoid using the term 'theoretical models': each approach can be 'theoretical', in the sense that it tests a theory (or hypothesis) to explain the impact of natural enemies on their prey.

5.3.2 Observing natural enemies and their prey

Many ecologists have used observations on natural enemies and their prey to assess the impact of natural enemies on prey populations, particularly the populations of pests. Ecologists observe changes in the abundance of either the prey species or the natural enemy species, or the mortality caused by natural enemies.

The best known example of observation of insects and their natural enemies is 'classical' biological control. The many forms of biological control are discussed in Chapter 10: 'classical' biological control is the term often used to describe the introduction of non-native (or 'exotic') natural enemies to control insect pests, which are, themselves, also exotic species. Two examples are shown in Fig. 5.9. In each case the introduction (and therefore increase in the abundance) of the natural enemy results in a decline in the abundance of the pest species. The implication is clear: the natural enemy is responsible for the decline in the pest species. Generally, however, there are no 'control' plots to assess whether the natural enemy introduction is responsible for the decline in prey numbers (Kidd & Jervis 1996). Nevertheless, successful cases of biological control dramatically demonstrate the potential impact of natural enemies. Moreover, classical biological control is not fundamentally different from the action of

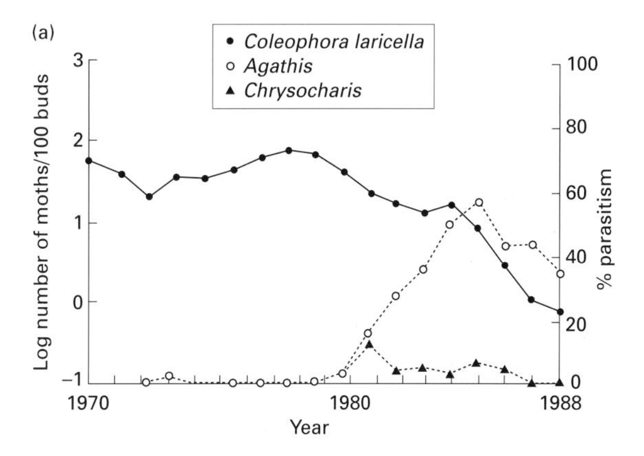

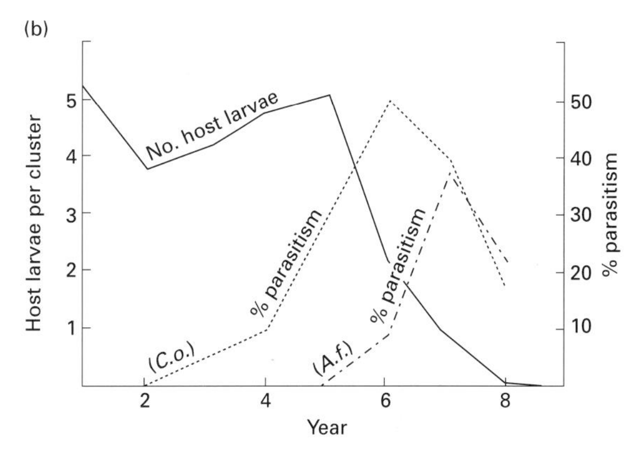

Fig. 5.9 Two examples of successful biological control of insect pests by natural enemies. (a) The introduction of two parasitoids, *Agathis pumila* and *Chrysocharis laricella*, to control the larch casebearer, *Coleophora laricella*. (From Ryan 1990.) (b) The introduction of *Cyzenis albicans* and *Agrypon flaveolatum* to control the winter moth in Canada (where 'year' is the number of years from which data were recorded, starting one year before the appearance of *C. albicans*. (From Embree 1966.)

indigenous natural enemies on their indigenous prey species (Waage 1992). In other words, 'classical' biological control demonstrates the generally 'unseen' impact of natural enemies.

Observations of the abundance of insects and their potential natural enemies can also be used to try to discover which natural enemy species are important in reducing the numbers of particular insect species. One situation where field observations, augmented by laboratory studies, have played a particularly important role is in the study of polyphagous predators (i.e. predators that consume a range of prey species). Agricultural pests, such as cereal aphids, are preyed upon by a wide range of arthropod predators. Some of these predators are aphid-specific, such as syrphid larvae and coccinelids; others, such as many ground-dwelling beetles, are polyphagous (consuming both aphids and other invertebrates).

Although aphid-specific predators are conspicuous and consume large numbers of prey during aphid outbreaks, simple regressions between the numbers of cereal aphids and aphid-specific predators suggest that these predators take advantage of aphid outbreaks but are incapable of preventing them (Edwards *et al.* 1979) (Fig. 5.10). In contrast, regressions between the numbers of cereal aphids and the diversity of predatory and other arthropods in cereal fields suggest that polyphagous predators can prevent aphid outbreaks (Potts & Vickerman 1974) (Fig. 5.11). Unfortunately, these types of observation fail to quantify the importance of polyphagous predators and, moreover, regressions between insects and their natural enemies can occur both by the natural enemy causing changes in prey numbers and by the natural enemy simply following changes in prey numbers (Kidd & Jervis 1996). Nevertheless, this type of study may usefully identify which types of predator merit further research.

A more useful form of observation than merely measuring the abundance of natural enemies and their prey is the estimation of mortality caused by natural enemies. Such observations are, however, much more difficult to make. For example, studies on the impact of parasitoids usually include observations on percentage parasitism. However, the technique used to quantify percentage parasitism can dramatically affect the estimate (Van Driesche 1983). Parasitism in natural populations can be measured by first taking a sample of the prey population and then either (a) dissecting each individual to detect parasitoid eggs or larvae within them or (b) rearing the sample to see how many parasitoids emerge. The most common technique used is rearing, but this typically underestimates parasitism in comparison with dissection by 12–44% (Day 1994). These underestimates are due to the relatively higher mortality of parasitized individuals because of disease and other factors during the rearing process. However, dissection can also lead to underestimates of parasitism because a proportion of parasitized hosts die of oviposition trauma and, in some species, after being fed upon by female parasitoids (Jervis *et al.* 1992). Day

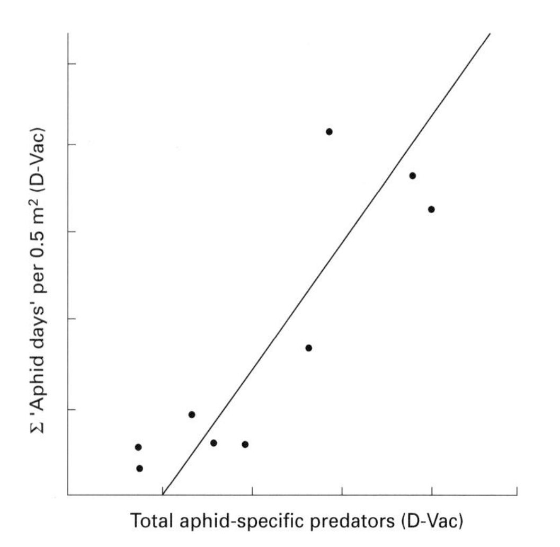

Fig. 5.10 The relationship between the abundance of aphid-specific predators and the abundance of cereal aphids. (From Edwards *et al.* 1979.)

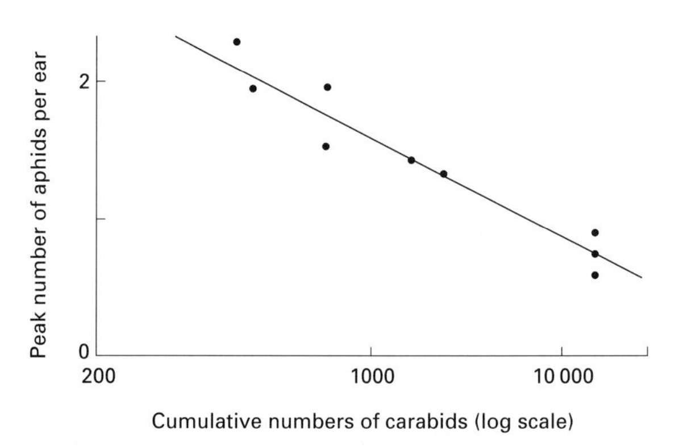

Fig. 5.11 The relationship between the abundance of polyphagous predators and the abundance of cereal aphids. (From Edwards *et al.* 1979.)

(1994) argued that to give the most comprehensive results both methods should be used concurrently. Kidd and Jervis (1996) reviewed some of the more complex ways of assessing parasitism.

Despite the problems associated with measuring parasitism accurately, the measurement of the impact of predators presents an even more challenging task because vertebrate and many invertebrate predators completely consume their prey and leave little or no evidence that they have done so. In some cases, measuring mortality caused by predators can be relatively straightforward. For example, the number of pupae of a univoltine (i.e. single generation per year) forest pest (such as the winter moth; see section 5.3.4) killed by predators can be calculated from (a) the number of insects entering the pupal stage, (b) the number of pupae killed by parasitoids, disease and unknown causes (e.g. weather) and (c) the number of pupae successfully emerging as adults. In some cases sampling of the pupal stage might also reveal evidence of feeding by particular types of predators and provide separate estimates of mortality caused by, for example, shrews and ground-dwelling beetles.

This simple calculation of predation is possible because the population does not, of course, increase during the pupal stage and because of the non-overlapping (discrete) generations of insect. Predation is much more difficult to assess in multivoltine insects, such as cereal aphids, with overlapping generations.

In such cases, an alternative to measuring mortality is to calculate a predation 'index' of some kind. For example, the research described above on the impact of polyphagous predators on cereal aphids led to more detailed work to identify which of the polyphagous predators found in cereal fields have a significant impact on aphid numbers. This research involved measuring the density of different insect predators during the period when aphid numbers were increasing and dissecting samples of predators to calculate the proportion that had consumed cereal aphids (Sunderland & Vickerman 1980). Different predators were then ranked according to an index calculated by multiplying predator density by the proportion of individuals in that species that had consumed aphids (Table 5.1) This then led to detailed studies on the species identified as potentially being the most important predators.

Field predation rates can also be estimated by direct observations of predation in the field, a technique unsuited to most predatory groups except perhaps spiders (e.g. Sunderland *et al.* 1986). In addition, serological methods can be used to measure predation. Serological techniques are based on the production of antibodies in rabbits and other mammals against antigens of prey species. These antibodies are then used to detect the presence of particular prey species in the gut contents of predators. One of the most commonly used techniques is ELISA or enzyme-linked immunosorbent assay (Sunderland 1988; Kidd & Jervis 1996). Sunderland *et al.* (1987), for example, calculated a predation index for different species as predators of cereal aphids based on enzyme-linked immunosorbent assay (ELISA), $P_g d/D_{max}$, where P_g is the percentage of predators testing positive in cereal aphid ELISA tests, D_{max} is the number of days over which cereal aphid antigens are detected in the predator and d is the density of the predator. Use of this index suggested that spiders were the most important predators of cereal aphids.

5.3.3 Density-dependent and density-independent mortality factors

Observations of the mortality caused by natural enemies have also been used to identify which, if any, natural enemies 'regulate' the abundance of particular insect species and, more generally, in the debate on the importance of regulation by density-dependent mortality. Density-dependent factors whose proportional impact varies according to population size or, more accurately, population 'density', which may be defined as population size in a given area. Density-dependent factors include predation and competition, and their impact may be manifest by a proportional change in birth rate or mortality rate as the population density changes (Fig. 1.13). In contrast, the proportional impact of density-independent factors is not affected by population density (see also Chapter 1).

The idea that the action of one or more density-dependent factors is needed to regulate animal population abundance (e.g. Nicholson 1933, 1957, 1958) is widely accepted despite the often acrimonious debate between ecologists (Dempster & McLean 1998). A number of population ecologists, most, notably, working on insects, have argued that density-independent factors are responsible for regulating insect densities (e.g. Andrewartha & Birch 1954; Milne 1957a, b; Den Boer 1991).

Table 5.1 Assessment of the importance of different polyphagous species as predators of cereal aphids by the calculation of a 'predation index'. (From Sunderland & Vickerman 1980.)

	Proportion containing aphid remains at aphid densities of:					
	1–1000 m^{-2} (increase phase)	1000–1 m^{-2} (decrease phase)	Limit 1000 m^{-2} (increase plus decrease phases)	Number examined	Mean density of predators (for aphid increase 1–1000 m^{-2})	Predation index*
Demetrias atricapillus	0.253	0.136	0.230	113	1.23	0.311
Agonum dorsale	0.236	0.336	0.257	653	1.28	0.302
Forficula auricularia	0.278	0.165	0.220	236	0.61	0.170
Tachyporus chrysomelinus	0.051	0.054	0.052	346	2.39	0.122
Tachyporus hypnorum	0.024	0.034	0.260	778	4.50	0.108
Bembidion lampros	0.082	0.167	0.093	989	1.23	0.101
Amara familiaris	0.033	0.019	0.027	255	1.47	0.049
Amara aenea	0.034	(0.00)	0.034	176	1.42	0.048
Nebria brevicollis	0.086	(0.00)	0.085	531	0.48	0.041
Notiophilus biguttatus	0.040	0.013	0.031	228	0.67	0.027
Asaphidion flavipes	0.048	0.000	0.044	114	0.31	0.015
Amara plebeja	0.016	0.000	0.014	147	0.88	0.014
Harpalus rufipes	0.053	0.056	0.054	147	0.14	0.007
Pterostichus melanarius	0.161	0.073	0.101	346	0.03	0.005
Loricera pilicornis	0.007	0.019	0.011	442	0.27	0.002
Calathus fuscipes	(0.000)	0.098	0.085	47	0.02	0.000

* Values in column 1 multiplied by values in column 5.
Parentheses denote sample size <10.

Put briefly, the fundamental argument for the importance of density dependence is that populations would increase indefinitely unless their fluctuations were somehow regulated by density-dependent mortality (or natality, emigration or immigration). The simplest population models demonstrate this (see section 5.4). The counter-argument is that in natural (as opposed to theoretical) systems, density-independent factors, particularly weather, are largely responsible for causing, and even limiting, the fluctuations of insect populations. As Andrewartha and Birch (1954) wrote in relation to the classic study of weather on thrips, *Thrips imaginis* (Thysanoptera), on rose: 'not only did we fail to find a density-dependent factor but . . . there was no room for one'. This remark was prompted by the finding (in multiple regression analysis) that weather explained over 80% of the variation in the abundance of thrips (Davidson & Andrewartha 1948a, b).

As well as stimulating the development of theoretical models, the density-dependent debate in the 1950s stimulated the growth of the 'life table' approach to the study of natural populations of insects (see below). What better way to determine whether or not density-dependent factors regulate insect abundance than to quantify the impact of different mortality factors and examine the relationship between their impact and population density?

5.3.4 Life table studies

Probably the most significant development in the detection of density dependence in insect populations has been the use of life tables and *k*-factor analysis (Varley 1947). One of the best-known examples of life table studies is the study by Varley and Gradwell (1968) and Varley *et al.* (1973) of the winter moth on oak trees in Wytham Wood in England (Fig. 1.14). Those workers monitored the abundance of adult and larval winter moth for nearly 15 years and measured the impact of pupal predation, parasitism by *Cyzenis albicans* and other parasitoids, and pathogen (microsporidian) infection. They also quantified the effect of what they called 'winter

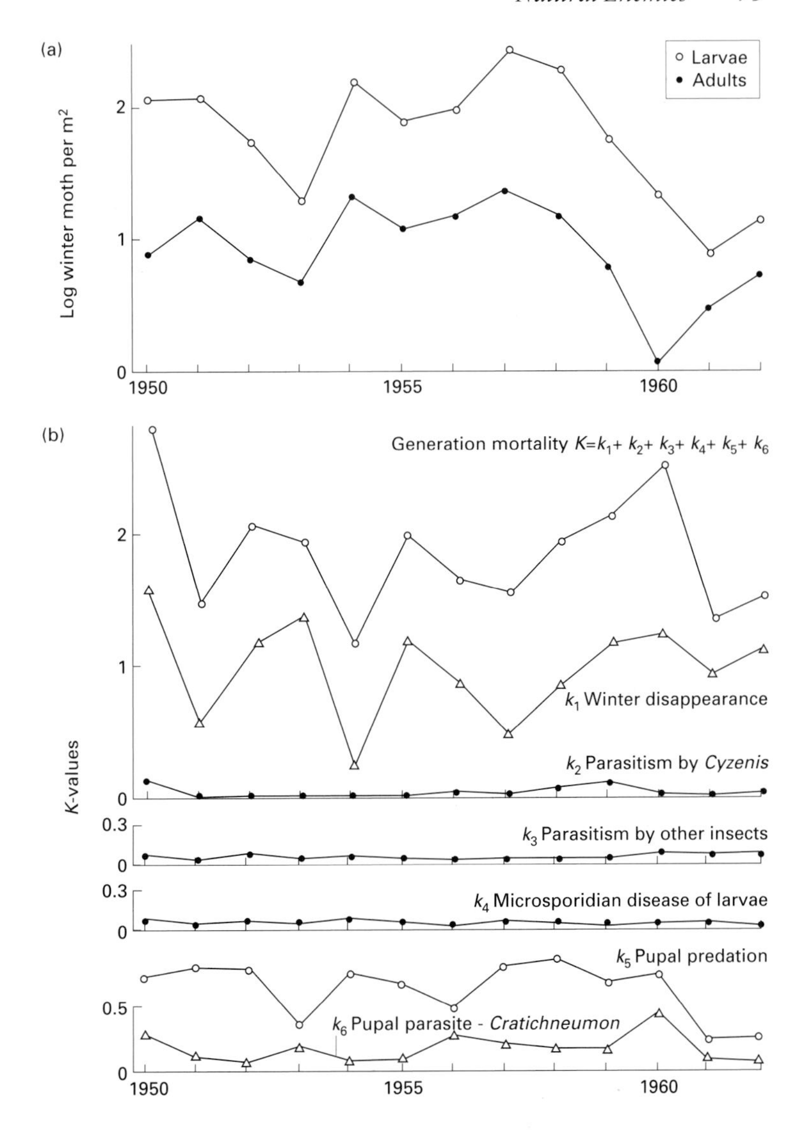

Fig. 5.12 Abundance (a) of winter moth larvae and adults in Wytham Wood, England 1950–62 and changes in mortality; (b) expressed as total generation mortality, *K*, and *k*-values for separate sources of mortality. (From Varley *et al.* 1973.)

disappearance', mortality between the adult and late larval stage.

The impact of each mortality factor in *k*-factor analysis is expressed as a *k* value:

$$k_i = \log N_i - \log N_{i+1} \quad (5.1)$$

where N_i and N_{i+1} are the densities of the population before and after mortality k_i.

The *k* value of each factor is calculated (as k_1, k_2, k_3, etc.) and total mortality, *K*, equals $k_1 + k_2 + k_3$, etc.

If the values of all *k* factors are plotted together with total mortality (*K*) for each year of the study, then the contribution of each *k* factor can be evaluated. The factor that makes the greatest contribution to total mortality is known as the key factor. This can be assessed visually or evaluated by plotting individual *k* factors against total mortality, the mortality factor with the greatest slope being the key factor (Podoler & Rogers 1975). For the winter moth, the key factor is winter disappearance (Fig. 5.12).

Density dependence is detected by plotting the values of each *k* factor against population density. A positive slope indicates density dependence, a negative slope indicates inverse density dependence and an absence of a significant relationship indicates density independence. This demonstrated that pupal predation acts as a density-dependent factor on the winter moth (Fig. 5.13).

Thus, k-factor analysis has shown that the size of the fluctuations of the population of winter moth, one of the most important defoliators of oak and other deciduous trees in Europe, is primarily determined by 'winter disappearance' but the population is regulated by pupal predation. The largest component of 'winter disappearance', although it includes adult, egg, and larval mortality, is mortality of young winter moth larvae, thought to be mainly determined by the degree of synchrony (or coincidence) between larval emergence and bud burst. However, studies elsewhere on this insect have produced different conclusions, and for an alternative interpretation of winter moth population dynamics in Wytham Wood, stressing the importance of density-independent factors, the reader is referred to Den Boer (1986, 1988), and the reply by Latto and Hassell (1987).

Detailed instructions on the construction of life tables and k-factor analysis have been given by Varley *et al.* (1973) and Kidd and Jervis (1996). Several reviews have been published summarizing the many insect life table studies that have now been carried out (Dempster 1983; Stiling 1987, 1988; Cornell & Hawkins 1995). Dempster (1983) analysed 24

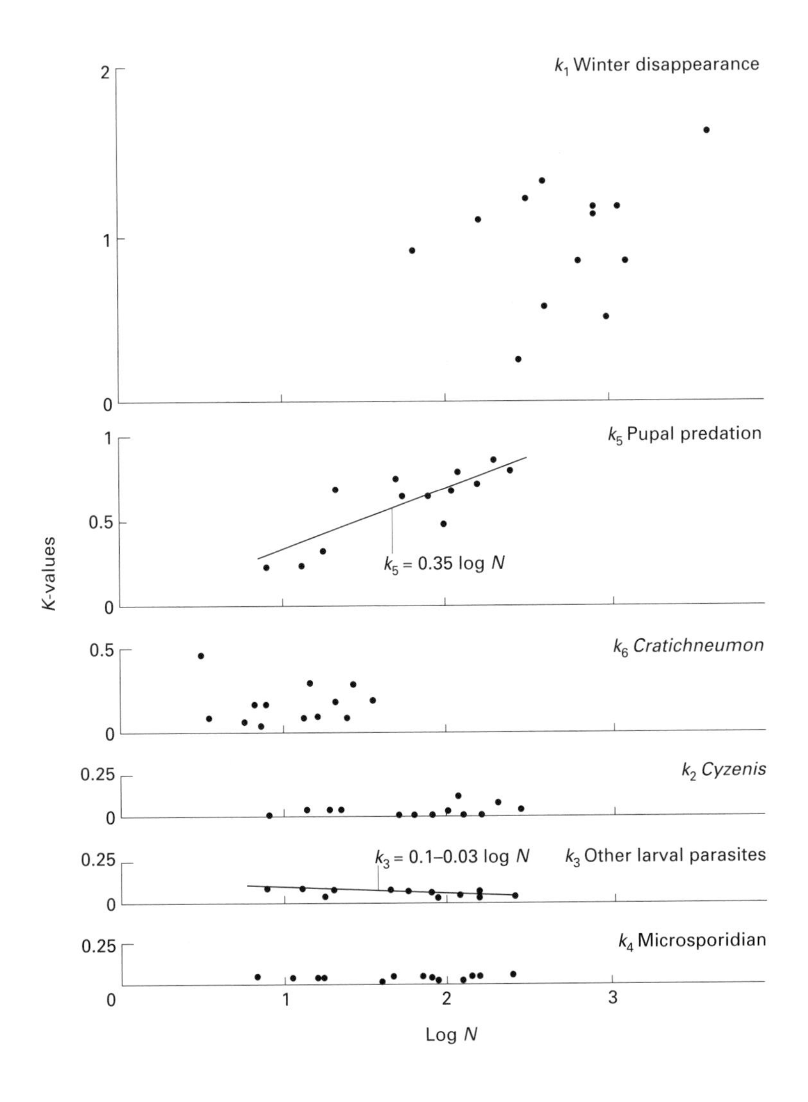

Fig. 5.13 Relationships between different winter moth mortalities and population density. (Source as Fig. 5.12.)

datasets and found evidence of density dependence by natural enemies in only three cases. Stiling (1987) examined life table data for 58 species of insect but found evidence of density-dependence in only about half of them. In approximately half of the cases where density-dependence was detected (27% of all cases) parasitism, predation or pathogens were responsible.

Unfortunately, there is a growing realization that there are many problems associated with *k*-factor analysis (e.g. Dempster 1983). First, both the estimation of key factors and the detection of density dependence by regression analysis are statistically flawed. In both cases the variables being analysed are not independent; for example, *k* values are calculated from, and therefore are not independent of, population densities. Alternative tests for the detection of density dependence have been proposed but these may be unnecessarily strict, potentially 'missing' density-dependent mortality factors (Hassell *et al.* 1987). There are several other 'technical' problems with *k*-factor analysis. For example, 'traditional' *k*-factor analysis assumes that each *k* factor acts in sequence, not in parallel. The order in which different mortalities are analysed may not seem to be important, but Putman and Wratten (1984) found that when they reversed the order of larval starvation and predation in the cinnabar moth life table analysis, predation, rather than larval starvation, became the apparent key factor.

Perhaps more important than these methodological problems, is the current concern that there are ecological reasons why *k*-factor analysis is flawed. One of the first potential problems of *k*-factor analysis was the realization that density-dependent factors did not always act immediately—their impact on a population could be delayed. These so-called delayed density-dependent factors are not immediately apparent from a regression of *k* values against density. However, time sequence plots can reveal the action of delayed density-dependent factors through the spiral patterns they cause (Fig. 5.14).

A further ecological problem with *k*-factor analysis is that it may take many generations to detect density-dependence (Hassell *et al.* 1989). However, during the course of a study, the key factor and the factor, or factors, responsible for regulating the population may have changed. Nor does *k*-factor analysis on its own reveal the potentially important influence of spatial density dependence (Hassell 1985) (section 5.4.7). Mortality factors may also interact, causing effects on populations that are not additive (section 5.5).

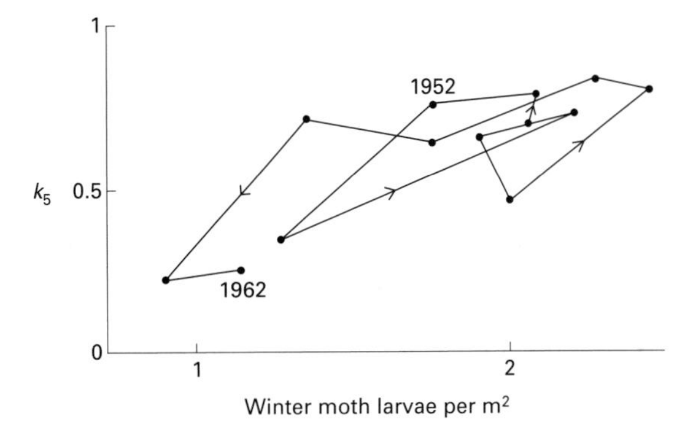

Fig. 5.14 Time series plot of pupal predation of the winter moth, the spiral form suggesting a delayed density-dependent component to this mortality. (Source as Fig. 5.12.)

Some of the problems associated with life tables and *k*-factor analysis are so severe that it has been referred to as a methodological 'straitjacket' (Putman & Wratten 1984): *k*-factor analysis clearly should not be used alone to try to understand the population dynamics of insects. However, life tables can be a useful way of comparing, for example, the susceptibility of different types of insect to predation or parasitism. Hawkins *et al.* (1997), for example, compiled life tables for 78 insect herbivores and found that leaf miners suffer the greatest levels of parasitism, and gallers, borers and root-feeders the least. In contrast, exophytic (externally feeding) herbivores experience the greatest level of mortality caused by predators and pathogens.

5.3.5 Experiments on natural enemies and their prey

As an alternative to observing natural populations, many ecologists have manipulated either natural enemy or prey species in experiments to quantify the impact of natural enemies on their prey. The experimental approach takes many forms but most experiments involve the exclusion of natural enemies and a comparison between the abundance of the prey species in exclusion and control 'treatments'. Natural enemies may be excluded by cages or barriers (Kidd & Jervis 1996). Cages may be placed over the whole plant, over single branches (sleeve cages) or leaves, or even small parts of a leaf (e.g. 'clip-cages'). Barriers to prevent access of natural enemies include greased plastic bands around tree trunks or branches. Exclusion experiments may be used to quantify the impact of the whole natural enemy complex of a single prey species or may focus on the impact of one, or a few, species of natural enemy (Kidd & Jervis 1996). The latter, so-called partial (as opposed to total), exclusion experiments include the use of mesh or gauze cages where the mesh size allows access of some natural enemies but not others. Alternatively, cages may be placed over plants to exclude, for example, hoverflies (Diptera: Syrphidae) and parasitoids but with their sides raised slightly above the ground to allow ground beetles to move in.

Morris (1992), for example, excluded predators from colonies of the aphid *Aphis varians* feeding on fireweed (*Epilobium angustifolium*) as part of a complex experiment designed to investigate the effects of predation, interspecific competition (with flea beetles—Coleoptera: Chrysomelidae) and water availability to the host plant. Morris found that coccinellid and syrphid predators had the greatest impact on the abundance of the aphid: in their absence, aphid numbers increased by 10% per day. The other factors had an insignificant impact on aphid abundance.

There are many problems associated with exclusion experiments but the main drawback of these experiments is that the cages may affect the microclimate of the enclosed plant, potentially affecting the performance of the prey species (Kidd & Jervis 1996). There are various possible solutions to this problem such as 'inclusion' cage experiments. In these experiments, known numbers of predators or parasitoids are placed inside the cages (e.g. Dennis & Wratten 1991). Even here, however, the unnatural microclimate may lead to misleading results (Kidd & Jervis 1996).

Despite their problems, exclusion experiments can, if carefully done, provide useful information on the impact of natural enemies. Other experimental techniques include the use of insecticides to remove natural enemies or even their physical removal (Kidd & Jervis 1996). An alternative experimental approach is to manipulate prey numbers instead of natural enemies. This is particularly useful for measuring the impact of natural enemies in different habitats. Watt (1988, 1990), for example, manipulated the abundance of pine beauty moth, *Panolis flammea* (Lepidoptera: Noctuidae), to measure, amongst other things, the role of natural enemies in different forest habitats. Run in parallel with natural enemy exclusion experiments, this approach identified the role of natural enemies in preventing pine beauty moth outbreaks on Scots pine (Fig. 5.15). A further alternative approach is to use artificial prey. For example, Speight and Lawton (1976) used laboratory-reared *Drosophila* pupae to assess the effect of weed cover in cereal fields on the impact of carabid beetles.

Experiments are also carried out to do more than quantify the impact of natural enemies. One interesting example is a 'convergence experiment' (Nicholson 1957). Convergence experiments involve the manipulation of insect populations to artificially high or low densities to determine whether the population returns to an equilibrium density. Although there are many technical problems associated with this type of experiment (Kidd & Jervis 1996), it can be useful in detecting density-dependent factors missed in life tables and k-factor analyses. Gould *et al.* (1990), for

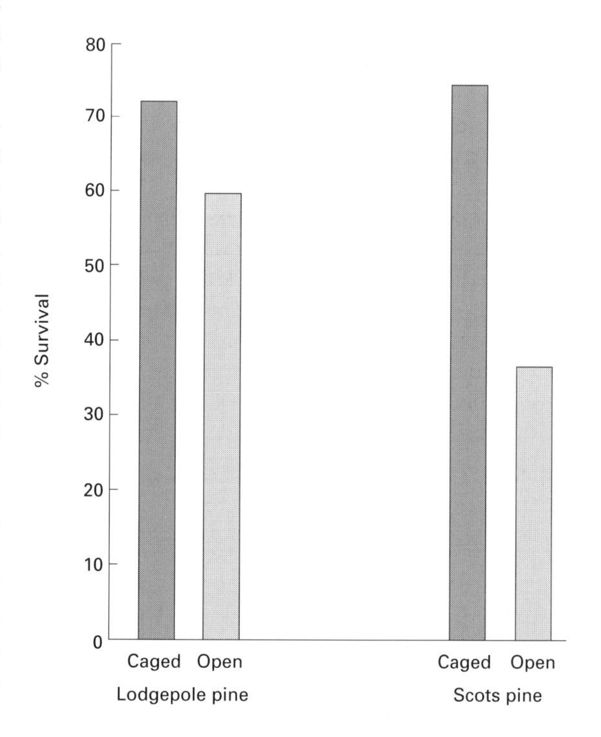

Fig. 5.15 The survival of pine beauty moth, *Panolis flammea*, larvae on different host plants in predator exclusion cages and exposed to predation. (From Watt 1989, 1990.)

example, discovered density-dependent mortality in gypsy moth, *Lymantria dispar* (Lepidoptera: Lymantriidae) populations caused by two parasitoids by artificially increasing the numbers of gypsy moth to provide a range of densities.

Experiments have also demonstrated how natural enemies can restrict the spread of insect outbreaks. For example, outbreaks of the western tussock moth, *Orgyia vetusa* (Lepidoptera: Lymantriidae) are known to persist for over 10 years, with little tendency to spread into suitable adjacent areas. One possible explanation for this is that natural enemies, more mobile than the moth, which has flightless females, disperse out from the edges of the outbreak and create a zone around the outbreak area where the ratio of natural enemies to tussock moth are particularly high. This 'predator diffusion' hypothesis was tested by establishing experimental populations of tussock moth along a transect from the edge of a tussock moth outbreak into suitable adjacent habitat where there were few tussock moths (Brodmann *et al.* 1997; Maton & Harrison 1997). The impact of natural enemies was assessed by measuring the numbers of eggs and larvae attacked by an egg parasitoid and four species of tachinid larval parasitoids. As predicted, parasitism was greatest in the zone immediately surrounding the outbreak, and it was concluded that the parasitoid did indeed restrict the spread of tussock moth outbreaks. Interestingly, this hypothesis was stimulated by 'reaction-diffusion' models, which predict that insects can become patchily distributed in uniform habitats as a result of interactions between mobile natural enemies and their relatively sedentary prey (Maton & Harrison 1997). This is a particularly good example of the power of experimentation and of how we need both models and experimentation to solve problems in insect ecology.

5.4 Modelling predator–prey interactions

5.4.1 The Lotka–Volterra model

No other aspect of theoretical population ecology has attracted more attention than predator–prey modelling. (In this context we include parasite–host models and host–pathogen models.) We cannot attempt to cover the whole history and complexity of models that have been developed to describe insect populations, and present only some of the main developments in modelling here. We are aware that many insect ecologists, particularly applied ecologists, are very sceptical of population models. Nevertheless, models are a useful way of illustrating complex interactions in ecology. A useful starting point for considering the development of population models is the theory of predator–prey interactions. These interactions have been explored to try to explain, for example, the population cycles of many insects such as several forest pests. Figure 5.16 illustrates the cycles of the pine looper moth, *Bupalus piniaria* (Lepidoptera: Geometridae), in the UK (see also Chapters 1 and 10).

The first population models to describe predator–prey interactions, the Lotka–Volterra equations were independently produced by Lotka (1925) and Volterra (1931). Such is the lasting impact of these publications that Hutchinson (1978) described Lotka's 1925 book as 'one of the foundation stones of contemporary ecology'. Hutchinson (1978) and Kingsland (1985) have described the history of early population models and their use, and Gillman and Hails (1997) have provided a thorough introduction to the mathematics of population models.

Lotka and Volterra first chose a model to describe the population dynamics of the prey population growing in the absence of predation. They used the logistic equation, probably the first population model, produced by Verhulst in 1838 to describe human population growth. The logistic equation is derived from the assumption that the rate at which a population grows (dN/dt) depends in some way on

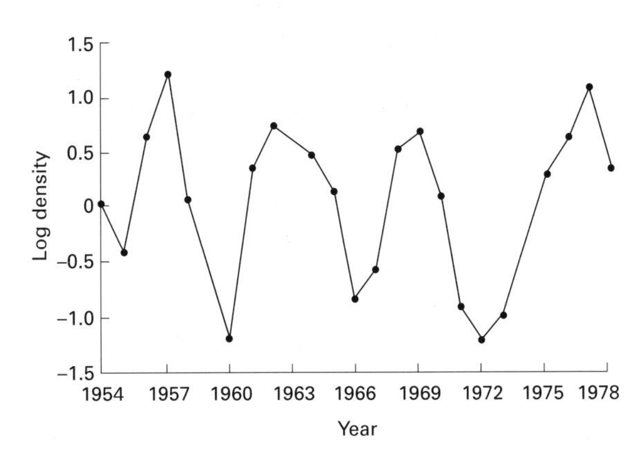

Fig. 5.16 Abundance of pine looper moth, *Bupalus piniaria*, pupae in Tentsmuir forest, Scotland 1954–78. (From Barbour 1985.)

the size of that population (N). This can be described mathematically by the equation

$$\frac{dN}{dt} = f(N). \qquad (5.2)$$

This equation has many solutions (Hutchinson 1978) but the simplest is

$$\frac{dN}{dt} = bN \qquad (5.3)$$

which can also be written (after integration) as

$$N = e^{bt}. \qquad (5.4)$$

These equations describe unrestricted, or 'exponential', population growth and, as natural populations cannot behave in this way indefinitely, it has to be rejected as being ecologically unrealistic. Nevertheless, the unrestricted rate of increase, b in the above equations, is used widely as r (or r_m) in population models and elsewhere, and is referred to as the 'intrinsic rate of increase' or the 'innate capacity for increase'.

As eqn 5.3 has to be rejected as being unrealistic, the next most simple solution to eqn 5.2 is

$$\frac{dN}{dt} = bN + cN^2. \qquad (5.5)$$

If we replace b by r (to conform with the usual writing of this equation) and rearrange by replacing c by $-r/K$ this gives the logistic equation

$$\frac{dN}{dt} = rN\left(1 - \frac{N}{K}\right) \qquad (5.6)$$

or (by integration)

$$N = \frac{K}{(1 + e^{-rt})}. \qquad (5.7)$$

The difference between the logistic model (eqn 5.6) and the exponential model (eqn 5.3) is that the unrestricted population rate of increase (r) is multiplied by the term $(1 - N/K)$ to give the actual rate of increase. When $N = 0$, $(1 - N/K) = 1$, and the population grows at an unrestricted rate. Eventually, $N = K$ and, because $(1 - N/K) = 0$, population growth is zero and the population remains at K. Thus the logistic equation describes the growth of a population towards an asymptote K, a population ceiling or the 'carrying capacity' of the habitat occupied by the population (see Chapter 4).

Lotka and Volterra used the logistic equation to represent the growth of a prey population as (where r_1 is the unrestricted rate of increase of the prey population):

$$\frac{dN}{dt} = r_1N\left(\frac{K - N}{K}\right). \qquad (5.8)$$

Second, the effect of predation on the prey was modelled by assuming that the rate of increase of the prey population would be reduced in proportion α to the abundance of predators (P) multiplied by the abundance of prey (N):

$$\frac{dN}{dt} = r_1N\left(\frac{K - N}{K}\right) - \alpha PN. \qquad (5.9)$$

Lotka and Volterra modelled the predator populations by assuming that the predator population would exponentially decline in the absence of prey:

$$\frac{dP}{dt} = -r_2P. \qquad (5.10)$$

They also assumed that the predator population would increase in proportion (β) to the abundance of predators (P) multiplied by the abundance of prey (N):

$$\frac{dP}{dt} = -r_2P + \beta PN. \qquad (5.11)$$

Taken together to represent the interactions between predator and prey, the Lotka–Volterra equations (eqns 5.9 and 5.11) can produce cycles in prey numbers (Fig. 5.17). These cycles are similar to those observed in many insect species (e.g. Fig. 5.16) and imply that they can be caused by predators. However, cycles are only produced under particular conditions of the Lotka–Volterra model (Gillman & Hails 1997) and the Lotka–Volterra equations have been widely criticised for their lack of realism. Nevertheless, their contribution to theoretical ecology as a stimulus for further research cannot be ignored (e.g. May 1974).

5.4.2 Discrete population models and density dependence

Population models can be either continuous or discrete (Gillman & Hails 1997). Continuous models are suitable for populations with overlapping generations and continuous reproduction, and are best described by differential equations such as those considered above. Discrete population models are suitable for insects with synchronous regular repro-

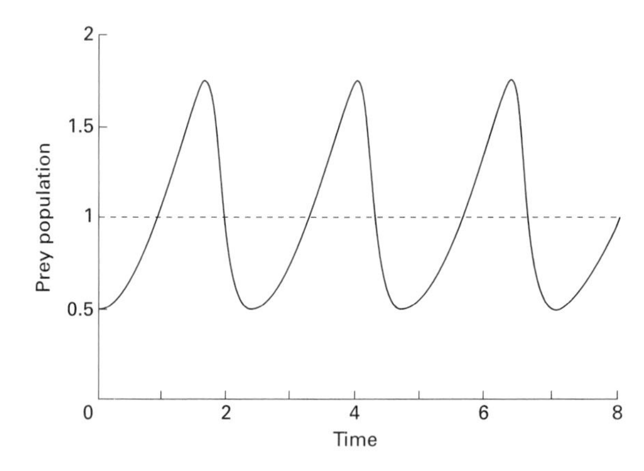

Fig. 5.17 Changes in the size of a prey population according to particular conditions of the Lotka–Volterra equation. (From May 1974b.)

duction such as univoltine (one generation per year) species. Discrete models are best described by difference equations and are much easier to construct and understand than models based on differential equations.

Gillman and Hails (1997) described the mathematics of difference equation models. In summary, they usually take the form of an equation describing the abundance of a population (N) at time t, usually written as N_t, as a function of the size of the population in the previous year or generation (N_{t-1}). For example, a population model for an insect with a fecundity of 100 eggs per female (and a sex ratio of 1 : 1) could be described by the equation

$$N_t = 50N_{t-1}. \quad (5.12)$$

This is a 'first-order' difference equation because abundance at time t is related to abundance one time interval previously ($t - 1$). Second-order difference equations relate abundance to two time intervals; for example,

$$N_t = 50N_{t-1} + 50N_{t-2}. \quad (5.13)$$

The exponential and logistic population models described above as differential equations have discrete equivalents. The discrete exponential model of population growth can be written as

$$N_t = \lambda N_{t-1} \quad (5.14)$$

and the discrete logistic equation as

$$N_t = \lambda N_{t-1}\left(\frac{1 - N_{t-1}}{K}\right). \quad (5.15)$$

This equation is also often written as

$$N_t = \lambda N_{t-1}(1 - \alpha N_{t-1}) \quad (5.16)$$

where $\alpha = 1/K$, or as

$$N_t = \lambda N_{t-1} - \frac{\lambda N_{t-1}^2}{K}. \quad (5.17)$$

One advantage of using discrete population models is that they can be readily used to demonstrate the potential effects of density dependence (section 5.3.3) on insect populations.

The complexity of apparently simple models such as eqns 5.15–5.17 can be demonstrated by changing the values of λ (Hassell *et al.* 1976; Gillman & Hails 1997). Using the examples in Fig. 5.18, where K = 100 and 200, increasing the value of λ to 3.1 results in oscillations between two population densities, or a 'two-point limit cycle' (Fig. 5.18b), and a further increase in the value of λ to 3.5 results in four-point limit cycles (Fig. 5.18c). If λ is increased still further, to 4.0, this results in irregular population fluctuations or 'chaotic dynamics' (Fig. 5.18d). Chaotic behaviour has received much attention in the dynamics of populations and other systems. The main message from models producing chaotic dynamics is that it may be wrong to assume that irregular population fluctuations are necessarily caused by irregularly acting factors, such as climatic variability: they may be caused by density-dependent factors.

One of the problems with the models described in this section is that, even if they can produce population behaviour of the types observed in nature (Fig. 5.16), they are oversimplified and say little about the causes of the population behaviour. This problem may be overcome by incorporating realistic values for model variables (such as λ) but nevertheless many population ecologists have felt the need to construct more complex models. The next section considers the most significant development in modelling predator–prey dynamics after the Lotka–Volterra models. Later sections look at some attempts to model the interactions between insects, their natural enemies and other factors.

5.4.3 The Nicholson–Bailey model

Returning to models that are designed to examine the

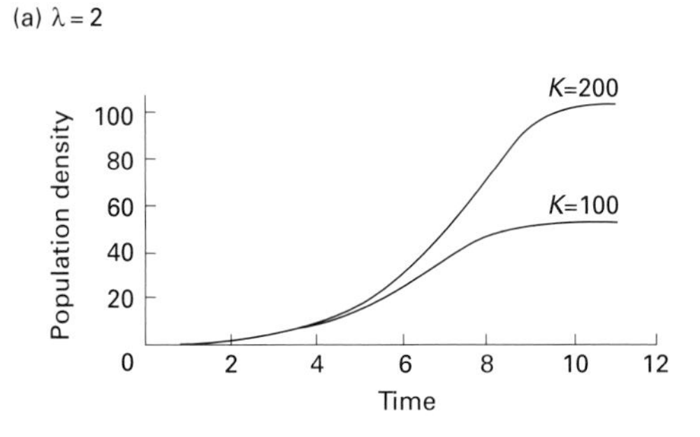

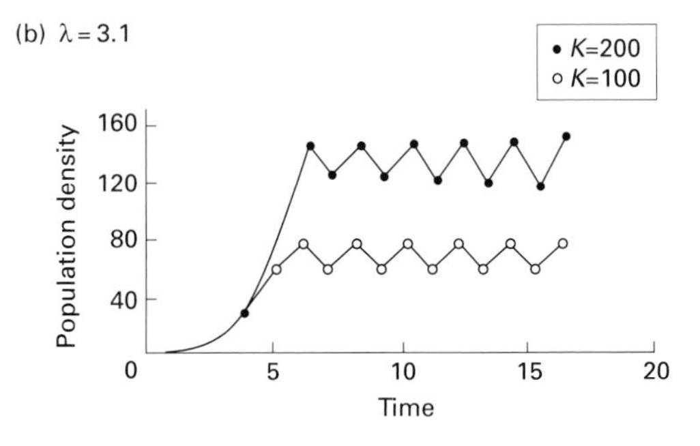

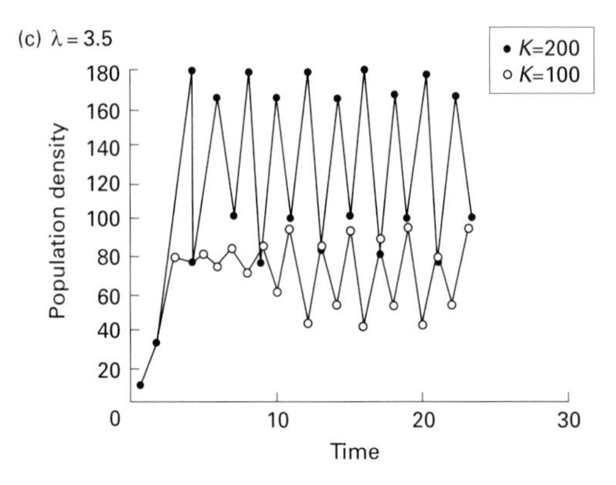

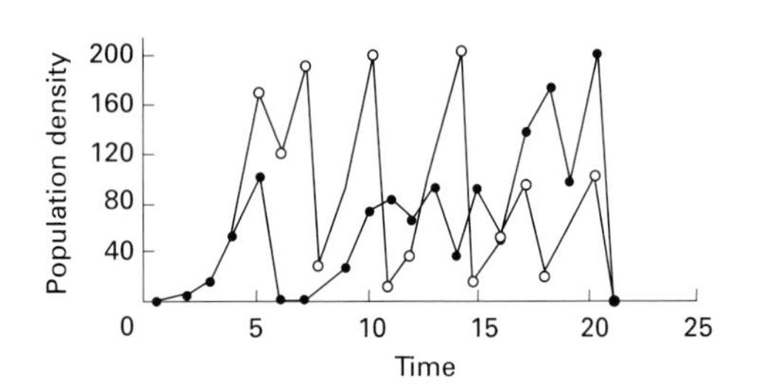

Fig. 5.18 The effects of varying λ and K in the discrete logistic equation. (From Gillman & Hails 1997.)

interactions between insects and their natural enemies, the first significant model to describe the population behaviour of insects and parasitoids was constructed by Nicholson and Bailey (1935). Those workers criticised the predator–prey models of Lotka and Volterra for several reasons including the incorrect assumption that there is an instantaneous response to encounters between natural enemies and their prey. Nicholson and Bailey pointed out that the response of natural enemies, particularly parasitoids, can take some time to occur. For example, the effect on the population density of adult parasitoids will take one generation to occur.

Nicholson and Bailey attempted to incorporate a greater degree of reality into their model. The important assumptions of the aspects of the Nicholson and Bailey model are as follows (Hassell 1981; Kidd & Jervis 1996; Gillman & Hails 1997):

- predators or parasitoids search randomly for their prey;
- predators can consume an unlimited number of prey (once located);
- parasitoids have an unlimited fecundity;
- only one adult parasitoid will emerge from a parasitized host;
- the area, described by Nicholson as the 'area of discovery' a, searched by a predator or parasitoid is constant (see below);
- the rate of increase of both the hosts (λ_N, in the absence of predation or parasitism) and the predators or parasitoids (λ_P) is known;
- generations of both the natural enemy and its prey are completely discrete and fully synchronized.

It should be noted that the area of discovery referred to above is more or less equivalent to the proportion of the habitat searched by the natural enemy during its lifetime and is used as a measure of the natural enemy's searching efficiency.

Using these assumptions, Nicholson and Bailey produced the following pair of discrete population models to describe the dynamics of the host (N) and parasitoid or predator (P):

$$P_{t+1} = \lambda_P N_t (1 - e^{-aP_t}) \tag{5.18}$$

$$N_{t+1} = \lambda_N N_t e^{-aP_t} \tag{5.19}$$

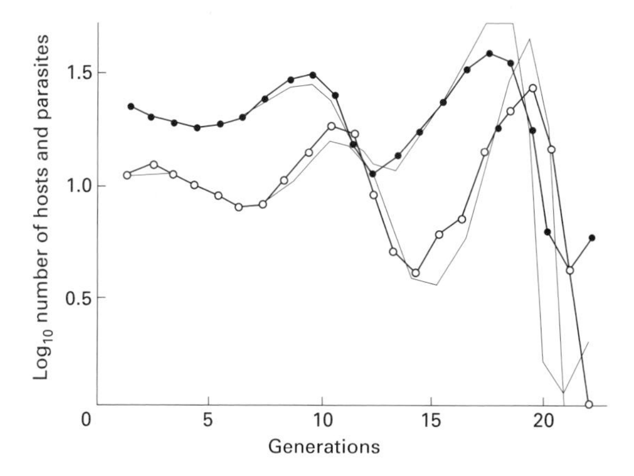

Fig. 5.19 Changes in abundance of greenhouse whitefly, *Trialeurodes vaporarium* (•—•), and its parasitoid *Encarsia formosa* (○—○) and as predicted by the Nicholson–Bailey model (solid lines without symbols). (From Burnett 1958 and Hassell & Varley 1969.)

It should be noted that the terms $1 - \exp(-aP_t)$ and $\exp(-aP_t)$ are the probability of a host being attacked and not being attacked, respectively, and are derived from the Poisson distribution, which is a probability density function that describes random processes such as the random encounters between natural enemies and their prey. The reader is referred to Gillman and Hails (1997) for a discussion of probability density functions and their use in models such as the above.

The Nicholson–Bailey equations predict that there are equilibrium densities for both the natural enemy and its prey. The equilibrium densities depend upon the particular values of λ and a. However, the equilibria are unstable: even the slightest disturbance leads to cycles of increasing size and the extinction of the parasitoid (see, e.g. Hassell 1978) (Fig. 5.19).

5.4.4 Making the Nicholson–Bailey model more realistic

The lack of stability in the Nicholson–Bailey model indicates that it has failed to capture some aspect of the interactions between natural enemies and their prey, which usually persist for much longer than predicted by the Nicholson–Bailey model. Some of the considerable amount of research to produce a more realistic model is considered below.

One aspect missing from the Nicholson–Bailey model (which is present in the Lotka–Volterra model) is density dependence in the equation that describes the population growth of the host. The incorporation of a density-dependent term in the Nicholson–Bailey model does increase the stability of the interaction between predators and their prey (Beddington *et al.* 1975) (Fig. 5.20). This, of course, implies that there is nothing inherently stable in the interactions between insects and their natural enemies. However, since insects and their natural enemies do persist together for many generations, many theoretical ecologists have tried to develop predator–prey models that are inherently stable.

5.4.5 Handling time and functional responses—the effects of prey density

One of the drawbacks of the Nicholson–Bailey model is that it is unrealistic to assume that predators have unlimited appetites and that parasitoids can produce an unlimited number of eggs (Hassell 1981). In addition, it is also unrealistic to assume that the time available to natural enemies for searching for prey is not related to prey density (Holling 1959). Holling predicted that as prey numbers increase, and therefore more prey are eaten or parasitized, then the predator or parasitoid spends an increasing amount of time in activities such as capturing, killing and eating prey. Holling called this the 'handling time' of a natural enemy. The consequence of an increase in handling time as prey density increases is that the number of prey captured does not show a linear increase as is implicit in the Nicholson–Bailey model.

The relationship between the number of prey killed or parasitized per natural enemy and prey density is known as the functional response, and Holling (1959) described three different types:

- Type I—response is linear up to a plateau (Fig. 5.21a);
- Type II—response rises at a decreasing rate (Fig. 5.21b);
- Type III—response is sigmoid (Fig. 5.21c).

A Type II functional response arises because, first, as mentioned above, predators spend an increasing amount of time handling their prey and a decreasing amount of time searching for prey as prey density increases. Holling described the Type II functional response curve with the following equation:

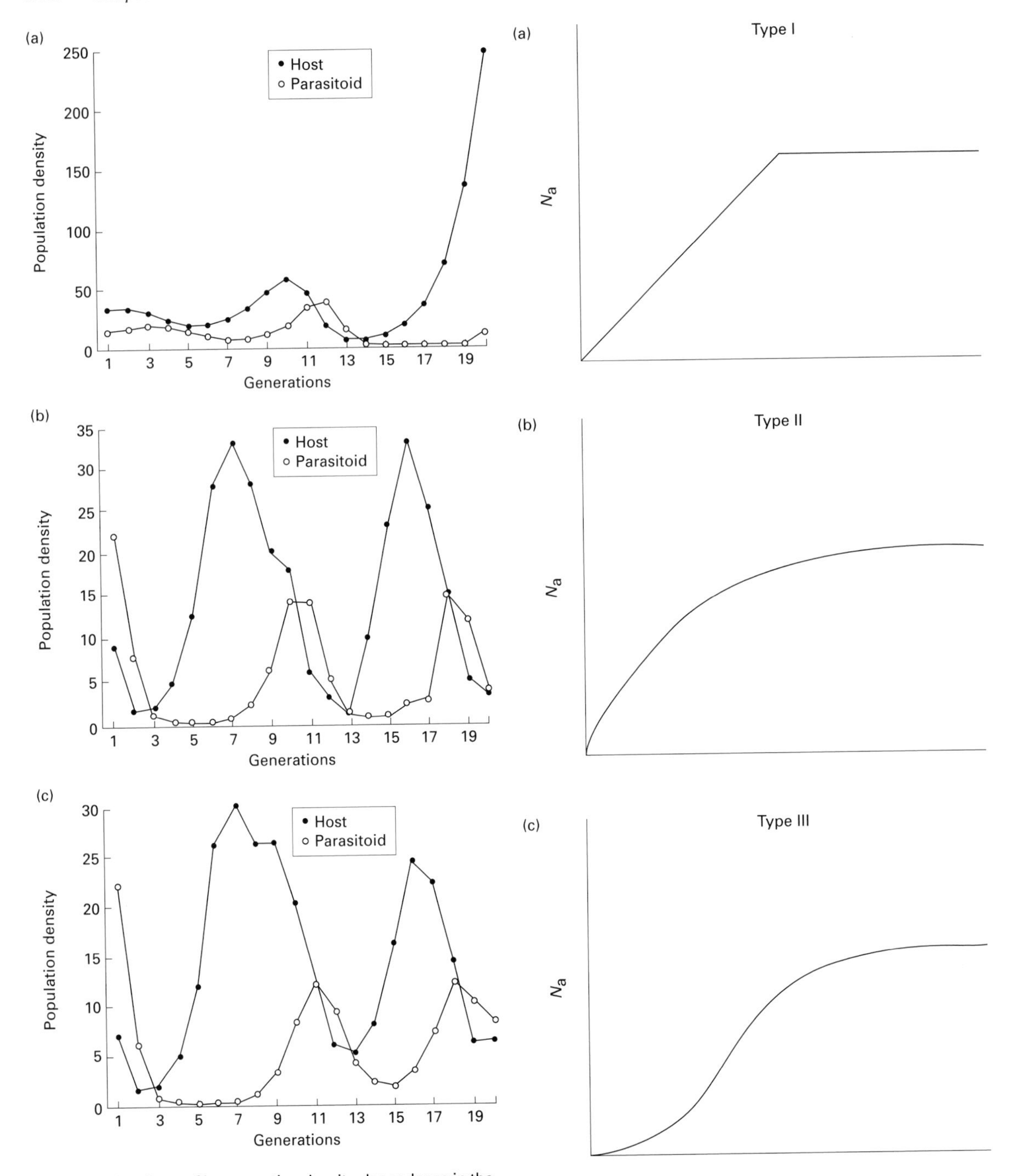

Fig. 5.20 The effects of incorporating density-dependence in the Nicholson–Bailey model (a) without density-dependence; (b) with density-dependence, resulting in cyclical oscillations with upper and lower limits (limit cycles); (c) increased density-dependence, resulting in oscillations with decreasing size which approach an equilibrium. (From Kidd & Jervis 1996.)

Fig. 5.21 Three types of functional response—the relationship between the number of prey killed per predator (or parasitized per parasitoid) (N_a) and prey density: (a) Type I, (b) Type II and (c) Type III. (After Holling 1959.)

$$N_A = \frac{aT_T N_o}{(1 + aT_H N_o)} \quad (5.20)$$

where N_A is the number of prey attacked, N_o is the density of prey, T_T is the time available for searching, T_H is the handling time and a is the searching efficiency or attack rate of the predator.

Holling developed this equation by carrying out an experiment with a human 'predator'. This predator was blindfolded and placed in front of a 0.9 m (3 foot) square table on which were pinned a variable number of paper discs 4 cm in diameter. The number of discs found by tapping and removed from the table in a minute was recorded at a range of densities. The above equation, derived from this experiment, is therefore known as Holling's disc equation.

As the disc equation shows, the Type II functional response depends upon the predator's rate of attack, the time available for the predator to encounter its prey, and handling time. One aspect not explicitly included in the disc equation, but nevertheless important in shaping the functional response, is hunger (Holling 1966).

The Type III functional response curve can arise in two linked ways, i.e. through learning and through switching. At low densities of a particular prey species, a predator may be considered to be naïve and inefficient at capturing their prey. However, as the density of the particular prey increases and the predator encounters it more frequently, the predator is likely to learn how to find and capture this species of prey more efficiently. Eventually, however, the effects of increasing handling time with increasing prey density have the same effect as they have in the Type II functional response and produce the sigmoid relationship between attack rate and host density.

The Type III response can also occur as a result of predator switching, that is, the phenomenon whereby predators swap from one prey item to another as it becomes more profitable to do so. For example, polyphagous predators may switch from prey species A to prey species B as the density of species A decreases and the density of species B increases. In these circumstances a predator might show a Type II response against both prey species together, but a Type III response against them separately.

Laboratory studies have shown that most predators and parasitoids have a Type II functional response (although it should be noted that some experimental procedures may be responsible for producing a Type II response (Van Roermund 1996)), a significant number have a Type III response, and the Type I response is very rare (Fig. 5.22). Holling considered that the Type II response was most typical of invertebrate predators and the Type III response most typical of vertebrate predators. However, that is clearly not universally true (Fig. 5.22).

Exceptions to these categories of functional response curve have been noted, and some researchers have suggested a reclassification of func-

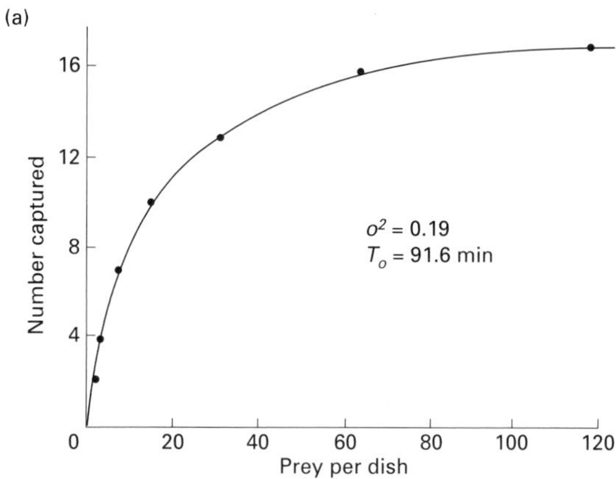

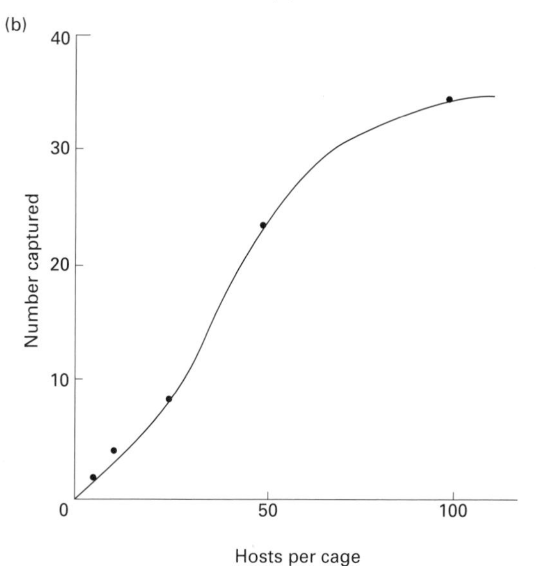

Fig. 5.22 Examples of functional responses: (a) concave Type II response for predation by the coccinelid *Harmonia axyridis* on *Aphis craccivora*; (b) sigmoid Type III response for parasitism of the aphid *Hylopteroides humilis* by the braconid wasp *Aphidius uzbeckistanicus*. (From Mogi 1969; Dransfield 1975; Hassell 1981.)

tional responses (e.g. Putman & Wratten 1984). One interesting exception is the response of predators to prey with defensive behaviour such as many sawflies. When threatened, sawfly larvae rear backwards, a particularly effective means of defence in these communally feeding species because they all react together. In an experiment on the sawfly *Neodiprion pratti banksianae* (Hymenoptera: Diprionidae) and a pentatomid bug predator, Tostowaryk (1972) found a humped functional response: when the prey reached a certain density, their defensive response to the predator was enough to bring about a decrease in attack rate. Despite this type of exception (which was noted by Holling (1965)), Holling's classification has largely stood the test of time and is widely accepted as a useful basis for describing the functional response of most predators and parasitoids.

The impact of the functional response on the interaction between predator and prey depends upon the type of the response (Hassell *et al.* 1977; Hassell 1978, 1981; Kidd & Jervis 1996). The inclusion of a Type II response in the Nicholson–Bailey model is theoretically destabilizing, but, as handling time is a small fraction of the total time available, the size of the destabilizing effect is likely to be small. A Type III response is theoretically stabilizing because it results in density-dependent predation, over lower prey densities.

Natural enemies show, in addition to their functional response, a numerical response. Whereas the functional response is the change in the attack rate of an individual predator or parasite in response to changing prey density, the numerical response is the change in the number of natural enemies in response to changing prey density. The total response is a combination of the individual functional response and the numerical response (Kidd & Jervis 1996). For example, a sigmoid (Type III) total response can be derived from a Type II functional response and a Type II numerical response (Fig. 5.23).

Holling (1965) produced a simple model to explore the consequences of different types of functional response, and this is considered below because it forms a useful basis for modelling the impact of multiple factors on the dynamics of insect herbivores.

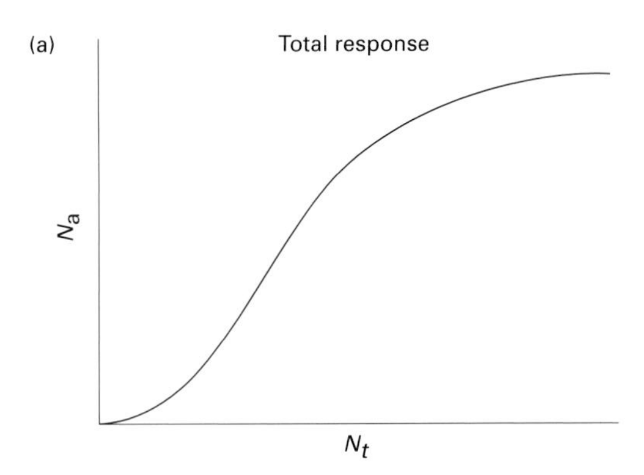

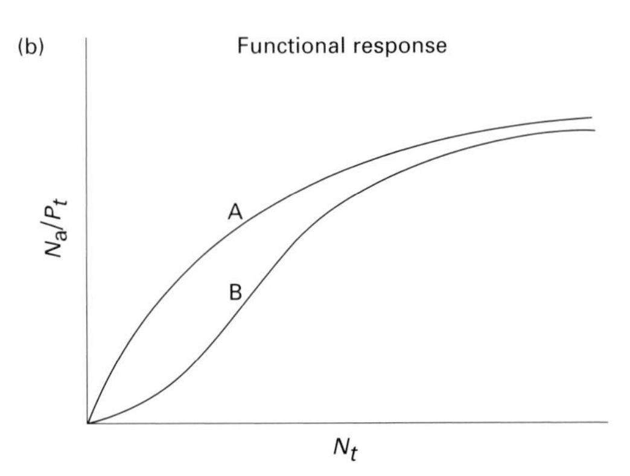

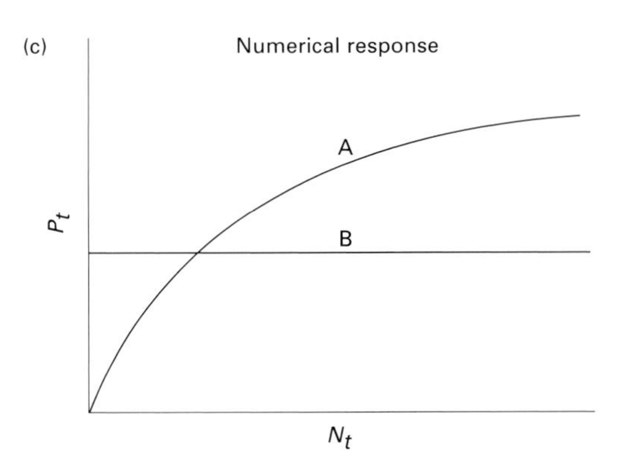

Fig. 5.23 Alternative ways of producing a sigmoid Type III total functional response (a), the combination of the individual functional response (b) and the numerical response (c), where N_a is the number of prey killed by predators, P_t is the density of predators and N_t is the density of prey. The same total functional response may be obtained from individual response A or B with numerical response A or B, respectively. (From Hassell 1978.)

5.4.6 Mutual interference—the effects of natural enemy densities

Another possible influence on predator–prey dynamics is mutual interference between predators (or parasitoids) (Rogers & Hassell 1974; Hassell 1981; Van Alphen & Jervis 1996). For example, many different types of natural enemy have been observed to disperse following an increase in encounters between other natural enemies (of the same species) or after detecting previous parasitism in hosts. There are also many examples of aggressive and host-marking behaviour in female parasitoids. These observations suggest that the searching efficiency of parasitoids and predators declines as the density of parasitoids and predators increases. Laboratory studies have confirmed this (Hassell 1978) (Fig. 5.24).

Mutual interference may be classified as either direct interference, resulting from behavioural interactions between natural enemies, or indirect interference caused by, for example, superparasitism, decreasing fecundity or a shift in sex ratio (Visser & Driessen 1991).

Hassell and Varley (1969) produced an empirically derived model for the effect of interference:

$$\log a = \log Q - m \log P_t \tag{5.21}$$

where a is searching efficiency (as discussed above), m is the slope of the line and $\log Q$ is the intercept (see Fig. 5.16). Q is therefore also equivalent to the attack rate in the absence of interference (Van Alphen & Jervis 1996).

This equation can be incorporated in the Nicholson–Bailey model and shows that the effect of mutual interference on predator–prey dynamics depends on the value of m and the prey rate of increase in the absence of all factors other than predation (λ_N) (Hassell & May 1973; Hassell 1981). An increase in m leads to an increase in stability, and an increase in λ_N leads to less stable interactions. The other term in the model above, Q, affects the equilibrium level but not stability.

Hassell *et al.* (1976) pointed out that it is difficult to predict the impact of interference on particular predator–prey interactions because, although the value of m can be determined from laboratory studies, the value of λ_N is difficult to acquire. It should be noted that λ_N is not the same as fecundity but includes the effects of all mortality factors other than the interaction being modelled (predation or parasitism).

The role of interference in predator–prey models has been criticised by several researchers (e.g. Free *et al.* 1977; Hassell 1981). The major criticism is that laboratory studies, conducted in atypically homogeneous environments at atypical natural enemy densities, have exaggerated the value of m and hence the role of mutual interference. Field studies have also suggested that mutual interference is generally not strong (Cronin & Strong 1993).

5.4.7 Aggregative responses by natural enemies to prey distribution

Most prey populations under natural conditions have

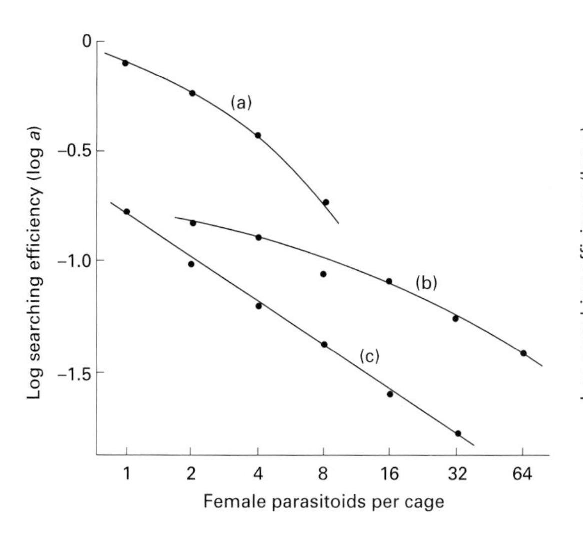

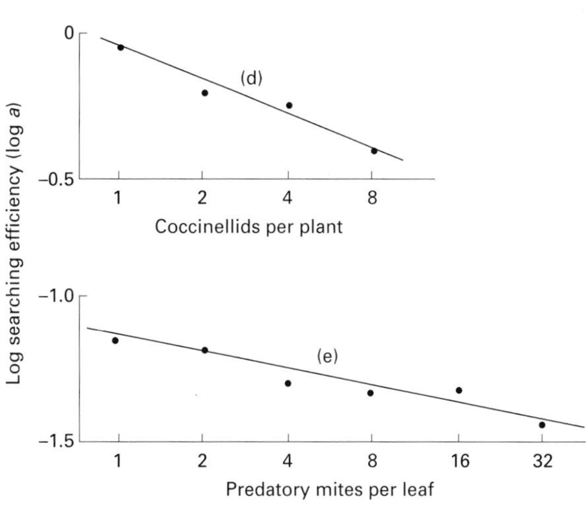

Fig. 5.24 Examples of decreasing searching efficiency of predators and parasitoids as their density increases: (a) *Pseudeucoia boche* (Bakker *et al.* 1967); (b) *Encarsia formosa* (Burnett 1958); (c) *Nemeriti canestens* (Hassell 1971); (d) *Coccinella semtempunctatus* (Michelakis 1973); (e) *Phytoseiulus persimilis* (Fernando, unpublished in Hassell 1981). (From Hassell 1981.)

a clumped distribution but the Nicholson–Bailey and related models implicitly assume that predators and parasitoids will not respond to this pattern of distribution (Hassell 1981). Instead, these models assume that natural enemies spend the same amount of time in each (equal-sized) patch of habitat irrespective of the density of prey in that patch. Many studies have shown that this is not true and that natural enemies show an aggregative response to the density of their prey, spending a disproportionately large amount of time in patches with high prey densities (Fig. 5.25).

Spatial variation in host density and the response of natural enemies to it can be modelled by running a modified Nicholson–Bailey model separately for a series of host patches (Hassell 1981). In addition, the distribution of the predator or parasitoid among these patches is modelled using the following equation (Hassell & May 1973):

$$\beta_i = c\alpha_i^{\mu} \qquad (5.22)$$

where β_i is the distribution of a predator or parasitoid in patch *i*, *c* is a constant, α_i is the distribution of prey in patch *i*, and μ is an aggregation index. The value of μ varies from $\mu = 0$, corresponding to random search, to $\mu = 1$, where natural enemies are evenly spread among all patches, and to $\mu = \infty$ (infinity), where all predators or parasitoids are concentrated in the highest host density patch and other patches are complete refuges.

The effect of incorporating the aggregation equation in the Nicholson–Bailey model depends on the value of μ (the natural enemy aggregation index), the prey distribution and the prey rate of increase as follows (Hassell & May 1973; Hassell 1981):

• an increase in aggregation by natural enemies in high-density host patches leads to an increase in stability;

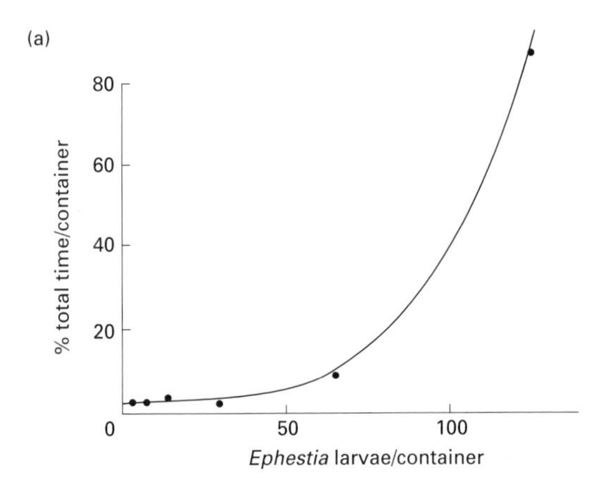

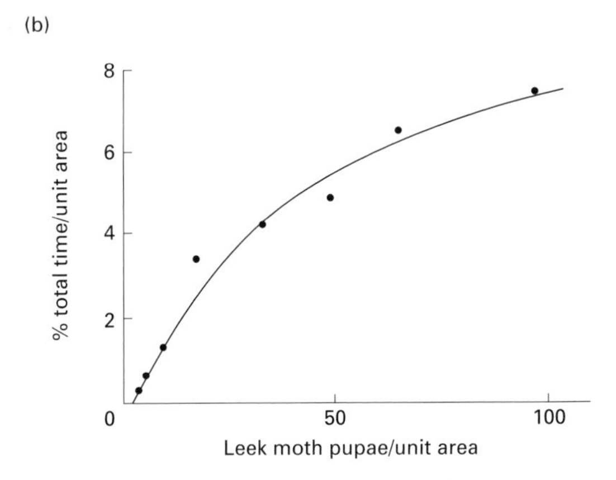

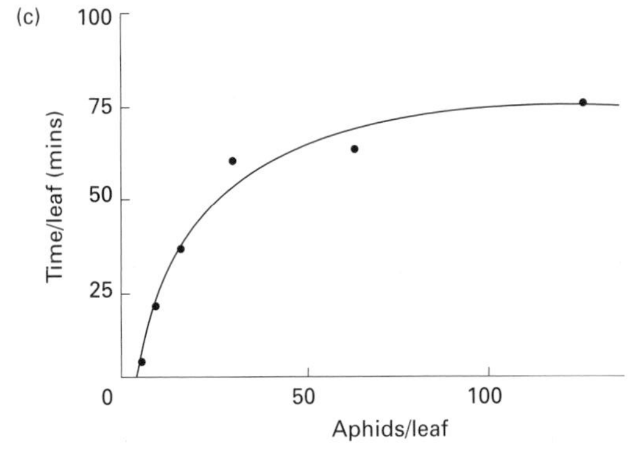

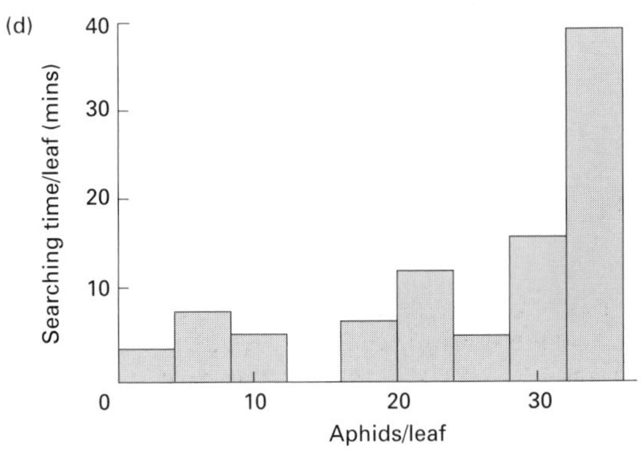

Fig. 5.25 Examples of aggregative responses in parasitoids and predators: (a) *Nemeritis canescens*, an ichneumonid parasitoid of flour moth larvae, *Ephestia cautella* (Hassell 1971); (b) *Diadromus pulchellus*, an ichneumonid parasitoid of leak moth pupae, *Acrolepia asseciella* (Noyes 1974); (c) *Diaereliella rapae*, a braconid parasitoid of the aphid *Brevicoryne brassicae* (Akinlosotu 1973); (d) *Coccinella septempunctata*, a predator of *B. brassicae* (Hassell, unpublished in Hassell & May 1974). (From Hassell & May 1974.)

• the more clumped the host distribution, the more likely the interaction is to be stable, but if the host is more or less evenly distributed amongst patches, the interaction will be unstable, irrespective of the degree of aggregation by the natural enemy;
• an increase in the prey rate of increase tends to decrease stability.

In other words, when natural enemies tend to aggregate in patches where the density of their prey is highest, they will tend to encounter prey at a higher rate than if they were searching randomly. This results in patches with low densities of prey becoming partial refuges from predation or parasitism, and it is this that contributes towards stability (Hassell 1981). This can be illustrated by comparing the relationship between searching efficiency (*a*) and predator density (P_t) for natural enemies with (a) a fixed aggregation strategy ($\mu = 1$), (b) a random search strategy ($\mu = 0$) and (c) an optimal foraging strategy (Fig. 5.26). When predators are following a fixed aggregation strategy, i.e. they are distributed in direct proportion to prey density, their searching efficiency is greater than under a random searching strategy while predator densities are low. However, as predator densities increase, searching efficiency decreases. This effect is similar to the effect of interference on searching efficiency discussed above and has been referred to as 'pseudo-interference' (because it may be wrongly attributed to mutual interference (Free *et al.* 1977)). 'Pseudo-interference' is the result of aggregation of large numbers of natural enemies, not mutual interference, although both are likely to occur in these circumstances.

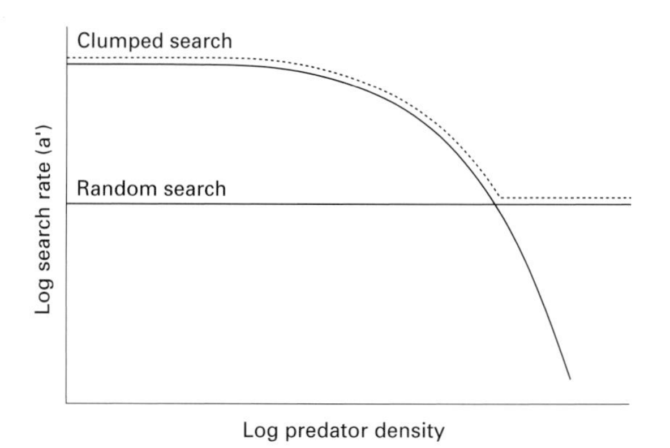

Fig. 5.26 The effects of different searching strategies on the relationship between searching efficiency and predator density. The broken line shows the switch between clumped and random search at high densities that an optimally-foraging or 'prudent' predator would adopt. (From Hassell 1978.)

Real predators and parasitoids are, however, unlikely to remain in high-density host patches when the number of predators or parasitoids reaches a level where their rate of prey capture is poorer than if they were adopting a random searching strategy. Instead, they may adopt an optimal foraging strategy (e.g. Comins & Hassell 1979). Optimal foraging behaviour has mostly been studied with vertebrate predators in mind but it has relevance to invertebrate predators and parasitoids too. Optimal foraging theory predicts that an optimal predator should forage preferentially in patches that are rich in food items and only forage in less profitable patches when the availability of good-quality patches is low (e.g. Royama 1970). In addition, an optimal predator should remain in a patch until its rate of capture falls below the average rate of food capture in the habitat as a whole. Thus in the case illustrated in Fig. 5.26, optimally foraging predators may be predicted to change their foraging strategy to minimize the effect of an increase in predator density.

Comins and Hassell (1979) constructed a model for optimally foraging predators and parasitoids. They found that the outcome was qualitatively similar to the model (Hassell & May 1974) developed for natural enemies with a fixed aggregation strategy.

The model of Hassell and May (1973), discussed above, thus appears to be the best of its kind in describing the population dynamics of natural enemies and their prey. The key aspects of that model are that it is spatially explicit and incorporates the aggregative behaviour of natural enemies. This leads to spatial density dependence: natural enemies aggregate in areas of high prey density and cause a disproportionate amount of mortality in those patches. This approach to modelling is important not only because it started the development of other spatially explicit models (see below) but also because it encouraged insect ecologists to include the spatial dimension in considering the temporal fluctuations in the abundance of insects.

5.4.8 The development of spatial models of natural enemy–prey interactions

One significant development in modelling the spatial dimension of the interactions between predators, parasitoids and their prey has been the use of cellular automata models (Comins *et al.* 1992; Gillman & Hails 1997). The cellular automata approach involves modelling the dynamics of a grid of cells linked by inter-generational dispersal and has many ecological applications. In the model constructed by Comins *et al.* (1992), the interaction between parasitoids and their prey in each of the cells in a square grid of width *n* was modelled with the Nicholson–Bailey model and dispersal was modelled with the following rules:

- in a dispersal phase, both a fraction of the adult hosts and a fraction of the adult parasitoids leave the cell in which they emerged and the rest remain and reproduce in it;
- the dispersing hosts and parasitoids move to the eight cells adjacent to the cell from which they emerged in equal numbers (in most other similar studies dispersing hosts and parasitoids have been distributed over the whole grid according to a specified rule);
- hosts and parasitoids emerging from the cells at the edge of the grid return to the grid in so-called reflective boundary conditions (for example, individuals that would have moved south-east from the middle of the east side of the grid move one cell south and individuals that would have moved south-east from the south-east corner of the grid stay in that cell).

In small grids (of a width of 10 cells or less) the parasitoid and its host become extinct within a few hundred generations. However, with a larger grid size (of width 15–30 cells), the model produces three types of behaviour: 'spirals', 'spatial chaos' and 'crystal lattices' (Fig. 5.27). Most importantly, in each case, the model predicts long-term persistence of the host and parasitoid, despite the unstable nature of the basic Nicholson–Bailey model.

5.4.9 Natural enemy–prey models —conclusions

As pointed out in section 5.2.1, it is easy to be sceptical about the value of mathematical models in insect ecology. Nevertheless, population models have given us an insight into how insects interact with predators and parasitoids. How else could the relevance of laboratory and field experiments on predators and parasitoids be demonstrated without the use of population models?

Lotka and Volterra showed that it was possible to model the interactions between predators and their prey, and since then ecologists such as Nicholson, Holling, Hassell, Comins and others have worked to inject ecological realism into population models. These ecologists have shown that the long-term persistence of insects and their natural enemies may arise as a result of the behaviour of the natural enemies (aggregating in high-density patches of their prey), other density-dependent factors acting on the prey populations, or by the dispersal behaviour of natural enemies and their prey (section 5.4.8). Some of the 'analytical' approaches to modelling natural enemy–prey interactions not dealt with here have been discussed by Jervis and Kidd (1996) and Gillman and Hails (1997).

In the final part of section 5.4 we will consider a slightly different approach to modelling. This approach is particularly important for modelling the interaction of multiple factors in population dynamics.

5.4.10 Holling's population models: multiple equilibria

Holling (1965) developed an approach to modelling, widely used since (by him and others) that demonstrates complex interactions in a simple way.

Holling's 'model' is essentially a description of the birth rate and mortality rate of a prey population. He first calculated the birth rate for different densities of the prey population. At low densities of the insect prey species birth rate is low because male and female insects have problems locating each other. (This is often referred to as the Allee effect because it was first suggested by Allee (Allee *et al.* 1949).) As insect densities rise, birth rate increases sharply but then declines as a result of a reduction in the food supply, e.g. pine foliage for a pine-feeding moth larvae. Holling calculated the mortality rate (NEC) required to balance this varying birth rate in order to (theoretically) stabilize the population (Fig. 5.28). He then superimposed a mortality rate on top of this birth rate to describe the effects of a predator with a Type II functional response. This mortality rate (ACT) shows a decline in percentage mortality (Fig. 5.28) as a result of the decline in searching efficiency as prey density increases.

When NEC equals ACT there is a theoretical equilibrium density (for the prey population). Such an equilibrium exists in Fig. 5.28. However, the equilibrium point is unstable: when the prey density is less than the theoretical equilibrium, the prey population

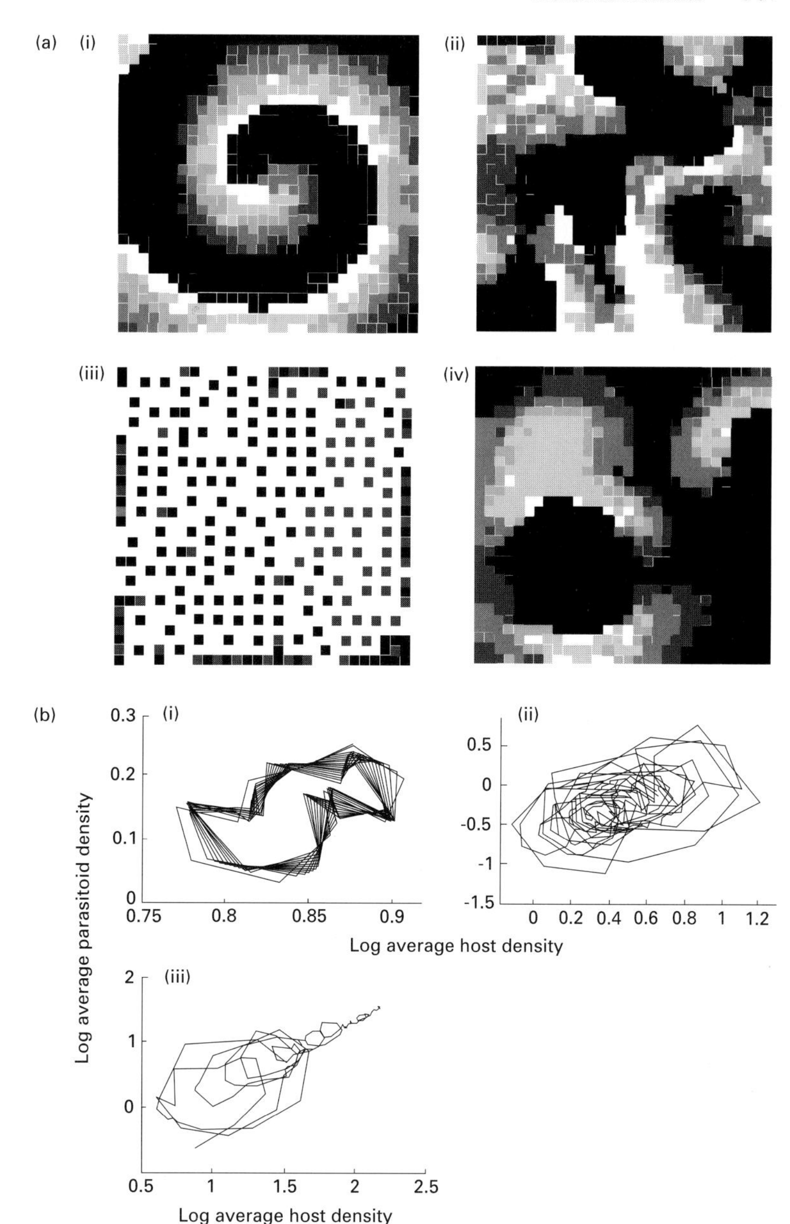

Fig. 5.27 Dynamics of an insect host and its parasitoid in a spatial grid of 30 by 30 cells linked by dispersal with Nicholson–Bailey model dynamics ($\lambda = 2$) in each cell (apart from iv—see below): (a) instantaneous maps of population density with different levels of density of host and parasitoid represented by different levels of shading (black, empty patches; dark shades becoming paler, patches with increasing host densities; light shades, patches with increasing parasitoid densities) resulting in (i) 'spirals', (ii) 'spatial chaos', (iii) 'crystalline structures', (iv) highly variable spirals from Lotka–Volterra model dynamics; (b) host–parasitoid density ('phase plane') plots showing dynamics over time with the same parameters as (a). (From Comins *et al.* 1992 and Gilman & Hails 1997.)

will decrease still further because mortality as a result of predation exceeds the birth rate; when the prey density is greater than the equilibrium the population increases because the birth rate always exceeds the mortality rate.

However, when this analysis was repeated with a Type III functional response, Holling could demonstrate a stable equilibrium. Figure 5.28 shows the mortality rate associated with a Type III functional response. This produces three equilibrium densities, two of which are unstable and one of which is stable. When the prey population density is less than the first equilibrium point it will decrease, for the reasons described above in relation to the effects of the Type II functional response. When the population density lies between the first and third equilibria it will tend to move towards the second, and therefore stable, equilibrium density. The Type III functional response

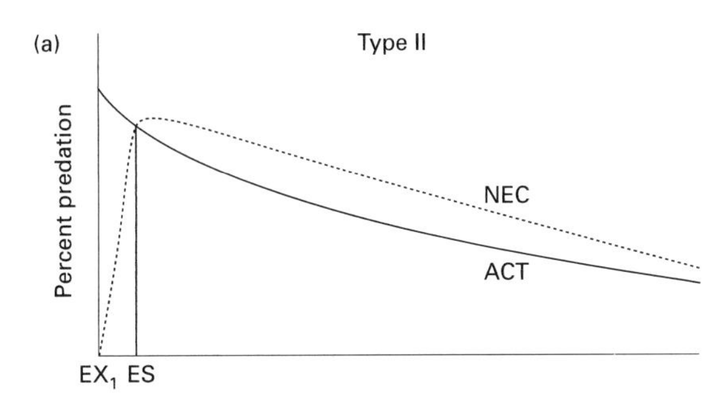

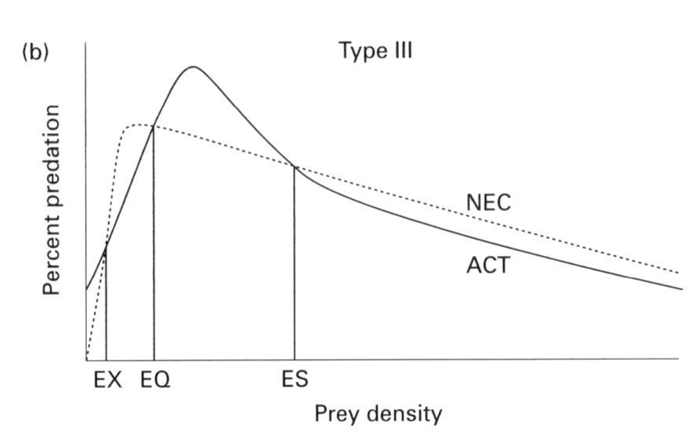

Fig. 5.28 Holling's (1965) representation of the consequences of different types of functional response for population regulation. In both graphs the relationship between predation and prey density is represented by ACT (actual percentage predation) and NEC (the mortality required to stabilize the prey population). A Type II functional response (a) results in one unstable equilibrium (labelled as EX, ES because it may be considered as either a threshold density for population extinction (EX) or an 'escape threshold' (ES)). A Type III functional response (b) also results in a stable equilibrium (EQ). (From Holling 1965.) See alternative graphical analysis of the interactions between different factors to create population equilibria in Figs 5.29, 5.32 and 5.33.

therefore leads to a range of prey population densities within which the predator regulates the density of the prey. When the population density is greater than the third equilibrium point it will tend to rise and can therefore be referred to as the 'escape' or 'outbreak' equilibrium.

In other papers, Holling and coworkers (e.g. Peterman *et al.* 1979) developed the above approach by calculating the year-to-year population growth rate or 'recruitment rate', R, of a population as

$$R = \frac{N_{t+1}}{N_t} \tag{5.23}$$

where N_{t+1} and N_t are prey population densities in years $t + 1$ and t, respectively.

Plots of R against population density or 'recruitment curves' can be used to explore the population dynamics of the spruce budworm, *Choristoneura fumiferana* (Lepidoptera: Tortricidae) (Peterman *et al.* 1979). This insect is one of the most important pests of boreal forests, periodically killing large numbers of balsam fir (*Abies balsamea*) and white spruce (*Picea glauca*) in North America (Plate 5.1 opposite p. 158).

In their analysis of spruce budworm population dynamics, Holling and coworkers first constructed a simulation model to describe the interactions between the budworm, its host plant and its natural enemies. Because of its complexity, they ran this simulation model for a single year at a time over a range of population densities, N_t, thus estimating population size, N_{t+1}, the following year. The recruitment rate was then calculated by using the above equation and plotted against population density: theoretical equilibrium densities occur wherever $R = 1$ (and therefore $N_{t+1} = N_t$). They are stable if $R > 1$ below the equilibrium density and $R < 1$ above it (and unstable if the opposite occurs). The population densities around a stable equilibrium are sometimes described as a 'domain of attraction' (Peterman *et al.* 1979) because while within it the population density is 'attracted' to the stable equilibrium. For the same reason a stable equilibrium is often referred to as an attractor.

The spruce budworm population can first be considered with only the effect of starvation included as a mortality factor. Starvation mortality increases gradually, then rapidly as a result of a sharp decline in the amount of foliage of the host plant as larval densities increase (Fig. 5.29a). Consequently, the recruitment curve gradually declines from a maximum level at low densities and results in one equilibrium point, P_1, at high budworm density (Fig. 5.29b).

The effect of parasitism on spruce budworm is greatest at low budworm densities and gradually decreases as budworm numbers rise (Fig. 5.29c). Thus parasitism has a marked effect on recruitment rate at low budworm densities but a minor effect on high densities and there is no increase in the number of equilibrium points (Fig. 5.29d). Predation by birds has a complex effect on budworm mortality and, hence, on the population recruitment rate. As a result of their Type III functional response, mortality rises with prey population density to a low density maximum after which it declines (Fig. 5.29e). This has a marked effect on recruitment rate at low

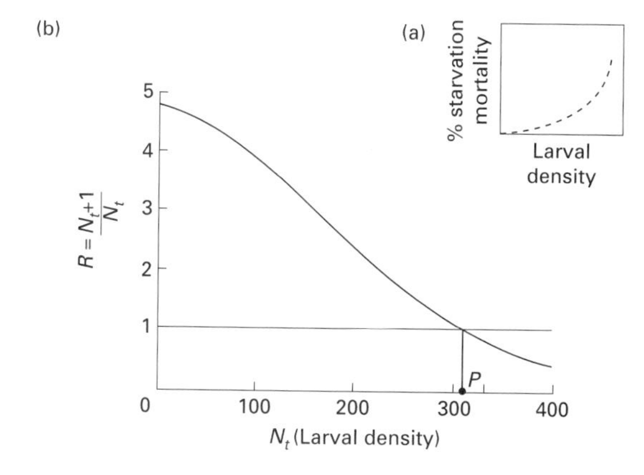

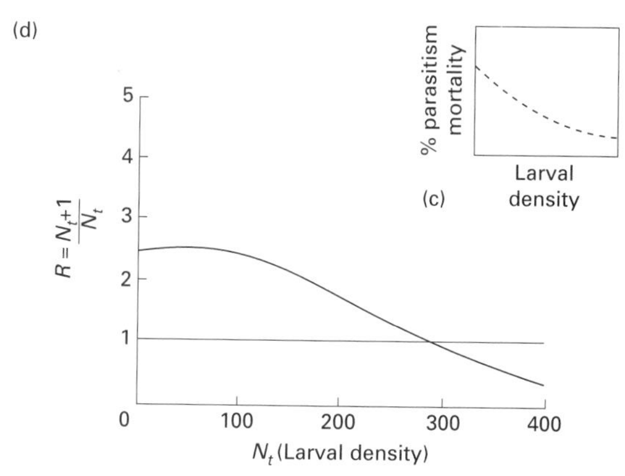

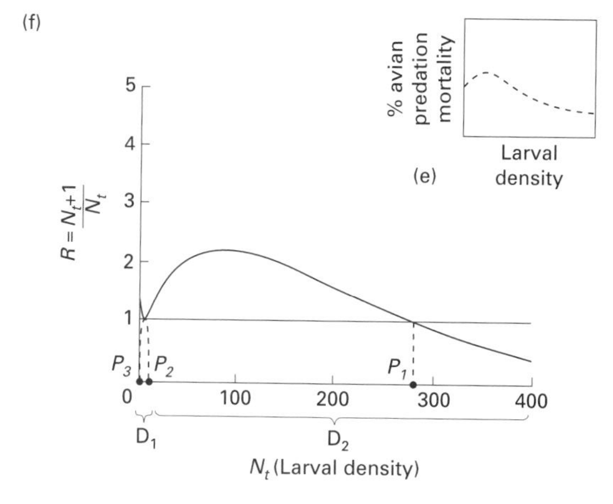

Fig. 5.29 Recruitment rate curves for the spruce budworm (where recruitment rate $R = (N_{t+1})/N_t)$, and N_t and N_{t+1} are population density in successive years. See text for details. (From Peterman *et al.* 1979.)

budworm densities and results in two additional equilibria, P_2 and P_3 (Fig. 5.29f).

P_1 and P_3 are stable equilibria and P_2 is an unstable equilibrium. Together they create two 'domains of attraction' (Peterman *et al.* 1979) with the two stable equilibria, or 'attractors', at the centre of each and the unstable equilibrium forming the boundary between these domains of attraction. (This is sometimes referred to as the occurence of 'alternative stable states'.) This produces a similar result to the first of Holling's models described above but with parasitism included as well as predation, even though the former makes little difference to the dynamics of the system.

One of the advantages of this approach to modelling is that it can be used to assess the impact of additional factors such as weather, which may bring about variability in either the birth rate or the mortality rate of the budworm population. The effect of these additional factors depends upon the relative positions of the theoretical equilibrium densities and, therefore, the strength of the attractors within each domain of attraction. If the domain of attraction is small, i.e. it occurs over a narrow range of insect densities, the insect population is unlikely to be regulated within it. In contrast, regulation is more likely within a wide domain of attraction. Thus the position of the unstable equilibrium is particularly important in defining the width of the first, and most important from an applied point of view, domain of attraction.

It is no coincidence that this approach to modelling was developed with the spruce budworm, one of the most important pests of temperate forests. It may explain the dynamics of this and other pests, which, after being at innocuous densities for a long time, suddenly erupt to damaging 'outbreak' densities. For example, climatic conditions favourable to the pest in a particular year may push its population out of the lower domain and cause a subsequent outbreak. These climatic conditions need not last for more than one year because the population has now been released from a predator-mediated domain of attraction and will now rise within the upper domain of attraction to the high-density equilibrium.

We will see in a later section that the spruce budworm model has been taken further than this to include the effects of changing forest conditions on the dynamics of the spruce budworm populations.

It should be noted that the results of some studies suggest that the occurrence of multiple equilibria

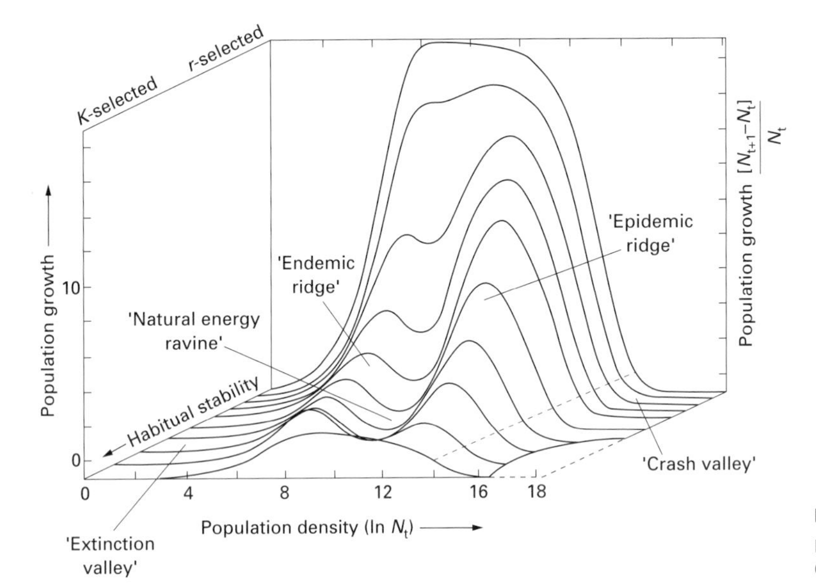

Fig. 5.30 A synoptic model of population growth. (From Southwood & Comins 1976.)

may be widespread. Jones *et al.* (1993), for example, showed that the combined impact of a generalist staphylinid predator (*Aleochara bilineata*) and a specialist cynipid parasitoid (*Trybliographa rapae*) produced multiple equilibria (or 'alternative stable states') in the population dynamics of the cabbage root fly (*Delia radicum*).

We end these sections on theoretical models of insect population dynamics by considering the 'synoptic' model of Southwood and Comins (1976) because it essentially takes the Holling recruitment rate model one step further by considering the relationship between population growth and density for insects with a range of life history strategies from *r*-selected to *K*-selected species. Southwood and Comins presented population growth *R* with the equation

$$R = \frac{(N_{t+1} - N_t)}{N_t}. \quad (5.24)$$

They argued that *r*-selected insects with their high dispersal capabilities and high rate of increase are unlikely to be affected by natural enemies and will usually have the type of relationship between population growth and population density shown in Fig. 5.30. *K*-selected insects have contrasting characteristics, poor dispersal capabilities and low rates of increase, and will also be unaffected by natural enemies as a result of their strong defences against them.

'Intermediate' species (that is, species with life history characteristics between *r*-selected and *K*-selected species) are, however, likely to be affected by natural enemies. These species have neither the high dispersal characteristics and high rates of increase of *r*-selected species nor the defences of *K*-selected species. Southwood and Comins argued that these species would be relatively immune to the effects of natural enemies at low and high densities but would be susceptible to the potentially regulating effect of natural enemies at intermediate densities. They suggested that intermediate species might have three equilibrium densities, much in the same way as in Holling's spruce budworm model. Taken together, the relationship between population growth rate and population density for the full spectrum of *r*-selected to *K*-selected produces a three-dimensional model of population behaviour (Fig. 5.30). Southwood and Comins aptly named parts of this three-dimensional model as extinction valley, endemic ridge, epidemic ridge and crash valley. From the point of view of regulation by natural enemies, the most important area is the natural enemy ravine (which makes up part of Holling's lower domain of attraction).

One of the strengths of the synoptic model is that it encompasses a range of population behaviours, which is what we experience in the real world. It predicts that natural enemies will be a significant influence only in certain types of species and that their influence is likely to vary in strength. Unfortunately, this approach has not been developed as much as it merits.

5.5 Synthesis: the impacts of natural enemies and other factors on insect population dynamics

5.5.1 The 'top-down' vs. 'bottom-up' debate

In a still influential paper written nearly 40 years ago, Hairston *et al.* (1960) argued that because the world is covered with a profusion of green plants, herbivores are having little impact on their host plants, particularly in natural ecosystems. They concluded that herbivores must be kept at insignificant levels by natural enemies. The many instances of successful biological control, observations and experiments on natural enemies and some of the theoretical models discussed in this chapter have supported this 'top-down' view of population dynamics.

However, several ecologists have also argued that plants play the dominant role in determining herbivore numbers because they set a resource limit, or ceiling (Dempster & Pollard 1981). Insect numbers cannot increase past the number theoretically sustained by this ceiling and will be increasingly affected by intraspecific competition as they approach the ceiling. Dempster and Pollard (1981) argued that there was little evidence to support the models that predict regulation around an equilibrium density (see examples above). One of the examples they gave was the cinnabar moth, *Tyria jacobaeae* (Lepidoptera: Arctiidae), whose abundance parallels the biomass of its host plant, ragwort (Fig. 5.31a). The population dynamics of many species appear to fit the 'ceiling' model, sometimes in subtle ways. For example, Auerbach (1991) found that natural enemies played an insignificant impact in the population dynamics of the aspen blotch miner, *Phyllonorycter salicifoliella* (Lepidoptera: Gracillariidae) and concluded that the abundance of this insect each year approached a 'ceiling' set by the numbers of young leaves available to the adult stage during oviposition. In this case, therefore, the ceiling was not set by the total amount of foliage that might be wrongly assumed to be available to this insect, but only the number of nutritious, young leaves. Dempster and Pollard (1981) extended their argument to include the abundance of natural enemies, suggesting that their numbers too are determined by a ceiling set by the abundance of their insect hosts (Fig. 5.31b). There are, however, many

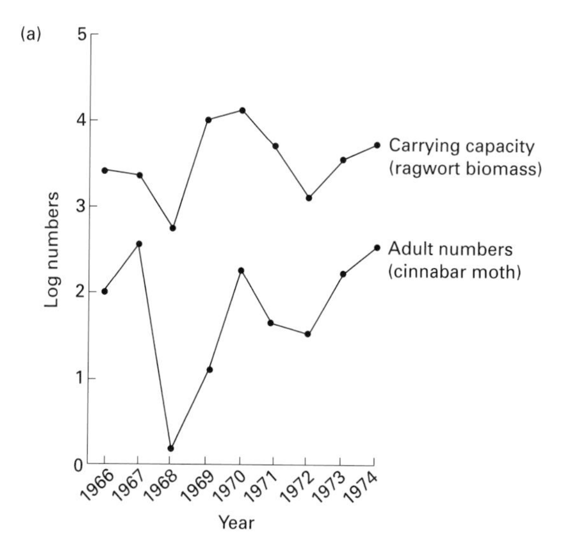

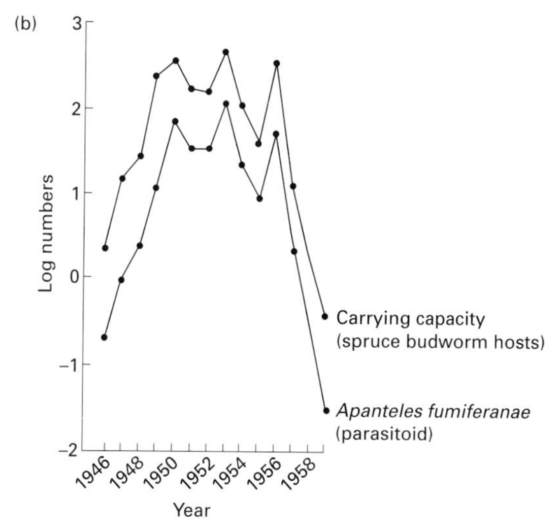

Fig. 5.31 The influence of resource limitation on herbivores and natural enemies: (a) abundance of cinnabar moth *Tyria jacobaeae* adults and biomass of its host plant, ragwort, the determinant of the carrying capacity of its habitat; (b) abundance of the parasitic wasp *Apanteles fumiferanae* and its host spruce budworm (*Choristoneura fumiferana*). (From Dempster & Pollard 1981.)

examples showing that abundance is not always determined by resource limitation.

Another way that the interaction between herbivores and their host plants may regulate insect numbers is through density-dependent changes in birth and mortality rates brought about by herbivore-induced changes in plant quality. This possibility was first suggested by Haukioja (1980), who argued that the population cycles of insects such as the autumnal moth, *Epirrita autumnata* (Lepidoptera: Geometridae) in Scandinavia are the result of an insect-induced deterioration in the quality of birch as the abundance of the autumnal moth increases. The topic of induced plant responses is discussed more fully in Chapter 3, and its role in the dynamics of the example of the larch budmoth is discussed later in this chapter.

There are other ways in which the host plant may play a significant role in determining insect numbers through, for example, phenological asynchrony and stress-mediated changes in plant quality (Chapter 3).

Fig. 5.32 Model of the combined role of different factors to determine equilibrium densities in insect herbivores: (a) separate relationships between the mortality caused by predation and competition and herbivore density; (b) combined mortality due to predation and competition; (c) the effects of host plant quality and herbivore density on herbivore birth rate; (d) the combined role of mortality and birth rate. Note that, even when predation maintains the herbivore population below its carrying capacity, variation in food quality still causes variation in equilibrium density (N_1, N_2 and N_3). N_1, N_2, N_3 and K are theoretically stable equilibria and u is an unstable equilibrium. (From Hunter 1997.)

The value of insect-resistant crop varieties (Chapter 10) also shows that 'bottom-up' factors can determine insect numbers. Life tables (see section 5.3.4) may be used to assess the relative importance of natural enemies and plant factors as causes of insect mortality. Cornell and Hawkins (1995) examined 530 life tables for 124 insect herbivores and found that natural enemies were a more frequent cause of mortality for immature herbivores (48% of cases) than plant factors (9%). However, they suggested that life table studies have probably underestimated the importance of plant factors.

5.5.2 The additive effects of insect–plant and insect–natural enemy interactions

The 'top-down' vs. 'bottom-up' debate was presented above as a choice between the regulation of insect numbers by natural enemies and by the host plant, respectively. Clearly, neither model applies in all cases, but nor is it true that either natural enemies or the host plant determines insect numbers. It is more likely that both natural enemies and the host plant

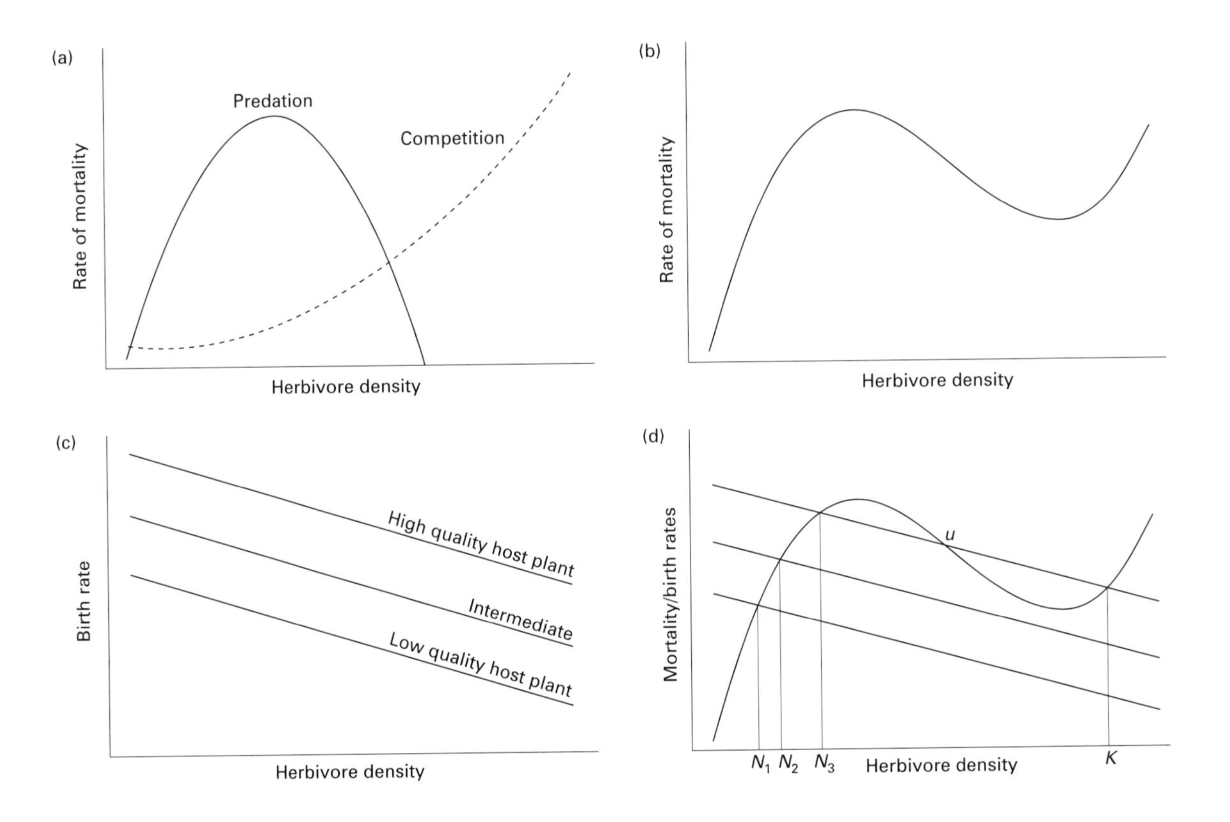

jointly determine the abundance of most insect herbivores (Hunter & Price 1992; Bonsall *et al.* 1998).

The combined effect of natural enemies and the host plant may be 'additive' or 'synergistic'. In the former case, mortality caused by natural enemies and variation in natality or mortality caused by the host plant add up to determine insect numbers. In the latter case, the impact of natural enemies on their prey is influenced by some other factor such as the host plant or other natural enemy species. The potentially additive effect of natural enemies and the host plant is shown in Fig. 5.32.

We must therefore conclude that density-dependent and density-independent factors both determine insect abundance, and that top-down and bottom-up forces are both important. The model above shows how, in theory, they interact. The following sections look at a few examples where insect ecologists have tried to tease apart the dynamics of insect populations, using a range of approaches, to identify how different factors together determine insect numbers.

5.5.3 Spruce budworm

Holling's model (and Southwood and Comin's synoptic model) provides an explanation for the irregular outbreaks of agricultural and forest pests. However, Holling and others have noted that pest outbreaks are more likely under some conditions than others.

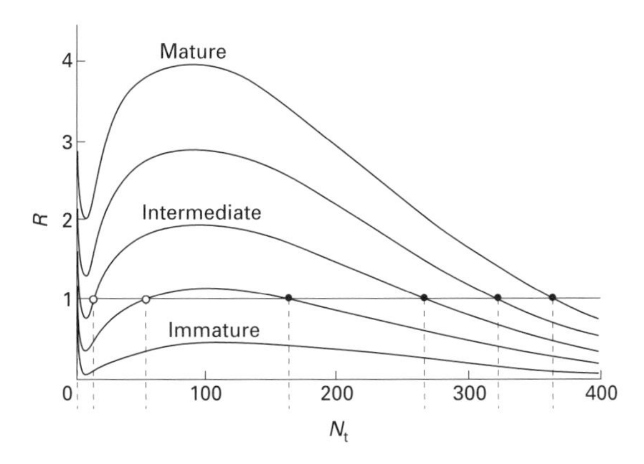

Fig. 5.33 Spruce budworm recruitment curves for different forest ages (immature, intermediate and mature). Stable equilibria are shown as solid circles and unstable equilibria as open circles. Fig. 5.29 demonstrates how single recruitment curves are generated. (From Peterman *et al.* 1979.)

Spruce budworm outbreaks (Plate 5.1, opposite p. 158), for example, become more likely as forests age. Holling argued that this is because the spruce budworm recruitment curve changes as forests age (Fig. 5.33). In young, or 'immature', forests the recruitment rate is low because forest conditions are relatively unfavourable for the spruce budworm. As the forest ages, it becomes more suitable for the spruce budworm but outbreaks are prevented by predation unless large numbers of moths immigrate into the forest (Fig. 5.33; 'intermediate' forest). Eventually, however, the ability of predators to prevent an outbreak disappears (Fig. 5.33; 'mature' forest). Holling's model therefore demonstrates that a complex interaction between different factors, including both top-down and bottom-up factors, determines the likelihood of a spruce budworm outbreak.

5.5.4 Larch budmoth

The larch budmoth, *Zeiraphera diniana* (Lepidoptera: Tortricidae), shows 8–10 year population cycles in the Alps (see Fig. 1.12, Plate 1.3, opposite p. 158). Several factors have been suggested to cause these cycles, such as parasitism and induced resistance (Benz 1974; Baltensweiler & Fischlin 1988). Several parasitoids attack the larch budmoth; different species appear to attack the budmoth at different stages of the population cycle (Delucchi 1982). Feeding damage by the larch budmoth results in a decline in larch needle quality and this is thought to cause an increase in larval mortality. Larch budmoth damage also results in a decline in the quantity of available foliage.

Van den Bos and Rabbinge (1976) constructed a simulation model to try to explain the cycles of the larch budmoth. They incorporated the three factors mentioned above in their model: parasitism, changes in host plant quantity and induced changes in host plant quality. Van den Bos and Rabbinge's model, which closely simulated actual population fluctuations when all three factors were included (Fig. 5.34a), suggested that all factors were needed to explain the population cycles. They further demonstrated this by excluding each factor in turn from their model. Without parasitism in the model the population cycled but to a much less marked degree (Fig. 5.34b); without food quantity, the population extremes were greater (Fig. 5.34c); without an induced response affecting food quality, the popula-

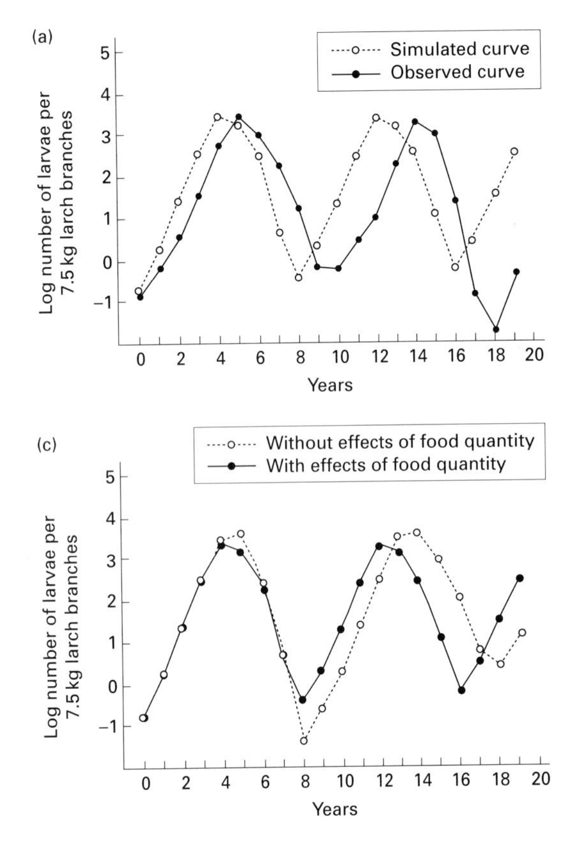

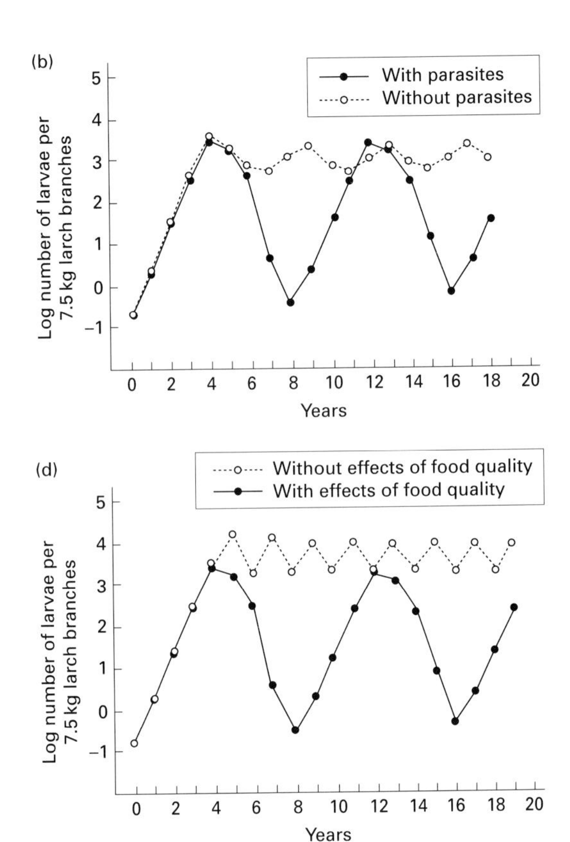

Fig. 5.34 The larch budmoth simulation model: (a) simulated and observed population density; (b) simulated population densities with and without parasitism; (c) simulated population densities with and without effects of food quantity; (d) simulated population densities with and without effects of food quality. (From Van den Bos & Rabbinge 1976.)

tion fluctuated around a higher density than usually recorded (Fig. 5.34d).

5.5.5 Lime aphid

Although the production of detailed models of the population dynamics of univoltine insects such as the spruce budworm and the larch budmoth is not easy, there have been a few attempts to analyse the even more complex dynamics of multivoltine insects such as aphids. One such example is Barlow and Dixon's (1980) research on the lime aphid, *Eucallipterus tiliae* (Hemiptera: Aphidae) (Plate 5.2, opposite p. 158). The fluctuations of this aphid were monitored for 8 years on lime (*Tiliae* × *europaea*) trees in Glasgow, Scotland. The lime aphid has 4–5 generations per year from fundatrices, which emerge from eggs in April–May, through to oviparae, which lay these eggs in September–October. The generations of the lime aphid rapidly overlap and the population assessments carried out by Barlow, Dixon and coworkers (Fig. 5.35) suggest very complicated population dynamics with no consistent pattern of within-year fluctuations in abundance. However, patterns in the dynamics of the lime aphid emerge when the trends in the abundance of fundatrices and oviparae are examined. There is a clear positive relationship between the number of oviparae in one year and the number of fundatrices the following year. In contrast, there is a negative relationship between the number of fundatrices in one year and both the number of oviparae later that year and the number of fundatrices the following year (Fig. 5.36).

Barlow and Dixon constructed a simulation model to try to explain the within-year and between-year fluctuations in lime aphid numbers. They considered various factors including:

- predation by the two-spot ladybird, *Adalia bipunctata* (Coleoptera: Coccinellidae);

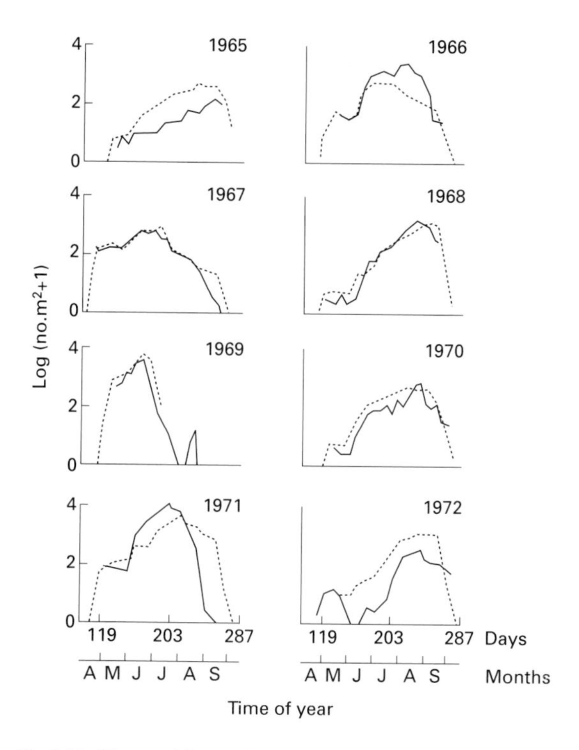

Fig. 5.35 Observed fluctuations in abundance of the lime aphid from 1965 to 1972 (solid lines) and simulated population density (broken lines). (From Barlow & Dixon 1980.)

- predation by the black-kneed capsid, *Blepharidopterus angulatus* (Hemiptera: Miridae);
- effects of plant quality (specifically, seasonal variation in plant nitrogen concentration) on growth, development and reproduction;
- effects of temperature on emergence of fundatrices in the spring and the development rate of aphid nymphs;
- effects of aphid density (measured in different ways; see below) on reproduction and emigration.

The resultant model was very successful in simulating the within-year population fluctuations of the lime aphid (Fig. 5.35). Barlow and Dixon therefore then examined the role of each of the major factors affecting the lime aphid, in much the same way as Van den Bos and Rabbinge had for the larch budmoth, by removing each factor in turn from their model. They also examined the role of each factor by running the model with each of these main factors alone. They assessed the impact of the removal or inclusion of each factor by looking at the effect it had on the relationship between the year-to-year changes in the numbers of fundatrices. The factors investigated in this way included the effect of adult density on flight (and hence emigration), the effect of nymphal density on flight, and predation.

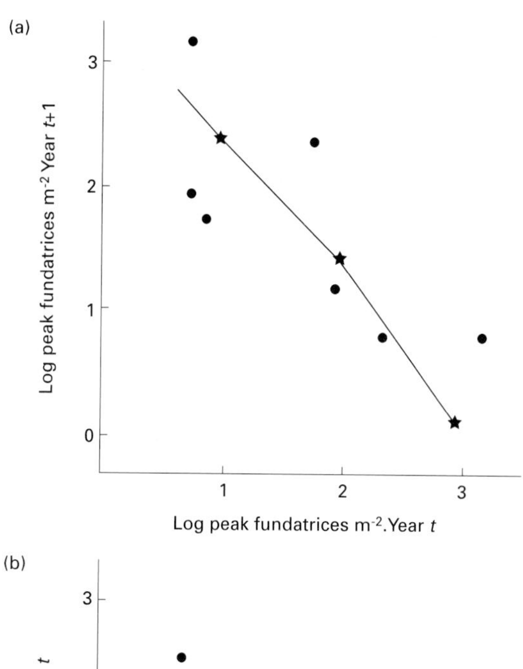

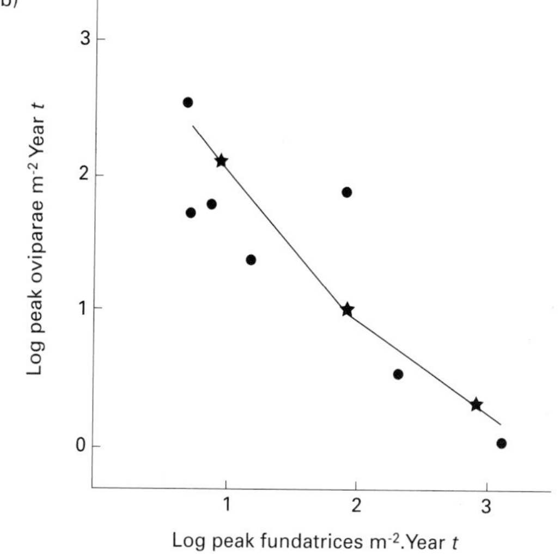

Fig. 5.36 Relationships between: (a) lime aphid densities in successive years (numbers of fundatrices, aphids which emerge from eggs in the spring); and (b) different aphid generations within the same year (fundatrices and oviparae, aphids which lay eggs in the autumn). Observed data shown as open circles, simulation model results shown as solid lines with stars. (From Barlow & Dixon 1980.)

When each factor was removed in turn, some factors, such as the effect of current aphid density on emigration, were found to have little impact on the year-to-year dynamics of the lime aphid (Fig. 5.37).

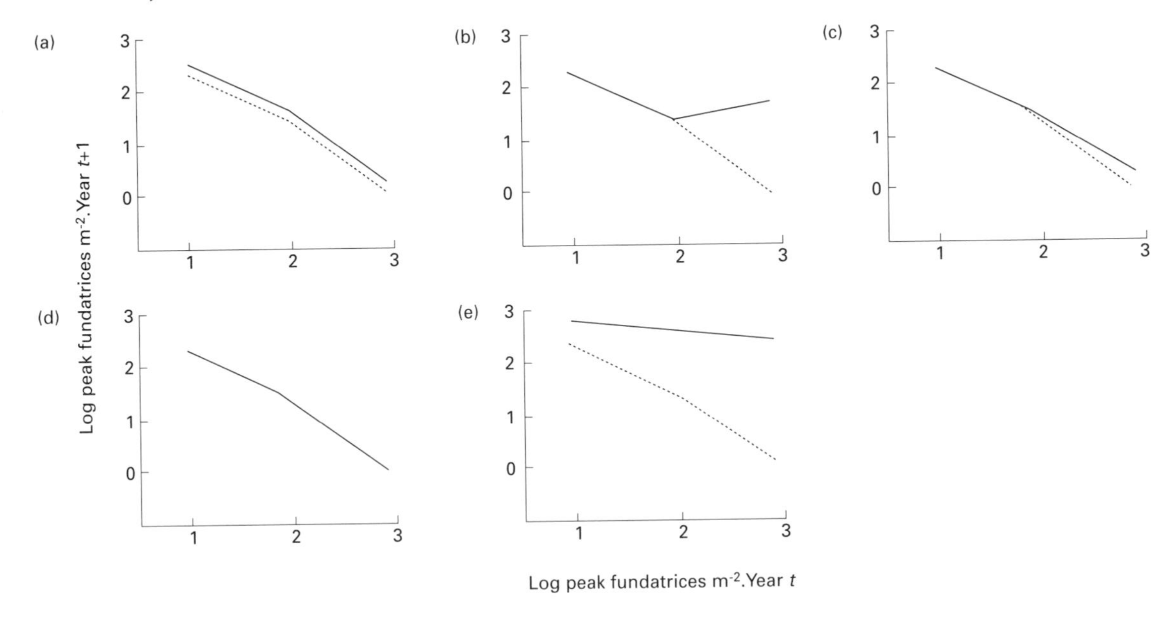

Fig. 5.37 The effect of removing different factors from the lime aphid model on the relationship between lime aphid (fundatrix) abundance in successive years: (a) impact of adult density on flight; (b) impact of nymphal density on flight; (c) changes in adult weight; (d) impact of cumulative aphid density on flight; (e) predation. In each case the new relationship is shown as a solid line and the original relationship (as Fig. 5.28a) as a broken line. (From Barlow & Dixon 1980.)

Two factors did have an impact: the effect on emigration of aphid density experienced during the nymphal stages and predation. However, the former appears to affect the lime aphid only at high densities (Fig. 5.37b), whereas the latter (predation) appears to have an effect at all densities (Fig. 5.37e).

When each factor was included alone in turn (Fig. 5.38), predation was found to be the only factor that had a significant effect on the year-to-year dynamics of the lime aphid. No other factor than predation was capable of regulating aphid densities alone (Fig. 5.38e). However, predation had this regulatory effect only at low and medium densities: at high densities, predation alone was incapable of regulating lime aphid numbers.

From their simulation studies, Barlow and Dixon concluded that different factors interact to regulate lime aphid numbers, and that the relative importance of these factors varied according to the abundance of the lime aphid. This conclusion is very similar to that of Holling and others developed by different approaches (see earlier in the chapter).

5.5.6 Insect–plant–natural enemy interactions

The above examples show how different factors (predation, parasitism, competition, damage-induced changes in host plant quality) together determine insect numbers. In each case, the overall effect of these factors appears to be an additive one. However, there are an increasing number of examples of synergistic interactions involving all three trophic levels (plants, herbivores and natural enemies). In particular, there are several ways in which the host plant can influence the impact of natural enemies of herbivores (Price *et al.* 1980).

Natural enemies use both insect (prey) and plant cues as part of the process of host location. Although parasitoids may use visual and acoustic cues, chemical cues appear to be the most common. Kairomones and synomones are both allelochemicals (chemicals that send information from one species (the emitter) to another (the receiver)) (Dicke & Sabelis 1988). A kairomone results in a response in the receiver that benefits only the receiver, a synomone benefits both the emitter and the receiver, and an allomone benefits only the emitter. Pheromones, in contrast, are chemicals used in communication between individuals of the same species. Most parasitoids use volatile

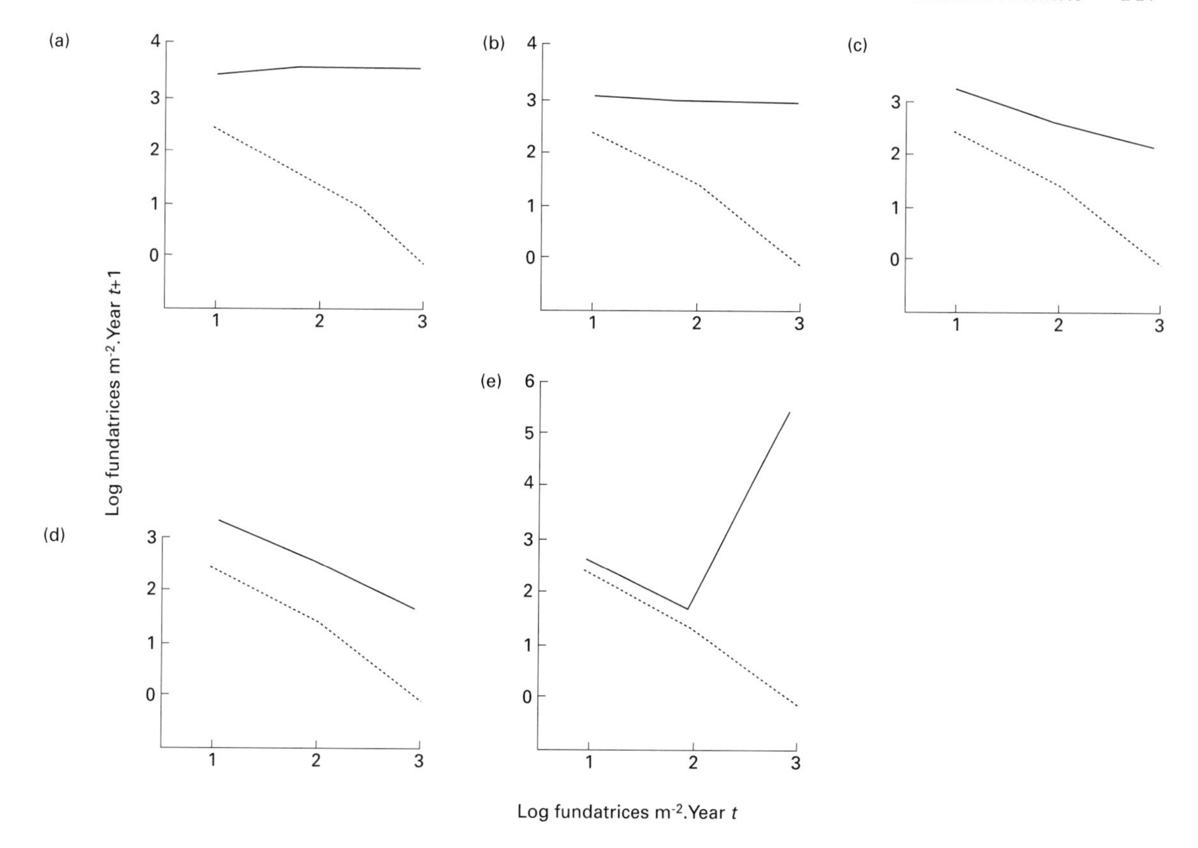

Fig. 5.38 The effect of different factors from the lime aphid model acting in isolation on the relationship between lime aphid (fundatrix) abundance in successive years: (a) impact of adult density on flight; (b) impact of nymphal density on flight; (c) changes in adult weight; (d) impact of cumulative aphid density on flight; (e) predation. In each case the new relationship is shown as a solid line and the original relationship as a broken line. (From Barlow & Dixon 1980.)

kairomones or synomones to locate the habitats where their hosts are located (Van Alphen & Jervis 1996). Chemicals that assist parasitoids in habitat location may be emitted from the insect or the plant (kairomones and synomones for the parasitoid, respectively). The former include chemicals emitted from frass (insect faeces), during moulting or feeding, sex pheromones and aggregation pheromones (Van Alphen & Jervis 1996). Several studies have demonstrated that plant odours attract predators and parasitoids. For example, the parasitoid *Leptopilina boulardi* (Hymenoptera: Eucoilidae) is attracted by ethyl alcohol, a chemical that is emitted from fermenting fruit, within which its prey *Drosophila melanogaster* (Diptera: Drosophilidae) feeds (Carton 1978).

Such is the dependence of parasitoids on plant cues that even a slight genetic change in the plant may result in a decline in parasitism. For example, Brown *et al.* (1995) demonstrated that very slight genetic changes in host plants can affect the impact of several natural enemies. They measured the effects of two different parasitoids and a mordellid beetle predator on the goldenrod ball gallmaker, *Eurosta solidaginis* (Diptera: Tephritidae), living on *Solidago altissima*, *Solidago gigantea* and host races derived from the two plant species. Observations on the behaviour of one species of wasp, *Eurytoma obtusiventris* (Hymenoptera: Eurytomidae) showed that it preferred to search on the ancestral plants than on the derived plant races and parasitism was much higher on the former (30.5%) than on the latter (0.4%). Although the host plant did not affect the mortality caused by the predator or the other parasitoid, total mortality caused by natural enemies was greater on the ancestral races.

This is an example of what is often referred to as an insect (such as *E. solidaginis* above) occupying 'enemy-free space' (Jeffries & Lawton 1984). The colonization of enemy-free space can have a dramatic impact on insect abundance. For example, the pine beauty moth, *Panolis flammea* (Lepidoptera: Noctuidae) (Plate 5.3, opposite p. 158) feeds on the native Scots pine (*Pinus sylvestris*) in Scotland but colonized lodgepole pine (*Pinus contorta*) after it was introduced as a plantation tree from Canada and the USA. Pine beauty moth is found in low numbers on Scots pine but regularly reaches 'outbreak' densities on lodgepole pine. This pattern occurs probably because there are fewer natural enemies in lodgepole pine plantations than in Scots pine forests and parasitic wasps, at least, are poorer at locating their prey on lodgepole pine (Watt *et al.* 1991).

One particularly interesting type of plant cue for natural enemies is the chemicals emitted by plants in response to insect damage (Dicke 1994). These chemicals, such as monoterpenes, lead to an increase in the activity of insect predators and parasitoids (see Chapter 3, section 3.3.7).

The physical structure of plants may also affect prey location or prey capture by natural enemies. For example, the parasitoid *Encarsia formosa* (Hymenoptera: Aphelinidae) is poorer at finding and killing greenhouse whitefly, *Trialeurodes vaporariorum* (Homoptera: Aleurodidae), on cucumber varieties with hairy leaves (Van Lenteren *et al.* 1995). This is one example of a potential pitfall of selecting plant varieties for resistance to insect pests. Another example is the reduction in parasitism and predation on the wild tomato variety PL 134417, which has glandular trichomes containing the methyl ketones 2-tridecanone and 2-undecanone (Kennedy *et al.* 1991; Farrar *et al.* 1992). This variety is resistant to pests such as the tobacco hornworm, *Manduca sexta* (Lepidoptera: Sphingidae), *Helicoperva* (*Heliothis*) *zea* (Lepidoptera: Noctuidae) and the Colorado beetle, *Leptinotarsa decemlineata* (Coleoptera: Chrysomelidae). However, predators and parasitoids such as *Archytas marmoratus* and *Eucelatoria bryani* (Diptera: Tachinidae) are less effective on PL 134417 than on tomato varieties that are susceptible to pest insects. These tachinid parasitoids both attack *H. zea* but have different life history characteristics. *E. bryani* lays its eggs directly into the host, but *A. marmoratus* gives birth to special larvae, called planidia, which attach themselves to passing *H. zea* larvae, penetrate the host cuticle and develop in the host pupae. The glandular trichomes kill *A. marmoratus* planidia and the larval development of both species of parasitoids is deleteriously affected by the methyl ketones consumed by their prey. Experimental removal of the trichomes eliminates the negative effects on these and other natural enemies. Predation and parasitism on hybrid plants with trichomes which do not contain the methyl ketones is intermediate between the susceptible and PL 134417 varieties, indicating that the negative effect of the glandular trichomes (in PL 134417) is partly physical and partly chemical (see also Chapter 10).

Other aspects of the physical structure of host plants that may affect insect natural enemies include bark texture and bark hardness, both of which influence the density of parasitic wasps that attack the bark beetle, *Ips typographus* (Coleoptera: Scolytidae) on spruce (Lawson *et al.* 1996). Another example of the effect of plant structure on natural enemies is the effect of gall size on parasitism of galling insects. Stiling and Rossi (1996) found that parasitism of the gall midge *Asphondylia borrichiae* on *Borrichia* decreased as gall size increased, largely because the most abundant parasitoid was largely unable to parasitize insects within large galls. As gall size was determined by plant clone and site conditions, these factors also indirectly determined parasitism.

Perhaps the simplest plant–insect–natural enemy interaction is the indirect effect of plant quality on predation. Several entomologists have suggested that insects on relatively poor-quality host plants tend to suffer higher rates of predation, because of their slower development rates, than insects on better-quality host plants (e.g. Price *et al.* 1980). Thus predation may be greater on an insect species feeding on one host plant species, or variety, than another. Haggstrom and Larsson (1995) tested this 'slow growth–high mortality' hypothesis by comparing the mortality caused by predators of the willow-feeding leaf beetle *Galerucella lineola* (Coleoptera: Chrysomelidae) on *Salix viminalis* and *Salix dasyclados*. *G. lineola* develops faster on *S. viminalis* than on *S. dasyclados*, and, in support of the slow growth–high mortality hypothesis, total predation during the larval period and the daily larval predation rate was found to be greater on *S. dasyclados* than on *S. viminalis*. Loader and Damman (1991) obtained similar results in a study of the cabbage-white butterfly *Pieris rapae* (Lepidoptera: Pieridae) on collard plants experimentally treated to be either rich or poor in leaf nitrogen concentration. Not all studies have given support to the slow growth–high mortality hypothesis. Leather and Walsh (1993), for example, found that predation of pine beauty moth larvae was greater on faster-growing larvae on better-quality host plants. Leather and Walsh compared predation on different provenances of lodgepole pine, and as different provenances are known to emit different mixtures of monoterpenes, perhaps predators were responding to the chemical composition of the host plants of their insect prey.

As discussed in Chapter 3, insects feeding on previously damaged plants may, amongst other effects, experience prolonged development times. Consequently, they may also suffer greater levels of predation (as discussed above). Herbivore damage may result in an increase in the mortality caused by natural enemies in other ways too (Karban & Baldwin 1997). For example, studies on plants with extrafloral nectaries, plant structures known to attract ants and therefore protect them against plant-feeding insects, have shown that the amount and quality of nectar may increase after they are damaged (Stephenson 1982; Smith *et al.* 1990). This should result in an increase in predation by ants attracted to extrafloral nectaries, although this has not been demonstrated conclusively (Karban & Baldwin 1997).

A particularly interesting example of a plant–herbivore–natural enemy interaction is the oak-gypsy moth–GMNPV interaction. GMNPV, or gypsy moth nuclear polyhedrosis virus, is a major natural enemy of gypsy moth, *Lymantria dispar* (Lepidoptera: Lymantriidae). The damage caused by gypsy moth larvae is known to affect the chemical composition of the foliage of its host plants and, in turn, the gypsy moth (Rossiter *et al.* 1988, and see Chapter 10). However, the chemical composition of oak foliage also affects GMNPV (Keating *et al.* 1990; Schultz *et al.* 1990). High foliar tannin concentration (particularly hydrolysable tannin concentration) inhibits GMNPV, increasing gypsy moth resistance to this pathogen (Fig. 5.39). Thus what may appear to be separate interactions (a) between the gypsy moth and its host plants, and (b) between the gypsy moth and one of its natural enemies, is a more complex interaction involving all three. This interaction has been explored in a model by Foster *et al.* (1992).

5.5.7 Natural enemies: competing natural enemies and other influences on their impact

In addition to the plant factors discussed above, the impact of natural enemies is influenced by a variety of factors, including interactions with other natural enemies. Ants, for example, can reduce the impact of parasitic wasps. Itioka and Inoue (1996) found that the presence of the honeydew-foraging ant *Lasius niger* (Hymenoptera: Formicidae) reduced the rate of parasitism of the red wax scale insect, *Ceroplastes rubens* (Hemiptera: Coccidae).

Only a few studies have actually quantified the effects of interactions between different natural enemy species on insect herbivores. Rosenheim *et al.* (1993), for example, studied the predatory commu-

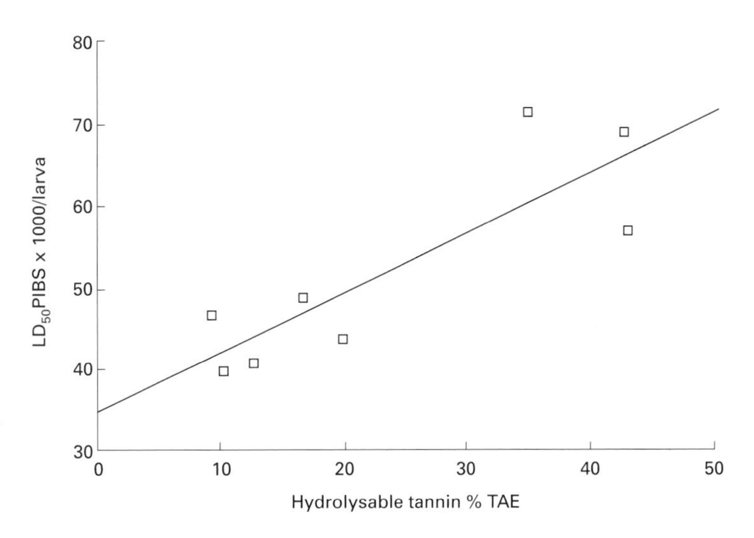

Fig. 5.39 Relationship between the concentration of hydrolyzable tannins in oak leaves from eight different locations and the susceptibility of gypsy moth larvae to GMNPV on leaves from the same locations (as measured by LD_{50} tests). (From Schultz *et al.* 1990.)

nity associated with the aphid *Aphis gossypii* (Hemiptera: Aphidae). They found that generalist hemipteran predators (*Zelus* and *Nabis* species) consumed larvae of the green lacewing, *Chrysoperla carnea* (Neuroptera: Chrysopidae), major predators of *A. gossypii*, under both natural and experimental conditions. Experiments showed that predation of lacewings by either hemipteran species could result in aphid outbreaks.

Species apart from other natural enemies can affect insect–natural enemy interactions. For example, endophytic fungi, which are thought to defend their host plants from herbivores, may also have a detrimental effect on the natural enemies of these herbivores. Bultman *et al.* (1997) found that parasitoids reared on fall armyworm, *Spodoptera frugiperda* (Lepidoptera: Noctuidae), that had fed on tall fescue infected by the endophytic fungus *Acremonium coenophialum* had a lower survival rate compared with those reared on fall armyworm that had fed on uninfected grass. However, as the endophytic fungus also has a direct negative effect on the herbivore (through the alkaloids produced symbiotically by the grass and the fungus), it is unclear what the net effect of the fungal infection on the herbivore might be.

The presence or absence of one species of herbivore may influence the action of natural enemies on another herbivore species because its presence promotes the activity of a natural enemy of the second herbivore. Evans and England (1996) showed that the presence of pea aphids can lead to an increase in alfalfa weevil parasitism, probably because parasitic wasps benefit from the honeydew that the aphids provide. Evans and England also showed that the addition of a predator (the seven-spot ladybird, *Coccinella septempunctata*) led to increased predation of alfalfa weevil but also led to a decrease in weevil parasitism, presumably because the predator also reduced the number of aphids.

Many other external factors may affect the relationship between natural enemies and their prey. Drought stress is often considered to have an indirect effect on insect herbivores (see Chapters 3 and 10). Drought stress may also have an effect on natural enemies. Godfrey *et al.* (1991) found that natural enemies were more abundant in drought-stressed than in irrigated corn (*Zea mays*). What caused this is unknown but it may be that drought-stressed plants emit chemicals similar to those emitted after they are attacked by herbivores and known to attract natural enemies (section 5.5.6).

Air pollution may also have a detrimental effect on natural enemies. Gate *et al.* (1995) found that parasitism of *Drosophila subobscura* (Diptera: Drosophilidae) by the wasp *Asobara tabida* (Hymenoptera: Braconidae) was reduced by exposure to elevated levels of ozone, although not to NO_2 or SO_2 fumigation. Similarly, elevated atmospheric CO_2, which can affect insect herbivores through a reduction in the nitrogen concentration of their host plants (Chapter 3), may also influence the impact of natural enemies. To date, only a few, inconclusive experiments have explored the effects of elevated atmospheric CO_2 on the action of natural enemies (Roth & Lindroth 1995; Bezemer *et al.* 1998).

There are many habitat characteristics that can affect the interaction between insects and their natural enemies. Soil type and condition (e.g. whether dry or waterlogged) can affect the abundance of soil or soil surface-dwelling predators in forests and other habitats. Their absence from certain areas may be one of the reasons why forest pest outbreaks start in particular parts of forests (Mattson & Addy 1975).

Forest fragmentation may lead to an increase in the severity of pest outbreaks because of a decrease in the impact of natural enemies. Roland *et al.* (1997) studied the effect of parasitism on the forest tent caterpillar, *Malacosoma disstria* (Lepidoptera: Lasiocampidae), a major defoliator of North American boreal forest dominated by trembling aspen and balsam poplar. The major parasitoids of *M. disstria* are the flies *Patelloa pachypyga* and *Arachnidomyia aldrichi* (Diptera: Tachinidae and Sarcophagidae). *P. pachypyga* lays its eggs on aspen foliage and *A. aldrichi* lays its eggs directly on the cocoon of *M. disstria*. Parasitism by *P. pachypyga* (but not by *A. aldrichi*) is lower in fragmented than in continuous forest (Roland & Taylor 1997; Roland *et al.* 1997). The lower abundance or efficiency of the tachinid parasitoid in forest fragments may be due to a preference for the relatively cool humid conditions of continuous forest stands. Some parasitoids are known to prefer such conditions; others prefer warm dry microclimates (Weseloh 1976). Other factors may be at least partly responsible for the prolonged outbreaks of *M. disstria* in fragmented forest. For example, virus transmission may be lower in forest patches than in continuous forest (Roland & Kaupp 1995) and the fecundity of *M. disstria* may be greater in fragmented than in continuous forest (Roland *et al.* 1997).

Some trees experience more damage by pests in

urban habitats, such as roadsides, than in forested habitats (Speight *et al.* 1998). This may be due to a range of factors, including differences in the abundance of natural enemies. Hanks and Denno (1993a), for example, found that generalist predators, such as phalangids, earwigs and tree crickets, were less abundant on trees along roadsides and in parking lots than in forested areas. Studies on experimental cohorts of the armoured scale insect, *Pseudaulacaspis pentagona* (Hemiptera: Diaspididae), showed that predator-induced mortality was greater on forest trees, and Hanks and Denno concluded that natural enemies and plant water relations (indirectly affecting scale insect performance through host plant quality; see Chapter 3) jointly explained the greater abundance of scale insects in disturbed habitats.

The above two examples are given to illustrate the effect of habitat characteristics on the impact of natural enemies. It should not be assumed from these examples that all forest pests are similar to *M. disstria*. The outbreaks of other major forest pests, such as the spruce budworms *Choristoneura fumiferana* and *Choristoneura occidentalis* (Lepidoptera: Tortricidae) are most severe in continuous forests (e.g. Swetnam & Lynch 1993). It should also be pointed out that not all studies have found that insects are more abundant on trees in urban than in natural (or seminatural) habitats (e.g. Nuckols & Connor 1995).

Perhaps the most important habitat feature from the point of view of the impact of natural enemies is plant diversity. There are several reasons why monocultures are more susceptible to pest outbreaks than mixed-species agricultural or forest crop stands but the main reason appears to be that an increase in plant diversity leads to an increase in the diversity and abundance of natural enemies. For example, Schellhorn and Sork (1997) found that the abundance of natural enemies (coccinellids, carabids and staphylinids) of collard-feeding pests was greater in polycultures (collards plus weeds) than monocultures (collards alone). Such is the importance of plant diversity for promoting the action of natural enemies that it forms an important basis for many pest management strategies such as intercropping (see Chapter 10).

5.6 Conclusions

The above examples show how different factors—predation, parasitism, the amount and quality of the food resource—interact to determine insect numbers. We could add many more examples showing how different factors interact to determine insect numbers.

These examples provide ample evidence for the argument that to understand the fluctuations of insects we need to carry out detailed studies of their population dynamics, through analysis of natural populations, experimentation and modelling. As we have tried to show in this chapter, ecologists have been trying to explain the dynamics of insects and their natural enemies using models for over 70 years. We hope that we have demonstrated that even some simple models have provided insight into natural enemy–prey interactions. In recent years, insect ecologists have uncovered more and more of the complex interactions involving insects, their host plants, their natural enemies and other factors. Surprisingly few models have been constructed to explore these interactions. Those that have been built (see section 5.4) have surely demonstrated that we should not ignore the complexity of insect population dynamics and that we can effectively use models to explore these dynamics.

However, it may be said that even after detailed research, we do not always arrive at a complete understanding of the population dynamics of the species studied across its full geographical range. A good example of how some species, at least, have different population dynamics in different parts of their geographical range is the winter moth. As discussed above, work on the winter moth, *Operophtera brumata* (Lepidoptera: Geometridae), in Wytham Wood in England showed that it was regulated by pupal predation and that one of its parasitoids, *Cyzenis albicans* (Diptera: Tachinidae) had no significant role in its population dynamics (section 5.3.4). In Canada, however, it was the introduction of this parasitoid that brought about the dramatic reduction of the winter moth (Fig. 5.9). Subsequent studies on the natural enemies suggested that parasitism and predation acted together to cause this reduction in winter moth densities, but the fact remains that the dynamics of this insect vary across its geographical range. Its dynamics also appear to vary from host to host. In the UK, the winter moth has recently emerged as a pest of Sitka spruce and heather (Plate 5.4, opposite p. 158). Hunter *et al.* (1991) failed to find that any of the factors that significantly affected the winter moth on oak affected it on Sitka spruce. Its dynamics on heather also appear to be different (Kerslake *et al.* 1996).

The studies discussed above show that that insect populations tend to be regulated by the interaction of a multiplicity of factors. Are there therefore a million types of population dynamics? The answer is probably no (Lawton 1992). The number of types of population dynamics is probably fairly small, even if more than one type of behaviour can be found in a single species. The reason for this conclusion is that underlying the complexity of population dynamics is an appealing simplicity. The work of Holling and others has demonstrated how the complex interactions of different factors can be represented in a simple way (Figs 5.20–5.22). They showed how a single species can exhibit different types of dynamics, but, more importantly, they also showed what might cause a species to move from one type of dynamics to another (section 5.4.5). Clearly, we need much more research on the impact of natural enemies and other factors on insect herbivores but this research must increasingly focus on the complex interactions (involving insects, their host plants, their natural enemies and other factors) determining insect abundance.

Chapter 6: Evolutionary Ecology

There are many ecological processes that are impossible to understand without considering evolution. Ecologists are rather prone to examining species of animals and plants at one point in time, and often also in one point in space. In fact, these species consist of a series of populations that may differ genetically and hence possess very different ecological properties (Bradshaw 1983). Populations of insects, as with all other taxa, are subject to natural selection, and all the topics that we discuss in the book, whether they be insects and climate, insects and plants, or insects and other animals or pathogens, are concerned with the results of evolution. Evolution is essentially about adaptation to environment, or fitness (Bulmer 1994), and the ways in which organisms develop, reproduce and die have been shaped by natural selection (Nylin & Gotthard 1998). Fitness considers the relative ability of individuals or their progeny to survive under various constraints imposed by the abiotic and biotic environment in which they find themselves. In this chapter we discuss some of the mechanisms of evolutionary ecology pertaining to insects, emphasizing particular examples to illustrate the complexities achieved as solutions to the problem of maximizing fitness.

6.1 Life history strategies

Because insects comprise such a vast number of species, it is not surprising that we can discover an enormous range of life history types. Even within a species, different individuals in different places or times may exhibit distinct life histories. Larvae of butterflies, for example, are adapted for feeding and growth, whereas their adults are specialized for dispersal and reproduction. The former activity may take some time, depending on food supply and climate, whereas the latter stage may be very much shorter. Adult aphids of a single species may vary in the incidence of winged forms (alatae), again depending on food supply and time of year, whereas other aphids alter the incidence of males and hence sexual reproduction in their life histories. Some adult insects never reproduce at all, as is the case with worker bees. Many species may produce enormous numbers of offspring, but others only a very few. Whatever the life history strategies evolved by insects, these may also vary within a species in different localities.

The life history strategy selected by evolution for a particular set of ecological circumstances thus varies tremendously. Some insects have evolved extremely rigid and non-adaptable strategies, as in the case of stick and leaf insects (Phasmida), where many species are host plant specialists, wingless as adult females, and reproduce only asexually. Others, however, are much more flexible in their life history patterns within one locality. The pea aphid, *Acyrthosiphon pisum* (Hemiptera: Aphididae) and its wasp parasitoid, *Ephedrus californicus* (Hymenoptera: Aphidiidae) both vary their life history strategies (in terms of variable numbers of offspring, life span and development time) in response to variations in the quality of the aphids' host plant (Stadler & Mackauer 1996).

To understand these complexities, it is helpful to consider them under discrete headings; here, we shall discuss survivorship, longevity, reproductive strategies, and the occurrence of different forms of the same species, polymorphisms.

6.1.1 *r*- and *K*-selection

All organisms have to make an evolutionary 'choice' when it comes to determining whether to produce very large numbers of offspring in whom very little investment is made per individual, or alternatively to have rather few, but to ensure that each one has the best possible chance of surviving to adulthood. Ecologists have pigeon-holed each strategy into *r*- or *K*-selection. Table 6.1 provides some of the characteristics of both types, though it must remembered that they are merely two opposing ends of a theoretical spectrum of reproductive strategy, with many intermediates being found in the real world.

Characteristic	*r*-selected	*K*-selected
Body size in general	Small	Large
Body size male vs. female	Larger females	Smaller females
Colonization ability	Opportunistic	Non-opportunistic
Development rate	High	Low
Dispersal ability	High	Low
Egg size	Small	Large
Fecundity	High	Low
Investment in each offspring	Low	High
Longevity	Short	Long
Occurrence in succession	Early	Late
Population density	Fluctuating, sometimes widely	Stable
Intraspecific competition	Often 'scramble'	Often 'contest'
Role of density-dependence	Unimportant	Important
Population 'overshoot'	Frequent	Rare
Sex ratio	Female biased	Normal

Table 6.1 Some characteristics of *r*- and *K*-selected insects. (From MacArthur & Wilson 1967; Southwood 1977; McLain 1991; Maeta *et al.* 1992).

From the table, it is clear that *r*-selected insects are likely to be unpredictable in their ecologies, to undergo 'boom and bust' population cycles, to produce large numbers of offspring, and essentially to over-exploit their habitats and leave them in ruins! Classic examples include locusts and aphids (see Chapter 1).

K-selected insects, on the other hand, are more often regulated by density-dependent factors such as intraspecific competition and natural enemies, and they have relatively few offspring and avoid over-exploiting their habitats. Some species are characterized by having giant eggs, low fecundities and prolonged life spans (Maeta *et al.* 1992). Examples include tsetse flies, carpenter bees, and many (though by no means all) forest Lepidoptera.

r-selected species often dominate in habitats that are relatively hard to find in time and/or space, or exist for only short periods of time. Thus, for example, work in the Apennine region of Italy found that 70% of dung beetle (Coleoptera: Scarabaeoidea) species living in cattle dung exhibited *r*-selected characteristics (Carpento *et al.* 1996). Dung pats are scattered fairly randomly around grassland, and remain fresh and hence suitable for colonization by beetles for only a short time, so that insects wishing to exploit them have to have evolved efficient and rapid searching mechanisms. *r*-selected insects also tend to be early colonizers of new or disturbed habitats, by virtue of their essentially opportunistic and pioneer characteristics. As mentioned above, aphids and their relatives, leafhoppers (Hemiptera: Aphididae and Cicadellidae), for example, are highly mobile and have high reproductive potentials—they are typically *r*-selected, and in one set of experimental manipulations they appeared early in the recolonization of apple trees previously emptied of their insect fauna by insecticide spraying (Brown 1993).

As with most things ecological, life is never simple, and some species may show traits of both *r*- and *K*-selection at different times or in different locations. A single species may exhibit *r*- or *K*-selection traits depending on its population density. Low population densities of the bean weevil, *Acanthoscelides obtectus* (Coleoptera: Curculionidae) showed *r*-selected characteristics of high fecundity, earlier age at first reproduction and higher growth rates when compared with high-density populations, which appeared to be more *K*-selected (Aleksic *et al.* 1993). In the case of the mosquito *Anopheles messeae* (Diptera: Culicidae), populations showed a seasonal conversion from *r*-selected traits to *K*-selected ones in periods of so-called population 'prosperity' (Gordee & Stegnii 1988), when resources were plentiful. Some of the characteristics mentioned in Table 6.1 will now be considered in greater detail.

6.1.2 Survivorship

Survivorship (or perhaps more accurately fatalityship) describes the rates at which organisms die as they progress through a generation. In theory,

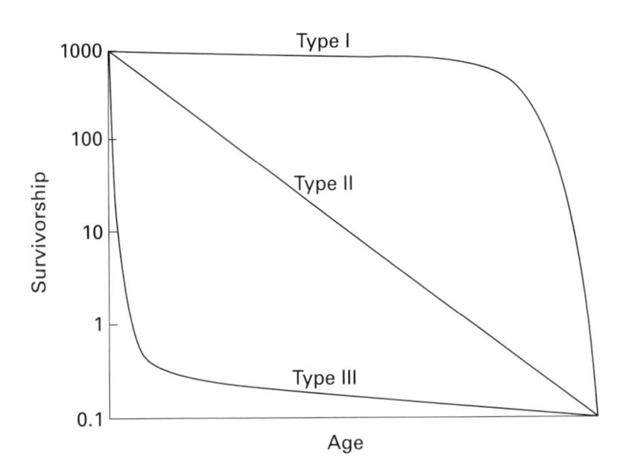

Fig. 6.1 Theoretical survivorship curves: Type I is where mortality is concentrated at the end of the maximum life span, Type II is where mortality rate stays constant with age, and Type III is where most mortality occurs at an early stage in the life cycle. (From Begon *et al.* 1986, after Slobodkin 1962.)

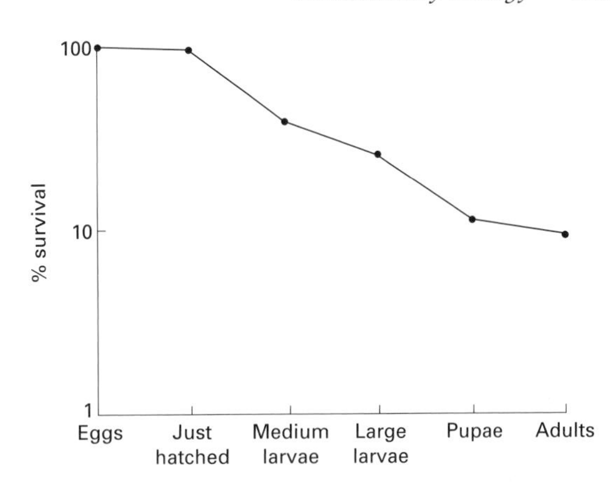

Fig. 6.2 Survival of lesser cornstalk borer, *E. lignosellus* (mean of three generations). (From Smith & Johnson 1989.)

animals including insects exhibit one of three types of survivorship curve, shown in Fig. 6.1.

These distributions may have limited real-world use, as most insects do not really conform to theory. Nevertheless, there is merit in considering a basic difference between a species or population wherein most of the individuals die at a very young age, compared with those where most individuals live to be old (Type III vs. Type I). The Type II survivorship curve describes a situation where no particular life stage shows significantly higher mortalities than any other. This pattern, where constant mortality occurs across ages, is more characteristic of birds and other vertebrates rather than insects (Strassman *et al.* 1997). However, it should be noticed that the *y*-axis is normally depicted on a logarithmic scale; back-transformation reveals that population declines illustrated by survivorship curves are largest in early stages even when the curve is a straight line (Type II). What really counts is the steepness of the slopes over a particular life stage. Three examples are given, one for each type of survivorship curve.

Figure 6.2 shows the mean percentage survival over three generations of several life stages of the lesser cornstalk borer (LCB), *Elasmopalpus lignosellus* (Lepidoptera: Pyralidae) (Smith & Johnson 1989). The larvae of this species can cause serious economic damage when they tunnel in the leading shoots of maize and other crops, causing dead-heart symptoms. Like many members of its family, LCB is not particularly host plant specific, and in this particular example, the so-called cornstalk borer is in fact acting as a pest of peanuts in Texas. From the graph, it is clear that no life stage through a generation exhibits particularly greater mortality than any other. There is just a suggestion that eggs and pupae suffer least, having the shallowest gradients. This is often the case, where quiescent, immobile, and relatively defenceless forms such as these are placed by the previous life stage (ovipositing females or late stage larvae) in safe and protected habitats. In fact, for borers such as LCB, the larvae are also fairly well protected from external influences including high levels of parasitism and disease, reflected in the roughly Type II curve.

Aphids more often than not show complex and specialist life history strategies, optimized for reproductive and host exploitation (see section 6.2). *Sitobion avenae,* the grain aphid (Plate 6.1, opposite p. 158), is a notorious pest of winter and spring cereals in western Europe, which, given the right weather conditions (dry and warm) when the crop enters the milk-ripe stage after flowering, can cause losses to wheat extending to several tonnes per hectare. All aphids at this time of the year are parthenogenetically reproducing females, each of whom has a maximum life span in the laboratory of around 10–12 weeks (Thirakhupt & Araya 1992). Under these ideal conditions, most females survive the nymphal stage and enter adulthood with little significant mortality (Fig. 6.3). Heavy mortalities occur only when the postreproductive stage is reached after a couple of months. It is important to realize that this type of survivorship

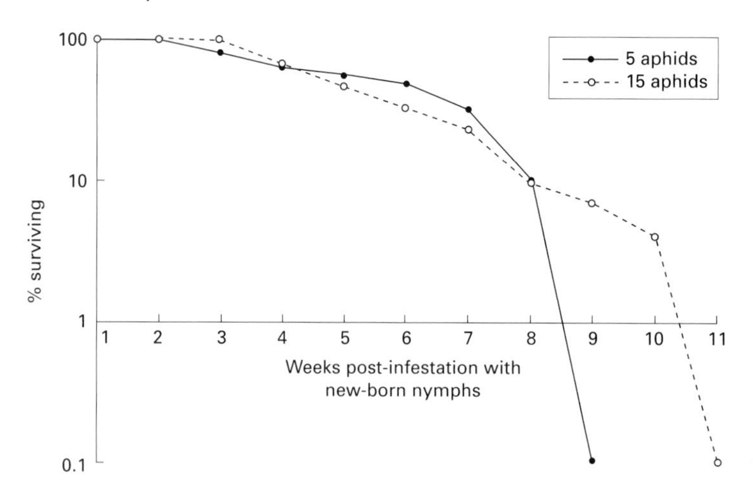

Fig. 6.3 Survivorship curves of two colonies of the cereal aphid, *Sitobion avenae*. (From Thirakhupt & Araya 1992.)

is unlikely to be encountered in the field, because natural enemies are likely to exact a toll. However, the importance of parasitoids, predators and pathogens in the life of aphids and their relatives is often limited (see Chapter 10), so that Type I survivorship curves may still operate. Even parasitoids themselves follow this type of strategy on occasion. Adults of *Cotesia flavipes* (Hymenoptera: Braconidae), a parasitoid of another pyralid moth related to the cornstalk borer discussed above, also show a Type I survivorship curve, where most individual females lived until the end of their normal life span, in this case a mere 24 h (Wiedenmann *et al.* 1992).

Finally, some insects produce enormous numbers of offspring, a high proportion of which are lost early in life. Scale insects (Hemiptera: Coccidae) provide some good examples. The horse-chestnut scale, *Pulvinaria regalis*, may lay several thousand eggs, according to the size of the female (see Fig. 6.11, below). As Fig. 6.4 shows, mortality during the egg stage is fairly minor (M.R. Speight, unpublished observations), but when the eggs hatch to produce first-instar nymphs called crawlers, very large losses may occur. The mobile crawlers disperse away from egg masses from which they originated on the main trunks of both host and non-host trees, and it is their job to locate leaves of limes, horse-chestnuts or sycamores on which older nymphs will feed during the summer months. Many crawlers merely move to the foliage of the trees on which they hatched, but the majority are blown off their hosts and lost on the wind, very few indeed ending up by chance on a new suitable host tree (Speight *et al.* 1998). These huge dispersal losses are

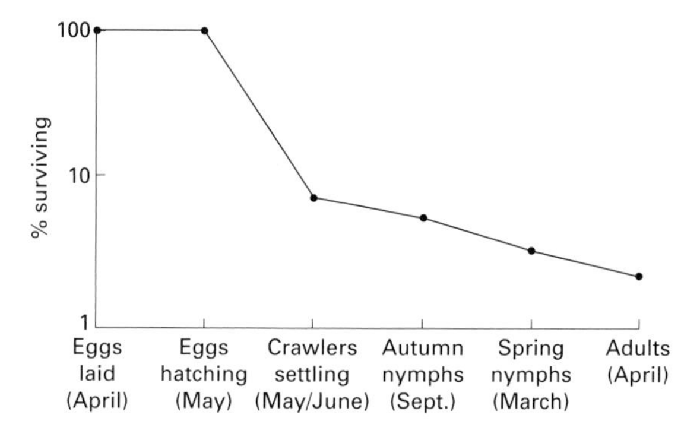

Fig. 6.4 Survival of horse-chestnut scale, *Pulvinaria regalis*, through an annual generation (M.R. Speight, unpublished observations).

analogous to those seen in the case of sessile marine invertebrates that consign their young larval stages to the vagaries of ocean currents, in the hope that at least a few will colonize new and distant substrates.

So, evolution has produced several types of survivorship strategy wherein the rates of mortality vary according to the life stage. An added complication, however, is that different life stages of a species may show tendencies towards various shapes of survivorship curve within that life stage, so that it is important to treat eggs, larvae and adults, for example, as separate stages wherein survival may show different patterns. Larvae of the caddisfly, *Rhycacophila vao* (Trichoptera: Rhyacophilidae), predate other invertebrates in small streams in Canada (Jamieson-Dixon & Wrona 1992). Their first winter is spent as the first three instars, developing to the fourth instar by

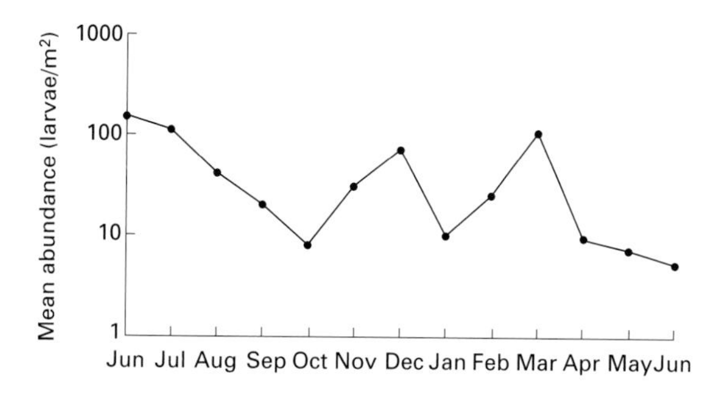

Fig. 6.5 Survival of a cohort of larvae of the caddisfly, *Rhyacophila vao*. (From Jamieson-Dixon & Wrona 1992.)

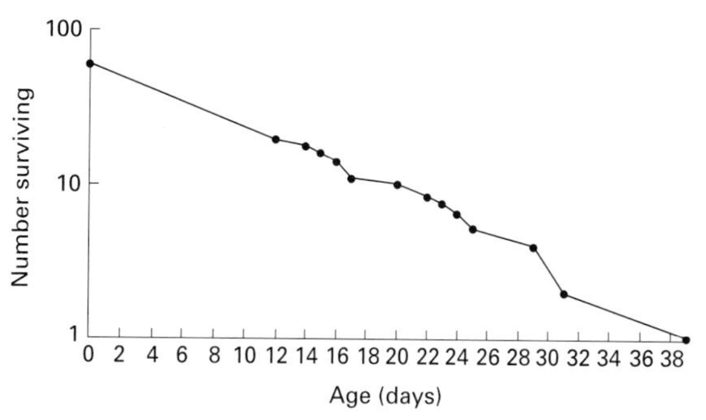

Fig. 6.6 Survival of adult female damsel fly, *Pyrrhosoma nymphula*. (From Bennett & Mill 1995.)

summer and the fifth by autumn. The second winter is spent as fifth instars, with new adults appearing in the summer of the second year. Figure 6.5 illustrates survivorship of a cohort of larvae in the third instar in late spring, and shows an exponential decline in numbers until winter. Higher mortalities in the active months are not surprising; losses through predation, competition and possibly starvation are likely to operate when the stream temperatures are relatively high, and a species such as this with a long larval stage through seasonally variable climatic conditions would be expected to show a Type III survivorship curve.

Insect species with a much shorter time spent in a particular life stage are less likely to exhibit this type of survivorship. Remaining with fresh-water insects, though this time the terrestrial adults, the damselfly, *Pyrrhosoma nymphula* (Odonata: Zygoptera), has a total adult life span of around 20 days in the UK (Fig. 6.6) (Bennett & Mill 1995). Within this time frame, the insects are reproductively active for a mean of 6.7 days only, and before this, young adults of both sexes remain away from water until sexually mature. It is thought that this tactic has evolved to protect them from predation before they have had a chance to reproduce, and the survivorship curve reflects the fact that prereproductives survive as well as, if not better than, mature adults.

6.1.3 Longevity

The length of time an insect spends in a particular life stage, and indeed the total time taken for one generation, depends on a wide range of factors that the species or population has encountered during its evolution. As seen in Chapter 1, the majority of insect species (about 80%) belong to the Endopterygotes, where the specialized larval stage has evolved to feed and grow, freeing the adult to specialize in its turn, this time for the major purposes of dispersal and reproduction. In general terms, growth demands a longer time span than the latter two activities, and hence it is not surprising to find that the majority of an insect's life is spent in the larval stage. Even the Exopterygote insects, whose nymphs frequently resemble their adults both morphologically and ecologically, tend to have protracted juvenile stages. Mayflies (Ephemeroptera) are an extreme example, where the mainly aquatic nymphs may spend a year or more growing slowly, whereas the adults, true to the order's name, may live only a matter of hours or days.

Obviously, the time spent as a larva or nymph is dictated by the amount of food available, and the climatic conditions, especially temperature, to which the insect is subjected. Poor food supply necessitates a longer juvenile period, as a minimum size must be attained before pupation or metamorphosis can proceed. In a simple example, some of the longest lived of all insects are larvae of certain species of longhorn beetle (Coleoptera: Cerambycidae) and goat moth (Lepidoptera: Cossidae) (Plate 6.2, opposite p. 158). Though living in tropical climates, the quality of food consumed by these wood borers is extremely low, being especially deficient in organic nitrogen. Larval periods of 10 years or more have been reported before the larvae are eventually able to pupate. A more complex example concerns pea aphids, *Acyrthosiphon pisum* (Hemiptera: Aphididae). This species has been shown to take longer from birth

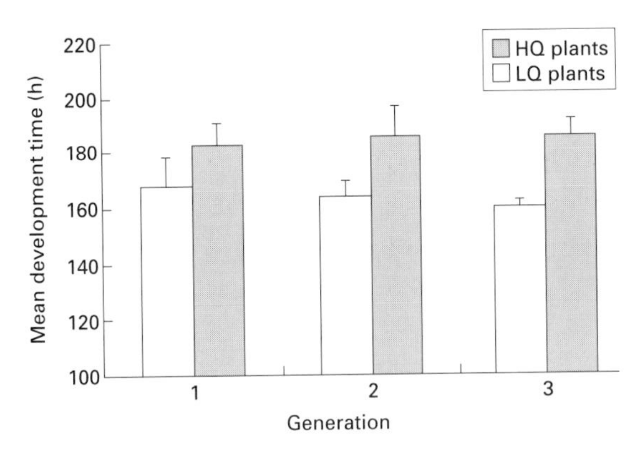

Fig. 6.7 Development time of pea aphid, *Acyrthosiphon pisum*, from birth to adulthood on high-quality (HQ) and low-quality (LQ) plants. HQ plants were grown in nutrient solution, and LQ plants in deonized water. (From Stadler & Mackauer 1996.)

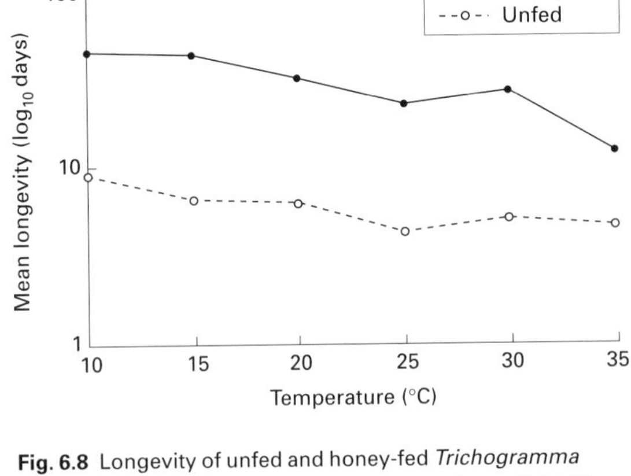

Fig. 6.8 Longevity of unfed and honey-fed *Trichogramma platneri* in relation to temperature. (From McDougall & Mills 1997.)

to adulthood when feeding on low-quality plants (Fig. 6.7) (Stadler & Mackauer 1996), but in compensation, the weight of adults produced on this poor food supply decreased. Thus, the insects seem to have evolved a life history strategy whereby they are able to compromise between lengthened nymphal periods and optimal adult size. Though the number of embryos produced decreases in smaller females (see section 6.1.4), at least the risk of long nymphal periods before reproduction is lessened.

The longevity of adult insects may also be related to food supply, and to other environmental factors such as temperature. In Fig. 6.8 it can be seen that adult *Trichogramma platneri* (Hymenoptera: Trichogrammatidae), a tiny parasitoid of lepidopteran eggs, live for less time as temperature rises (see also Chapter 2), but at all temperatures, those wasps provided with honey lived very much longer than those with no food (McDougall & Mills 1997). In fact, providing them with eggs of their host insect, in this case codling moth, *Cydia pomonella* (Lepidoptera: Tortricidae), made no difference to longevity, indicating the crucial importance of energy sources in the vicinity of pest outbreaks (see also Chapter 10).

6.1.4 Fecundity

The number of offspring produced by an organism is fundamental to its fitness. If too few are produced, then not enough may survive to continue the lineage; too many may waste resources and result in

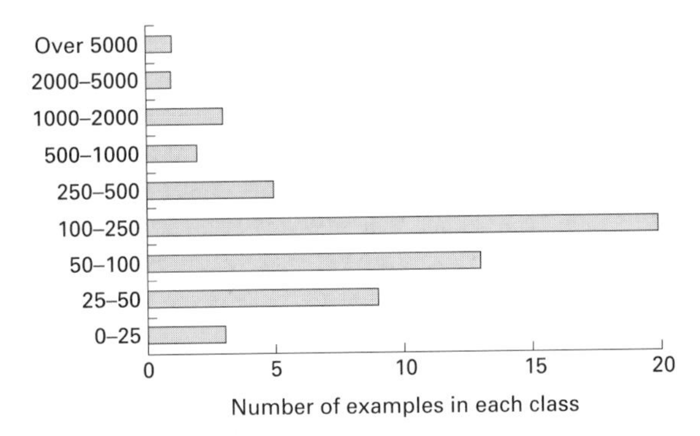

Fig. 6.9 Distribution of fecundities reported in 57 cases, combining all insect groups.

intense competition amongst relatives in the next generation. Moreover, current resources may limit reproduction below optimal levels. Some species may rely on their offspring to disperse, whereas others merely expect them to remain where their mother placed them and to convert food to growth as efficiently as possible. Fecundity is a measure of the number of offspring produced per female, and as might be expected, it varies widely between insect species.

Figure 6.9 shows a summary of 57 reports of insect fecundities from the recent literature. The median of this survey is 142 offspring per female, but it can be seen that some species produce very many more than this, with the maximum reported of over 5200 from a

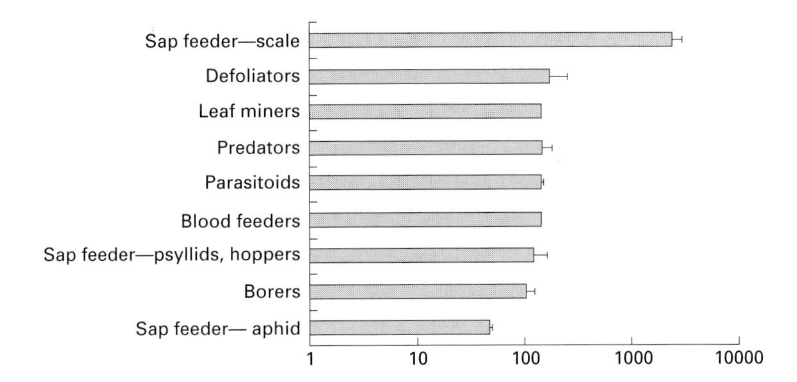

Fig. 6.10 Mean fecundities for various functional insect groups (± SE). (Various sources.)

scale insect (Hemiptera: Coccidae) (Lo 1995). At the other end of the spectrum, a mere 10 eggs per female was recorded for a scavenging dermestid beetle (Coleoptera: Dermestidae) (Iwasaki *et al.* 1994). From a functional point of view, it is useful to see if certain guilds of insects exhibit common fecundities. The same 57 cases are shown in Fig. 6.10, classified according to ecology. It is significant to note that sap-feeders appear at both ends of the fecundity range, with aphids showing very low numbers as opposed to scale insects, which show the highest fecundities on average of any insect group. Also noteworthy is the fact that many predators, parasitoids and herbivores all fill in the mid-range of fecundities.

Fecundities also vary considerably within a species or population, depending on various physiological and ecological parameters. Paramount amongst these is the size of the adult female. Larger females of a given species would be expected to be able to invest more in their offspring than smaller ones, assuming a fixed proportion of the adult is allocated to each. The size of adult insects can vary considerably within a species, related to food quantity and quality available while she was a nymph or larva (see Chapter 3). Fecundity therefore may be considered to correlate with resource availability (Speight 1994). Figure 6.11 illustrates just how variable fecundity can be within a species, according to the size of the female. In this case, the scale insect *Pulvinaria regalis* (Hemiptera: Coccidae) was sampled from urban trees in one small part of Oxford, UK. Maximum fecundity was recorded as nearly 3000 eggs, with a minimum of a mere 20 or so. It should be noted that the best fit to the relationship is not linear, but of a sigmoid nature, indicating a maximum potential dictated by insect physiology (Speight 1994).

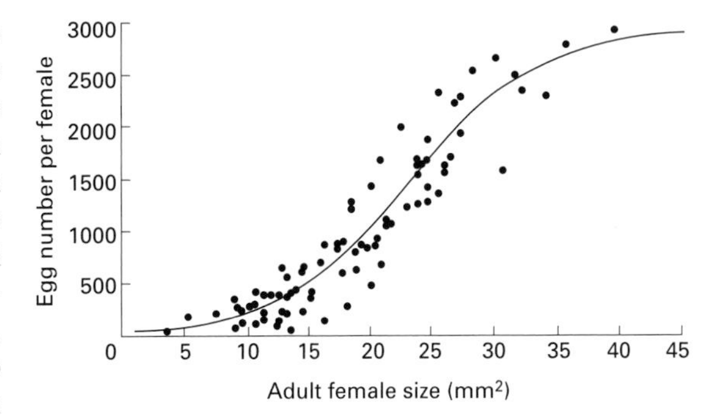

Fig. 6.11 Relationship between adult female size and fecundity in horse-chestnut scale, *Pulvinaria regalis*. (From Speight 1994.)

Fecundity may also vary through the life of one individual female, especially if, unlike the scale insect described above, she lays eggs over a prolonged time period rather than all at once. The weevil *Larinus latus* (Coleoptera: Curculionidae) feeds on thistles (*Onopordum* spp.) in Europe, and because the larvae destroy the seeds before they are able to enter the seed bank in the soil, this species is being considered as a biological control agent (Briese 1996). As can be seen from Fig. 6.12, adult female weevils have a high survival rate for the first 30 days or so of life (Type I curve, lx), and during that time their fecundity, though low overall, peaks after a week or so, and declines gradually thereafter. Even old adults are still able to lay a few eggs.

Finally, the investment made by a female insect in her offspring may be a function not merely of the number of offspring, but also of the fitness of each individual. This may be measured by the size of eggs or newborn nymphs, again frequently related to adult size or even to the ecological role to be played by the eggs after being laid. In the latter case, some eggs laid

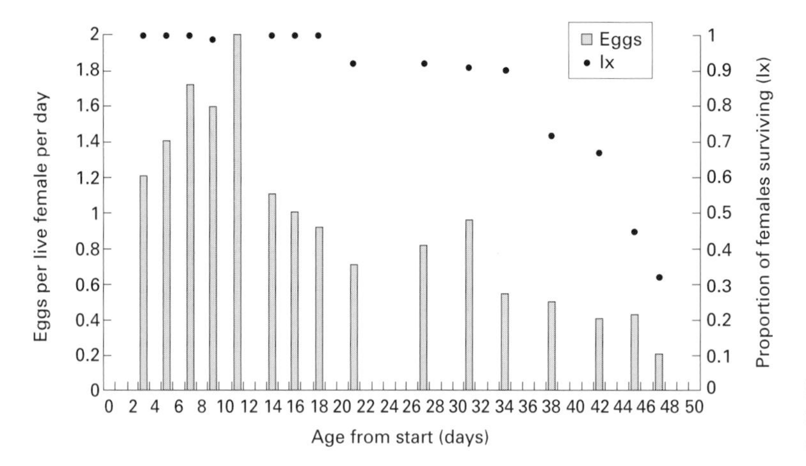

Fig. 6.12 Age-specific fecundity and mortality of the weevil *Larinus latus*. (From Briese 1996.)

later in a temperate season may have to overwinter in diapause (see Chapter 2), rather than hatch immediately, there being too little time left for the resulting juveniles to complete their development before the onset of winter. The cricket *Allonemobius fasciatus* (Orthoptera: Gryllidae) exhibits what has been termed 'bet hedging', by producing more diapause eggs and fewer eggs that hatch immediately, as the season progresses (Bradford & Roff 1993). In Fig. 6.13 it can be seen that female crickets that experienced constantly changing environmental conditions showed a virtually complete swap from direct-development eggs to diapause ones within a time scale of around 30–40 days, whereas those kept under constant conditions did not exhibit such a marked transition in reproductive strategy (Briese 1996). This life history plasticity, as it is termed (Nylin & Gotthard 1998), enables insects to perform trade-offs on such factors as optimal body size, where the fitness advantages of large size, especially high fecundity, are compared with the disadvantage of a long development time (and hence higher risks of mortality).

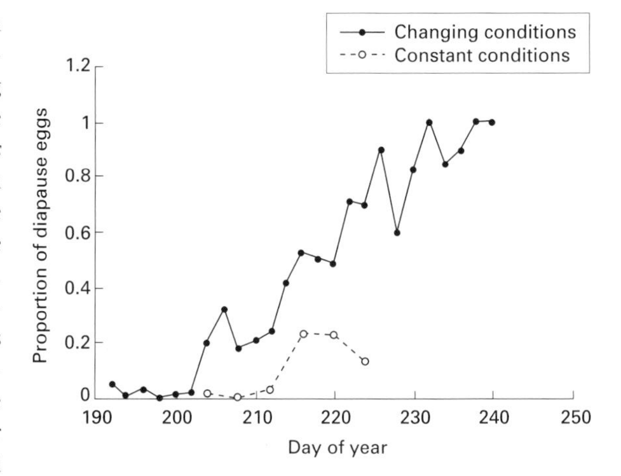

Fig. 6.13 Average proportion of diapause eggs produced by female crickets under two environmental conditions. (From Bradford & Roff 1993.)

Undoubtedly, if an insect lays more eggs in her lifetime, the likelihood is that she will be able to invest less in each one so that the chances of survival of the subsequent juveniles may be reduced. This phenomenon is well illustrated by the Hessian fly, *Mayetiola destructor* (Diptera: Cecidomyiidae). This insect is a serious pest of cereals such as wheat in various parts of the temperate world, where the larvae tunnel into the growing stems of the crop and cause it to fall over. Work in Australia by Withers *et al.* (1997) has shown that as the clutch size of female adults, measured as the number of eggs laid per plant, increases, then the survival of the arising offspring declines markedly (Fig. 6.14). Clearly, a smaller clutch size would enhance the survival probability of the offspring, and Withers *et al.* predicted that when host plants are abundant and the likelihood of the survival of the population is high, then Hessian flies will lay fewer eggs, but invest more in each. Only when host plants are scarce are larger clutches produced, which are distributed more thinly, giving at least some individuals a chance of survival in the next generation.

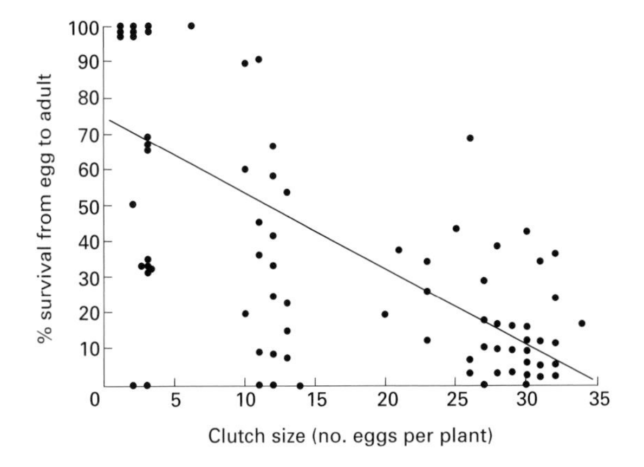

Fig. 6.14 Percentage survival of Hessian flies from egg to adult emergence as influenced by clutch size per plant ($r^2 = 0.50$). (From Withers *et al.* 1997.)

6.2 Sexual strategies—optimizing reproductive potential

As discussed above, many insects appear to maximize their reproductive potentials, and this is manifested by the evolution of mechanisms that increase their fecundities. Two examples are the evolution of flightlessness and of asexuality. Both mechanisms fulfil other purposes too, which will also be discussed here for convenience.

6.2.1 Flightlessness

Loss of the ability to fly has occurred in nearly all orders of winged insects, many times within most of these, and perhaps hundreds of times in the Hemiptera and Coleoptera (Andersen 1997). Figure 6.15 shows the major orders of insects and the per-

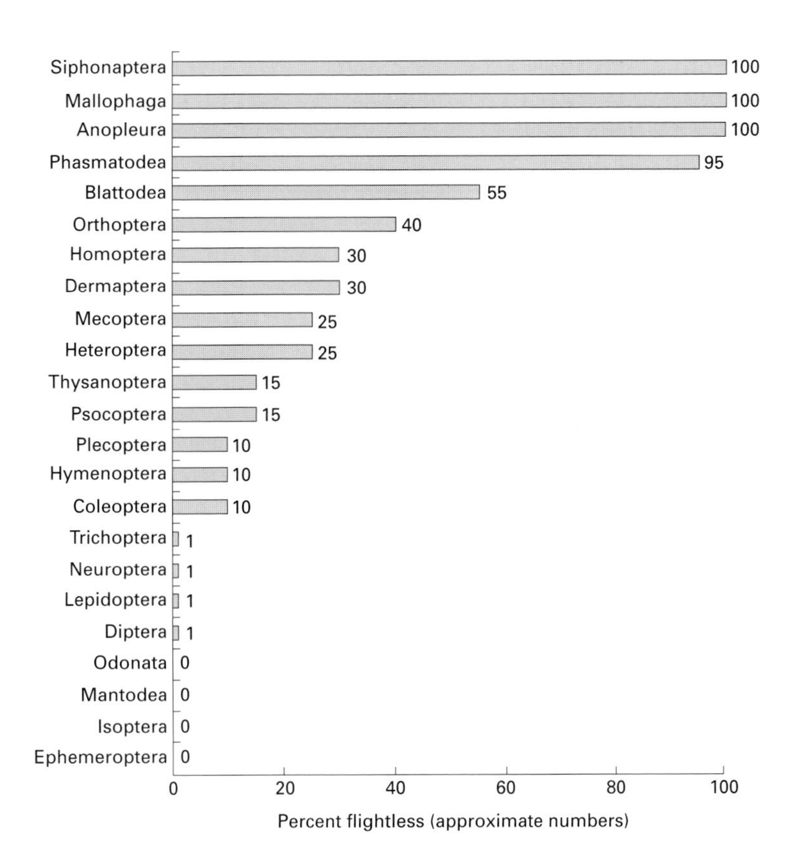

Fig. 6.15 Percentage of species within temperate insect orders that are flightless. (From Wagner & Liebherr 1992.)

centages of each that exhibit flightless adults (Wagner & Liebherr 1992). As all pterygote insects can be considered, by their very name, to possess fully functional wings ancestrally, the appearance of flightlessness must be a secondary event. As Wagner & Liebherr pointed out, the evolution of wings is heralded as the most important event in the diversification of insects, yet as the figure shows, the loss of wings and flight has occurred in almost all the pterygote orders. Loss of the ability to fly seems in part to be related to the stability of the habitat in which insects live (Hunter 1995). It is argued that if dispersal from one habitat to another is not required for the long-term survival of populations, because a single habitat has predictable and long-lasting abiotic and biotic factors to which the species is well adapted, then there is no need to disperse (Wagner & Liebherr 1992). Roff (1990) has reviewed reports in the literature of the occurrence of flightlessness in insect groups from various habitats. These include woodland, desert, the surfaces of the sea and snow, caves, bird and mammal body surfaces, and hymenopteran or termite nests. Woods, deserts and caves, for example, are clearly permanent ecological fixtures, at least until fragmentation of forests, for example, because of human activities occurs on a large scale, whereupon it might be expected that flightless species are the first to be endangered. Habitats with less than average incidence of flightless species are predictably changeable or ephemeral ones, such as the seashore and fresh water (both its margins and the water body itself) (Roff 1990). So, why should insects lose their supposedly most important trait as soon as it is no longer required? Clearly, there must be a cost involved in having full wings and using them to fly; this cost can be related to reproductive potential.

There is undoubtedly a trade-off between flight capability and reproductive success (Denno 1994). It is significant that where adults are flightless, it tends to be the females that exhibit this trait rather than the males (Wagner & Liebherr 1992). One order that contains many flightless species is the Phasmida (stick and leaf insects) (Plate 6.3, opposite p. 158). Many female adults have wings that are very much reduced or, in some cases, completely absent, yet many also have males with full wings, who fly readily. The Lepidoptera are not renowned for a high incidence of flightlessness, but in woodland habitats several families, such as the Lymantriidae and Geometridae, do have flightless species. In the case of the latter family at least, it has been shown that those species with reduced wings show a significantly higher maximum fecundity than fully winged species (Fig. 6.16) (Hunter 1995).

Many species of Hemiptera exhibit both flying and flightless morphs within one species. A relatively simple scenario of this type exists in northern European populations of the water strider, *Gerris thoracicus* (Hemiptera: Gerridae). Both morphs start off as long-winged forms that are able to fly before the reproductive period, but one form histolyse (breakdown) their flight muscles at the start of reproduction, and hence lose the powers of flight (non-flyers) (Kaitala 1991). The other form maintain their ability to fly throughout their adult life (flyers). In experiments, both morphs were maintained in one of two regimes, one where food (insect prey) was abundant, and the other where it was limited to three *Drosophila* per adult per day. Figure 6.17 shows that no matter what the food regime, the non-flyer morph had a higher fecundity and reproductive rate (measured as number of eggs laid per day) compared with the flyers, though both variables were also significantly affected by the quantity of food available to female insects (Kaitala 1991). The ability to change reproductive strategy via alterations in ecology or morphology (known as phenotypic plasticity) according to environmental conditions enables an animal to perform 'trade-offs' between such things as longevity, fecundity and reproductive rate. In the example of water striders, flightless females are able to be more reproductive at

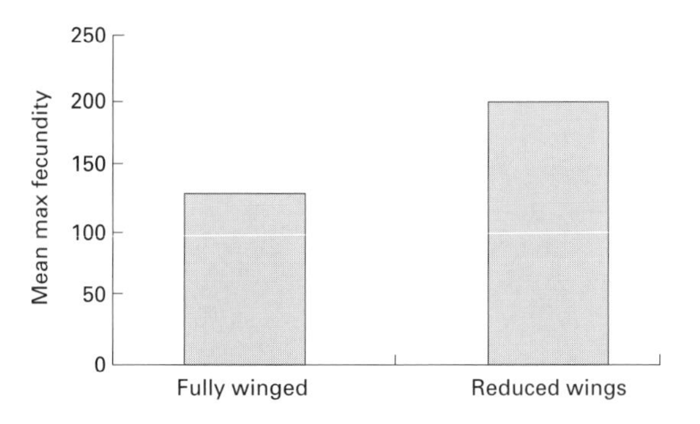

Fig. 6.16 Simple comparison of mean maximum reported fecundity of woodland geometrid moths according to their winged status. (From Hunter 1995.) (Differences between means significant at $P=0.05$ level.)

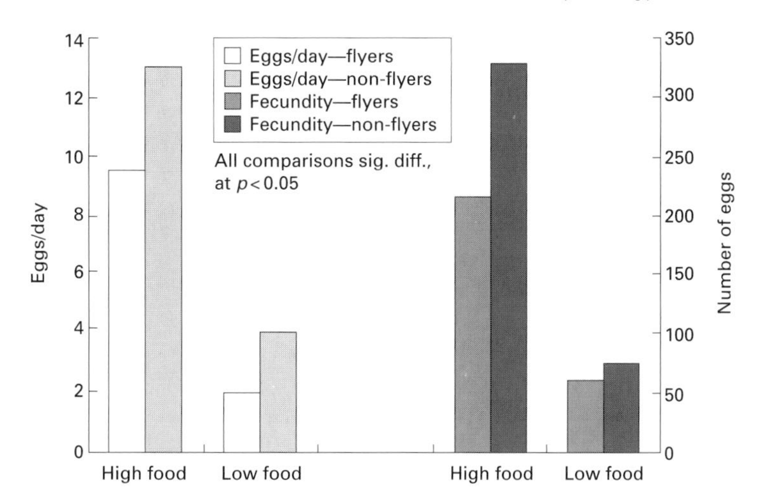

Fig. 6.17 Reproductive rates and fecundities of the water strider, *Gerris thoracicus*. (From Kaitala 1991.)

low food levels, but clearly are unable to move between distant and unpredictable habitats.

Other Hemiptera show more complex phenotypic plasticities, via changes in the occurrence of winged (alate) and wingless (apterous) morphs in response to variations in food supply and/or seasons. Aphids (Hemiptera: Aphididae) such as the grain aphid, *Sitobion avenae*, produce apterous offspring when food is both abundant and high in organic nitrogen, as occurs during the milk-ripe stage of cereal growth, when the grain is forming in early summer. Embryos can be found developing in the bodies of relatively young female nymphs, and their volume increases exponentially as their mother matures. Third instar nymphs are to be found around the fourth day of life, and they become adult on the seventh day in apterous morphs, whereas alatae do not become adult until the eighth day (Newton & Dixon 1990). Not only is there a trade-off between being able to fly and development time, but, as can be seen from Fig. 6.18, the embryos of apterous females grow faster. Also, the total fecundity of apterae is greater than that of alatae. Hence, winged morphs initially produce smaller and fewer offspring than non-winged ones, though alate females are able to catch up with their apterous sisters after 4 days of reproduction (Newton & Dixon 1990). It is clear that if the habitat is able to provide sufficient quantities of high-quality food, then a better strategic 'bet' for aphids is to remain wingless. Only when this food declines as the season progresses, or intraspecific competition intensifies, is there a need to leave, and hence having wings becomes an advantage (Leather 1989).

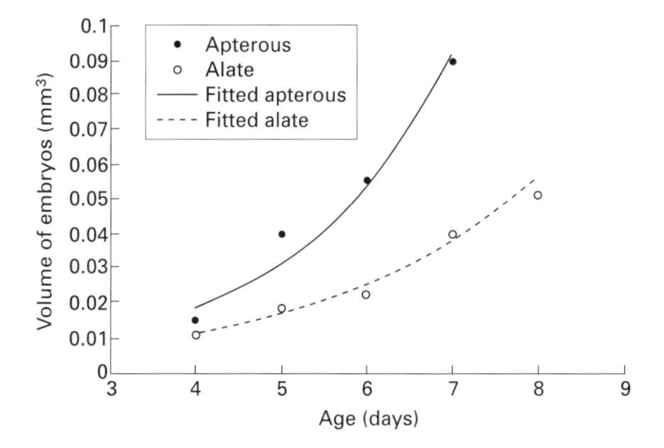

Fig. 6.18 Volume of largest embryos in relation to age of alate and apterous mothers of *Sitobion avenae*. (From Newton & Dixon 1990.)

One final problem is that insects with wings as adults not only show reduced fecundities, but they tend also to exhibit longer development times. In habitats where there is only a very limited summer season in which to complete maturation, such as in high polar situations, there is an increased incidence of flightlessness (Strathdee & Bale 1998). A variety of insects, including aphids and crane flies (Diptera: Tipulidae) show reductions in wings or their complete absence at higher latitudes.

6.2.2 Asexuality

Many insects reproduce, in part or wholly, by asexual means. Males are unknown or non-functional, and offspring are produced parthenogenetically without fertilization. Parasitic wasps, aphids, stick insects and sawflies, for example, all have asexual species, or morphs of species. The evolution of sex in any organism is of course a process by which genetic diversity, hybrid vigour and outbreeding are maintained or promoted. Sexual reproduction is said to enable organisms to adapt to changing environments, to move into new niches, to avoid parasites, and to act as a 'sieve' to remove deleterious genes and mutants (Fain 1995). However, it may be that these normally admirable traits are, on occasion, not desirable or adaptively significant, i.e. in some situations, males may be counter-productive or simply a waste of resources! Let us take a theoretical example where normally a population consists of roughly equal numbers of males and females. Fifty per cent of this population is necessarily non-productive, in that the males consume resources, or merely take up space, without producing offspring. Reducing the number of males, or eradicating them altogether, should in theory dramatically increase the reproductive potential of the population; in fact, double it (Keeling & Rand 1995). All that would be lost would be genetic diversity, itself not always a desirable trait, especially if hard-earned (in evolutionary terms) specialization such as host-specific parasitism or herbivory are to be maintained unchanged in offspring. Hamilton *et al.* (1990) suggested that the main reason for the evolution of sexual reproduction is linked to resistance to parasites and pathogens, via a continuing adaptability in the host, but this supposed advantage cannot operate in all cases, otherwise asexuality in higher animals, including insects, would not be so widespread.

A large number of species, again in the Hemiptera, reproduce predominantly, if not exclusively, by asexual means. Aphids have a widespread tendency to lose the sexual phase of reproduction (Blackman & Eastop 1984), and they frequently exhibit cyclical parthenogenesis where their life history strategy has evolved to separate sexual reproduction from the maximization of biomass and offspring production. Both activities involve specializations, ecologically and morphologically, to gain efficiency. This also allows host plant alternation, a phenomenon known as heteroecy. An excellent example of an aphid that undergoes both cyclical parthenogenesis and heteroecy is the black bean aphid, *Aphis fabae* (Hemiptera: Aphididae) (Plate 6.4, opposite p. 158), whose life cycle is illustrated in Fig. 6.19 (Blackman & Eastop 1984). Some of the terms relating to different aphid morphs need explanation: 'fundatrix' is the first parthenogenetic generation; 'virginoparae' are alate or apterous females who produce live female offspring parthenogenetically; 'gynoparae' are alate parthenogenetic females who migrate to the primary host and produce egg-laying sexual female offspring. It should be noted that all offspring are produced viviparously, allowing the shortest possible generation times.

Thus it can be seen that aphids such as this are able to maximize their exploitation of the host plant at the best times of year, when the food quantity and quality is at its height, and weather conditions allow for rapid development rates. No wastage of resources by non-productive males is allowed. Sexual reproduction prepares the population for winter, providing the scope for genetic mixing so that next year's generations will, if required, be able to cope with environmental changes. On another point, eggs tend to be most resistant to winter cold, and are thus again an insurance policy for the aphids.

Some aphids take this to even further extremes; for example, the green spruce aphid, *Elatobium abietinum* (Hemiptera: Aphididae). The British population never has males (anholocyclic life cycle), whereas that in Continental Europe uses sexual reproduction in autumn (holocyclic life cycle), to produce the overwintering adult stage. The onset of autumn and then winter in the UK does not appear to be a sufficiently strong stimulus to trigger the production of males. There is also an advantage in overwintering in the more cold-susceptible adult stage (see Chapter 2), in that reproduction can ensue at the earliest possible opportunity in the following spring. Judging by the seriousness of *Elatobium* as a pest of spruces in the UK, the loss of sexual reproduction has not hampered their ability to fully exploit their host.

6.3 Life history variations with region

As illustrated with *Elatobium* above, within a species' range, the dominant life history strategy may also vary according to the geographical location of a particular population, based predominantly on the local environmental conditions. For example, worker castes in the ant *Trachymyrmex septentrionalis* (Hymenoptera: Formicidae) differ in their mean sizes (Beshers & Traniello 1994). Colonies in Long Island, New York State, have larger individuals on average

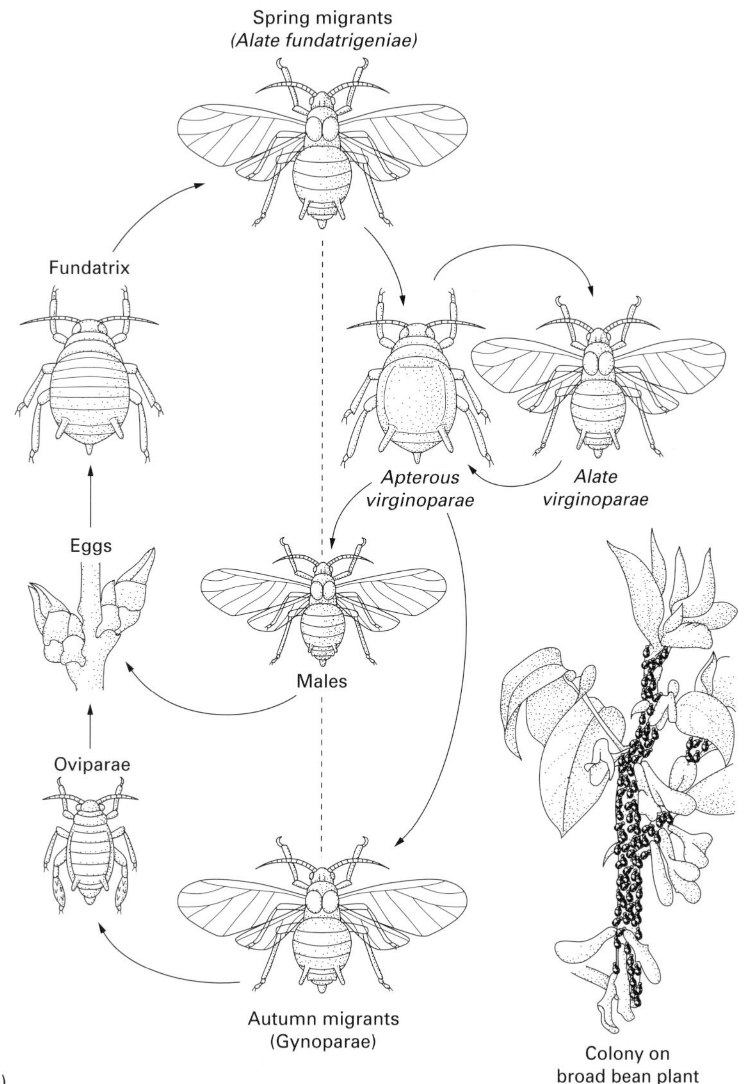

Fig. 6.19 Life cycle of the black bean aphid, *Aphis fabae*. (From Blackman & Eastop 1984.)

than those in Florida. It is suggested that the northerly colonies have evolved a life history strategy adapted to survive temperate winters, whereas those in the subtropical south have experienced different selection pressures and have adapted for rapid colony growth in the absence of climatic constraints. In Western Europe, the grayling butterfly, *Hipparchia semele*, shows significant regional differences in several life history features, including longevity, fecundity, egg-laying rates and egg size (Garcia-Barros 1992). This species feeds on grasses as a larva, and is commonly found in heathland, sand dunes and along the edges of cliffs. In the north of the region, there is a higher concentration of egg production early in the life of the adult female, and smaller eggs are produced, relative to butterfly populations in Mediterranean areas. Adult females also live longer in the latter region. Adverse climatic conditions expected in more northerly climes mean that adult females must reproduce faster and earlier before the shorter warm seasons are over. Of course, these examples describe extremes of what is likely to be a continuum of strategies where a species range is continuous, or clinal. Where a single species is necessarily split into more isolated populations by natural barriers or man-made ones, demes may be generated where population characteristics may be discontinuous (see section 6.5).

6.4 Coevolution

Darwin did not use the word coevolution in *The Origin of Species* (1859), but he did use the word coadaptation to describe reciprocal evolutionary change between interacting organisms (Thompson 1989). For example, in discussing the evolution of interactions between flowers and pollinators, Darwin wrote, 'Thus I can understand how a flower and a bee might slowly become, either simultaneously or one after the other, modified and adapted in the most perfect manner to each other, by continual preservation of individuals presenting mutual and slightly favourable deviations of structure' (p. 95, *The Origin of Species*). As we have come to expect, Darwin provided us with a fundamental description of, and a mechanism for, interactions between organisms that result in reciprocal changes in traits (morphology, behaviour, physiology) over evolutionary time, and his observations have generated the field of study that we now call coevolution.

It is important that we have a clear understanding of the concept of coevolution before we apply it to the ecology of insects. As Janzen (1980) has pointed out, misuse of the concept of coevolution is common, and he offers this definition: 'an evolutionary change in a trait of the individuals in one population in response to a trait of the individuals of a second population, *followed by an evolutionary response by the second population to the change in the first*' (emphasis ours). It is this reciprocal change (A affects B which affects A again) that distinguishes coevolution from simple adaptation by organisms to the abiotic and biotic environment. For example, an insect herbivore that has the ability to detoxify certain secondary metabolites in the tissues of its host plant is not necessarily 'coevolved' with that plant. The secondary metabolites might be present for a variety of reasons (Chapter 3), or the herbivore may have had its detoxification mechanisms in place before encountering the host plant in question. Janzen (1980) also offered a definition for 'diffuse coevolution' in which the populations of A or B or both (above) are actually represented by an array of populations that generate a selective pressure as a group. Potential examples include some mimicry complexes among multiple species of butterfly or some insect pollinator interactions with flowering plants: the floral adaptations of most flowering plants probably represent the result of selection from a number of different pollinator species. In both kinds of coevolution (diffuse and 'pairwise'), the key requirement is repeated bouts of reciprocal genetic change specifically because of the interaction. As reciprocal bouts of genetic change are not easily observed in natural populations, the study of coevolution requires careful detective work combining the disciplines of ecology, systematics, and genetics (e.g. Hougen-Eitzman & Rausher 1994; Iwao & Rausher 1997). More complex subdivisions of the coevolutionary process, particularly subdivisions of diffuse coevolution, are possible (Thompson 1989, 1997) but are beyond the scope of this text. Interested readers are strongly encouraged to read Thompson's (1994) excellent text on the subject.

Reciprocal evolutionary change can occur between populations of organisms that interact with one another in a variety of ways (see for example Fig. 4.2). Competitive, parasitic, and mutualistic interactions can all provide grist for the coevolutionary mill. In the following sections, we describe potential examples of antagonistic coevolution (insect herbivores on their host plants) and mutualistic coevolution (flowering plants and their insect pollinators, and ant–plant mutualisms).

6.4.1 Antagonistic coevolution: insect herbivores and plants

In Chapter 3, we described in some detail the myriad of physical and chemical traits exhibited by plants that can influence the preference (host choice) or performance (growth, survival, and reproduction) of insect herbivores. We were careful to point out that simply demonstrating that a plant trait exhibits antiherbivore activity is not sufficient evidence to conclude that the trait is an evolved defence. Many physical and chemical features of plants appear to serve a variety of functions, and tight evolutionary associations between specific plant traits and specific insect herbivores are difficult to demonstrate unequivocally. However, there is an almost devout belief among many insect ecologists that the diversity of chemical structures found in the tissues of plants represents in large part the result of selection by insect herbivores. If this is true, we might expect to find that a subset of these selective events has been reciprocal and, hence, coevolutionary. This is the basis of what has been described as the 'evolutionary arms race' between insect herbivores and their host plants. In its simplest form, the arms race can be characterized as follows: one or more individuals within a plant population develops a new, genetically based defensive trait by random mutation or recombination. Individuals with this trait suffer lower levels of herbivory than their conspecifics. Low levels of herbivory are associated with higher rates of survival or fecundity, and the proportion of individuals carrying the novel defence increases over time by the process of natural selection. Subsequently, within the herbivore population, one or more individuals develop a genetically based ability to breach the novel plant defence. These insects are at an advantage over their conspecifics, and so the ability to breach the plant defence spreads through the insect herbivore population. At some future point, yet another novel plant defence emerges, and the process of 'escalation' begins again.

An important parallel to this coevolutionary process was suggested by Ehrlich and Raven (1964), who argued that coevolution between plants and insects could lead to speciation events and adaptive radiation of taxa. This has subsequently come to be known as 'escape-and-radiation coevolution' (Thompson 1989). Ehrlich and Raven envisaged five steps to this type of coevolution, as follows.

1 Plants produce novel secondary compounds through mutation and recombination.

2 The novel chemical compounds reduce the palatability of these plants to insects, and are therefore favoured by natural selection.

3 Plants with these new compounds undergo evolutionary radiation into a new 'adaptive zone' in which they are free of their former herbivores. Speciation results in a new taxon of plants that share chemical similarity.

4 A novel mutant or recombinant appears in an insect population that permits individuals to overcome the new plant defences.

5 These insects then enter their own new 'adaptive zone' and radiate in numbers of species onto the previously radiated plant species containing the novel compounds, thereby forming a new taxon of herbivores.

Ehrlich and Raven's view of coevolution, then, results in speciation of both plants and insects rather than a tight coevolved interaction between a specific species of plant and a specific insect herbivore.

Ehrlich and Raven (1964) amassed an extraordinary amount of information to support their view of escape-and-radiation coevolution. Using the associations between major taxa of butterflies and their host plants, they argued convincingly that most butterfly taxa are each restricted to using a few related families of plants. Examples of major butterfly–host associations include Papilionidae on Aristolochiaceae, Pierinae on Capparidaceae and Cruciferae, Ithomiinae on Solanaceae, and Danainae on Apocynaceae and Asclepiadaceae (Table 6.2). In addition, when taxonomic associations between butterflies and their hosts are 'broken', it may be because the plants are phytochemically similar. For example, the plant families Rutaceae and Umbelliferae share some attractant essential oils (Dethier 1941), which may be responsible for an apparent shift of some Rutaceae-feeding groups of *Papilio* to Umbelliferae. Ehrlich and Raven's paper has had a profound impact on the fields of evolutionary and population ecology, and is a 'must read' for any serious student of insect ecology.

What evidence is there that coevolution in general is a powerful evolutionary force, and that Ehrlich and Raven's hypotheses in particular are correct? Surprisingly, few studies have emerged in full support of escape-and-radiation coevolution (Smiley 1985; Thompson 1989), perhaps because detailed and unequivocal phylogenies of both phytophagous insects and their host plants have been slow to emerge. One possible example in support of Ehrlich and Raven was presented by Berenbaum (1983), who described the diversification of coumarin-containing plants that followed the increasing modification of the coumarin molecule. Simple coumarins are small molecules (Fig. 6.20a) found in many families throughout the plant kingdom. A modification of this molecule, hydroxycoumarin, occurs in about 30 plant families. Linear furanocoumarins (Fig. 6.20b) are found in eight plant families, but most genera are within the Rutaceae or Umbelliferae. Finally, the angular furanocoumarins (Fig. 6.20c) occur in only 13 plant genera (including two in the Leguminaceae and eleven in the Umbelliferae). One striking fact is that the diversity of plant species containing coumarin-based molecules increases as the complexity of the molecule increases (Fig. 6.21). Genera containing angular furanocoumarins are the most speciose, followed by genera containing linear furanocoumarins,

Table 6.2 Some relationships between butterfly taxa and plant taxa discussed by Ehrlich and Raven (1964). (From Price 1996.)

Butterfly family	Subfamily	Tribe or lower taxon	Location	Plant family or higher taxon
Pieridae	Dismorphiinae			Leguminosae (exclusively)
	Coliadinae			Leguminosae (mostly)
	Pierinae	Pierini	Temperate	Cruciferae
			Tropical	Capparidaceae
		Euchloini	Temperate	Cruciferae
			Tropical	Capparidaceae
Nymphalidae	Ithomiinae			Solanaceae
	Danainae			Apocynaceae
				Asclepiadaceae
	Morphinae	11 Genera		Monocotyledons
		Morpho		Dicotyledons
	Satyrinae			Graminae, Cyperaceae
	Charaxinae			Ranales and others

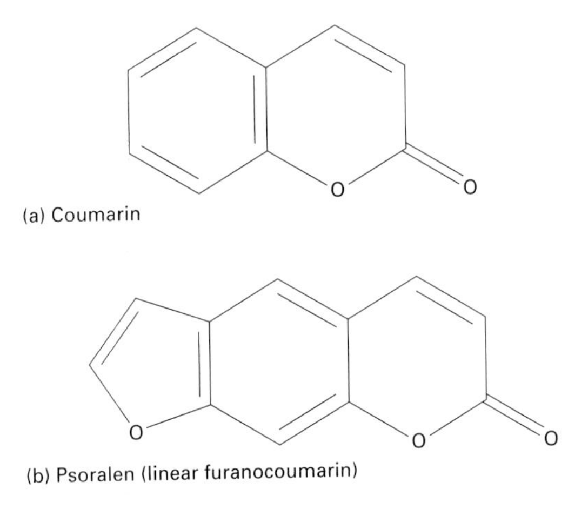

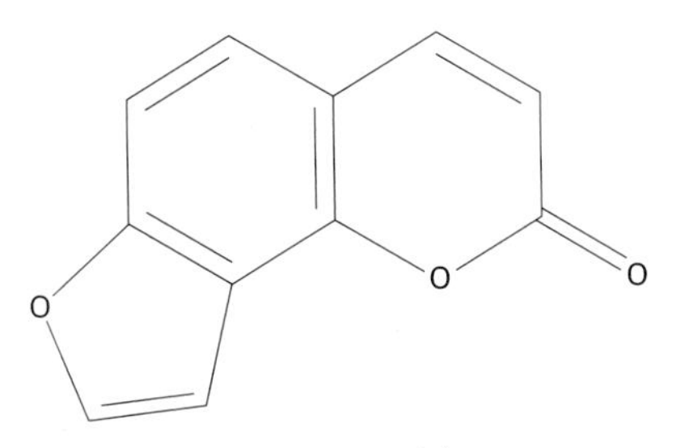

Fig. 6.20 Increasing complexity of (a) the basic coumarin structure found in many plant families, through (b) linear furanocoumarins found in eight plant families, and (c) angular furanocoumarins found in two plant families.

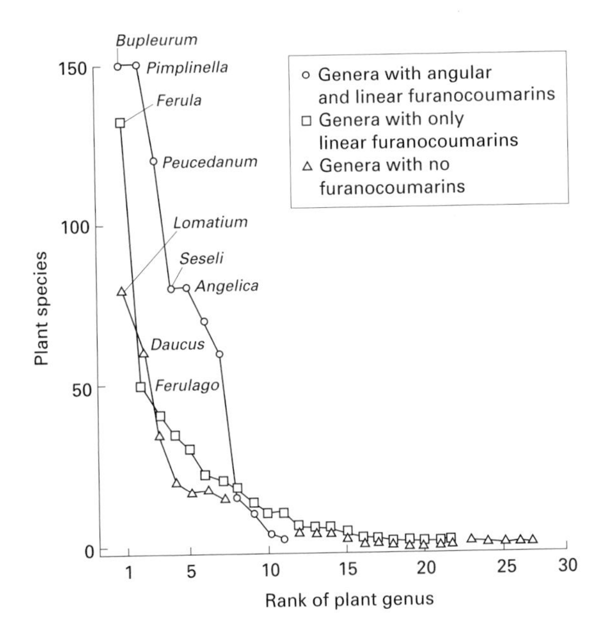

Fig. 6.21 Number of the plant species per genus in genera of Umbelliferae with different chemistries. There are generally more species per genus in those genera whose species contain angular and linear furanocoumarins. (After Berenbaum 1983; from Strong *et al.* 1984.)

and so on. These data suggest that the evolution of phytochemical complexity may indeed shift plants into a new 'adaptive zone' where radiation can occur. Even more striking, the number of species of insect herbivore associated with each chemical type also increases with chemical complexity (Fig. 6.22). The diversity of *Papilio* butterfly species (Plate 6.5, op-

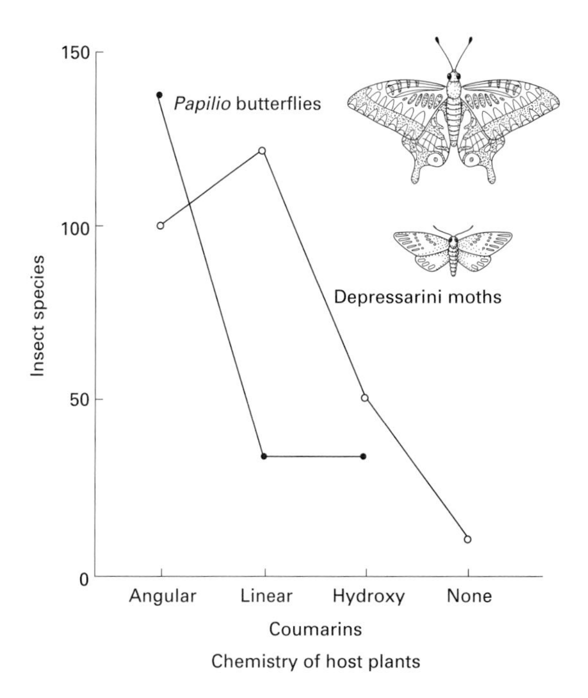

Fig. 6.22 Number of species in two lepidopteran groups associated with Umbelliferae differing in their coumarin-based defences. (After Berenbaum 1983, from Strong *et al.* 1984.)

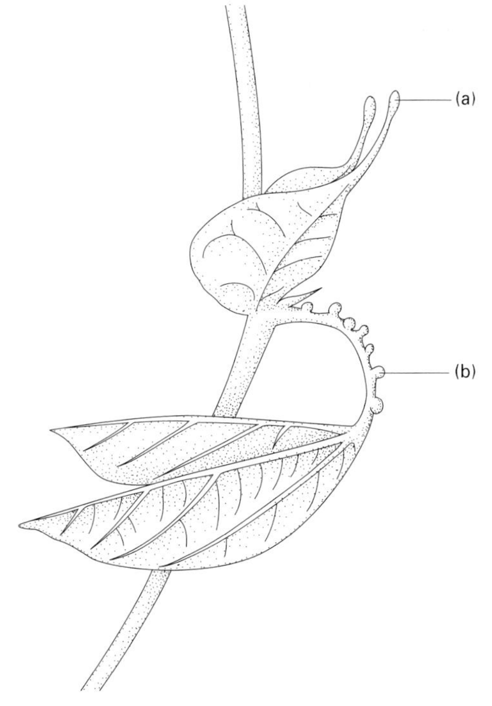

Fig. 6.23 (a) Mimetic Heliconius eggs on *Passiflora*, and (b) extrafloral nectaries. (From a photograph in Gilbert & Raven 1975, reprinted from Strong *et al.* 1984.)

posite p. 158) and moths in the tribe Depressarini increases with the complexity of the coumarin structure. This would suggest that insects that have overcome each of the new coumarin forms have entered their own 'adaptive zone'.

Not everyone agrees that the evolution of furanocoumarin-containing plants, and their associated insect herbivores, provides support for Ehrlich and Raven's hypotheses. Thompson (1986) has argued that differences in the way that species are lumped and split in different phylogenies can change the conclusions of the analyses. Moreover, 'parallel cladogenesis' (in this case, the congruent development of taxa of insect herbivores with taxa of their host plants) is not sufficient evidence to demonstrate that escape-and-radiation coevolution has operated. Secondary plant metabolites can serve a large variety of functions (Chapter 3), and the elaboration of chemical structures, and the radiation that may follow, need not be insect generated. If insects were not the major force leading to plant diversification, then adaptation and radiation by insect herbivores onto plants with novel compounds would be simple evolution, not coevolution. Although recent advances in molecular techniques are likely to provide us with much more accurate insect and plant phylogenies (Farrell *et al.* 1992; Funk *et al.* 1995), the precise mechanisms leading to parallel cladogenesis may always be open to question.

Evidence of tight coevolution at the species level between insect herbivores and their host plants is also not as common as we might once have thought. Spectacular examples do exist, such as the evolution of egg-mimic structures on *Passiflora* vines that are thought to deter *Heliconius* butterflies from laying eggs (Gilbert & Raven 1975; Fig. 6.23). There seems little doubt that the egg mimics are a defence directed at the butterflies and that the energy-efficient flight and inspection responses of the butterflies have resulted from selective pressures imposed by the vine (Gilbert 1975). With such enthralling natural history stories, it is not surprising that many insect ecologists embraced coevolution as a dominant force in insect–plant interactions, and were less careful in applying the concept to their own data than they should have been (Janzen

1980). In reality, we might expect tight coevolution between insect herbivores and their host plants to be relatively rare for several reasons. Perhaps most importantly, many different kinds of organisms can attack plants, including bacteria, fungi, viruses, nematodes, molluscs, mammals, birds, and reptiles. We should not expect perfect adaptation nor counter-adaptation between insects and plants when such a diverse group of consumers have the potential to injure many plant species. Second, different insect species on the same host plant can exert selection pressures in different directions. For example, tannins appear to deter leaf-chewing insect herbivores on red oak (*Quercus rubra*) in Pennsylvania while, at the same time, attracting gall-forming and leaf-mining insects (Hunter 1997). Clearly, the advantages of producing tannins will vary depending on the relative densities of leaf-chewing and endophagous herbivores. Third, it is not clear that most insects exert much of a selective force on most plants most of the time (Strong *et al.* 1984). Rather, strong selection imposed on plants by insects may be relatively rare and episodic in natural systems, and the links between episodic selection and plant diversification are not well established. Finally, related to the second and third points just mentioned, the 'insect–plant interface' is highly variable in space and time. Different plant (and insect) populations separated in space are under different selective pressures and unlikely to be on exactly the same 'coevolutionary trajectory' at the same time (Thompson 1997). Similarly, a single population of insects and plants may interact to varying degrees at different times depending upon variation in the biotic and abiotic environment.

What is clear is that attitudes have changed concerning the 'evolutionary arms race' between insect herbivores and their hosts. In 1964, Ehrlich and Raven suggested: 'The plant–herbivore interface may be the major zone of interaction responsible for generating terrestrial organic diversity.' In 1984, 20 years later, Strong *et al.* concluded: 'Coevolution most certainly does not provide a general mechanism to explain the contemporary structure of phytophagous insect communities.' An additional 14 years further on still, we are perhaps cautiously exploring the variety of conditions under which coevolution is likely to occur, the different mechanisms that can be involved, and with a clearer view of the language that we should use in the process (Thompson 1994, 1997).

6.4.2 Mutualistic coevolution

There is little doubt that mutualistic associations are common in insects. Table 6.3 provides an estimate of the number of species of insects in the British Isles that are involved with mutualistic interactions. Overall, about 36% of the British insect fauna have at least one mutualist, and many have more than one (Price 1996) (see also Table 6.4 later in this chapter). Although any estimation technique is open to criticism, it seems clear that mutualism is a pervasive feature of the ecology and evolution of insects. The success of some insect groups is almost certainly dependent upon their associations with other species. For example, flagellate protozoan or bacterial symbionts in the guts of termites (Isoptera) have permitted this order of insects to exploit a high-cellulose diet that is unsuitable for most other insect species (Bignell *et al.* 1997). The association is sufficiently well developed that many termites have modified guts with anaerobic microsites to enhance cellulase activity. Presumably, this association has facilitated the radiation of termites into the wide variety of environments in which they are found today (see section 6.4.5 for more details). The role of termites in terrestrial ecosystems is considered in detail in Chapter 7.

The 'intraorganismal' mutualisms of termites and other insects, where the associated species actually live within the body of the insect, almost inevitably require adaptive adjustments by both species in the association, and are therefore amongst the most reliable examples of coevolution (see also section 6.5). Many groups of Hemiptera (cicadas, plant-hoppers,

Table 6.3 An estimate of the numbers of insects involved with mutualistic micro-organisms in the British Isles.

Order	Species	Number with mutualists
Orthoptera	39	8
Phthiraptera	308	308
Thysanoptera	183	183
Heteroptera	411	288
Homoptera	976	891
Lepidoptera	2233	1116
Coleoptera	2844	709
Hymenoptera	6224	2874
Diptera	3190	811
Siphonaptera	47	47

Based on data from Buchner (1965) and Price (1996).

aphids, etc.) have specialized organs called mycetocytes in which mutualistic micro-organisms live. The presence of these organs in even very primitive Hemiptera has led some researchers to suggest that the whole taxon was able to radiate onto plants and suck plant juices because their symbionts provided essential nutrients absent from the plants (Buchner 1965; Price 1996). If true, this would be one of the most vivid examples of adaptive radiation driven by coevolution. Aphid–symbiont interactions are considered in more detail later in this chapter. Nutrition mutualists are also found in phytophagous Lepidoptera and parasitic fleas (Siphonaptera) and lice (Phthiraptera).

One fascinating 'endosymbiotic' relationship that has gained increasing attention in recent years is the association between some parasitic wasps (Hymenoptera: Ichneumonidae and Braconidae) and polydnaviruses (Edson *et al.* 1980). Ichneumonids are endoparasites of other insects (Chapter 4) and, as such, must deal with the immune systems of their hosts. The ichneumonids may have adopted viral mutualists that suppress the cellular immune response of their host insect. The polydnaviruses (or virus-like particles, VPs) are held within the lumen of the oviducts of female wasps. VPs enter the host insect, often a lepidopteran larva, when the female wasp oviposits within the body of the host. The VPs have been detected in host haemocyctes (Han *et al.* 1996), where they apparently attack the immune system. Because (a) successful parasitism appears much reduced in the absence of VPs, and (b) the VPs replicate primarily in the wasps, the association appears truly mutualistic. To what degree the association has led to coevolution and parallel cladogenesis is as yet unclear. Fleming's (1992) review provides an excellent introduction to the subject.

Mutualisms between insects and other taxa that are not intraorganismal (i.e. one mutualist does not live permanently within the other) may be less tightly coevolved. In fact, while searching for examples of tight coevolution that included insects for this text, we were struck by the repetition of four or five well-known examples (e.g. yucca moths, *Heliconius* butterflies, fig wasps, etc.) from one review to the next. Although we commit the same sin below, we suggest that the lack of diversity of examples of tight coevolution among free-living mutualists historically chosen by researchers reflects a paucity of such relationships in natural systems. Most free-living insects, or their free-living mutualists, are probably not in tight coevolutionary relationships with one another. There are, of course, some exceptions described below. But we believe that the best examples of the coevolution of insects with mutualists are intraorganismal (see above).

'Looser' mutualisms between individuals or populations of insects and their symbionts, resulting from diffuse coevolution, are perhaps more common than tight species-to-species coevolution. The pollination of most insect-associated angiosperms (flowering plants) probably falls into this category. Many flowering plants use insects such as bees as 'flying genitalia' for reproduction, providing nectar rewards in return for the transfer of gametes. However, the majority of flowering plants are visited by more than one species of insect and most insect pollinators visit more than one species of plant. Diffuse coevolution may be the best way to view the adaptive changes in flowers and insects that benefit both the pollinator and pollinated species. Roubik (1992) has suggested that there is very little tight coupling between bees and flowering trees in the New World tropics, and has found over 200 pollen species within bee colonies at a single site! As Feinsinger (1983) remarked, 'most plants and pollinators move independently over the landscape, not matched in pairs'. Indeed, many pollinating insects appear adapted specifically to account for spatial and temporal variation in flowering plants. Inoue and Kato (1992) provided compelling evidence that variation in the body size of bumblebees, both within a single colony at one point in time and from the start to the end of the pollinating season, allows the bees to visit a wide array of plant species. Their studies of five species of *Bombus* in Japan have demonstrated remarkable plasticity in proboscis length and head capsule width within a species. Although many pollination biologists have reported the averages of morphological features in insect pollinators that match the morphology of flowers (e.g. Fig. 6.24), it is the variation around these averages that provides the best evidence for diffuse coevolution between flowers and pollinators (Figs 6.25 & 6.26).

Of course, there are a few examples of insect pollinator–plant mutualisms that are highly specialized and, presumably, tightly coevolved. Gilbert (1975, 1977, 1991) has suggested that the pollination of *Anguria* and *Gurania* vines by *Heliconius* butterflies represents the result of reciprocal evolutionary change (Plate 6.6, opposite p. 158). Male vines flower

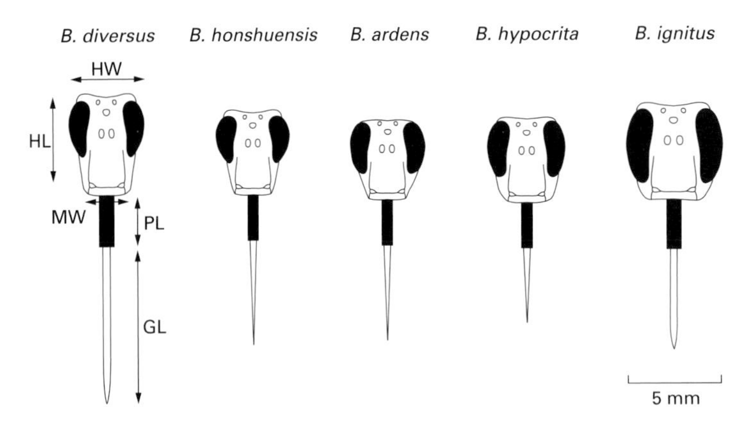

Fig. 6.24 Average sizes of the heads of workers of five bumblebee species in Japan. These averages hide the considerable variation in morphology that occurs within species. (From Inoue & Kato 1992.)

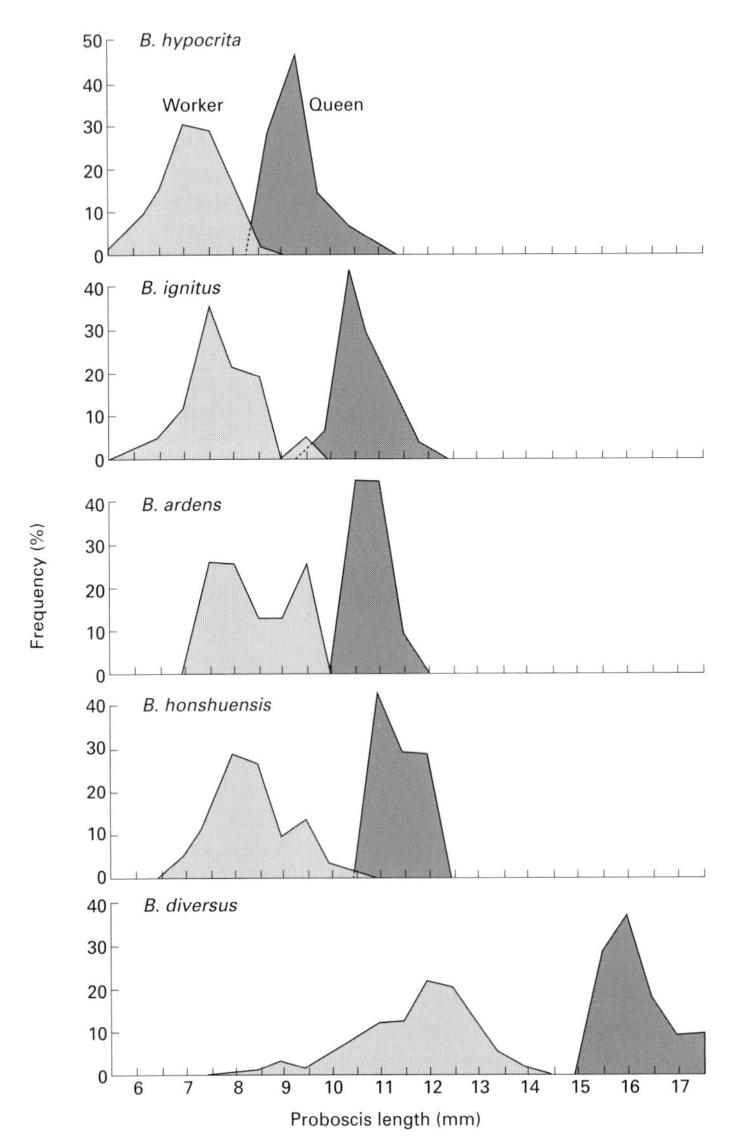

Fig. 6.25 Frequency distributions of proboscis length of five bumblebee species. (From Inoue & Kato 1992.)

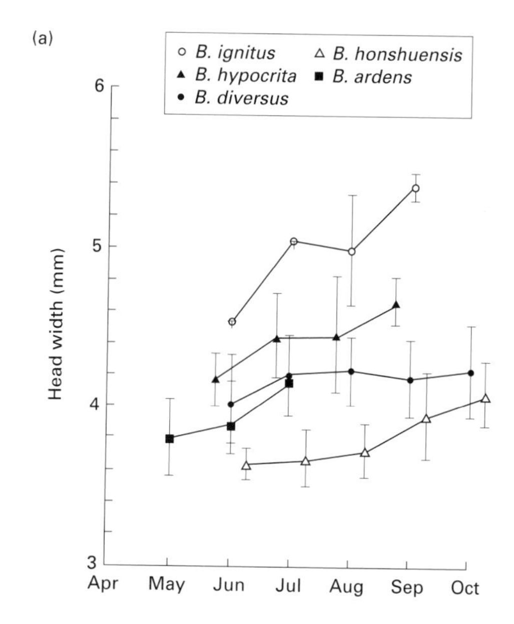

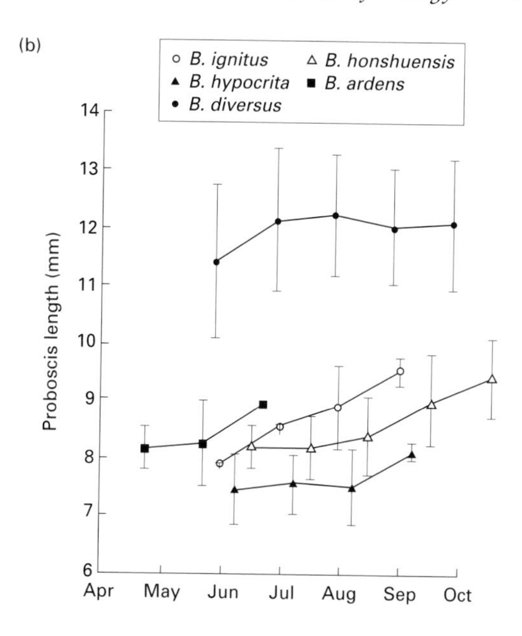

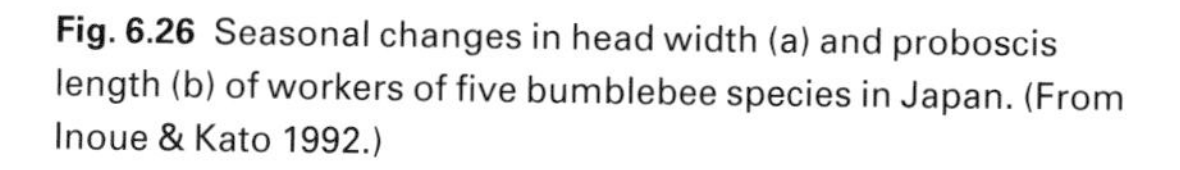
Fig. 6.26 Seasonal changes in head width (a) and proboscis length (b) of workers of five bumblebee species in Japan. (From Inoue & Kato 1992.)

throughout the year, even when females are not in flower, providing a constant source of nectar (and pollen) for the long-lived butterflies. The combination of longevity and the ability of *Heliconius* to learn specific routes through the forest provides an evolutionary incentive for the vines to offer a consistent reward to visitors. In fact, a single male inflorescence of *Anguria* may produce about 100 flowers consecutively. Even though each flower lasts only a day or so, the inflorescence will attract *Heliconius* for several months. Butterflies increase the probability of seed set and are predictable visitors over extended periods while the plants provide nutritious nectar and pollen that presumably improve both the fecundity and longevity of *Heliconius*.

Arguably the most remarkable example of plant–pollinator coevolution is that of the fig wasps (Hymenoptera: Agaonidae) and fig trees (*Ficus*) (Galil & Eisikowitch 1968). With few exceptions, each species of fig in the genus is pollinated by a different species of agaonid wasp, and the near-perfect match between fig species and wasp species is an astonishing example of parallel radiation. Figs produce false fruits (Fig. 6.27) within which male and female flowers are produced. Initially, an opening at the top of the inflo-

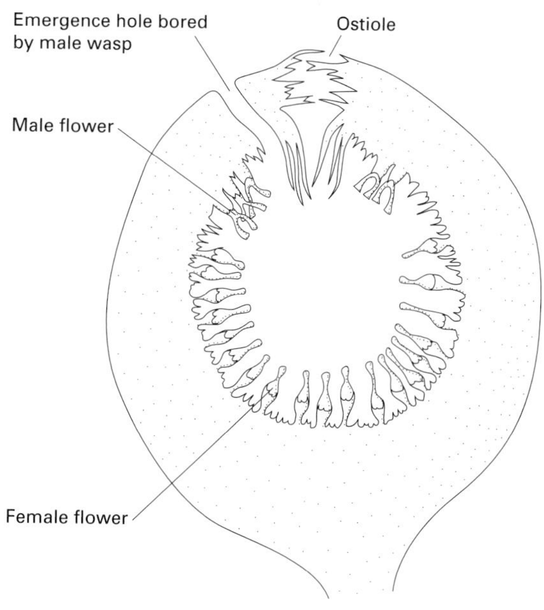

Fig. 6.27 Cross-section of a fig inflorescence. (After Galil & Eisikowitch 1968; Price 1996.)

rescence, called the ostiole, opens in synchrony with female flowering. Female wasps enter through the ostiole and pollinate some female flowers while ovipositing in, and pollinating, others. Flowers that receive pollen only (usually long-styled flowers) develop to produce seeds whereas flowers that receive both wasp eggs and pollen (usually short-styled flowers) become galls with developing wasps inside. Male wasps, wingless with modified, elongated abdomens, emerge from the galls first and mate with females while they are still within the galls. The males then bore exit holes in the side of the inflorescence. When the mated females finally emerge from their galls, it is in synchrony with male flowering. Female wasps collect pollen from the male flowers and store it in specialized pouches on the coxae of their front legs. They then leave the inflorescence through the hole bored by the males, and fly in search of another fig to start the process over again. Almost every aspect of inflorescence morphology and phenology (timing of development) appears adapted to the morphology and phenology of the fig wasps (and vice versa!). The intricacy of the coevolved relationship may arise, in part, because the fig wasps are indeed 'endosymbionts' for at least a portion of their life cycle, living as they do within the false fruits of the fig trees. As we suggested earlier, intraorganismal mutualisms are likely to be more tightly coevolved than those of free-living organisms and there are many variations in the degree of coevolution between pollinators and their hosts. The pollination of yucca plants by yucca moths, an interaction that appears to vary with species and environment (Pellmyr *et al.* 1996; Thompson 1997) is a particular case in point.

6.4.3 Ant–plant associations I: myrmecophytism

In Chapter 3, we described a special kind of 'plant defence' in which certain plants provide incentives (food and shelter) to ants. The ants, in turn, are thought to remove insect herbivores from the plants and so reduce tissue damage. Ant–plant associations have arisen independently in many plant taxa, and are often thought to represent good examples of coevolution. For example, Janzen (1966, 1967) described the close association between species of *Acacia* in Central America and ants in the genus *Pseudomyrmex*. So-called 'Beltian bodies' on the leaf tips of *Acacia* provide protein rewards to foraging ants. In addition, hollow swollen thorns provide shelter ('domatia'), and extrafloral nectaries provide sources of sugar (see Fig. 3.15). Plants that provide such rewards for ants are termed 'myrmecophytic', or 'ant-loving'.

Plants in the genus *Cecropia* in central America also provide domatia and food rewards for ants. In this case, the extrafloral nectaries produce glycogen, a highly branched polysaccharide that is normally found in animal tissues such as our own liver and muscle. Glycogen can be broken down rapidly by animals to release glucose. The production of an animal sugar by a plant is a remarkable adaptation that is hard to explain in any context except the benefits that must be received by attracting foraging ants. It seems obvious that ants benefit from the relationship, but what about the plants? In one study, Vasconcelos and Casimiro (1997) have shown that *Azteca alfari* ants influence the foraging of leaf-cutting ants (*Atta laevigata*) among *Cecropia* species. Leaf-cutting ants cut foliage from plants and use the leaf fragments to grow fungus gardens within their nests (Plates 6.7 & 6.8, opposite p. 158). Although the leaf-cutters are actually fungivorous, they can be viewed as herbivores from the plant's perspective. In the presence of the predaceous *A. alfari*, the leaf-cutting ants are forced to utilize less-preferred species of *Cecropia*.

Similarly, Fonseca (1994) demonstrated a clear benefit to plants as a result of hosting ants. Experimental removal of *Pseuomyrmex concolor* from the myrmecophytic plant species *Tachigali myrmecophila* resulted in a 4.3-fold increase in herbivore densities, and a 10-fold increase in levels of leaf damage. Leaf longevity was two times higher on plants occupied by ants than on unoccupied plants, and apical growth was 1.6-fold higher on occupied plants.

Myrmecophytic plants may gain additional advantages from the ants that they attract beyond the removal of insect herbivores. In several cases, ants have been shown to prune back plants adjacent to the one on which they live. Pruning of neighbours may reduce competition between their host plant and surrounding plants for light and nutrients. However, ants may also benefit by pruning. Studies of *Triplaris americana* plants in Peru and their associated *Pseudomyrmex dendroicus* ants have shown that pruning activity reduces invasions by *Crematogaster* ants by physically separating adjacent plants. *Crematogaster* invasions inhibit *Pseudomyrmex* feeding activity, and the invaders can steal *Pseudomyrmex* broods and usurp their nests (Davidson *et al.* 1988). This example illustrates the general point that we have to be careful when we make assumptions about the benefits of apparent mutualisms. In this case, pruning of neighbours may actually be more important for the ants than for their plant hosts.

Indeed, the fact that *Crematogaster* ants will readily utilize the nests of *Pseudomyrmex* on *T. americana* suggests that the mutualism between the plant and *Crematogaster* may not be especially tight. We do not really know whether the plant benefits more from the presence of one ant species than the other. Previous examples described in this chapter have taught us to be suspicious of inferring coevolution by simply observing current ecological relationships, and ant–plant associations should be no exception. Are ant–plant mutualisms really tightly coevolved? Certainly, there is reason to doubt that ant–plant associations have arisen by escape-and-radiate coevolution (above). For example, the *Leonardoxa africana* (Leguminosae: Caesalpinioideae) complex in Africa is composed of four closely related plant species. DNA sequence data have shown that the interactions of two of the four plant species with the ants *Aphomomyrmex afer* and *Petalomyrmex phylax* appear to have arisen independently and are not the result of cospeciation (Chenuil & McKey 1996). They further suggest that specific ant–plant associations originate by 'ecological fitting of preadapted partners', not escape-and-radiate coevolution. As ecologists and evolutionary biologists become increasingly aware of the value of molecular techniques, we will be able to test rigorously some of our preconceived notions of coevolution.

Even if ant–plant associations are not all tightly coevolved, they provide dramatic examples of mutualism and adaptation, and are remarkably common. In the Pasoh Forest Reserve of Penninsular Malaysia, for example, 91 of 741 woody plants examined had extrafloral nectaries (12.3% of species and 19.3% of vegetation cover). The nectaries occurred in 47 genera and 16 plant families. The interactions between the plants with extrafloral nectaries and ants appeared to be rather facultative and non-specific (Fiala & Linsenmair 1995). A continuum of myrmecophytism, from weakly facultative to obligate, is shown by the palaeotropical tree genus *Macaranga* (Euphorbiaceae) (Fiala & Maschwitz 1991). All *Macaranga* species provide food for ants in various forms (extrafloral nectaries or fat bodies), but the obligate myrmecophytes start producing food rewards at a younger age and may produce more. In addition, only the obligate myrmecophytic *Macaranga* offer nesting spaces (or domatia) inside internodes in the stem of the plant, which become hollow because of the degeneration of the pith. Non-myrmecophytic trees retain solid stems with a compact and wet pith. The stem interior of some 'transitional' species remains solid, but the soft pith can be excavated: provision of nesting sites may be the most important step toward obligate myrmecophytism in this genus of trees.

As might be expected, some insect herbivores have developed a variety of adaptations to counter the effects of ants on plants. For example, although larval densities of *Polyhymno* sp. (Lepidoptera: Gelechiidae) are lower on ant-defended individuals of *Acacia cornigera* in Mexico than on ant-free individuals, larvae gain some protection from the ant *Pseudomyrmex ferruginea* by constructing sealed shelters made from the pinna or pinnules of *Acacia* leaves. As a result of this protection, the larvae of *Polyhymno* can sometimes reach densities that can defoliate and kill their host plants (Eubanks *et al.* 1997). Likewise, the Brazilian savanna shrub *Caryocar brasiliense* has nectaries that attract ants, and oviposition by the butterfly *Eunica bechina* (Lepidoptera: Nymphalidae) is deterred by the presence of real and experimental (rubber) ants. Although larval mortality is strongly affected by ant visitation rates to plants, the butterfly larvae have adopted a remarkable defence to reduce predation pressure. They build stick-like structures from their own frass (faecal pellets) and retire to the end of these frass chains when harrassed by the ants (Fig. 6.28; Freitas & Oliveira 1996).

6.4.4 Ant–plant associations II: myrmecochory

A second form of ant–plant association that can be described as a mutualism is myrmecochory—the dispersal of seeds by foraging ants (see Fig. 6.29). In the same way that plants can exploit pollinators as flying genitalia, so they can also exploit ants as mobile gardeners. By providing rewards for ants on the surface of seeds, some plants can benefit from the movement of their propagules across distances greater than would otherwise be possible.

Nakanishi (1994) distinguished between three kinds of myrmecochory. First, there is myrmecochory with 'autochory'. In this case, the seed pods of plants release the seeds 'explosively' so that seeds are already deposited some distance from the parent plant, and are then dispersed over greater distances by foraging ants. Second, there is myrmecochory

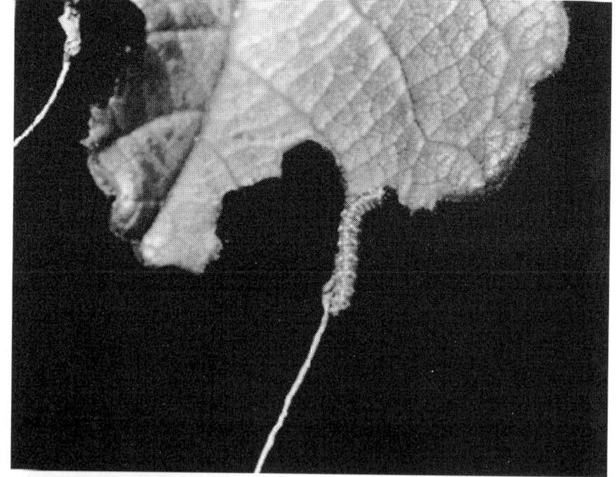

(a)

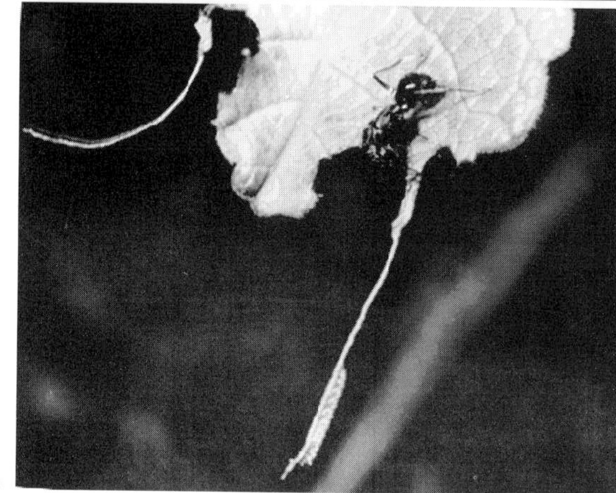

(b)

Fig. 6.28 Larvae of *Eunica bechina* (a) build frass-chains as they feed, and (b) retreat down these chains when harrassed by ants. (Photograph from Freitas & Oliveira, 1996.)

Fig. 6.29 Elaiosome on the seed of *Croton priscus*. (Courtesy of I. Sazima.)

with vegetative reproduction: plants reproduce both vegetatively and sexually, with ants transporting the sexual propagules (seeds). Finally, there is 'pure' myrmecochory, where the only dispersal of seeds away from the parent plant relies on ants. The last of these depends most heavily on ants to disperse seeds away from the parent, and we might expect plants that rely on pure myrmecochory to provide the greatest reward to their ant mutualists.

How do plants reward ants for seed dispersal? Ant-dispersed seeds generally have detachable protrusions on their surfaces called elaiosomes (Fig. 6.29). Elaiosomes are high in lipids and fatty acids, and can also contain proteins. Foraging ants return seeds to their nest, remove the reward, and the seeds are left to germinate, often underground in soil that has been aerated by ant activity. It has been suggested that the chemistry of elaiosomes influences their attractiveness to ants and subsequent ant behaviour. For example, a comparison of the chemistry of the elaiosomes from three species of *Trillium* found significant variation in their protein, neutral lipid, and fatty acid contents (Lanza *et al.* 1992). Results suggested that oleic acid stimulated ants to pick up the seeds and that linoleic acid stimulated ants to carry the seeds to the nest. Variation in these compounds, and in lipid to protein ratios, appeared to explain the relative dispersal success of the three *Trillium* species. As an aside, it has been suggested recently that the capitula (small protrusion) on the eggs of some stick insects mimics the elaiosomes of seeds, and stimulates ants to collect them. Eggs that were buried in ants' nests were found to suffer lower rates of parasitism than unburied eggs (Hughes & Westoby 1992).

Do plants really benefit from elaiosome production and ant dispersal? Horvitz and Schemske (1994) found that seeds with rewards had 1.6-fold higher emergence and were dispersed on average three-fold farther than seeds without rewards. However, other biotic and abiotic factors also influenced seed dispersal and germination so that the value to the plant of elaiosome production is not likely to be equivalent in every environment. The tropical pioneer tree *Croton priscus* (Euphorbiaceae) has explosive seed capsules that are dispersed some distance (about 3 m) ballisti-

Fig. 6.30 Ant worker in the genus Pheidole removing an elaiosome from *Croton priscus*. (Photograph courtesy of I. Sazima, property of the Association of Tropical Biology, Kansas, USA.)

cally. However, they still produce elaiosome-bearing seeds (myrmecochory with autochory; Fig. 6.30). Research by Passos and Ferreira (1996) found that ants remove seeds at a rate of about 88% per day and move them 1–2.5 m farther than for autochory alone. Seedlings of *C. priscus* are often found on the refuse piles left by foraging ants, suggesting that ant dispersal does result in germination (Passos & Ferreira 1996). However, the dispersal of seeds by ants may not always be entirely beneficial. Bond and Stock (1989) showed that *Leucospermum conocarpodendron* seeds dispersed by ants after fire in the Cape fynbos ecosystem of South Africa were generally moved to nutrient-poor sites in comparison with seeds that were passively dispersed from the plant.

If elaiosome production and the dispersal of seeds by ants is the result of coevolution, it is probably diffuse rather than tight coevolution. There do not appear to be many examples of obligate relationships between a single species of plant and a single ant disperser. Rather, ants will disperse seeds from many plant species, including novel species in the environment. For example, Pemberton (1988) described myrmecochory between plants introduced to North America (e.g. leafy spurge, *Euphorbia esula*) and ants native to North America (e.g. *Formica obscuripes*). As there cannot have been a long-term evolutionary relationship between the introduced plant and native ant, this demonstrates the facultative nature of the mutualistic interaction.

6.4.5 Insect–microbe interactions

A large number of insects from many orders have evolved mutually beneficial associations with micro-organisms including fungi, bacteria and protozoans (Table 6.4). This coevolution has arisen many times, and is predominantly an adaptation to promote the utilization of difficult or nutrient-poor food resources, such as plant sap, wood and leaf material. The symbiotic micro-organisms may exist outside the insects' body, as with fungus-gardening ants mentioned above, inside the gut of the insect, as with certain termites, or inside the cells of the insect, as in the case of aphids. The mechanisms of this coevolution and its results will be discussed in detail as they occur in various insect groups.

Coleoptera

Various families of beetle have evolved complex symbioses with micro-organisms. Some such as the chafers and rhinoceros beetles (Coleoptera: Scarabaeidae) possess colonies of methane-producing bacteria in their hind-guts, which break down plant material consumed by the host (Hackstein & Stumm 1994). Weevils (Coleoptera: Circulionidae) have developed specialized organ-like structures called mycetomes, which contain enterobacteria (Campbell *et al.* 1992). In the case of the grain weevil, *Sitophilus oryzae*, a widespread pest of stored cereals, the symbionts occur in the ovarioles of the female (Grenier & Nardon 1994), from where they may be transferred to the offspring via the eggs. Woodworm, museum beetles and so-called cigarette beetles (Coleoptera: Anobiidae) are all able to feed on dead organic material, from wood to skins and even wool. They also possess mycetomes that open into the alimentary tract between the fore- and mid-gut, which in the case of wood boring species contain yeast-like symbionts including *Simbiotaphrina buchneri* and *S. kochii* (Noda & Kodama 1996). These yeasts are thought to aid in larval nutrition (Suomi & Akre 1993), and also to detoxifiy plant-derived toxins such as tannins (Dowd & Shen 1990). Museum beetles, *Anthrenus* spp., are notorious destroyers of preserved animal material from skins, hides, mummified bodies and insect collections (Fig. 6.31). Trivedi *et al.* (1991) examined the ability of *Anthrenus* larvae to digest wool fibre, a particularly difficult food source. The larvae (so-called 'woolly bears') bite off

Table 6.4 Summary of insect associations with micro-organisms.

Order	Sub-group	Micro-organism	Type of association	Location of symbiont
Hemiptera	Aphididae (most aphids)	Bacteria (Entobacteriaceae)	Endo-symbiotic–intracellular	Maternal tissues, embryos
	Aphididae (Cerataphidini)	Yeast-like	Endo-symbiotic–extracellular	Haemocoel
	Delphacidae (planthoppers)	Yeast-like	Endo-symbiotic–intracellular	
Dictyoptera	Blattodea (cockroaches)	Bacteria (Flavobacteria)	Endo-symbiotic–intracellular	Fat body
		Protozoa	Endo-symbiotic–extracellular	Gut
	Cryptocercidae (wood roaches)	Protozoa	Endo-symbiotic–extracellular	Hind-gut
Isoptera	Rhinotermitidae (lower termites)	Protozoa (flagellates)	Endo-symbiotic–extracellular	Hind-gut
	Higher termites	Bacteria and yeasts	Endo-symbiotic–extracellular	Hind-gut
	Mastotermes spp.	Bacteria	Endo-symbiotic–intracellular	Fat body
	Higher termites (macrotermes)	Fungi	Ecto-symbiotic	Within colony
Hymenoptera	Formicidae (ants–leaf-cutters)	Fungi	Ecto-symbiotic	Within colony
	Siricidae (wood wasps)	Fungi	Ecto-symbiotic	Larval tunnels, wood tissue
Coleoptera	Scarabaeidae (chafers, etc.)	Bacteria	Endo-symbiotic–extracellular	Hind-gut
	Curculionidae (weevils)	Bacteria	Endo-symbiotic–intracellular	Ovaries, other tissues
	Anobiidae (museum beetles, woodworm, etc.)	Yeast-like	Endo-symbiotic–intracellular	Fore- and mid-gut
	Platypodidae (ambrosia beetles)	Fungi	Ecto-symbiotic or extracellular	Larval chambers
	Scolytidae (bark beetles)	Fungi	Ecto-symbiotic or extracellular	Larval chambers

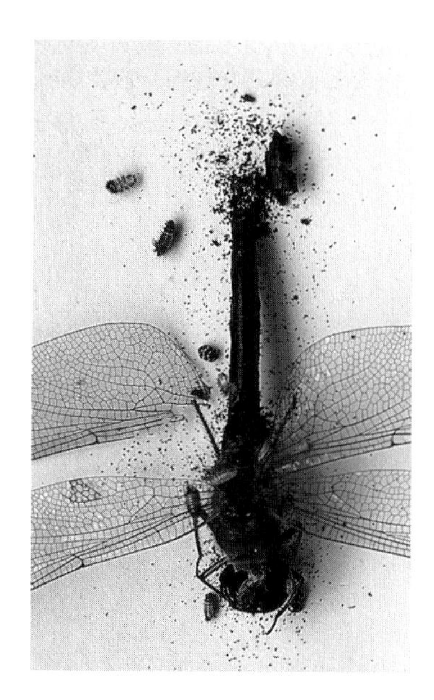

Fig. 6.31 Insect specimen destroyed by *Anthrenus*. Larvae and adults can be seen. (Courtesy of P. Embden.)

wool fragments in 20–100 μm lengths. Digestion of the outer scales of the wool occurs in their sack-like mid-gut. In the ileum, large numbers of bacterial cells help to digest keratin fibrils, resulting in complete structural disintegration of the wool fibre. Actively feeding amoeba-like protozoans gradually replace the bacterial flora in the posterior ileum and rectum, where digested wool components and microbial biomass become compacted because of the absorption of nutrient fluid from the hind-gut. The final compact faecal pellets appear as dust—anyone who has had an infestation of museum beetle in their prized collection of insects or shrunken heads will vouch for the efficiency of this symbiotic interaction!

One of the most widespread beetle–microbe symbioses involves two families of wood and bark borers, the bark beetles (Coleoptera: Scolytidae), and the ambrosia beetles (Coleoptera: Platypodidae). Both families contain large numbers of both tropical and temperate species that are frequently serious pests of trees, both standing and fallen or felled. The beetles themselves can kill trees, especially those that are already in some way debilitated by fire, windblow, drought or defoliation, by extensive larval boring between the bark and sapwood, which effectively ringbarks or girdles the host. Some species cause extra problems by way of the maturation feeding activities of young adults, before mating; Dutch elm disease is a classic example of this, which we discuss fully in Chapter 9. Almost all scolytids and platypodids have evolved intimate symbiotic associations with various types of fungi, which assist the beetles by breaking down wood material, aiding in the rotting process by rendering the host trees susceptible to further beetle attack, or, as in the case in some of the ambrosia beetles, providing a direct source of food for the growing larvae. In the main, the fungal symbionts grow externally to the beetles, using wood substrates for their own nutrition, but they rely on the beetles to carry them from host to host, so much so that some fungal species are unable to survive on their own. This feeding habit in platypodids is known as xylomycetophagy, where the ectosymbiotic 'ambrosia' fungi form the major part of the food of both larval and adult beetles (Beaver 1989). As mentioned above, the beetles themselves are frequently important forest pests, but the fungi can also cause serious wood degradation. Blue-stain fungi in the genus *Ceratocystis*, for example, can render timber unmarketable (Plate 6.9, opposite p. 158).

To vector their fungal symbionts efficiently, many species of beetle have evolved specialized structures called mycangia in which the fungi proliferate (Fig. 6.32) (Cassier *et al.* 1996). Mycangia have openings to the outside of the prothorax of adult beetles, ready to inoculate new host trees when the beetles tunnel into their bark and wood to deposit their eggs. Fungal taxonomy is, as always, complex and seemingly incomplete, at least to an entomologist, but various genera are commonly found in association with these mycangia, including *Graphium* spp. (Cassier *et al.* 1996), *Ceratocystiopsis* spp. (Coppedge *et al.* 1995), *Monacrosporium* spp. (Kumar *et al.* 1995), and *Ophiostoma* spp. (Fox *et al.* 1992). The coevolution of these highly successful beetle–fungus associations is clearly ancient; amber from the Dominican Republic, dated at 25–30 Myr old (see also Chapter 1), contains well-preserved adult platypodids wherein the mycangia are readily identifiable, still containing spores of symbiotic fungi (Grimaldi *et al.* 1994).

The role of these fungal symbionts in the life of the beetles has been investigated in detail. In Fig. 6.33, it can be seen that bark beetles in the genus *Ips*, globally widespread pests of pine and spruce especially, have significantly shorter development times when their fungi (*Ophiostoma* spp.) are present (Fox *et al.* 1992). Not only that, but as the figure shows, the length of the tunnels excavated by bark beetle larvae are significantly shorter when fungi are available, indicating that these symbionts reduce the need for larvae to feed extensively on what can only be described as a suboptimal diet. The number of beetles in a population that carry fungal symbionts, and the concentration of the microbe in each insect, can be variable. It was said that in the case of *Scolytus* spp., the beetle

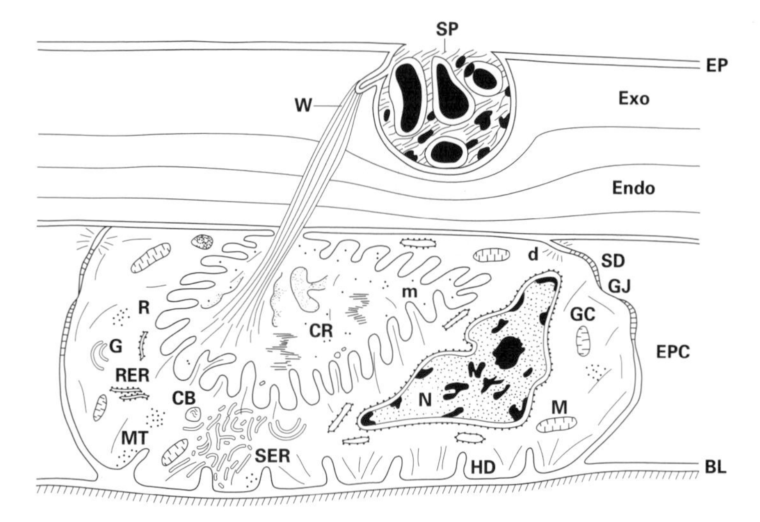

Fig. 6.32 Schematic drawing of a spheroidal cavity of the *Platypus cylindrus* mycangium connected to a glandular cell by a wick of micronobules. BL, basal lamina; CR, central reservoir; d, desmosome; Endo, endocuticle; EP, epicuticle; EPC, epithelial cell; Exo, exocuticle; G, Golgi apparatus; GC, glandular cell; GJ, gap junction; HD, hemidesmosome; LB, lytic body; M, mitochondria; MT, microtubules; m, microvilli; N, nucleolus; R, ribosome; RER, rough endoplasmic reticulum; SD, septate desmosome; SER, smooth endoplasmic reticulum; SP, spores of fungi; W, wick of microtubules. (From Cassier *et al.* 1996.)

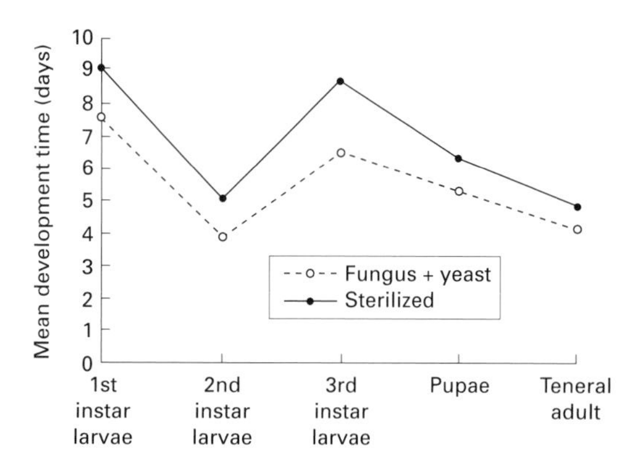

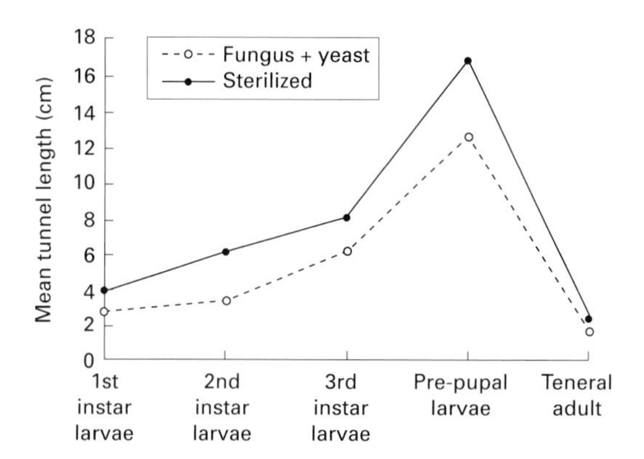

Fig. 6.33 Mean development time and tunnel length of various life stages of the bark beetle *Ips paraconfusus* with and without symbionts (all comparisons within a life stage significantly different at $P = 0.05$ level except adult). (From Fox *et al.* 1992.)

vectors of Dutch elm disease, one in two beetles visiting a new and healthy tree to maturation feed would carry the fungus *Ceratocystis* (C.J. King, personal communication). In another world-renowned genus of bark beetle, *Dendroctonus*, there is a highly significant relationship between the number of female beetles carrying the symbiotic fungus and both the weight of the beetles and their fat content (Fig. 6.34) (Coppedge *et al.* 1995). Both these factors are directly correlated with the survival of the beetles and their reproductive fitness.

Hemiptera

Symbiotic relationships between members of the order Hemiptera and micro-organisms have been studied most intensively in the aphids (Hemiptera: Aphididae). Most aphids possess bacteria in the genus *Bruchnera* (Wilkinson & Douglas 1995). These intracellular symbionts are located in the cytoplasm of structures known as mycetocytes, large cells in the aphids' abdomen (Fukatsu 1994). The association is thus known as 'mycetocyte symbiosis' (Douglas 1998). As well as occurring in the mycetocytes, for the definition to be upheld the micro-organisms have to be maternally inherited, and both the insect and the micro-organisms have to require the association. In fact, many groups of insects exhibit mycetocyte symbioses; most of these insects live on nutritionally poor or unbalanced diets such as vertebrate blood, wood or phloem sap (Douglas 1998). For aphids, this last diet is normally deficient in essential amino acids, which the bacteria are able to synthesize (Liadouze *et al.* 1995), and studies of aphid fossils using rRNA molecular clocks suggest that this association may have been established as much as 160–280 Myr ago. *Bruchnera*, which may represent as much as 2–5% of the total aphid biomass (Whitehead & Douglas 1993), is thought to be related to the well-known *Escherichia coli*, indicating a free-living and monophyletic origin of the symbionts (Harada & Ishikawa 1993).

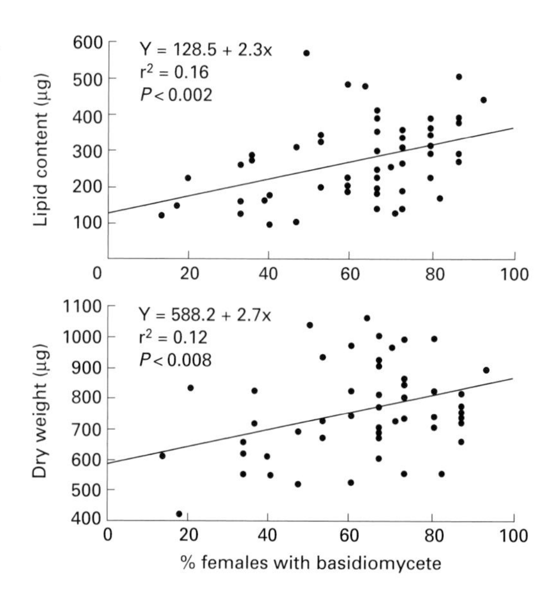

Fig. 6.34 Regressions of adult dry weight and lipid content of the bark beetle *Dendroctonus frontalis* against percentage of females carrying a basidiomycete fungal symbiont. (From Coppedge *et al.* 1995.)

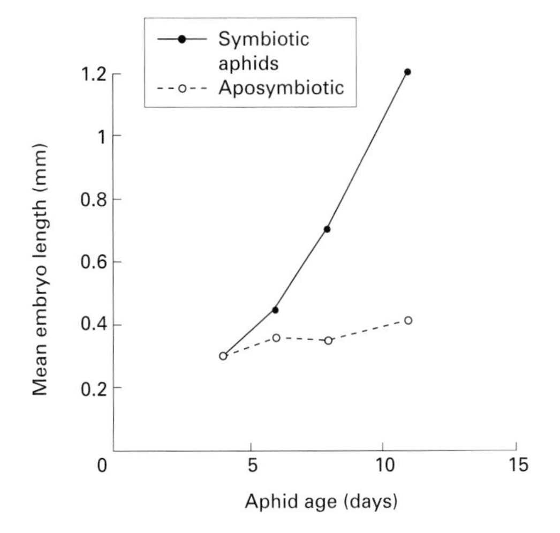

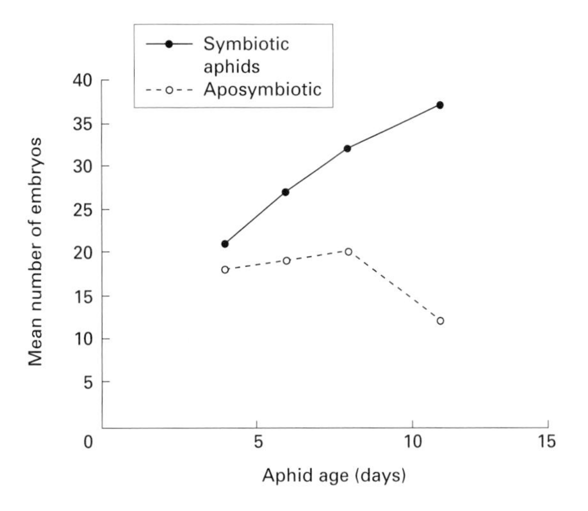

Fig. 6.35 Size and number of embryos in nymphs of the pea aphid, *Acyrthosiphon pisum*, with and without their endosymbiotic bacteria. (From Douglas 1996.)

Much research has been carried out to elucidate the advantages conferred on aphids by their bacterial symbionts, using the relatively simple system of treating the insects with antibiotics that kill the bacteria. Aphids with their symbionts removed in this way are termed aposymbiotic. Figures 6.35 and 6.36 show two examples of this type of study. Both examples used the pea aphid, *Acyrthosiphon pisum*, and showed that aposymbiotic aphids grow very slowly relative to normal ones, and that both the number of embryos and the size of each in developing females are also much reduced (Douglas 1996). In fact, it seems that embryos are more dependent on the symbiotic bacteria than are maternal tissues, probably because of limitations caused by the absence of certain amino acids such as tryptophan and phenylalanine. Further evidence of the nature of amino acid limitation in this context comes from the study by Liadouze *et al.* (1995), where it was found that the fresh body weight of aphids reared on bean plants was much reduced in aposymbiotic aphids when compared with symbiotic ones. Furthermore, aposymbiotic aphids have a significantly higher but unbalanced concentration of free amino acids such as asparagine, aspartic acid and glutamine. Aphids missing their bacterial symbionts are unable to maintain required and balanced levels of essential amino acids, with resultant losses in fitness.

Isoptera

The ability to digest cellulose is rare in most higher animals, including insects. This cellulytic capacity may be uncommon simply because it is rarely an advantage (Martin 1991), in that cellulose digestion is usually mediated by micro-organisms. Even termites (Isoptera) (Fig. 6.37), wherein some independent cellulose digestion may occur, rely in the main on a battery of microbial symbionts to break down the celluloses and hemicelluloses in their food, mainly dead wood and other plant tissues. Some termites have a complex and varied gut microflora, consisting of various bacteria and yeasts including *Bacillus*, *Streptomyces*, *Pseudomonas* and *Acinetobacter* (Schaefer *et al.* 1996). In others, such as *Coptotermes* spp., protozoans including *Pseudotrichonympha* are essential for wood degradation in the gut. Termites from which these protozoans had been eliminated showed a 30% reduction in their wood-attacking activities, though they were readily able to pick up new symbionts from fresh worker termites in the neighbourhood (Yoshimura *et al.* 1993). As all endosymbiotic microbes in termites inhabit the gut of their hosts, the insects usually lose their symbionts when they empty their guts before a moult (ecdysis) (Lelis 1992), and hence have to pick them up again in the next instar from surrounding termites. This is thought to be one of the reasons for the evolution of sociality in termites, which ensures that there is always someone close by from whom to reacquire the essential symbionts. It may thus be possible to trace the ancestry of termites back to primitive

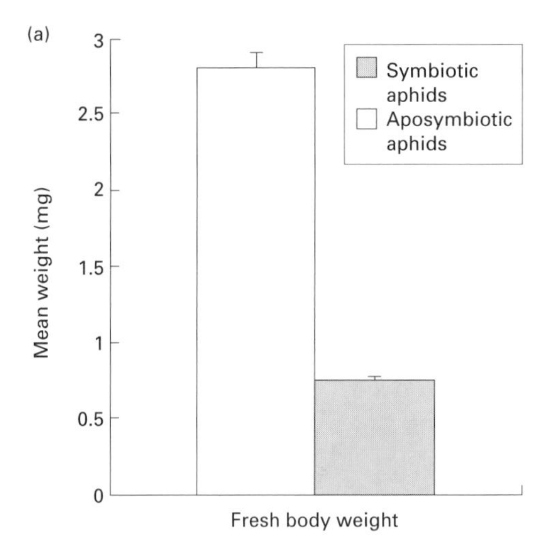

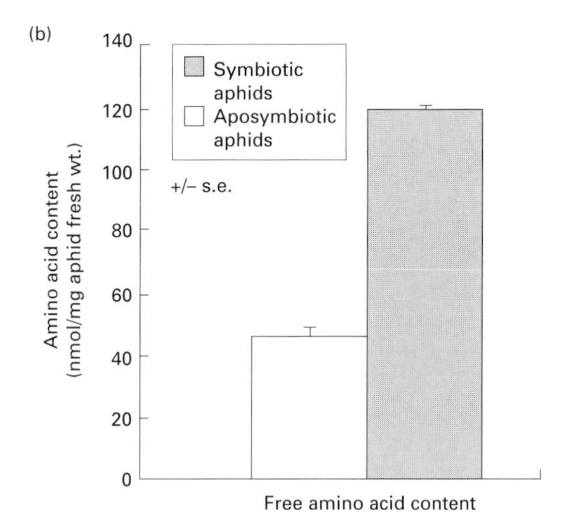

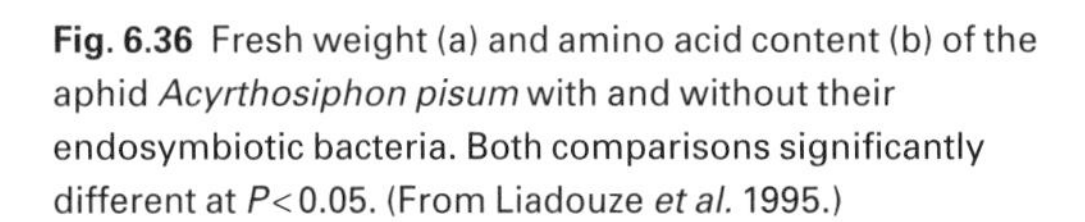
Fig. 6.36 Fresh weight (a) and amino acid content (b) of the aphid *Acyrthosiphon pisum* with and without their endosymbiotic bacteria. Both comparisons significantly different at $P < 0.05$. (From Liadouze *et al.* 1995.)

woodroaches (Dictyoptera: Cryptocercidae), which also have cellulytic protozoans in their hind-guts (Thorne 1991).

Termites in the family Rhinotermitidae, such as *Reticulitermes* spp., have been found to support at least 11 species of hypermastigote flagellate protozoans in their hind-guts, many of which seem to perform different functions under the general heading of wood degradation. In experiments in Japan, Inoue *et al.* (1997) fed termites with different diets and assessed the population densities of various protozoan species in their guts after 20 days. Figure 6.38 illustrates the effect on just two protozoan species, comparing different diets with the natural (freshly caught) situation. Wood and pure cellulose seem able to maintain high densities of protozoans, but xylan, a secondary compound found in wood material, alone reduces the numbers of one species significantly, and starvation virtually or completely wipes out the symbionts. Both termite and symbiont clearly require adequate mixtures of food derived from natural wood material.

Some Rhinotermitidae even have the facility to fix atmospheric nitrogen using bacteria in their hind-guts (Curtis & Waller 1995), but the most complex

Fig. 6.37 Worker termites consuming dead leaves.

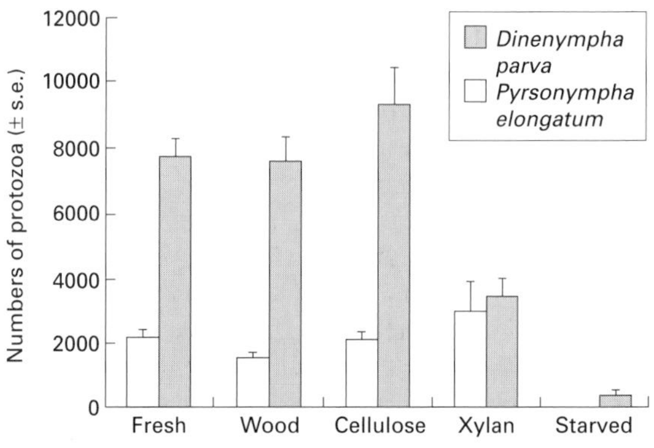

Fig. 6.38 Numbers of protozoans of two species in the hind-gut of the termite *Reticulitermes speratus* maintained on different diets for 20 days. (From Inoue *et al.* 1997.)

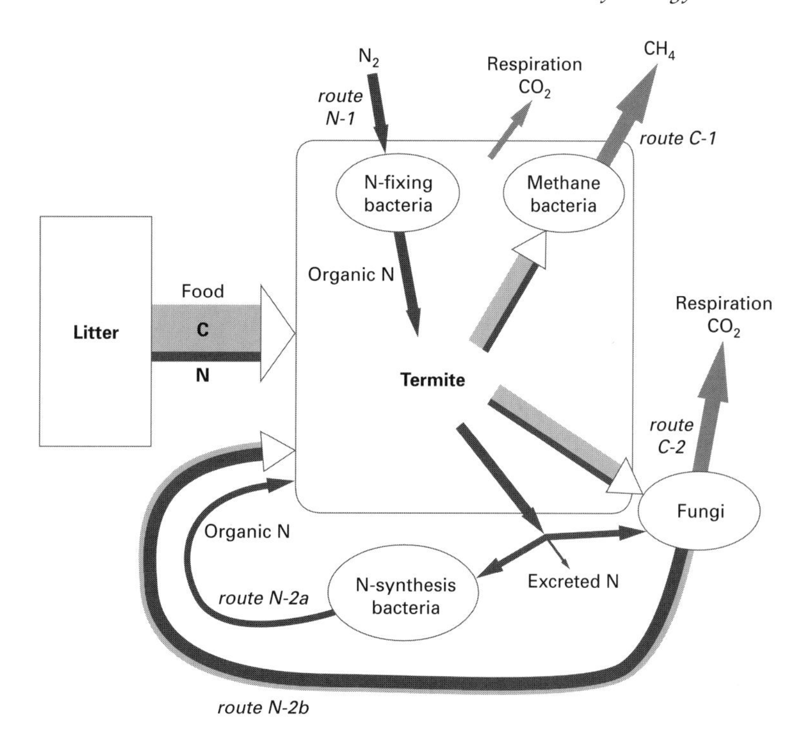

Fig. 6.39 Summary of possible C–N balance mechanisms used by various species of termites (see text for details). (From Higashi *et al.* 1992.)

insect–microbe relationships in the Isoptera are found in the so-called higher termites, such as *Macrotermes* spp., which culture symbiotic 'gardens' of the fungus *Termitomyces* (Bignell *et al.* 1994). Young worker termites eat dead plant material encountered during their foraging activities away from the colony, but also consume the conidia of this fungus when back in the nest. The conidia contain cellulase enzymes, which may help to break down plant tissues in the termite gut. Fungus gardens also break down composted plant debris directly, which older workers feed to the juveniles, the latter receiving food in which cellulose, hemicellulose, pectin and lignin are all already significantly degraded (Bignell *et al.* 1994).

One of the most complex problems in the evolutionary ecology of termites and their various symbionts involves the carbon to nitrogen balance of their food relative to their own tissues (Higashi *et al.* 1992). Dead wood contains less than 0.5% organic nitrogen, whereas termite tissues range from 8 to 13% (fresh weights). Wood is also very much richer in carbon-based celluloses and hemicelluloses, so that the C/N ratio can range from 350 to 1000, very much higher than that required by the insects. Termites have evolved two solutions to this problem; one involves adding N, the other, deleting C (Higashi *et al.* 1992). Figure 6.39 shows pathways that may solve these problems, and illustrates the roles of microbial symbionts. Those species that add N to their food may do so either by using nitrogen-fixing bacteria to fix atmospheric nitrogen (route N-1 in Fig. 6.39), or by relying on other bacteria that can synthesize amino acids from inorganic nitrogen excreted as urea by the insects themselves (route N-2a). Excreted N may also be used as a manure for termite fungal gardens, with the termites eating the fungi (route N-2b). Other termites get rid of C instead. Symbiotic methanogenic bacteria release excess C from wood material in the form of methane (route C-1), or once again, the fungi in the fungus gardens can export C to the atmosphere in the form of carbon dioxide during their respiration (Higashi *et al.* 1992). From these observations, it is hypothesized that two evolutionary strategies have been followed by termites. Some possess only C-eliminating symbionts, and live entirely inside wood, whereas the second group possess both types of symbionts and can deal with C and N balancing routes. These species are able to forage outside their nests, and have greater productivity and larger colonies. The fact that colonies can become larger when both C and

N routes are available also means that this group of termites can possess truly sterile worker castes, wherein the likelihood of having to take over reproduction is very low because of the large number of individuals in the colony.

Hymenoptera

The ant family Formicidae contains a tribe called the Attini, which are unique among the ants in that they depend on externally cultured symbiotic fungi for food for their larvae. The most advanced group of these insects are known as the leaf-cutter ants from genera including *Acromyrmex* and *Atta*, which together constitute 39 New World species (Cherrett *et al.* 1989) (Plates 6.7 & 6.8, opposite p. 158). Leaf cutters are able to utilize enormous quantities of plant material, not just leaves, but also stems, fruit and flowers cut from living plants, as substrates on which their fungal symbionts grow inside the ant colony. Because of these activities, leaf-cutter ants are considered to be extremely serious forest and farm pests in Central and South America. Indeed, Cherrett *et al.* (1989) suggested that they consume more grass than cattle, and Wint (1983) reported that up to 80% of defoliation seen in rain forests in Panama was caused by leaf cutters.

Clearly, this obligate symbiosis works very well; the basic biology is simple enough, in that homobasidiomycete fungi in the order Agricoles (Hinkle *et al.* 1994) produce hyphae that terminate in apical swellings called gongylidia (Maurer *et al.* 1992), which are eaten by adult ants and fed by them to their larvae. The gongylidia appear to have no function except to provide ant 'rewards' (Bass & Cherrett 1995), and occur grouped together in 'bite-sized' chunks called staphylae (Cherrett *et al.* 1989). The greater the number of gongylidia in an ant colony, the more workers can be supported (Fig. 6.40). This relationship predominantly works via the larvae, which use the staphylae as a protein-rich source of plant-derived nutrients. The insect uses the fungus to defeat physical and chemical plant defences, and the fungus gains because the chewing activities of the ants allow easy penetration of plant material by their hyphae (Cherrett *et al.* 1989).

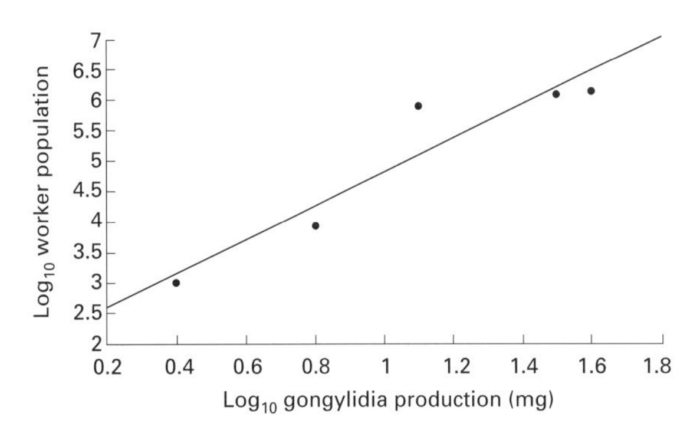

Fig. 6.40 Relationship between the productivity of symbiotic fungi (measured by gongylidia production on culture plates) and the maximum number of workers in leaf-cutting ant colonies. (From Cherrett *et al.* 1989.)

6.5 Sequestration of plant secondary metabolites

In Chapter 3, we described some of the secondary metabolites in plant tissue that can influence the foraging behaviour and feeding success of insect herbivores. At least some of those secondary metabolites are likely to serve a defensive function in plant tissue and, earlier in this chapter, we described potential evolutionary and coevolutionary associations between insects and plants that are chemically based. Here, we explore one further evolutionary response of insects to plant secondary metabolites—the ability of some herbivores to retain and utilize plant chemicals for their own defensive purposes (Rowell-Rahier & Pasteels 1992). This process is known as sequestration. Sequestered plant defences can make insect herbivores less attractive, palatable, or nutritionally valuable to parasitoids (Campbell & Duffy 1979; Greenblatt & Barbosa 1981), invertebrate predators (Kearsley & Whitham 1992; Dyer & Bowers 1996), and vertebrate predators (Brower & Brower 1964; Rothschild 1973). The number of plant toxins sequestered by individual insect herbivores can be dazzling. For example, Wink and Witte (1991) have identified 31 different quinolizidine alkaloids in the aphid *Macrosiphum albifrons* and 21 in *Aphis genistae* (Hemiptera: Aphididae). The concentrations of alkaloids in the tissues of these aphids can exceed 4 mg per g fresh body weight.

Of course, not all chemical defences found on or in the bodies of insect herbivores are sequestered from plants. For example, phytophagous leaf beetles in the genus *Oreina* (Coleoptera: Chrysomelidae) have a variety of defensive strategies (Pasteels *et al.* 1995;

Table 6.5). All beetles within the genus produce pronotal and elytral secretions that appear defensive in function. Most of the species secrete cardenolides, but these are synthesized *de novo* from cholesterol, and are not derived from the plants on which *Oreina* beetles feed. In contrast, *O. cacaliae* secretes nitrogen oxides of pyrrolizidine alkaloids (PAs) and no cardenolides. The PAs are derived from their host plants. Between these two extremes are other species of *Oreina* that can both synthesize and sequester their defensive secretions. These intermediate species can exhibit remarkable variation in their defences both within and among populations. This variation probably arises from differences in the local availability and toxicity of their host plants, but allows for considerable flexibility in defence.

We should stress that sequestration is an active process. Not all plant secondary metabolites are retained, and those that are sequestered can be modified to varying degrees. For example, the polyphagous grasshopper *Zonocerus variegatus* (Plate 6.10, opposite p. 158) sequesters only two (intermedine and rinderine) of the five pyrrolizidine alkaloids from the flowers of the tropical weed, *Chromolaena odorata*. About 20% of the PAs that are sequestered are converted to lycopsamine and echinatine by chemical inversion at one of the carbon bonds of the compounds (Biller *et al.* 1994). In other words, insects that make use of plant toxins for defence have the capacity to sequester certain compounds, excrete others, and sometimes modify those that they keep.

Moreover, not all compounds sequestered by insects from plant tissue are necessarily toxic or defensive. For example, poplar hawkmoth and eyed hawkmoth larvae (*Laothoe populi* and *Smerinthus ocellata*, respectively (Lepidoptera: Sphingidae)) sequester chlorophylls and carotenoids (pigments) from their hosts. The pigments are translocated to the integument, and provide for accurate colour matches between the insect larvae and the plants on which they are feeding (Grayson *et al.* 1991) (Plate 6.11, opposite p. 158). Although this sequestration is defensive, presumably helping to protect larvae from visually hunting predators, it does not rely on plant toxins. Similarly, the Lycaenidae, the second largest family of butterflies, appear to use excretion rather than sequestration as the dominant mode of coping with toxic host plants. However, many larvae sequester flavonoids from their hosts, which are later concentrated as pigments in the adults' wings, and are thought to play a role in visual communication (Fiedler 1996).

Table 6.5 Sequestered and autogenous defensive secretions in the chrysomelid beetle genus *Oreina*. (Data from Rowell-Rahier and Pasteels 1992 and Pasteels *et al.* 1995.)

Species	Sequestered secretions	Autogenous secretions
O. bifrons	—	Cardenolides
O. gloriosa	—	Cardenolides
O. speciosa	—	Cardenolides
O. variabilis	—	Cardenolides
O. speciosissima	Pyrrolizidine N-oxides	Cardenolides
O. elongata	Pyrrolizidine N-oxides	Cardenolides
O. intricata	Pyrrolizidine N-oxides	Cardenolides
O. cacaliae	Pyrrolizidine N-oxides	—

None the less, there is clear evidence that toxins sequestered from plants can act as powerful deterrents to their natural enemies. For example, the leaf beetle *Chrysomela confluens* (Coleoptera: Chrysomelidae) produces a salicylaldehyde-based defensive secretion that is effective against generalist predators. In an experiment by Kearsley and Whitham (1992), some larvae were 'milked' of their defensive secretions, whereas other larvae were left with their secretions intact. Only 7% of the unmanipulated larvae were attacked by ants, and none suffered serious injury. In contrast, 48% of milked larvae were attacked by ants, and two-thirds of these were killed. The salicylaldehyde secretion is produced from salicin, a precursor present in the leaves of host plants consumed by *C. confluens*. It is interesting to note that the conversion of salicin to salicylaldehyde liberates glucose, which may act as a source of energy for the larvae. If there are any costs associated with sequestration and secretion, they may be offset by the energy gained from the liberation of glucose. In this case, at least, larvae appear to obtain their defence 'for free'. In a similar study, Dyer and Bowers (1996) demonstrated that *Junonia coenia* (Lepidoptera: Nymphalidae) larvae sequester iridoid glycosides from their host plants. In their experiments, larvae fed on diets with high concentrations of iridoid glycosides, and individual larvae with high rates of sequestration, were more likely to be rejected by (and survive attack by) ant predators.

One of the most extensively studied cases of sequestration by an insect herbivore is that of the monarch butterfly, *Danaus plexippus* (Lepidoptera: Danaidae). Monarchs sequester cardenolides from their milkweed (*Asclepias*) host plants. Cardenolides are toxic, bitter-tasting steroids that attack the sodium–potasium–ATPase enzyme system of most animals, and they have been shown to protect adult (and perhaps larval) monarchs from their natural enemies (Brower *et al.* 1988). Cardenolides are sequestered by monarch larvae (Plate 6.12, opposite p. 158) that feed on *Asclepias*, then concentrated into the exoskeletal tissues of adults. The insensitivity of monarchs to cardenolides is thought to result from a single amino acid replacement (asparagine replaced with histidine) at position 122 in the alpha subunit of the Na^+, K^+, ATPase system (Holzinger & Wink 1996). To understand the cardenolide–monarch–predator interaction, we must appreciate that (a) *Asclepias* varies in the concentration of cardenolides in its tissues, and (b) the movement of monarchs in space and time dictates the type of *Asclepias* to which they are exposed and their subsequent level of protection from predators.

Asclepias species vary in toxicity at a variety of spatial scales. For example, *A. syriaca*, the species on which about 92% of larvae feed in the northern USA, exhibits large-scale variation in cardenolide concentration across its range from North Dakota east to Vermont, and south to Virginia (Malcolm *et al.* 1989; Fig. 6.41a,b). At very fine spatial scales, cardenolides within individual plants are inducible (Chapter 3), increasing in concentration over 24 h following defoliation by monarchs (Malcolm & Zalucki 1995). Induction therefore will generate small-scale variation among individuals within populations. In addition, small- and large-scale variation in cardenolide concentrations is superimposed upon seasonal varia-

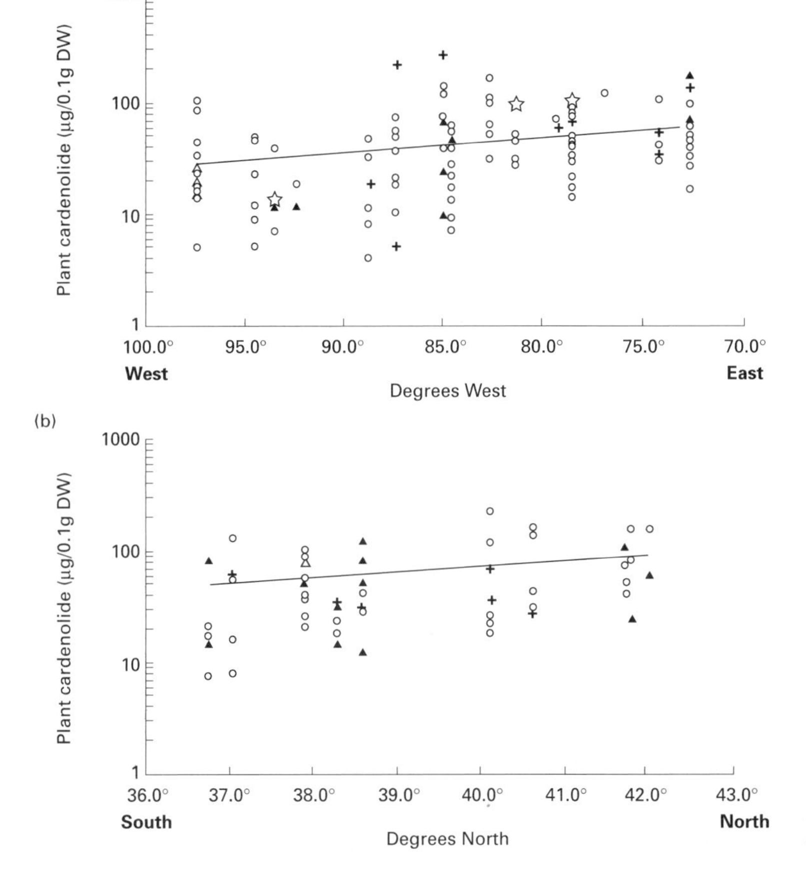

Fig. 6.41 (a) Longitude and (b) latitudinal variation in the cardenolide concentrations of *A. syriaca*. DW, dry weight. (From Hunter *et al.* 1996.)

Plate 1.1 Insects preserved in amber. (Courtesy of Oxford University Museum of Natural History.)

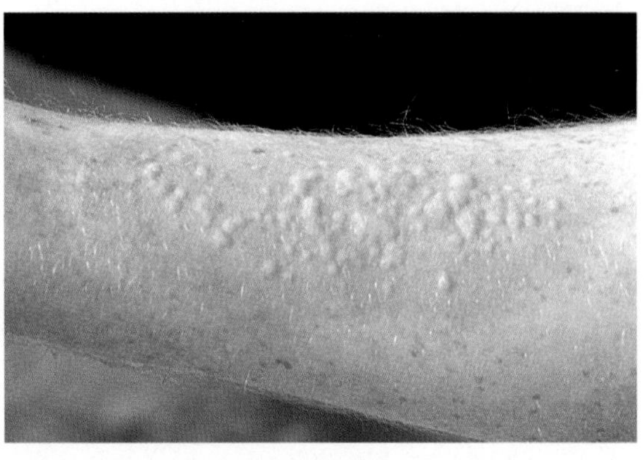

Plate 1.4 Skin rash caused by the urticating hairs of brown tail moth larvae. (Courtesy of P.D. Embden.)

Plate 1.2 Old logging road through secondary rainforest.

Plate 1.5 Fried grasshoppers for sale on the streets of Bangkok, Thailand.

Plate 1.3 Defoliation of Corsican pine by larvae of *Zeiraphera diniana*.

Plate 2.1 (*right*) Grassland stripped by armyworm (*Spodoptera* spp.) in Uganda. (Courtesy of D.J. Rogers.)

[*facing page 158*]

Plate 2.2 Gypsy moth larva.

Plate 3.1 Defoliators. Lepidopteran larva on tropical broadleafed tree, Borneo.

Plate 3.2 Defoliators. Saw fly (Hymenoptera: Symphyta) larvae on pine, UK.

Plate 3.5 (*right*) Gall forming. Pineapple galls produced by conifer woolly aphids (Hemiptera: Adelgidae) in spruce. (Courtesy of M.R. Mitchell.)

Plate 3.3 Sap feeder. Aphid feeding on sap of annual plant whilst giving birth. (Courtesy of P.D. Embden.)

Plate 3.4 Leaf mining. Microlepidopteran larva tunnelling between epidermises of hazel leaf. (Courtesy of P.H. Sterling.)

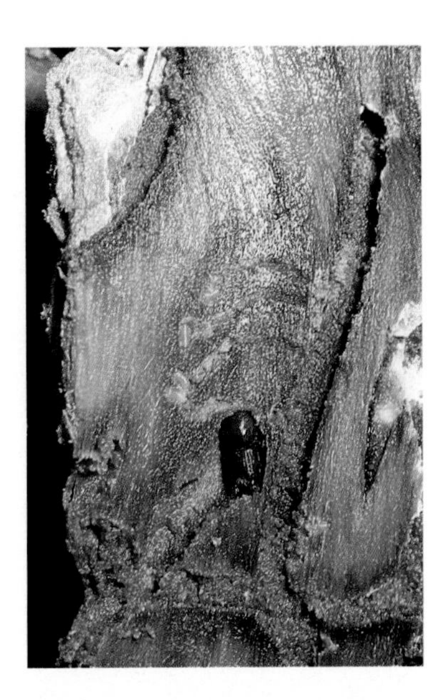

Plate 3.6 Bark boring. Adult female bark beetle (Coleoptera: Scolytidae) with maternal gallery, with larvae in separate tunnels.

Plate 5.1 Trees defoliated and killed by spruce budworm, *Choristoneura fumiferana* in Canada. (Courtesy of G.C. Varley.)

Plate 3.7 Resin pitched out of spruce bark as a result of bark beetle tunnelling.

Plate 5.2 Lime aphid, *Eucallipterus tiliae*. (Courtesy of P.D. Embden.)

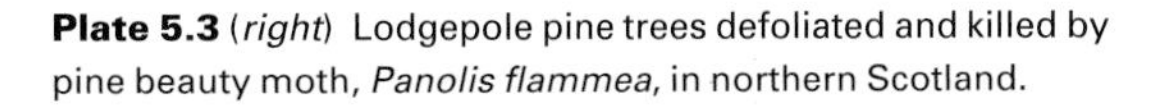

Plate 5.3 (*right*) Lodgepole pine trees defoliated and killed by pine beauty moth, *Panolis flammea*, in northern Scotland.

Plate 5.4 Spruce damaged by winter moth, *Operophtera brumata* in southern Scotland.

Plate 6.2 Cossid moth larva in tunnel showing exit hole.

Plate 6.1 Cereal aphids on wheat ear.

Plate 6.3 Adult female leaf insect (*Phasmida*) showing vestigial wings. (Courtesy of P.D. Embden.)

Plate 6.4 Black bean aphid, *Aphis fabae*, on field beans.

Plate 6.5 Swallowtail butterfly (Papillionidae). (Courtesy of P.D. Embden.)

Plate 6.7 Leaf cutter ant with leaf fragment. (Courtesy of P.D. Embden.)

Plate 6.8 Leaf cutter ant fungus garden. (Courtesy of P.D. Embden.)

Plate 6.9 Pine logs showing blue-stain fungus in cut ends after forest fire and subsequent attack by *Ips*, Queensland, Australia.

Plate 6.6 (*left*) Heliconius butterfly. (Courtesy of P.D. Embden.)

Plate 6.10 *Zonocerus* grasshopper. (Courtesy of P.D. Embden.)

Plate 6.11 Eyed hawk moth larva showing pigmented coloration. (Courtesy of P.D. Embden.)

Plate 6.12 Monarch butterfly larva.

Plate 6.13 Hornet exhibiting aposematic (warning) coloration. (Courtesy of P.D. Embden.)

Plate 6.14 Hornet clear-wing moth exhibiting mimicry of aposematic coloration. (Courtesy of P.D. Embden.)

Plate 9.1 Trees in English hedgerow with Dutch Elm disease. (Courtesy of D. Barrett.)

Plate 9.2 Cereal plant infected with barley yellow dwarf virus (BYDV). (©Kansas Department of Agriculture, USA.)

Plate 9.3 Pine tree in Japan showing classic wilt symptoms arising from attack by pine wilt nematode (PWN). (Courtesy of H.F. Evans.)

Plate 10.1 Horse chestnut scale egg masses on tree branch.

Plate 10.2 (*left*) Green spruce aphid.

Plate 10.3 Adult female cottony cushion scale with first instar cawlers. (Courtesy of M.R. Mitchell.)

Plate 10.6 Pine looper moth larvae.

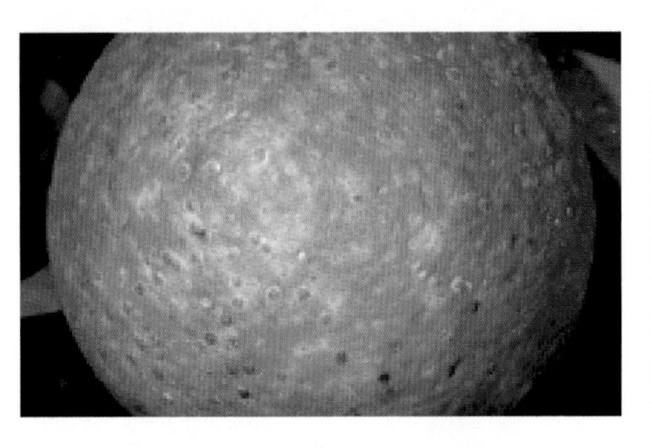

Plate 10.4 Citrus fruit infested with California red scale. (Courtesy of J.K. Clark, ©Regents, University of California, USA.)

Plate 10.7 Rice paddy in north Vietnam.

Plate 10.5 Browntail moth larva killed by nuclear polyhedrosis virus. (Courtesy of P.D. Embden.)

tion, and all levels of variability are likely to influence the monarch–enemy interactions because of diversity in the power of sequestered defences.

Perhaps the most striking example of variation in the cardenolide–monarch–milkweed interaction comes from studies of variation among milkweed species. Milkweeds used by monarchs in the southern USA have much higher concentrations of cardenolide than the species, *A. syriaca*, used by most larvae in the north (Hunter *et al.* 1996; Table 6.6). The migration of monarchs among species of *Asclepias* that vary in toxicity brings them in and out of 'zones of susceptibility' to vertebrate natural enemies. Monarchs become susceptible to vertebrate predators (e.g. mice, black-backed grosbeaks and black-headed orioles) if their total cardenolide concentration drops below 121 μg per butterfly (or 57 μg per 0.1 g dry weight of butterfly; Fink & Brower 1981). If we start a typical year in January, we can follow the movement patterns of monarchs, the chemistry of the hosts on which they feed, and changes in monarch cardenolide defences over time (Fig. 6.42). In January, almost the entire monarch fauna east of the Rocky Mountains is overwintering in high-altitude fir forests in central Mexico. These adults developed from larvae feeding in the northern USA on *A. syriaca*. Because *A. syriaca* has relatively low levels of cardenolides, and because adult butterflies apparently lose cardenolide as they travel south, the adults overwintering in Mexico in January are not well protected from predators, and suffer high rates of vertebrate predation (Malcolm & Zalucki 1993). This is seen in Fig. 6.42(A) where cardenolide concentrations in the butterflies are just at the threshold of susceptibility to vertebrate predators.

During March and April (Fig. 6.42(B)), the surviving adult monarchs migrate northwards, into the southern USA (e.g. Florida, Louisiana, and Texas) and produce the first larvae of the year. These larvae feed on southern species of *Asclepias* (e.g. *A. viridis*, *A. asperula*, *A. humistrata*), which have much higher cardenolide concentrations (Table 6.6), and the adult population arising from these larvae has high levels of sequestered cardenolides (Fig. 6.42(C)). These new spring adults, well protected against their vertebrate predators, then migrate to the northern USA (Fig. 6.42(D)). They begin breeding and ovipositing on *A. syriaca*, and produce another three generations.

Table 6.6 Cardenolide contents of the commonest *Asclepias* species used by monarch butterfly larvae in the southern and northern USA. (Data from Hunter *et al.* 1996 and sources therein.)

Asclepias spp.	Mean cardenolide content (μg/g leaf)
A. syriaca	50
A. humistrata	417
A. viridis	376
A. asperula	886

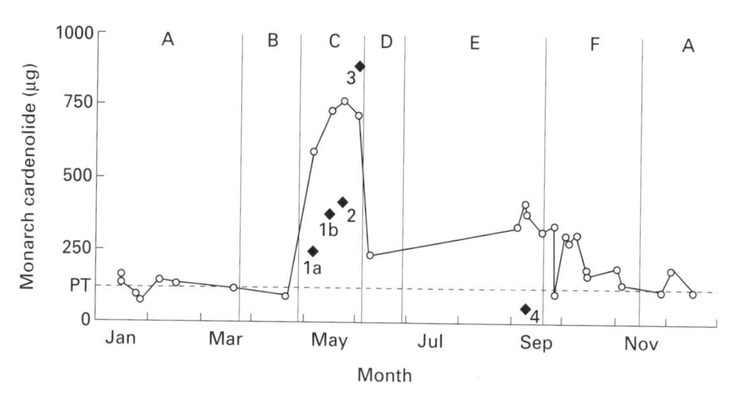

Fig. 6.42 Annual cycle of mean cardenolide concentrations (μg per butterfly) in populations of *D. plexippus*. Open circles are cardenolides in butterflies; solid diamonds are cardenolides in *Asclepias* hosts. (1a) *A. viridis* in Texas and Louisiana; (1b) *A. viridis* in Florida; (2) *A. humistrata* in Florida; (3) *A. asperula* in Texas; (4) *A. syriaca* in northern USA. Temporal zones are: (A) overwintering in Mexico; (B) northward spring migration to southern USA; (C) spring breeding in the southern USA; (D) spring generation arriving in northern USA; (E) summer breeding in northern USA; (F) autumn migration from northern USA to Mexico. (From Hunter *et al.* 1996.)

These northern generations (Fig. 6.42(E)) are less well protected from predation, although sequestration can concentrate cardenolides to levels higher than those in the host plant. However, as the adults produced in the north migrate south to Mexico for the winter (Fig. 6.42(F)), they lose cardenolide, becoming increasingly susceptible to predators (Fig. 6.42(A)), right-hand-side).

The migration of monarchs, then, exposes them to *Asclepias* species that vary in cardenolide content, and there is a dynamic interaction between monarchs, cardenolides, and vertebrate predators over the course of a single year.

Although most studies of the natural enemies of monarchs have focused on vertebrate predators, a recent study has shown that insect parasitoids are also influenced by the cardenolides sequestered by monarchs. The number of adult tachinid parasitoids (Diptera: Tachinidae) emerging from parasitized monarchs declined as the cardenolide concentration of the *A. syriaca* on which the larvae were feeding increased (Hunter *et al.* 1996). Because tachinid larvae hatch outside their hosts and then penetrate the cuticle, the reduced number of emerging parasitoids could have resulted from either reduced rates of penetration or lower levels of survivorship of parasitoids within host larvae. Interestingly, the probability of parasitism was unrelated to cardenolide concentration, and so the cardenolides did not protect larvae from tachinids directly. Rather, the sequestered compounds simply reduced the number of parasitoids emerging. This suggests that the cardenolide–monarch–parasitoid interaction has a greater effect on the third trophic level (tachinids) than the second (monarchs).

We pointed out above that the cardenolides sequestered by monarchs were concentrated into exoskeletal tissues. It makes intuitive sense that insects are more likely to survive attack by enemies if they concentrate their toxins in the tissues first encountered by those enemies: bitter tasting, or noxious compounds can produce a rapid rejection response by both vertebrate and invertebrate predators. Of course, over evolutionary time, we might expect concentration of toxins to tissues that maximize the inclusive fitness of the sequestering insect. For example, the painted grasshopper *Poecilocerus pictus* (Orthoptera: Pyrgomorphidae) sequesters cardenolides from the milkweed *Calotropis gigantea* (Pugalenthi & Livingstone 1995). Concentrations are particularly high in the metathoracic scent gland, which probably serves to warn predators even before contact is made. However, concentrations are also high in ovaries and eggs. The painted grasshopper may be providing protection to its unborn offspring (and neonate nymphs) from predators and parasites by including cardenolides with other egg provisions. This is an excellent example of a maternal effect (Rossiter 1994, 1996) in which the environmental experience of the parent (in this case, cardenolides consumed by the mother) is passed on to the offspring and influences their performance. Because the success of offspring is a primary component of fitness, it should be no surprise to see sequestered compounds concentrated in tissues that protect reproductive investment.

One common feature of insects that are protected by toxins (or venoms, such as bees and wasps) is bright 'aposematic' coloration. Aposemitism is the combination of warning coloration and toxicity to predators, and has been studied for years by entomologists and evolutionary biologists. The hypothesis, at its most basic, is that aposematic insects gain some protection from visually hunting predators that can learn to ignore distasteful prey. The subject is too broad to cover in detail here, but Owen's (1980) text provides a useful introduction to the topic. We restrict our comments to a couple of general points. First, there is no direct link between aposematism and sequestration (Pasteels & Rowell-Rahier 1991). It should be recalled that chemical defences in insects can be autogenous (self-generated) and so not all insects that display warning colours have sequestered compounds from plants. For example, although aposematism and chemical defence are both common in the Chrysomelinae (Coleoptera: Chrysomelidae) sequestration is the exception, not the rule, in this group of insects. Sequestration appears to be a secondary trait derived from aposematic, autogenously defended, ancestors.

Second, aposematic coloration has traditionally been considered to lead to two kinds of mimicry by other species of insect. Batesian mimicry is when a palatable mimic comes to resemble an unpalatable, aposematic model, and so gains protection from predators that have learnt to avoid the model insect. Mullerian mimicry, on the other hand, is when a series of unpalatable aposematic insects come to resemble one another, so reinforcing the message of unpalatability (or danger) to potential predators

(Plates 6.13 & 6.14, opposite p. 158). However, mimetic relationships do not always fall neatly into these categories, and can be difficult to untangle (Ritland & Brower 1993). For example, the viceroy butterfly, *Limenitis archippus* (Lepidoptera: Nymphalidae) has been considered a classic example of a Batesian (palatable) mimic of two aposematic danaid models (the monarch, *D. plexippus*, and the queen, *D. gilippus berenice*). However, experiments with captive red-winged blackbirds indicate that *L. archippus* may be as unpalatable as *D. gilippus*, and that the mimetic complex in Florida (USA) may be dynamic, shifting along a continuum from Batesian to Mullerian mimicry. Queens and monarchs feed on a variety of hosts, and therefore vary in palatability in space and time (above). The relative abundance of the three species varies in space and time also, and viceroy coloration varies from monarch-like to queen-like, with intermediate forms (Ritland & Brower 1993). The precise division of species among Batesian and Mullerian mimicry may be, in reality, more complicated than it first appears. There are many other kinds of mimicry in the insect world that we do not have space to consider here. Again, we direct interested readers to an introductory text on the subject (Owen 1980).

6.6 Deme formation and adaptive genetic structure

In previous sections of this chapter, we have described evolutionary responses of insects to their biotic and abiotic environment. It is important to recognize, however, that environmental variation generates a series of habitat mosaics, and that the evolutionary responses of insects occur in a patchy environment. Because the environment is patchy, insect populations can frequently become structured into discrete genetic groups or 'demes'. Demes can be thought of as groups of individuals that show marked genetic similarity because gene flow within the group exceeds gene flow among groups. Of course, restricted gene flow and deme formation can occur for many reasons, including stochastic (chance) events that isolate individuals from others in the population. However, deme formation can also result from adaptation to local environmental conditions, and this is called adaptive genetic structure (Mopper 1996).

We have seen in this and earlier chapters that host plant quality can vary markedly in space. Quality can vary among populations of plants, among individuals within a population, and even among different parts (e.g. branches) of the same plant (Denno & McClure 1983). If insect herbivores become locally adapted to such variation in their host plants, then adaptive genetic structure can develop. Measuring this adaptive genetic structure is important, because it helps us to understand interactions between insects and plants in ecological and evolutionary time, and may provide information on the routes of speciation.

The study of deme formation within populations of insect herbivores came to prominence in the early 1970s, when Edmunds was studying populations of black pineleaf scale, *Nuculaspis californica* (Hemiptera: Diaspididae) on its host *Pinus ponderosae*. In a paper describing the ecology of this insect–plant interaction, Edmunds (1973) noted that, 'scale populations apparently become adapted to specific host individuals' and his observations paved the way for a classic study of deme formation in this system that was later published in *Science* (Edmunds & Alstad 1978). In this study, scale insects from 10 infested trees were transferred onto 10 uninfested trees, and the performance of the insects compared between natal (original) and novel hosts. The survival of scale insects was higher on the original host tree than on trees to which they were transferred, and Edmunds and Alstad concluded that scales had become locally adapted to the individual tree on which they lived. Although there was some criticism of the methods and conclusions in this paper (e.g. Unruh & Luck 1987), more recent studies that compared variation in enzymes produced by scales ('allozyme markers') have confirmed that there is genetic structure in *N. californica* populations among trees, and among individual branches within trees (Alstad & Corbin 1990).

Genetic structure within populations of phytophagous insects now appears to be widespread (Mopper & Strauss 1998). Deme formation has been documented in such diverse taxa as Coleoptera, Hemiptera, Thysanoptera, Lepidoptera and Diptera. What factors could cause insect herbivores to exhibit genetic differentiation among individual plants, or even different parts of the same individual? Alstad (1998) described three possible sources of genetic structure (Table 6.7). The intrinsic local adaptation hypothesis suggests that, over evolutionary time, insects become adapted to the genotype of the particular plant on which they live, because plant genotype is the principal factor influencing plant quality (defences, nutritional quality) for insects. In contrast,

Table 6.7 Three hypotheses on the origin of genetic structure within insect populations.

1 *Intrinsic local adaptation hypothesis*
Genotypic variation among individual plants influences their quality as hosts for insects
Genotypic variation among individual insects results in some individuals out-performing others on particular host genotypes
Natural selection favours the insect genotype(s) most suitable for the genotype of each individual plant
Reduced gene flow or strong selection maintains differences in the genotypes of insects among individual plants
2 *Extrinsic local adaptation hypothesis*
Local variation in the environment (light, temperature, nutrients, water, etc.) generates variation in quality both within and among plants
Genotypic variation among individual insects results in some individuals out-performing others on particular host phenotypes or on parts of individual plants that vary in quality
Natural selection favours the insect genotype(s) most suitable for the phenotype of each plant or plant part
Reduced gene flow or strong selection maintains differences in the genotypes of insects among plants or plant parts
3 *Drift hypothesis*
The insect population is characterized by low levels of gene flow among plants or plant parts
Isolated groups of insects lose certain alleles whereas others become fixed because of random drift
The population as a whole develops genetic structure by non-adaptive processes

the extrinsic local adaptation hypothesis suggests that environmental effects on plant quality (e.g. nutrient availability, sun vs. shade, etc.) are dominant in determining the phenotypic traits of plants that are important to herbivores. These two hypotheses may appear similar because they both operate through phenotypic traits of the plant. However, the extrinsic local adaptation hypothesis allows for differences in quality within plants that are not genetically based. For example, if different branches of the same tree are growing in different environments (e.g. sun vs. shade) their qualitative differences can contribute to genetic structure in the insect population. Under both hypotheses, however, genetic structure is considered to arise from adaptive genetic change in the insect population. Finally, the drift hypothesis suggests that limited gene flow can result in genetic structure by chance alone. In other words, isolated groups of insects become genetically distinct simply by the random loss of alleles that are shared by the global population. In this last case, demes form not by adaptation, but by 'neutral evolution'.

All three hypotheses leading to genetic structure probably require a couple of conditions in common. First, the plant must be long-lived relative to the insect herbivore. The development of genetic structure by insects on annual plants seems unlikely because their hosts live for only one year. Few, if any, insect species are likely to have a sufficient number of generations within that year for distinct genetic structure to develop by either adaptation or drift: they will broadly resemble a cross-section of the entire population. In contrast, on long-lived tree species, phytophagous insects can produce literally hundreds of generations on the same individual plant, providing the potential for adaptation by the herbivore. Second, deme formation seems more likely when gene flow is low among the fragmented groups of individuals. If sexual reproduction is common among individuals from different groups, then they are unlikely to develop genetic dissimilarity. Restricted gene flow can occur for a number of reasons, including low rates of insect dispersal within and among plants, or phenological asynchrony (differences in developmental timing) among groups living on different hosts.

What evidence is there in support of adaptive vs. neutral theories of deme formation? Virtually no studies have compared the extrinsic vs. intrinsic local adaptation hypotheses (above). One reason for this is the very real difficulty in estimating the relative contributions of genotype and environment to variation in qualitative traits (e.g. nutrition, defence) of long-lived plants in natural populations (Klaper & Hunter 1998). Although new statistical techniques combined with molecular studies (e.g. microsatellite DNA markers) are making this feasible, it has been extremely difficult to say with any certainty whether differences among trees in natural populations are genetic or environmental. However, if we combine the intrinsic and extrinsic hypotheses together into an 'adaptive' model, we can compare it with the neutral (drift) model of deme formation. The adaptive model argues that, as insects repeatedly colonize a natal host, they become more successful at exploiting it and less successful at exploiting novel conspecific hosts. After several generations, natural selection produces fine-scale demic structure in the insect population. If this is

Table 6.8 Experimental tests of the adaptive deme formation hypothesis. (After Mopper 1996. Data from: 1, Edmunds and Alstad 1978; 2, Rice 1983; 3, Wainhouse and Howell 1983; 4, Karban 1989; 5, Hanks and Denno 1993b; 6, Unruh and Luck 1987; 7, Cobb and Whitham 1993; 8, Mopper *et al.* 1995; 9, Stiling and Rossi 1998; 10, Ayres *et al.* 1987.)

Insect	Host plant	Mobility	Deme formation
Black pineleaf scale[1] *Nuculaspis californica*	Ponderosa pine *Pinus ponderosae*	Sessile	Yes
Black pineleaf scale[2] *Nuculaspis californica*	Sugar pine *Pinus lambertiana*	Sessile	No
Beech scale[3] *Cryptococcus fagisuga*	Beech *Fagus sylvatica*	Sessile	Yes
Thrips[4] *Apterothrips secticornis*	Seaside daisy *Erigeron glaucus*	Sessile	Yes
Armoured scale[5] *Pseudaulacaspis pentagona*	Mulberry *Morus alba*	Sessile	Yes
Needle scale[6] *Matsuccoccus acalyptus*	Pinyon pine *Pinus edulis*	Sessile	No
Needle scale[7] *Matsuccoccus acalyptus*	Pinyon pine *Pinus edulis*	Sessile	No
Leaf-miner[8] *Stilbosis quadricustatella*	Sand-live oak *Quercus geminata*	Dispersive	Yes
Gall midge[9] *Asphondilia borrichia*	Sea oxeye daisy *Borrichia frutescens*	Dispersive	Yes
Geometrid moth[10] *Epirrita autumnata*	Mountain birch *Betula pubescens*	Dispersive	No

true, transplanted insects will fare worse on novel than on natal hosts. In contrast, the neutral model predicts that genetic structure arises from drift, and is not adaptive. If true, the performance of transplanted insects will not differ between novel and natal hosts.

Mopper (1996) and Mopper and Strauss (1998) have collected together most of the experimental tests of deme formation in insect herbivore populations. Of 10 field experiments reported by Mopper (1996), six support the adaptive model of deme formation and four do not (Table 6.8). Like many of the other topics covered in this book, the jury is still out on which forces, selective or stochastic, generate genetic structure in insect herbivore populations. However, compelling examples of adaptive deme formation do exist. Karban (1989) found that thrips, *Apterothrips secticornis* (Thysanoptera) displayed higher growth rates on natal plants (*Erigeron glaucus*) than on novel plants of the same species. His experiment was particularly well designed because all the thrips were kept in a common environment on novel hosts for several generations beforehand. By doing this, Karban made sure that differences in performance among hosts were due to the genotype of the insects and not to their immediate environmental history.

One interesting fact to emerge from Mopper's (1996) review is that deme formation may not be linked to the potential of insects to disperse. Three of the 10 insect species examined were dispersive, and two of these exhibited deme formation (Table 6.8). This is surprising because dispersal ability is likely to be related to gene flow, and high levels of gene flow should eliminate genetic structure. One possible explanation is that strong selection pressure maintains genetic structure, even in the presence of significant gene flow. Introduced, 'novel' genotypes might be removed rapidly if host quality exerts considerable selection on insect genotype. Of course, to demonstrate that selection is mediated by host quality, it is necessary to show that survival or fecundity of 'novel' insects is lower than that of 'natal' insects because of plant attributes. Remarkably few studies of deme formation in insects have attempted to show that mortality or fecundity is directly linked to plant traits. The importance of this is aptly demonstrated by an experimental study of deme formation in the leaf-mining moth, *Stilbosis quadricustatella* (Lepidoptera: Cosmopterigidae) on oak (Mopper *et al.* 1995; Fig. 6.43a,b; Fig. 6.44). Leaf-miners are excellent insects for such studies because they leave evidence of the causes of mortality behind in the mine. Plant-related mortality can be separated from predation or parasitism by examining damage (or lack thereof) to the mine.

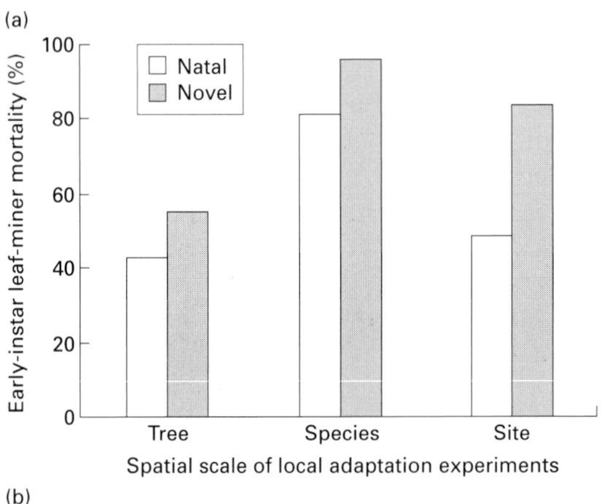

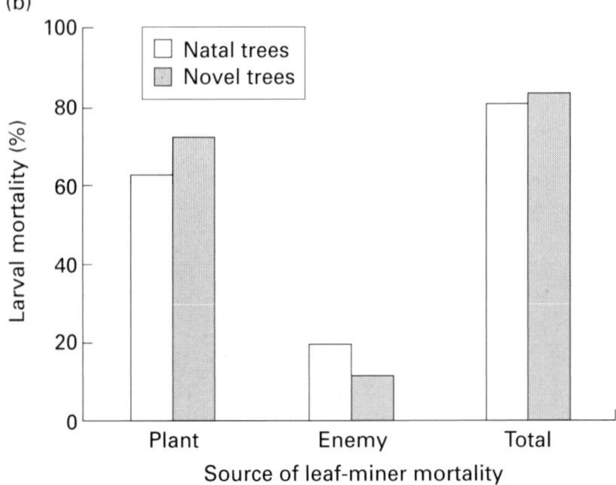

Fig. 6.43 (a) Experimental transfer of *S. quadricustatella* between natal and novel environments demonstrates genetic structure at three spatial scales (tree, species, and site). (From Mopper *et al.* 1995.) (b) Larval mortality of *S. quadricustatella* transferred between natal and novel host plants. Although host-mediated mortality is lower on natal hosts, predation is higher. Overall, levels of mortality do not differ between natal and novel hosts. (From Mopper *et al.* 1995.)

Mopper and her colleagues transferred *S. quadricustatella* among (a) natal and novel sites, (b) natal and novel oak species (*Quercus geminata* and *Q. myrtifolia*), and (c) natal and novel *Q. geminata* individuals within a site. They found that, as predicted by the adaptive deme formation hypothesis, there was significantly lower plant-mediated mortality in the natal treatment groups at all three spatial scales (site, species, individual). Apparently, there is genetic structure in the leaf-miner population at a variety of scales, and host quality is directly responsible for some larval mortality. However, despite reduced plant-mediated mortality on natal trees, rates of overall survival were not different. Mortality from natural enemies was actually lower on novel trees than on natal trees, and it compensated for the effects of tree quality on survival (Fig. 6.43b). If the investigators had not divided mortality among host and enemy effects, overall survival would have been identical, and the hypothesis of adaptive deme formation rejected. Further work is needed in this system to clarify the role of natural enemies over a period of years to examine the circumstances under which plants and predators might costructure the genetics of insect populations.

In several of the studies in which adaptive deme formation has been demonstrated, endophagous (internally feeding) insects have been used in the experiments. Galling and mining insects (Chapter 3) are intimately associated with their host plants, and it has been suggested that they are more likely to adapt to the quality of individual hosts over time than are insects on the plant surface. External feeders may be more prone to unpredictable environmental changes

Fig. 6.44 Two lepidopteran leaf mines (*Stilbosis quadricustatella*) on the sand-live oak, *Quercus geminata*. (Courtesy of S. Mopper.)

that limit the insects' responses to the 'constant' nature of their host plant. Endophagous insects may be somewhat 'buffered' from the environment. For example, adaptive deme formation has been demonstrated for the gall-forming midge, *Asphondylia borrichia* (Diptera: Cecidomyiidae) on sea oxeye daisy, *Borrichia frutescens* (Stiling & Rossi 1996, 1998). In this case, gall size is significantly larger on natal than on novel plants, and adult flies prefer natal to novel plants for oviposition. However, it is clear that we need many more studies before we can generalize about links between feeding mode and deme formation. Indeed, the study of genetic architecture in insect herbivore populations still suffers from a lack of experimental studies. Turning full circle, Alstad now appears to favour a neutral drift model, and not an adaptive model, for deme formation in the black pineleaf scale insect. He has concluded (Alstad 1998): 'after 20 years' field research with black pineleaf scale and ponderosa pine, the neutral drift and the extrinsic local adaptation hypothesis remain viable. The intrinsic local adaptation hypothesis is dead.' We need much more research to determine how often, and by what mechanisms, insect herbivore populations develop genetic structure.

6.7 Extreme ways of life

It should be clear by this stage in the chapter that evolution, through the proto process of natural selection, can result in adaptation by insects to a wide variety of environments. In this final section, we discuss some examples where insects have adapted to habitats that may be considered to be extreme, in that they present particularly difficult conditions for insects and hence demand very special modifications to ecology, physiology and behaviour. These extremes will be considered under the headings of salinity, temperature and toxic environments.

6.7.1 Salinity

As mentioned in Chapter 1, the immediate ancestors of insects, the proto myriapods, were terrestrial, and so any adaptations seen in the Insecta which involve the use of aquatic habitats must be considered to be secondary. Living in fresh water for at least part of the life cycle is a common feature for some insect orders, but it is perhaps initially surprising to find that water containing appreciable quantities of salt is almost entirely unexploited by insects. True sea water is defined as containing in the region of 35‰ (parts per thousand) dissolved sodium chloride. Water with 8–35‰ is known as brackish, and with anything below 8‰ is thought of as fresh. Almost all aquatic insects are to be found only in fresh water, but a few species such as brine flies have evolved the ability to withstand relatively high levels of salinity. *Ephydra hians* (Diptera: Ephydridae) is a species commonly found in warm saline–alkaline lakes in Canada, where its larvae graze on microbial mats consisting mainly of photosynthetic filamentous cyanobacteria (Schultze-Lam *et al.* 1996). Other species of *Ephydra* are thought to be fairly tolerant of salinity levels, but some brine flies may be so well adapted to these conditions that they are unable to tolerate lower salinities. *Paracoenia calida* is endemic to saline springs in northern California, with salinity levels of 22‰. It is suggested that its physiology has become highly adapted to these waters, and specializations for life under such severe and rare conditions may account for the endemism (Barnby & Resh 1988).

The sea covers somewhere in the region of 70% of the Earth's surface, but there are almost no insect species at all that could be considered to be marine, in the strict sense of the word. Certainly, many species can be found in the intertidal zone, in the strand line and in salt marshes, but in almost all of these cases, the insects that live there are able to do so by avoiding much contact with sea water. Along the North Sea coast of Germany, for example, 310 species of Lepidoptera from 29 families were found inhabiting salt marsh ecosystems (Stuening 1988), but most are found in salt-tolerant but terrestrial plant communities dominated by grasses in the upper reaches where there is little tidal influence. Even when the sea covers the marsh, insects such as the beetle *Bledius spectabilis* (Coleoptera: Staphylinidae) may be able to avoid coming into contact with water by constructing burrows into which they retreat. When the tide comes in, surface tension blocks the narrow neck of the burrow, preventing flooding (Wyatt 1986).

Thus, there are very few truly marine insects. Indeed, if 'marine' implies that the species lives *in* sea water, then there are none, though just a handful occupy the sea–air interface by exploiting the surface of the sea. These are the sea-skaters or ocean-striders, belonging to the genus *Halobates* (Hemiptera: Gerridae), of which only five species have colonized the open ocean (Andersen 1991), all of them in the tropics. These species would appear to have evolved from fresh-water gerrids, of which there are plenty of species, via a coastal, intertidal or estuarine habitat. The earliest fossils of sea-skaters are about 45 Myr old (Andersen *et al.* 1994), and show very similar characteristics to modern species. The adults are always wingless, presumably reflecting the stable nature of marine habitats (Andersen 1991), and they feed on dead bodies of arthropods floating on the surface of the sea. Eggs are laid on whatever hard substrates are available; floating seaweed and albatross tails have even been suggested! Some species have been found more than 150 km from shore (Cheng *et al.* 1990), but others are more coastal, especially in the nymphal stages. Figure 6.45 represents the distributions of nymphs and adults of *Halobates fijiensis* in relation to the coastline in Fiji, and it can be seen that most juveniles are found close to the seaward edge of mangroves, and generally confined to pools high on the shore when the tide recedes. Adults, on the other hand, can be found some distance out to sea in sheltered bays, though sparsely distributed (Foster & Treherne 1986). Thus potential competition for food and space between adults and offspring is likely to be reduced.

6.7.2 Temperature extremes

We discuss the multivarious effects of temperature on insects in Chapter 2. Exceedingly high temperatures that approach the point of protein denaturation must represent an upper limit to animal distribution, but, nevertheless, some insects live in very hot water. The brine fly *Paracoenia calida* mentioned above occurs in high densities in water that is not only very salty, but where the water at its spring source reaches 54°C

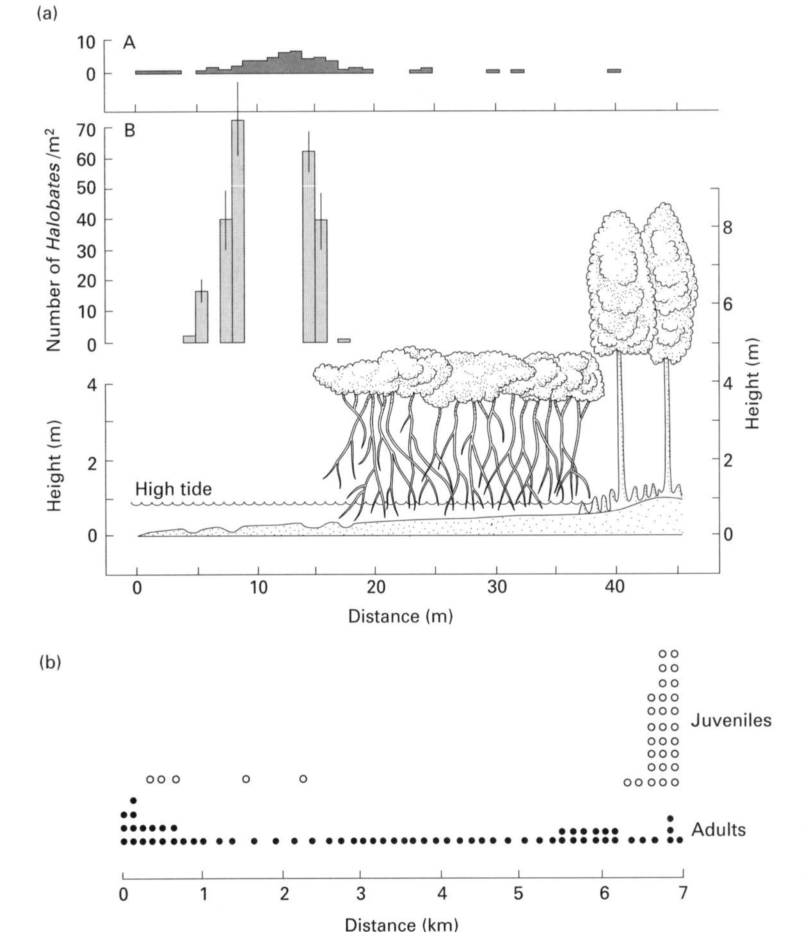

Fig. 6.45 (a) Distribution of juvenile *Halobates fijiensis* along upper shore and mangrove transects at (A) high tide, and (B) lowtide (± SE). Numbers counted in five 1 m squares every 1 m. (b) Distribution of adult and juvenile *H. fijiensis* along an open-water transect across Lucacala Bay, Fiji. (From Foster & Treherne 1986.)

(Barnby & Resh 1988). Eggs are laid near the source to remain warm, and the larvae, which are to be found in all but the hottest water, avoid the problem of low oxygen tensions in warm water by using atmospheric oxygen for respiration. Water temperatures can be responsible for the distribution of insects according to thermal gradients in hot springs (Fig. 6.46). In a study on streams supplied by hot springs in California (Lamberti & Resh 1985), no macroinvertebrates were to be found in water temperatures equal to or above 45°C, but the highest density (mainly midges (Diptera: Chironomidae)) occurred at 34°C, with maximum species richness at a much more tolerable 27°C. Both midges and caddisflies (Trichoptera) were commonest at sites that were several degrees below their lethal thermal thresholds, which suggests some leeway in the system should temperatures suddenly rise.

Temperature extremes are more common on land, where the buffering effects of water are absent. Deserts can range from scorchingly hot to well below freezing within a single 24 h cycle, and insects have adaptations to tolerate both extremes. According to Gehring and Wehner (1995), ants in the genus *Cataglyphis* (Hymenoptera: Formicidae) are amongst the most thermo-tolerant land animals known. They forage in the desert at temperatures above 50°C, when their thermal lethal temperatures are a mere 53–56°C, depending on species. It seems that the ants are able to perform this feat by accumulating special heat shock proteins in their tissues before venturing out.

The ability of many insects to survive periods of freezing by supercooling is described in Chapter 2, and it can be seen that temperatures several degrees below zero can often be survived by changes in cell chemistry induced by exposure to low but not lethal temperatures. One example of extreme cold survival comes from *Onychiurus*, a genus of springtail (Collembola) living in Spitsbergen. Monthly mean air temperatures in this region exceed 0°C for only 4 months of the year (June to September) (Strathdee & Bale 1998), and temperatures in the soil can be expected to be appreciably lower most of the time. Strictly speaking, of course, springtails are no longer considered to belong within the Insecta, but as hexapods their evolutionary ecology can be expected to parallel that of true insects. These animals are exposed to soil temperatures as low as –29.6°C, and temperatures that remain below zero for up to 289 days. In fact, the animals may be encased in ice for an astonishing 75% of the year (Coulson *et al.* 1995).

6.7.3 Toxic environments

A large number of chemicals derived from human activities of one sort or another are toxic to most animals, including insects. Many types of chemical are produced as by-products of manufacturing, agrochemical and power generation industries. These unwanted by-products include atmospheric pollutant gases and petroleum wastes in water and soil. Other 'toxic chemicals' are actually synthesized on purpose, such as fuels including petrol, and jet or

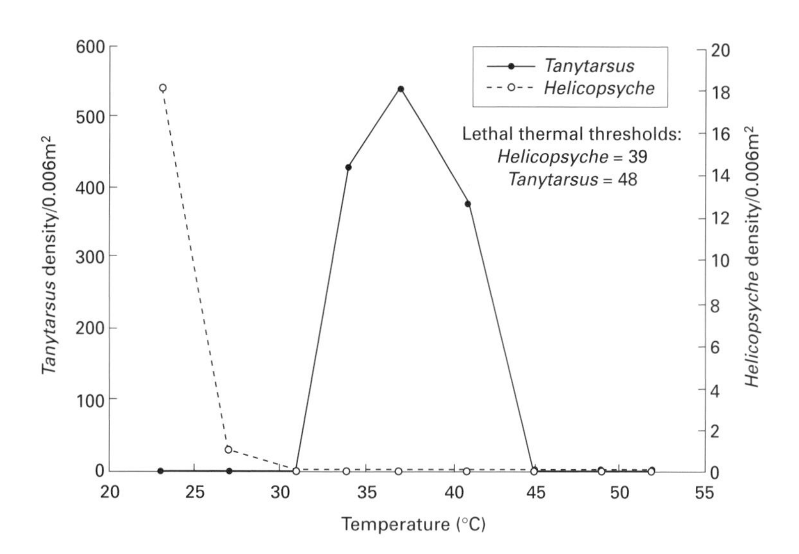

Fig. 6.46 Distribution of larvae of a chironomid midge, *Tanytarsus* sp. and a caddisfly, *Helicopsyche borealis*, according to the thermal gradient in a stream emanating from a hot spring. (From Lamberti & Resh 1985.)

even rocket fuel. Some of these compounds end up in the environment, where they can have dire consequences for insects and other animals. Not surprisingly, residues of pollutants including petroleum hydrocarbons, ammonia and metals in river sediments in Indiana, USA, were all toxic to *Chironomus* midge larvae (Diptera: Chironomidae) (Hoke *et al.* 1993). Even more understandable is the lethal effect of aircraft fuels. Chironomid adults of various species have been seen to be overcome by fumes during the refuelling of helicopters, and dead adult midges have actually been found clogging up the filters in the fuel lines of such aircraft belonging to the Irish Air Corps (O'Connor 1982). The evolution of fuel-tolerance in insects is somewhat hard to imagine, though there is, in fact, some evidence suggesting that not all insects are equally affected. In the case of standard, petroleum-derived jet fuel as used by the US Air Force, the order of decreasing susceptibility was earwigs, rice weevils, flour beetles, ladybirds, and finally cockroaches (Bombick *et al.* 1987)—cockroaches have apparently adapted to survive almost anywhere! Fuel-derived toxins may also have sublethal effects on insects, which, though the exposure may be rather short-lived, can have serious effects on population survival in the longer term. Undoubtedly, insect ecologies and physiologies have had little time to adapt, even if it were possible, to the presence of solid rocket fuel (SRF) components in their environment. None the less, honey bees, *Apis mellifera* (Hymenoptera: Apidae), in the vicinity of the Kennedy Space Centre, Cape Canaveral, Florida, are exposed to hydrogen chloride gas, a hazardous exhaust component of space shuttle launches. In experiments, Romanow and Ambrose (1981) found that HCl caused depressions in brood and honey production, elevations in aggressive behaviour by the bees, and increases in the incidence of both viral and fungal stress-related diseases. Quite how the evolutionary ecology of bees would handle this problem given time and sufficient selection pressure is uncertain.

Not all insects find petroleum hydrocarbons lethal, and in fact a very few have actually adapted to crude oil as a habitat. Maggots of the fly *Psilopa petrolii* (Diptera: Ephydridae) were found before World War II living in crude oil spillages in California, where they fed on other insects that fell into the oil (Erzinclioglu *et al.* 1990). It is not known how this insect avoids or tolerates the toxic effects of the oil,

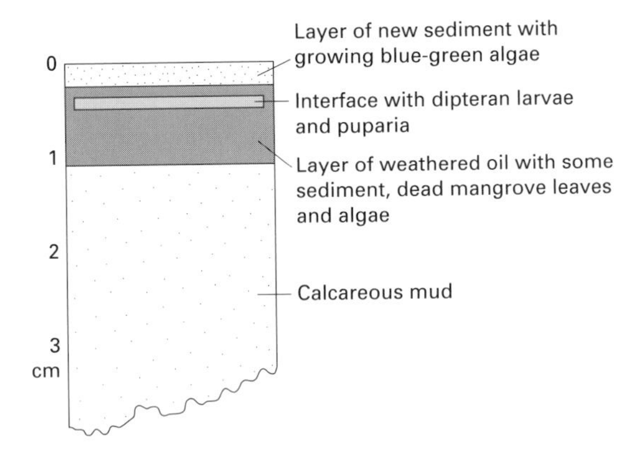

Fig. 6.47 Diagrammatic cross-section of soil profile from mangrove oil-spill in the United Arab Emirates, showing location of oil-tolerant fly larvae and pupae. (From Erzinclioglu *et al.* 1990.)

but here is clearly an unoccupied niche ripe for colonization. Much more recently, similar dipteran larvae have been discovered in intertidal sediments in the Arabian Gulf, living in the interface between a layer of oil derived from fuel oil spillages and a new thin overlying layer of blue–green algae (Fig. 6.47) (Erzinclioglu *et al.* 1990). To make matters worse, these oily sites also experienced periodic inundations of highly saline sea water.

Other anthropogenic toxins may have less direct effects. A wealth of literature describes the many effects of acid rain on animal and plant communities, either by affecting the insects directly or by changing the characteristics of their host plant. Many examples are available of both positive and negative influences (see Chapters 3 and 8), and only one is presented here. In NE France, many mountain streams have become acidified because of precipitation in their catchment areas depositing atmospheric pollutants such as SO and NO gases. pHs as low as 4.9 have been recorded, with associated reductions in bicarbonate ions and elevated aluminium concentrations (Guerold *et al.* 1995). The populations of certain aquatic insects were seriously depleted by these conditions. Mayflies (Ephemeroptera) were least abundant in the acidic waters, but stoneflies (Plecoptera) were the dominant insect group. The latter have adapted to more acidic conditions, and as environmental change progresses, it will be these types of insects that will survive more readily (see also Chapter 8).

Chapter 7: Insects in Ecosystems

7.1 What is ecosystem ecology?

So far in this book, we have concentrated on interactions between one or a few species of insects and their abiotic and biotic environments; our focus has been at the population and community level. In this chapter, we look at the importance of insects at the broader ecological scale of entire ecosystems.

We can define an ecosystem as all the interactions among organisms living together in a particular area, and between those organisms and their physical environment (Tansley 1935). The foundation of ecosystem ecology is that ecosystems can be represented by the flow of energy and matter from one subsystem to the next. Figure 7.1 shows how energy from the sun is trapped by plants, or 'primary producers', in both terrestrial and aquatic systems, and is passed on to primary consumers (herbivores), and then on to secondary, and tertiary consumers. Subsequently, the energy reaches the decomposers and, finally, is either 'trapped' (at least temporarily) in matter that is difficult to decompose or dissipated as heat. In fact, energy is lost to heat during each exchange between trophic levels (see below). Ecosystem ecologists study the movement of energy and matter among components in this way. Very often, they care less about the identity of a particular individual or species than its 'functional role' in the environment (decomposer, producer, etc.). Although insects are never primary producers, they are important primary, secondary, and even tertiary consumers in some ecosystems. Moreover, they are fundamental participants in the decomposition process in terrestrial and aquatic ecosystems.

When we look at a complex ecosystem such as a tropical forest or temperate lake, we know that all of the species in it are dependent on the availability of energy and matter (e.g. nutrients) flowing into the system. We ought to be able to chart the input and output of energy and materials, and see how they are divided up to support the diversity of organisms in the system. Specifically, we ought to be able to assess the importance of insects in the transformation and flow of energy and matter.

7.2 A few fundamentals of ecosystem ecology

Before exploring the role of insects in ecosystems, we should describe a few important rules or laws that govern the movement of matter and energy through systems. These rules set fundamental limitations on the structure and function of ecosystems and, by extension, the potential effects of insects on ecosystem processes.

7.2.1 Energy and ecosystems

Where does the energy come from to drive ecosystems? What is the source of all the energy required for organisms, including insects, to grow, move, reproduce, and repair damaged tissues? Energy exists in several forms: heat; radiant energy (electromagnetic radiation from the sun); chemical energy (stored in the chemical bonds of molecules); mechanical energy; and electrical energy. One fundamental rule governing energy, called the first law of thermodynamics, says that energy cannot be created or destroyed. It can be converted from one form to another, or moved from one system to another, but the total amount never increases or decreases. That means that when the universe was created about 15 billion years ago, the amount of energy existing then equals the amount of energy existing now. The energy has changed form over and over again in that time, but nothing has been gained nor lost.

So, if organisms cannot create the energy that they require to live, they have to transform it from some existing source. Plants (or, strictly speaking, 'autotrophs' including some bacteria and Protista) are critical to this process in almost all ecosystems. They trap energy from the sun, converting electromagnetic energy into chemical energy by photosynthesis. The chemical energy is stored in the molecules that make

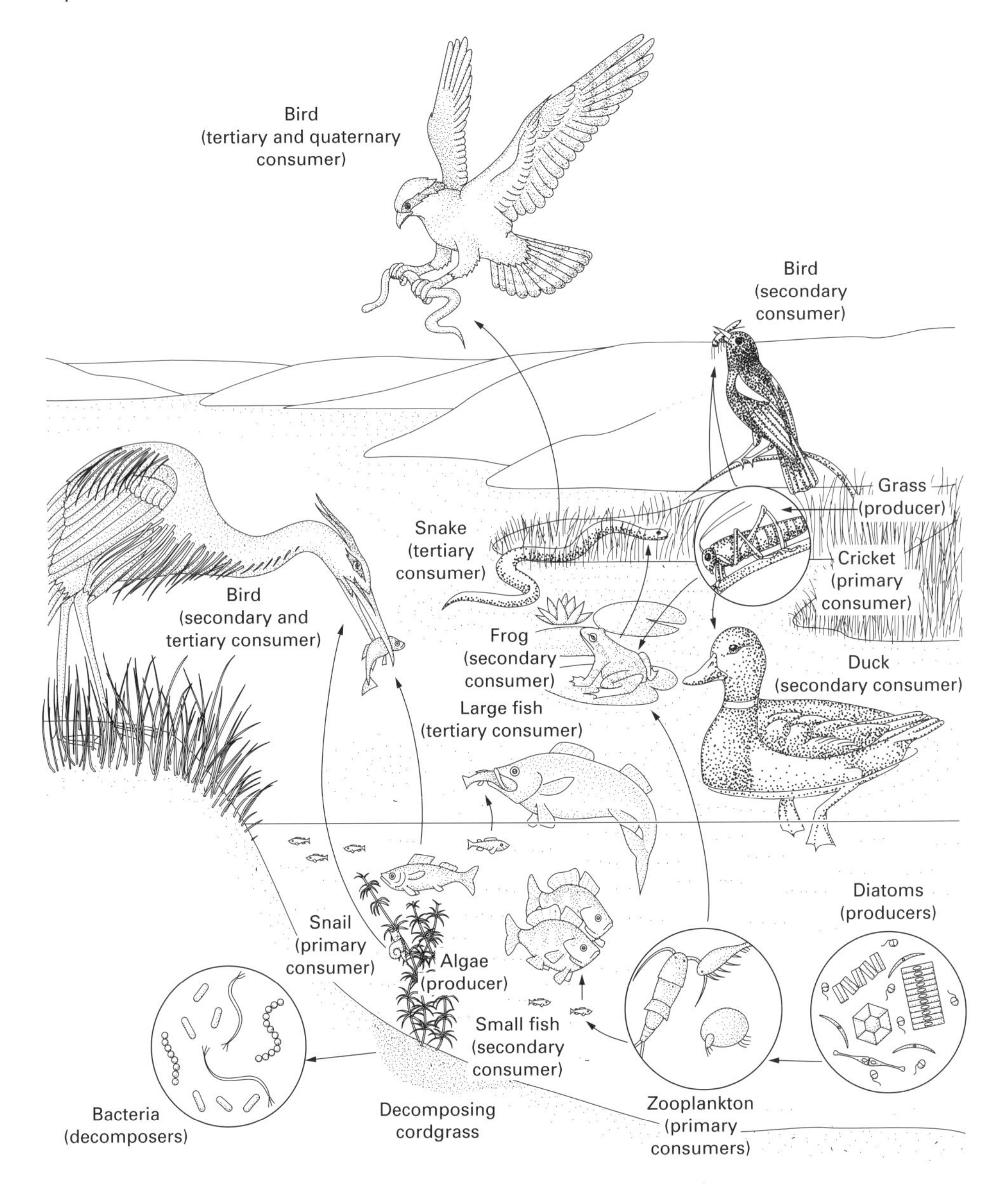

Fig. 7.1 Simplified food chains showing how primary producers in both aquatic and terrestrial systems trap energy from the sun. The energy is then passed through subsequent trophic levels. (From Raven *et al.* 1993.)

up plant tissue. Herbivores, including many of the phytophagous insects discussed in Chapter 3, can convert this chemical energy into the mechanical energy of flight. Secondary and tertiary consumers such as insect predators and parasitoids (Chapter 5) then utilize or convert the chemical energy stored in herbivores for their own growth, movement, and reproduction.

However, the transformation of energy from one form to another always involves some loss of 'useful' energy. For example, when a grasshopper transforms the chemical energy of plant tissue into mechanical energy to leap away from a mantid predator, some of the chemical and mechanical energy is changed into heat energy, and 'lost' to the plant–herbivore–

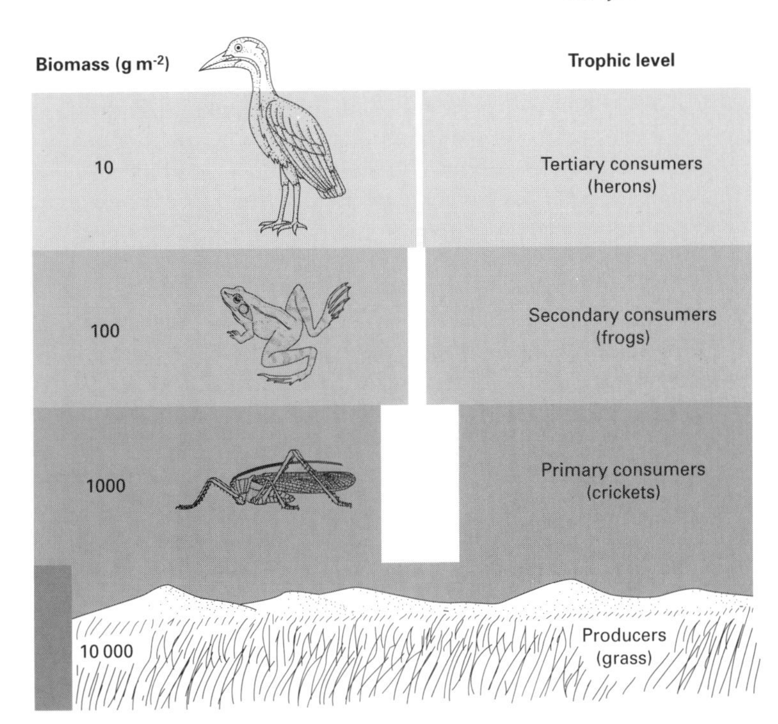

Fig. 7.2 Losses in useful energy during transfer among trophic levels. It should be noted that 10 000 $g\ m^{-2}$ of grass will support, on average, only 10 $g\ m^{-2}$ of tertiary consumers, placing fundamental limitations on ecosystem processes. (From Raven *et al.* 1993.)

predator system. That does not mean it disappears (remember the first law of thermodynamics); it is still in the environment, but in a form that biological life cannot readily reuse. This is the second law of thermodynamics: during the transformation of energy from one form to another, there is a continual loss of energy from 'useful' or reusable forms, to 'less useful' or non-reusable forms (heat). That means that the amount of useful energy available to do 'work' in the universe decreases over time, even if the total amount of energy stays the same.

The first and second laws of thermodynamics are therefore critical to the way that ecosystems are put together. First, the amount of energy entering the ecosystem sets a limit on the total productivity of the system, because the system cannot generate energy of its own. Second, the loss of useful energy to heat sets a limit on how many transformations, or retransformations, can take place. The losses to heat at each stage mean that, eventually, all the useful energy has been used up. This will set a limit to the number of different transformations or 'trophic levels' (from plants to herbivores to carnivores to decomposers) that can take place. In fact, the transformation of energy from one trophic level to the next is only about 10% efficient; about 90% of the 'useful' energy is lost between each trophic level (Fig. 7.2). Ecosystems are ultimately constrained by the first and second laws of thermodynamics. By extension, so are the insects that inhabit ecosystems.

7.2.2 Matter and ecosystems

The flow of matter through ecosystems differs fundamentally from the flow of energy. Energy flow through ecosystems is linear, and there is no recycling. It is dissipated in unusable forms into the environment during each stage of transformation until there is none remaining from the original pool. In contrast, materials such as nitrogen, carbon, phosphorus, potassium, and even water, flow through ecosystems in cycles. When matter cycles from the living world to the non-living physical environment and back again, we call this biogeochemical cycling. Other than sunlight and the occasional meteorite, the Earth is essentially a closed system. That means that matter cannot escape from the system and materials are reused and recycled both within and among ecosystems. Insects are important components of several biogeochemical cycles as well as mediators of energy transformation. In the rest of this chapter, we describe two examples of the importance of insects to the cycling of materials and the transformation of energy in ecosystems. Specifically, we will examine the role of insect decomposers in the carbon cycle and the importance of leaf-shredding insects to ecosystem processes in streams. We chose these two examples from many possible systems because they are both topical and moderately well understood.

7.3 Insects and the terrestrial carbon cycle

The carbon cycle (Fig. 7.3) describes how carbon flows from the physical environment to the living world, and back again. Figure 7.3 is a simplified diagram of the global carbon cycle, including the 'storage' of carbon in coal, oil, and natural gas that results from partial decomposition of plant and animal material. For our purposes, we concentrate here on the terrestrial components of the carbon cycle.

Carbon is incorporated from the physical environment into the living world by the process of photosynthesis; carbon dioxide from the atmosphere is captured by plants, and turned into organic compounds in living plant tissue using the energy of the sun:

$$6CO_2 + 12H_2O + \text{radiant energy} \rightarrow C_6H_{12}O_6 + 6H_2O + 6O_2.$$

About 0.03% of the world's atmosphere is carbon dioxide, and it acts as the major pool from which carbon enters the living world. Once the carbon is incorporated into plant tissue, it can follow several paths. First, some of it is used for the growth and reproduction of the plant itself. This costs energy, and the source of that energy is 'respiration'. We can think of this as the plant burning fuel, in this case carbon-containing molecules, to release the energy:

$$C_6H_{12}O_6 + 6H_2O + 6O_2 \rightarrow 6CO_2 + 12H_2O + \text{transformed energy}.$$

It should be noted that respiration causes the release of carbon dioxide, which then re-enters the atmosphere. This is one simple part of the carbon cycle: carbon dioxide is taken up by plants during photosynthesis, and so moves from the physical world to the living world; then, as plants respire and use the energy for growth and reproduction, carbon is returned to the physical world.

Second, some of the carbon captured by plants is consumed by primary consumers such as insect herbivores and, in turn, by predators that eat herbivores. At each trophic level, carbon that was originally captured by plants is returned to the atmosphere by respiration of organisms at that trophic level. For example, hawks that eat mice that eat grasshoppers that eat plants return a portion of the carbon originally captured by plants back to the atmosphere during respiration (Fig. 7.4).

7.3.1 Insect decomposers

Insects play their major role in the carbon cycle (and

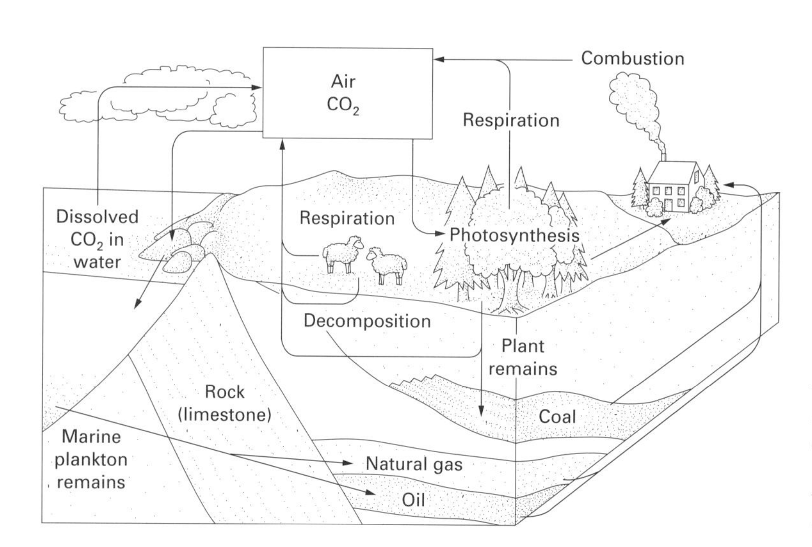

Fig. 7.3 A simplified diagram of the global carbon cycle. The text refers only to terrestrial components of the cycle. It should be noted that incomplete decomposition of animal and plant remains can lead to carbon 'storage' in the forms of coal, oil and natural gas. (From Raven *et al.* 1993.)

in the cycling of other materials) during the decomposition process. Dead and decaying plant and animal tissues or waste products serve as sources of food for many kinds of decomposers, including insects, mites, fungi, and bacteria. As these tissues are processed by decomposers, carbon dioxide is released into the atmosphere, completing the major part of the carbon cycle. Blow flies and flesh flies (Diptera: Calliphoridae and Sarcophagidae, respectively) are well-known insect decomposers whose larvae often feed within carrion or excrement. Likewise, carrion beetles (Coleoptera: Silphidae) in the aptly named genus *Necrophorus* excavate chambers beneath the dead bodies of small mammals. The buried carcass provides food for their offspring. Carrion beetles are physically strong enough that a pair of beetles can move a large rat several feet before finding a suitable burial site. Although this feeding habitat may seem distasteful to some, these insects provide a valuable service by removing dead animals and animal waste from the environment, and completing the cycling of matter between the living and non-living world. Similarly, many dung beetles (Coleoptera: Scarabaeinae) chew off portions of animal faeces, work them into balls, and can roll the dung considerable distances. They often work in pairs, with one adult beetle pulling from the front while the other pushes from behind. The dung ball is buried, and provides food and shelter for larvae of the next generation (Fig. 7.5). The scarab, *Scarabaeus sacer*, which populates the perimeter of the Mediterranean, was sacred to the early Egyptians, who drew parallels between dung-rolling and the mysterious daily movement of the sun across the sky.

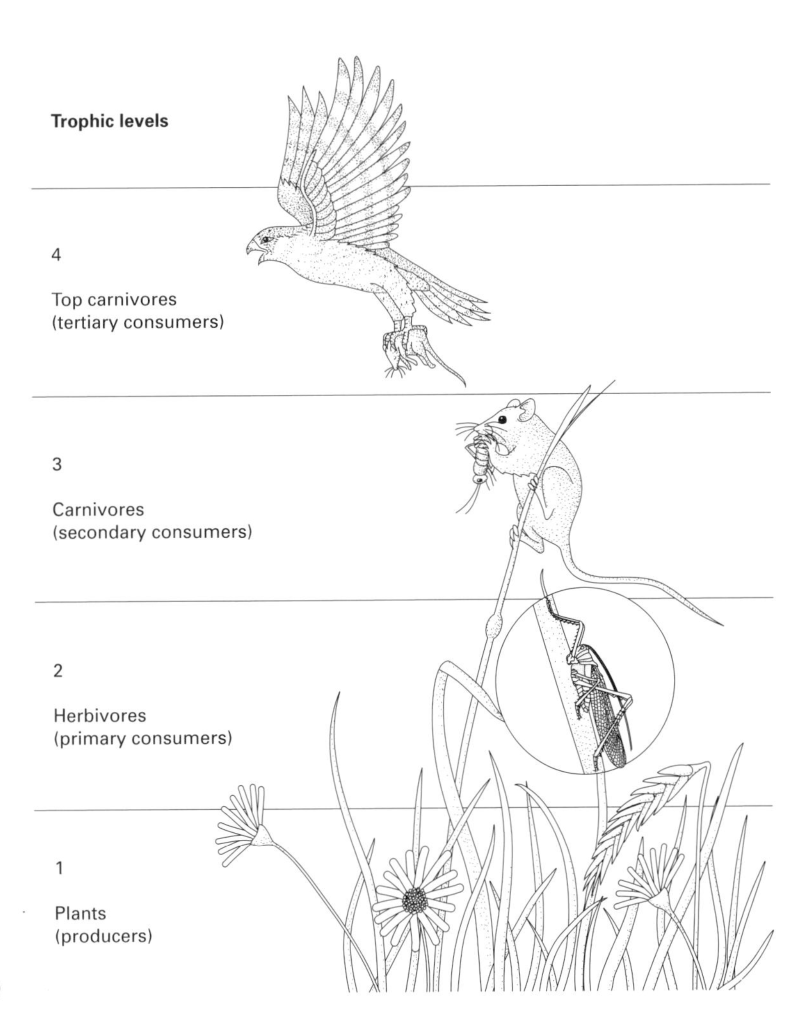

Fig. 7.4 Some of the carbon trapped by plants during photosynthesis is ultimately released during respiration by top carnivores (tertiary consumers) such as hawks. (From Raven *et al.* 1993.)

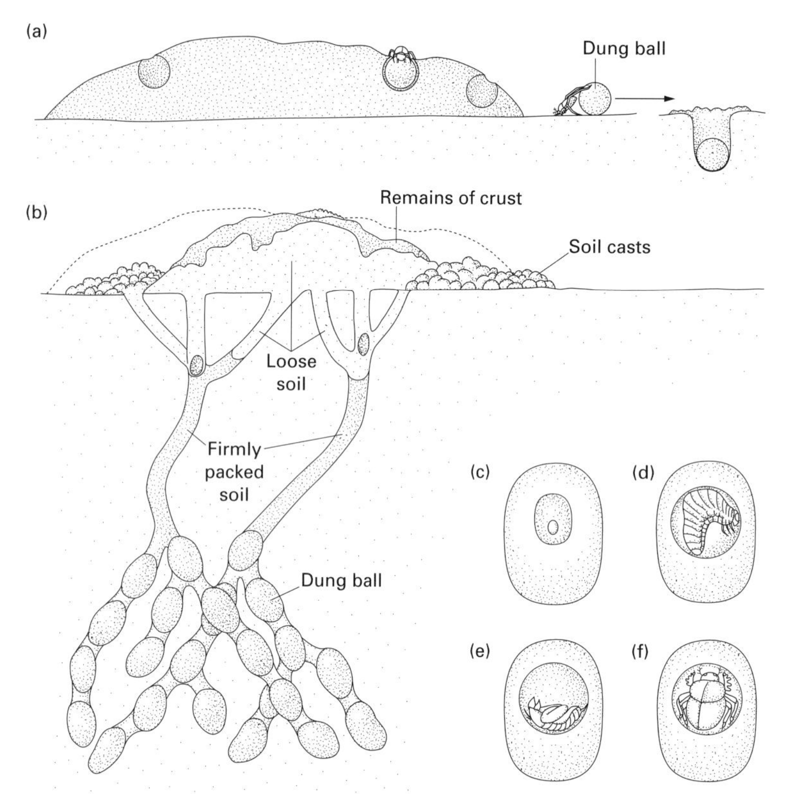

Fig. 7.5 Some dung beetles such as (a) *Garreta nitens*, roll dung for burial some distance from its source. Most (b) form nests below the dung pat where (c) eggs, (d) larvae, (e) pupae, and (f) adults develop. (From Waterhouse 1977.)

Dung beetles have actually been used as 'decomposition control agents' in Australia. Although Australia has about 250 native species in the subfamily Scarabaeinae, their foraging habits appear much better adapted to the small, dry dung of marsupials than the large, wetter dung of the cattle introduced by European settlers in the late 1700s. A single adult bovine drops an average of 12 dung pads per day and, if not removed, the dung can render inaccessible about one-tenth of a hectare per bovine per year (Waterhouse 1977). With 30×10^6 cattle in Australia, dung pads can put about 2.5×10^6 hectares of pasture land out of service each year, constituting a truly enormous loss to the dairy and beef industries. In addition, dung pads act as resources for insect pests such as the bush fly, *Musca vetustissima*, and the buffalo fly, *Haematobia irritans exigua* (Diptera: Muscidae). In 1967, dung beetles from Africa, with its extensive large mammal fauna, were introduced into Australia for the control of cattle dung. Dung beetles have spread widely in Australian pasture and have significantly reduced the build-up of cattle dung. Researchers continue to seek the most appropriate beetle species for both the removal of the dung itself and to control the insect pests that live within it (Kirk & Wallace 1990; Davis 1996). The battle is certainly not over. There is some evidence that modern grain feeds reduce the palatability of cattle dung for dung beetles (Dadour & Cook 1996). None the less, the dung beetle introductions provide compelling evidence of the importance of certain insects to decomposition processes in ecosystems.

Perhaps the most impressive decomposers in the insect world are the termites (Isoptera, Fig. 7.6). Although some termites feed on soil organic matter, most feed on dead and decaying plant material including leaf litter, roots, and woody debris (Whitford *et al.* 1988; Moorhead & Reynolds 1991). Dead plant material poses special problems to decomposers because it contains high proportions of cellulose and lignin, neither of which is readily digested by animals. In fact, amongst the insects, some cockroaches and higher termites in the subfamily Nasutitermitinae are the only taxa known to synthesize enzymes capable of

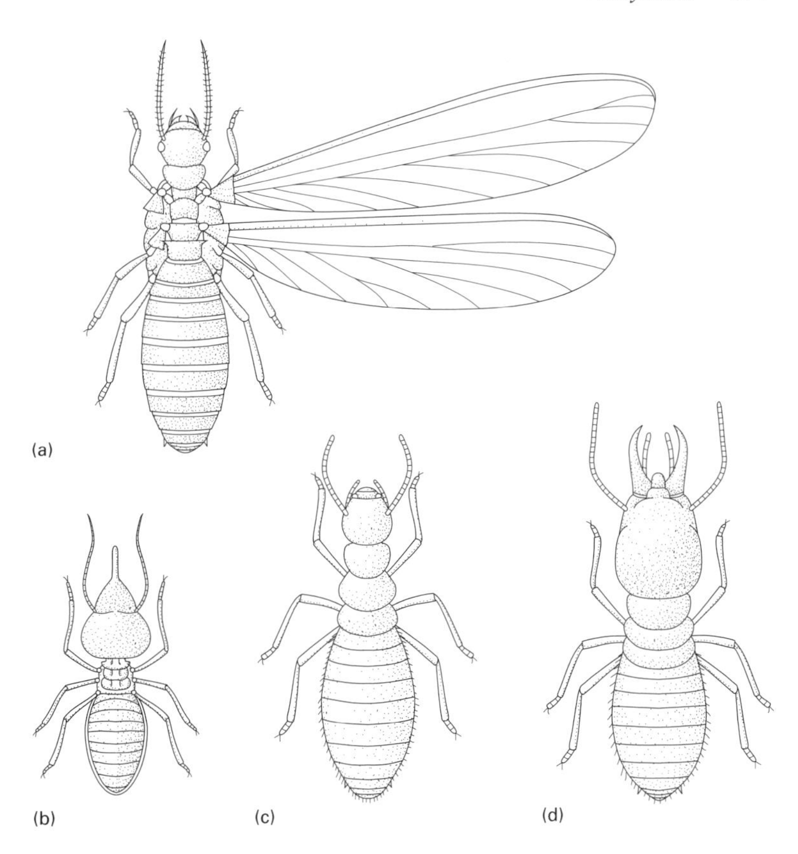

Fig. 7.6 Castes of termites. (a) Sexual stage of *Amitermes tubiformans*. (b) Nasutus of *Tenuirostritermes tenuirostris*. (c) Worker, and (d) soldier of *Prorhinotermes simplex*. (From Borror *et al.* 1981.)

degrading cellulose (Martin 1991). None the less, termites are efficient decomposers of plant material in desert (Whitford *et al.* 1988), savanna (Wood & Sands 1978), and forest ecosystems (Bignell *et al.* 1997) because of their symbioses with micro-organisms that live in the guts of termites and produce cellulolytic (cellulose-digesting) enzymes (see also Chapter 6, section 6.4.5). The gut symbionts of various termite groups include both flagellate protozoans (Yoshimura *et al.* 1993) and bacteria (Basaglia *et al.* 1992). One crucial feature of termites relevant to the carbon cycle is the occurrence of anaerobic microsites in termite guts. When plant material decomposes in the absence of oxygen, the end product is methane (CH_4) instead of carbon dioxide (CO_2). Termites therefore have the potential to recycle significant amounts of carbon to the atmosphere in two gaseous forms. Methane is one of the principal greenhouse gases, contributing about 18% to the effects of such gases on climatic variation (Anon. 1992).

There is some debate on the relative importance of termites to global fluxes of CO_2 and NH_4. We might expect termites to be an important component of the terrestrial carbon cycle, if only because they are very abundant in many forest ecosystems. Forests contain more organic carbon than all other terrestrial systems and at present account for about 90% of the annual carbon flux between the atmosphere and the Earth's land surface (Groombridge 1992). The literature now contains estimates of termite abundance and biomass from each of the three global blocks of tropical forest, and these data suggest that termites may be an order of magnitude more abundant than the next most abundant arthropod group, the ants. The highest biomass reported for termites, 50–100 g/m^2 in southern Cameroon forest (Eggleton *et al.* 1996), is greater than any other component of the invertebrate (or vertebrate) biota and may constitute as much as 95% of all soil insect biomass (Bignell *et al.* 1997). If direct carbon fluxes by insects are significant components of ecosystem processes, by far the greatest contribution will be made by termites. In addition to direct carbon emissions during decomposition, the Macrotermitinae also cultivate a fungus on dead plant material. The metabolic rate of this cultivated fungus and the quantity of CO_2 that it releases exceed the corresponding values for its termite gardeners (Wood & Sands 1978).

What are the effects of termites on the global carbon cycle? Because of technical difficulties in estimating termite densities, their rates of respiration, and the effects of the environment on CO_2 and NH_4 emissions, there are probably no truly reliable estimates (Bignell *et al.* 1997). However, data suggest that 20% of all CO_2 produced in savanna ecosystems results from termite activity (Holt 1987): termites can consume up to 55% of all surface litter in such systems (Wood & Sands 1978). Globally, the number is likely to be much lower. Based on the most reliable estimates to date (Bignell *et al.* 1997), termites probably contribute up to 20% of the global production of NH_4 and 2% of the global production of CO_2. Although these numbers may not seem dramatic, we should remember that these are the contributions of a single order of insects that represent less that 0.01% of terrestrial species richness. In fact, the global CO_2 produced by termites each year probably exceeds that taken up by northern hemisphere regrowth forests (Houghton *et al.* 1990). Incidentally, termite densities drop dramatically in clear-cut forests (Watt *et al.* 1997), and the practice of clear-cutting therefore has implications for global levels of NH_4 and CO_2 production.

7.4 Leaf-shredding insects and stream ecosystems

The role of insects in the flow of energy and matter through ecosystems is understood nowhere better than in stream systems. One of the reasons that ecosystem approaches have been so pervasive in stream entomology is the recognition that many individual species may be 'functionally redundant' (Lawton 1991), behaving in such similar ways that the presence or absence of a particular species matters less than the 'functional group' to which it belongs (Wallace & Webster 1996). Insect functional groups in streams (Cummins & Klug 1979) are somewhat analogous to the 'guilds' of insects described in Chapter 3, although there is even more emphasis on the mode of feeding by which resources are consumed. Functional groups include grazers, shredders, gatherers, filter feeders, and predators, and are based on their morpho-behavioural mechanisms for gathering food rather than taxonomic relationships. Although all of these functional groups are relevant to the pathways taken by energy and materials through stream ecosystems (Wallace & Webster 1996), we focus here on shredding insects because of their profound effects on stream food chains, particularly in forested stream environments.

Upland streams in forested areas receive much of their energy base from outside the stream itself. Although primary production by algae and other plants occurs in such streams, low light levels under forest canopies reduce the importance of this energy pathway compared with terrestrial systems (above). Instead, much of the energy base in streams is derived from litter inputs (leaves, twigs, woody debris, etc.) from the terrestrial environment. Of course, this energy source is still the result of primary production, it just comes from outside the stream (allochthonous production) as opposed to inside the stream (autochthonous production). For example, in the eastern USA, litter-fall into streams averages $600\,g/m^2$ per year, over half of which consists of leaves (Webster *et al.* 1995).

Shredders are the functional group of insect larvae (and sometimes other arthropods) that fragment (or 'comminute') the litter in streams into particle sizes that can be used by other invertebrate groups. Leaf-shredding insects are dominated by three insect orders: the caddisflies (Trichoptera) (Fig. 7.7), stoneflies (Plecoptera) (Fig. 7.8) and true flies (Diptera). It is important to point out, however, that many species in these three orders fall into other functional groups as well, such as grazers, filter feeders, etc. Some shredders appear to select leaf-litter tissue that has been colonized and partially decomposed (or 'conditioned') by fungi and bacteria (Cummins & Klug 1979). Indeed, shredders also ingest attached algae and bacteria along with litter tissue (Merritt & Cummins 1984), although they probably do not receive much of their energy requirements from such sources (Findlay *et al.* 1984). Instead, some shredding insects derive enzymes from microbial symbionts or ingested microbes that may be active in cellulose hydrolysis (Sinsabaugh *et al.* 1985), a similar strategy to that used by the termites in terrestrial systems described earlier in this chapter.

However, the major importance of shredding insects in stream ecosystems is their ability to turn the

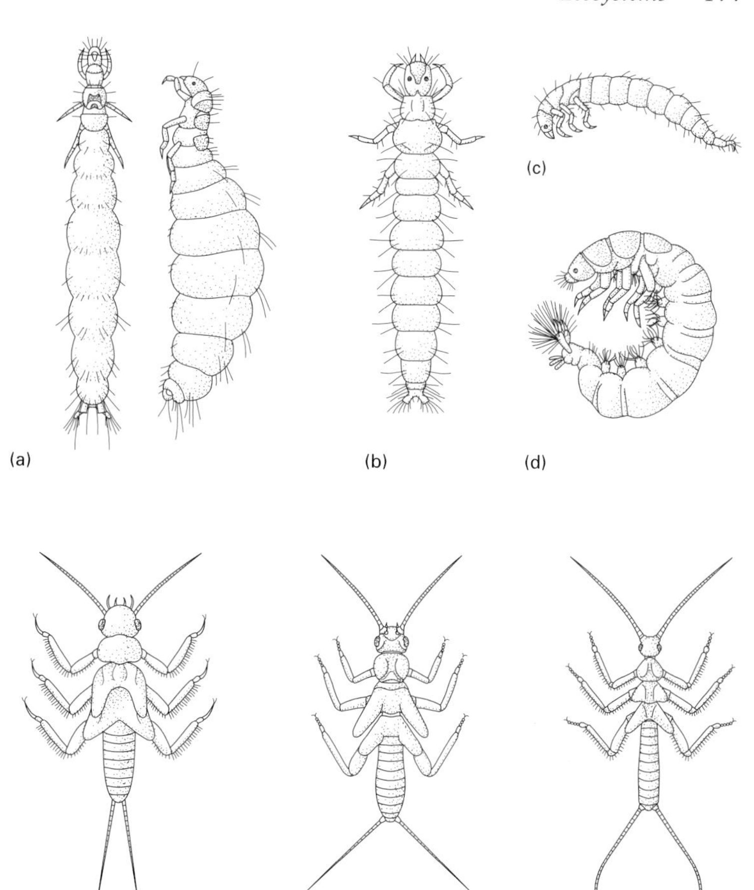

Fig. 7.7 Aquatic larvae in the order Trichoptera (caddisflies). Larvae are (a) *Hydroptila waubesiana*, (b) *Rhyacophila fenestra*, (c) *Polycentropus interruptus*, and (d) *Hydropsyche simulans*. (From Borror *et al.* 1981.)

Fig. 7.8 Aquatic larvae in the order Plecoptera (stoneflies). Larvae are (a) *Isoperla transmarina*, (b) *Nemoura trispinosa*, and (c) *Taeniopteryx glacialis*. (From Borror *et al.* 1981.)

coarse particulate organic matter (CPOM) of litter into fine particulate organic matter (FPOM) and dissolved organic matter (DOM) (Wallace *et al.* 1982; Meyer & O'Hop 1983). Shredders are not especially efficient feeders, assimilating only 10–20% of the material that they ingest. Most litter passes through the gut, emerging as fine particles or dissolved fractions in the faeces. FPOM and DOM are major sources of nutrition for gatherers, filter feeders (e.g. blackfly larvae in the dipteran family Simulidae) and microbes in streams (Cummins *et al.* 1973; Short & Maslin 1977; Wotton 1994; Fig. 7.9). In other words, if allochthonous (outside) litter is a major energy source for streams, shredding insects are the catalyst that makes that energy available to a wide variety of stream biota. Moreover, as both FPOM and DOM are readily transported downstream (Cuffney *et al.* 1990; Cushing *et al.* 1993), shredders make the products of litter decomposition available to organisms at some distance from the major sites of litter input. In addition, defecation by shredding insects that burrow periodically in stream sediments increases the organic content of those sediments by 75–185% (Wagner 1991). Finally, insect shredders also promote wood decomposition by scraping, gouging, and tunnelling into the woody debris (twigs, branches, stems) that falls into streams. Freshly gouged surfaces act as sites for microbial activity and subsequent decomposition (Anderson *et al.* 1984), and shredder activity has even been implicated in the collapse of highway bridges supported by untreated timber pilings.

In an experimental investigation of the role of insects in stream ecosystem processes, Wallace and colleagues from the University of Georgia eliminated more than 90% of the insect biomass in a southern

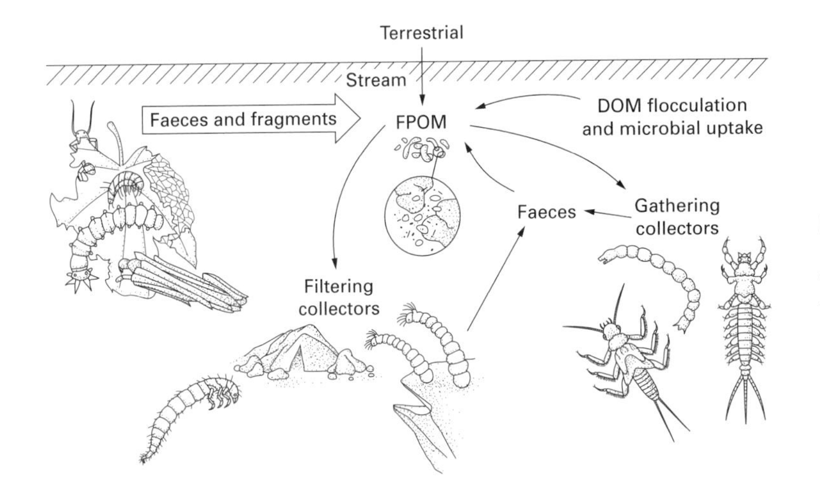

Fig. 7.9 Sources and users of fine particulate organic matter (FPOM) in temperate streams. It should be noted that insect shredders (top left) are a major source of FPOM. They provide both fragments of leaf litter and faeces that are necessary sources of food for filtering collectors and gathering collectors. (After Cummins & Klug 1979 and redrawn from Allan 1995.)

Appalachian stream by applying insecticide. Their manipulation significantly reduced the rate of leaf litter breakdown and the export of fine particulate organic matter in the stream (Wallace *et al.* 1982, 1991; Cuffney *et al.* 1990). Both litter decomposition and FPOM export recovered at the same time as the shredder community recovered from the insecticide treatment (Wallace *et al.* 1986; Fig. 7.10). Wallace's data suggest that insects accounted for 25–28% of annual leaf-litter processing and 56% of FPOM export over a 3 year period in this stream. Larval insects are clearly critical to ecosystem processes in streams, and we encourage interested readers to explore the literature on functional groups other than shredders and their effects on the flows of energy and matter (Wallace & Webster 1996).

In this chapter, we have presented two examples of the role of insects in ecosystem function. Perhaps the most important function that insects can serve in ecosystems is the transformation of plant material to animal material during primary consumption. We learned in Chapters 1 and 3 that plant-feeding insects are abundant and more speciose than other kinds of insect. As such, they provide a crucial resource base for a diversity of secondary consumers including birds, mammals, reptiles, amphibians, fish, other insects and, in some parts of the world, humans. Other potential roles of insects in ecosystems are less well characterized. They may, for example, regulate primary production in some systems (Schowalter *et al.* 1986; Dyer *et al.* 1993) by intense grazing pressure.

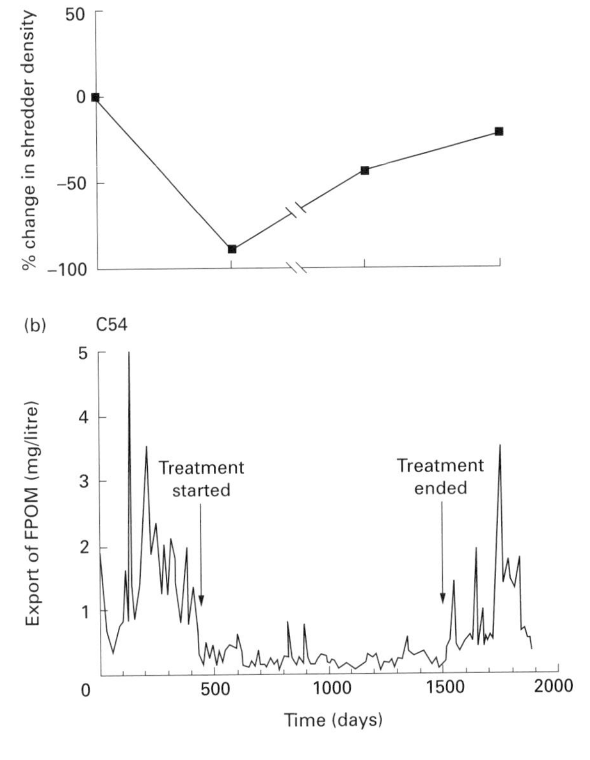

Fig. 7.10 Changes in (a) shredder density (%) and (b) export of fine particulate organic matter (mg/l) after insecticide treatment in a headwater stream in the southern Appalachians. Concentrations of FPOM return to normal when insect shredder density recovers from the treatment. (After Wallace *et al.* 1991; Whiles & Wallace 1995.)

In addition, by dropping leaf fragments and frass (faeces) to the forest floor, canopy-dwelling insects may return nutrients from the tops of trees to the litter and soil (Seastedt *et al.* 1988; Lovett & Ruesink 1995). Given that we understand only a little of the ecosystem ecology of insects, and the potential benefits that they provide, it would seem wise to conserve as much insect biodiversity as possible. The diversity and conservation of insects is the topic of the next chapter.

Chapter 8: Biodiversity and Conservation

8.1 Introduction

The word 'biodiversity' was first used (as 'BioDiversity', a contraction of 'biological diversity') in 1986 (Wilson 1997). It has become widely used since then. Perhaps the main reason for its popularity has been its widespread usage in relation to threats to the natural environment and the plants and animals in them (Gaston 1996a). This usage is understandable: it became clear during the 1980s that most of the plants and animals in the world remained undescribed and that deforestation, climate change and other factors threatened to cause massive extinction of species (Reid 1992; Mawdsley & Stork 1995).

At the species level, there is no doubt that insects make up most of the world's biodiversity. Of all the 1.7×10^6 species described, approximately 45 000 are vertebrates, 250 000 are plants and 950 000 (56%) are insects (Groombridge 1992). Of the estimated number of species in the world, some 64% are thought to be insects (Fig. 8.1). The most diverse order of insects is the Coleoptera (300 000 species, 24% of all species), followed by the Diptera (85 000, 7%), Hymenoptera (110 000, 8%) and Lepidoptera (110 000, 9%) (May 1988) (see also Chapter 1). Figure 8.1 shows estimates for the numbers of species in each of these orders, adjusted to take into account the fact that some insect groups are much better described than others. For example, 9% of all described species are Lepidoptera but only 3% of all described and undescribed species are thought to be from this order. However, even the adjusted estimate for Lepidoptera still leaves them eight times as speciose as vertebrates.

Insects are not only more numerically diverse than other organisms, they also perform a disproportionate number of functional roles: herbivory, predation, parasitism, decomposition, pollination, etc. The scale and breadth of insect diversity is well summed up by the title of Wilson's (1987) paper: 'The little things that run the world (the importance and conservation of invertebrates).' This chapter discusses the measurement of biodiversity, particularly insect diversity, natural patterns in insect diversity, threats to biodiversity and insect conservation.

8.2 Measuring biodiversity

8.2.1 Defining biodiversity

Biodiversity may be defined as 'the variability among living organisms from all sources including, *inter alia*, terrestrial, marine and other aquatic ecosystems and the ecological complexes of which they are part; this includes diversity within species, between species and of ecosystems' (The Convention on Biological Diversity 1992). Most other definitions also include a hierarchical element, typically genes, species and ecosystems, and many discussions of biodiversity implicitly or explicitly include ecosystem processes, such as nutrient flow (Noss 1990; Gaston 1996a). The inclusion of ecosystem processes in a definition of biodiversity seems to be both confusing and unnecessary, and is not followed here, but the measurement of species diversity within and between different functional groups of insects is discussed in section 8.2.6 and the critical role of biodiversity in maintaining 'ecosystem health' (through its effect on ecosystem processes) is considered in section 8.3.6.

The fundamental level of biodiversity is the species, notwithstanding the other levels of biodiversity and the taxonomic problem of what constitutes a species. Species diversity tends to be expressed as species richness, i.e. the number of species in a given area or habitat (but see below). Although the terms 'species richness' and '(bio)diversity' are often confused, species richness tends to be closely correlated with other measures of biodiversity (Gaston 1996b).

8.2.2 Species diversity, species richness and diversity indices

Species diversity may be measured as species richness or, more strictly, as a diversity index. Diversity indices

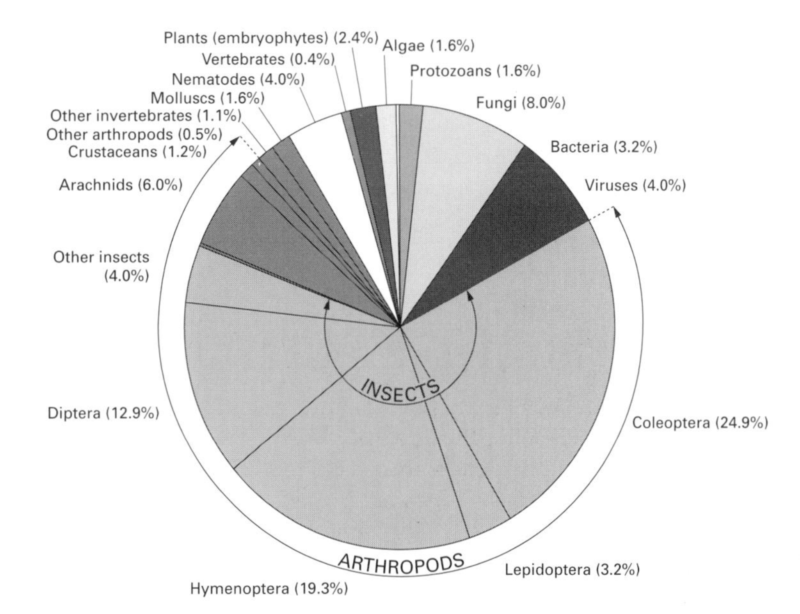

Fig. 8.1 Estimated proportions of species of insect and other groups of organisms of the global total. (From Groombridge 1992.)

measure both species richness and the evenness of their abundance. Magurran (1988) has provided a critical review of the many diversity indices that were developed during the 1960s and 1970s, and both her book and that of Krebs (1989) have provided excellent guides to calculating diversity indices such as the log series index α, Simpson's index, D, and the Shannon index, H'. Table 8.1 demonstrates the application of these and a few other indices for four ground-dwelling beetle communities.

Both Krebs and Magurran have urged caution in the use of biodiversity indices because of the controversy over which indices of diversity are the 'best'. Magurran recommended the calculation of the Margalef and Berger–Parker indices because they are simple and easy to interpret, the log series index $\grave{a}$ because of its wide use and theoretical robustness, and the Shannon index because of its widespread use and despite the widespread criticism it has attracted. Magurran also recommended the drawing of rank abundance graphs (Fig. 8.2).

However, the problem with diversity indices, no matter how strongly based in ecological theory, is that they provide descriptors of diversity whose values mean very little to most people. Indeed, Pielou (1995) considered that diversity indices 'are wholly unsuitable for measuring biodiversity'. In contrast, the number of species in a particular area, community or habitat is clearly a tangible entity (even if it is hard to estimate). Moreover, information on abundance of the species in a community is better interpreted graphically (Fig. 8.2) than through a complex index.

8.2.3 Calculating species richness

Although conceptually a simple measure, species richness may not be simple to assess accurately. The total number of plant species in most temperate habitats may be easy to measure but the total number of species in most insect groups in most habitats must be estimated by sampling. Several books describe sampling methods for different insect groups (e.g. Southwood 1978; Jermy *et al.* 1995). Sampling of insects (and other animals and plants) produces species-accumulation curves such as those shown in Fig. 8.3. The total number of species found by sampling rises rapidly at first then the rate at which new species are recorded gradually declines, eventually reaching an asymptote, given sufficient sampling. There are several statistical techniques for estimating species richness from sampling data. These techniques were reviewed by Colwell and Coddington (1994), who concluded that there have been insufficient studies on which to base a recommendation on which technique is best. Even so, one of the most frequently used techniques is the Jack-knife method (Colwell &

Table 8.1 Species diversity indices for ground beetles in four forests: Sitka spruce and birch sites in Moray (north-east Scotland) and Knapdale (south-west Scotland). (Data from Watt, Barbour & McBeath, unpublished. Note that only data from the 20 most abundant of the 81 species recorded are shown.)

	Moray spruce	Moray birch	Knapdale spruce	Knapdale birch
Philonthus decorus	0	246	0	39
Tachinus signatus	4	184	2	2
Abax parallelpipedus	0	0	6	167
Nebria brevicollis	2	149	0	0
Geotrupes stercorosus	0	0	2	91
Carabus problematicus	4	40	0	25
Leistus rufescens	20	22	5	6
Carabus violaceus	34	14	0	0
Calathus micropterus	4	38	0	0
Strophosomus melanogrammus	12	26	0	0
Barypeithes araneiformis	9	0	26	0
Catops nigrita	15	12	1	2
Olophrum piceum	7	17	1	0
Othius myrmecophilus	5	6	7	1
Pterostichus madidus	0	11	3	5
Catops tristis	3	13	1	1
Carabus glabratus	0	0	7	10
Tachinus lacticollis	3	12	2	0
Anthobium unicolor	2	10	0	0
Dalopius marginatus	3	9	0	0
Number of species	39	56	22	30
Number of individuals	187	905	78	394
α	1045	8766	396	3408
Margalef index	7.26	8.08	4.82	4.85
Berger–Parker index	0.182	0.272	0.333	0.424
$1 - D$ (Simpson's index)	0.931	0.850	0.854	0.751
Shannon index (of diversity)	3.142	2.538	2.478	1.953

Coddington 1994). The species richness of different sites (or the same site at different times) can be compared by the total number of species sampled or the Jack-knife (or other technique for extrapolation) estimate of species richness. The former approach is valid only when the same sampling intensity is used in each site or on each occasion. Statistical errors can be calculated for the Jack-knife and similar methods (Colwell & Coddington 1994).

A widely used alternative to comparing the species richness of different sites by extrapolation is rarefaction. In essence, if there are differences in the total number of individuals sampled in different sites, rarefaction can be used to estimate the number of species present in samples of the same size. Krebs (1989) provided examples of its use.

8.2.4 Attaching values to species

A major problem with the use of species richness (or, indeed, any diversity index) in relation to the conservation of biodiversity is that it ignores any value that may be attached to a species. In particular, it fails to differentiate between common species and rare, potentially threatened species. It has been suggested that weights be applied to different species. For example, Usher and Pineda (1991) recommended that, in assessing the diversity of plant species in a given area, the greatest weights should be given to native species in their characteristic habitats, lower weights should be given to native species not in their characteristic habitats, and the lowest weights should be given to alien (non-native or 'exotic') species. Alternatively, weights could be given to species according to their rarity or geographical range (Pielou 1995). These approaches have not, as yet, been widely used and they run the risk of producing species richness indices as difficult to understand as diversity indices. However, there may be much to be gained by separately presenting the species richness

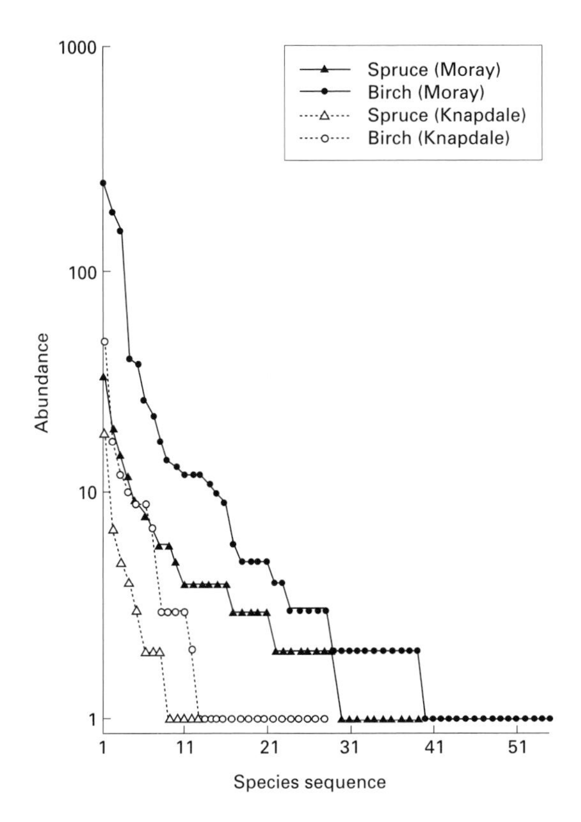

Fig. 8.2 Species abundance curves for ground-dwelling beetles in four Scottish forests: Sitka spruce and birch in Moray and Knapdale. (Source as Table 8.1.)

of clearly defined groups of species, such as native and alien species, or the numbers of species in different classes of rarity or risk from local, regional or global extinction. Probably the best example of classifying species according to rarity and risk of extinction is the Red List and Red Data Book (RDB) system, considered below in relation to insect conservation (section 8.5.2). Unfortunately, although valuable for focusing attention on particular threatened species, particularly vertebrates, the problem with applying systems such as the RDB classification to insects is that, in most parts of the world for most insect groups, we do not know which insect species are endangered. Thus, although some measure of the status of insects is desirable, it is, apart from well-studied groups in a small number of countries, usually impossible to achieve.

8.2.5 Comparing species composition

Insect ecologists frequently wish to compare the biodiversity of two or more sites with something more ecologically meaningful than species richness. This usually involves a comparison of species composition. This can be done in various ways but the simplest way is to compare the presence of different species. Figure 8.4 shows the number and composition of four communities of rove beetles (Coleoptera: Staphylinidae) in different sites in the coastal Sitka spruce forest of Vancouver Island (Winchester 1997). Winchester found that sites varied in their species richness and their species composition. Figure 8.4 summarizes variation in species composition in two ways: by the number of species shared between different (pairs of) sites and by the number of species unique to each site. This clearly shows how important the canopy and 'forest interior' sites are in terms of the total number of species found there and, more importantly, in terms of the numbers of species found only in each of these sites.

The approach illustrated above has two important limitations. First, its value is restricted to the comparison of a relatively small number of sites. A figure such as Fig. 8.4 could not easily be interpreted with more than four sites. Second, it ignores the abundance of different species in each site. This is not a problem if we are sure that sampling tells us what species exist in each site, but sampling, particularly of most insect groups, will often result in the collection of specimens that are simply dispersing through the site being sampled. The problem of these so-called 'tourists' has long been recognized (see section 8.3.5). To a large extent it can be overcome by avoiding sampling techniques that tend to collect large numbers of tourists and by concentrating on insect groups that do not disperse widely. In addition, ecologists can use measures of the abundance of different species to compare the insect communities in different sites.

Various statistical techniques are needed to overcome the problems outlined above. The simplest approach is to use a similarity index to compare different sites. There are a wide range of similarity indices available, reviewed with worked examples by Krebs (1989). These indices vary from those based on the presence or absence of species, such as the Jaccard and Sorensen coefficients, to those that are also based on the abundance of species. The latter include the Percentage Similarity (or Renkonen) index, which is notable for being easy to calculate, and the Morisita index, which is widely considered to be the best available. Table 8.2 shows an example of

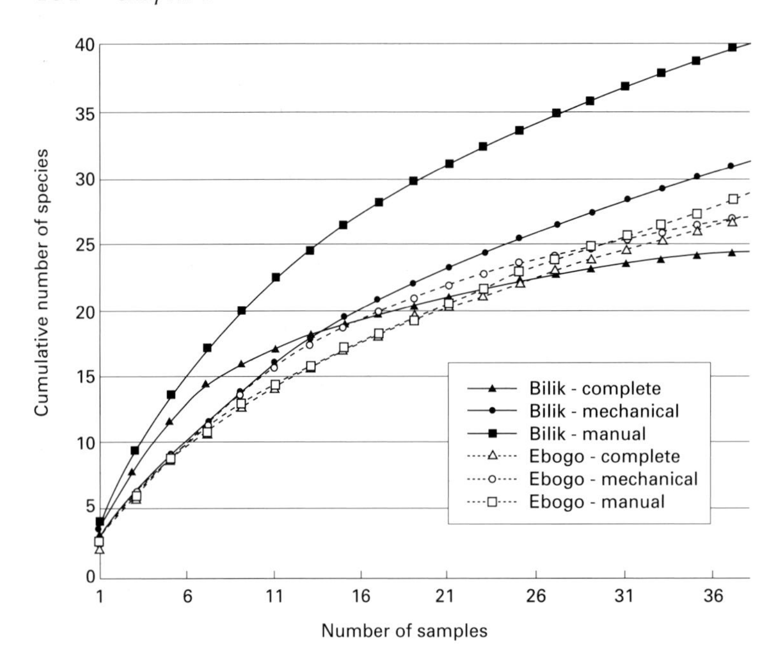

Fig. 8.3 Species accumulation curves for canopy-dwelling ants in six plantation plots in the Mbalmayo Forest Reserve in southern Cameroon: forest plots cleared by complete, partial manual and partial mechanical clearance in two locations within the Reserve (Bilik and Ebogo). (Data from Watt, Stork & Bolton, unpublished.)

Table 8.2 Morisita similarity index values based on Table 8.1 (plus additional data collected at the four forest sites). The analysis shows moderate similarities between both the different beetle communities at each forest (maximum for identical communities would be approximately unity) and the communities associated with the two tree species.

	Moray birch	Knapdale spruce	Knapdale birch
Moray spruce	0.67	0.00	0.00
Moray birch		0.00	0.00
Knapdale spruce			0.84

the use of the Morisita index on the data shown in Table 8.1.

Similarity analysis results in the production of tables with similarity values comparing pairs of communities. For small numbers of sites no further analysis may be needed but these tables are often difficult to interpret. In most circumstances, therefore, cluster analysis on the index values is carried out. There are, of course, a range of cluster analysis techniques available (Krebs 1989) but, because of their general utility, they can be found within statistical packages such as SAS or SPSS. The latter package was employed, for example, by Chey *et al.* (1998) in their comparisons of canopy arthropod communities in exotic tree plantations in Sabah, Borneo. Figure 8.5 shows the results of average linkage cluster analysis on termitic communities.

Further analyses of the composition of insect communities can be carried out and, although complex, these analyses can be particularly useful for identifying the underlying factors causing changes in species composition. These multivariate analytical techniques include Canonical Correspondence Analysis (CCA), Detrended Correspondence Analysis (DCA or DECORANA) and Two Way Indicator Species Analysis (TWINSPAN). It is beyond the scope of this book to describe these techniques but a brief description of some recent examples should illustrate their value. Dennis *et al.* (1997) used Canonical Correspondence Analysis to show how plant species composition, vegetational structure and degree of grazing by sheep and cattle affected the composition of the ground beetle community. Setala *et al.* (1995) used Canonical Correspondence Analysis and Detrended Correspondence

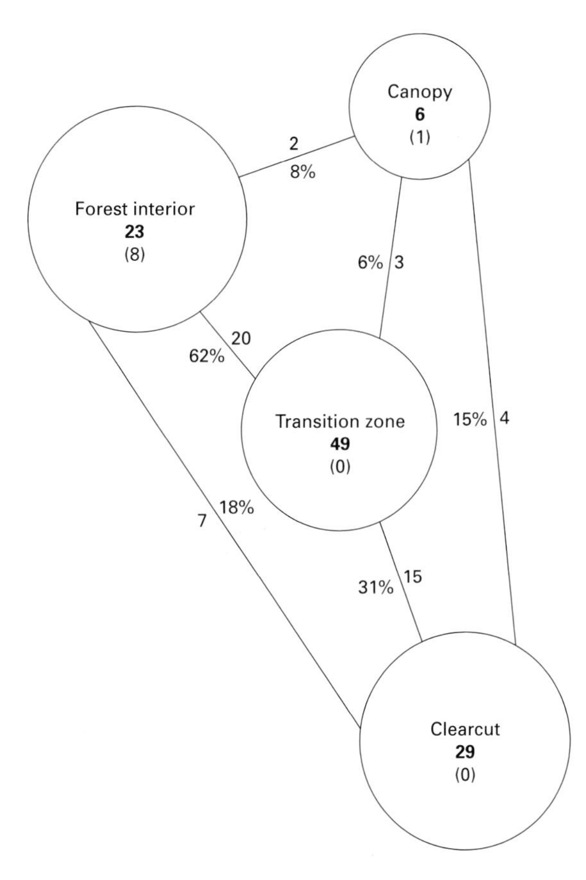

Fig. 8.4 Staphylinid species in different habitats in the Upper Carmanah Valley, Vancouver Island, Canada. The number of species in each habitat is given in bold, the number of species occurring in common in different habitats is given along the lines joining the habitats, and the number of species unique to each habitat is given within parentheses within the circles. (From Winchester 1997.)

Analysis to show how the age of a forest stand affected the collembolan communities living within Douglas fir stumps. Chey *et al.* (1997) employed TWINSPAN to investigate variations in forest Lepidoptera collected by light trapping in relation to forest management and tree species composition, and Collier (1995) used Detrended Correspondence Analysis and TWINSPAN to identify the importance of factors such as the amount of native forest in the riparian zone of streams and rivers on the aquatic macroinvertebrate community downstream. The techniques used in these examples may be complex but they can have an important practical value by identifying factors that are critical for the maintenance of the biodiversity of particular environments. The last of the four studies cited above, for example, identified the need to maintain native trees alongside streams and rivers.

8.2.6 From All Taxa Biodiversity Inventories to rapid biodiversity assessments

Anyone wishing to measure the biodiversity, in the 'complete' sense of the word, of a given area is faced with an enormous amount of work. Some measure of the amount of sampling and identification is given by studies of insects in forest canopies. Considering only beetle species, Erwin (1982) found 955 species of Coleoptera (excluding weevils, which were not sorted to species) in the canopy of only 19 trees in Peru, and Allison *et al.* (1993) reported 633 beetle species in a total of 4840 individuals in a study in Papua New Guinea. Basset (1991) sorted 51 600 arthropods in an Australian study, and Moran and Southwood (1982) sorted and identified nearly 42 000 arthropods in their study of insects associated with British and South African trees (see below). Few entomologists record the amount of time taken to acquire the information they report but Lawton *et al.* (1998) estimated that it took 3470 h to sample, sort and identify the butterflies (132 species), canopy beetles (342 species), canopy ants (96 species), leaf-litter ants (111 species) and termites (114 species) in a study of an area of about 10 ha in a forest in southern Cameroon.

Despite the enormity of the task, some ecologists have advocated the need to carry out a total inventory of all the plant and animal species in a given area. This approach, known as an All Taxa Biodiversity Inventory (ATBI), has been championed by Janzen (1997). However, the purpose of an ATBI is much more than an inventory. Janzen (1997) outlined some of the reasons for establishing ATBIs. They may provide:

- a baseline for measuring the impact of global change (i.e. climate change, land use change, etc.)—'a gigantic canary in the mine' (Janzen 1997);
- a standard for calibrating and developing methods of measuring biodiversity;
- an area for studying ecology;
- a source of species, genes and products for human use;
- an area for demonstrating the biological world to students (of all stages);
- a provider for ecosystem services such as carbon sequestration.

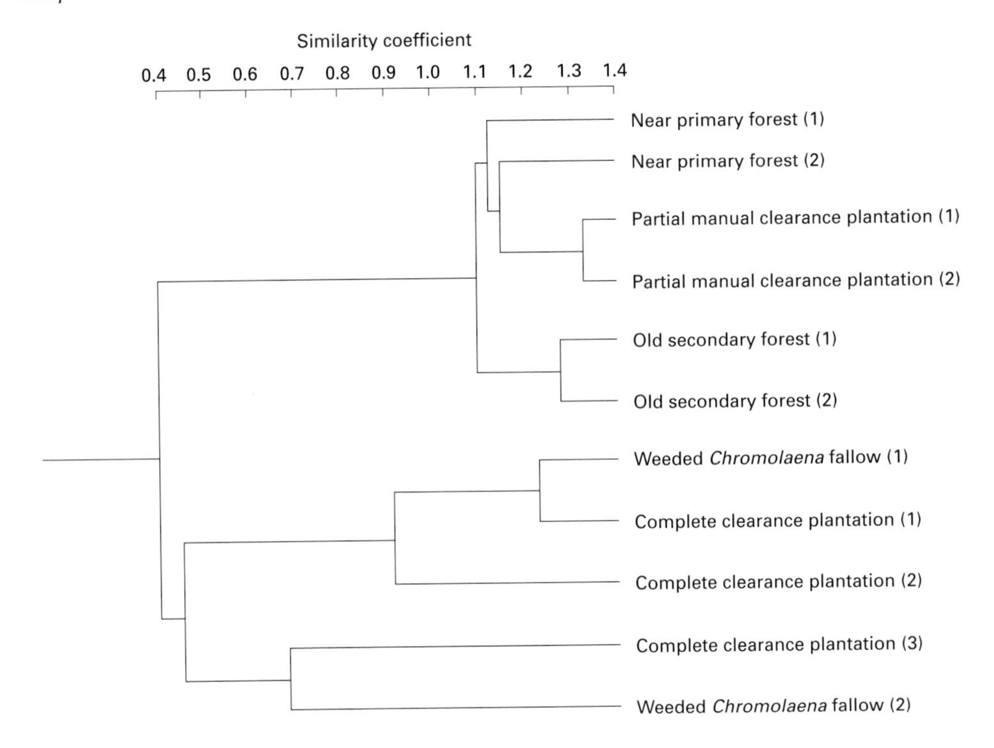

Fig. 8.5 Results of a cluster analysis showing the similarities of termite assemblages in the Mbalmayo Forest Reserve in Southern Cameroon (based on Morisita index values and using the UPGMA ('unweighted pair-group method using arithmetic averages') method (Krebs 1989)). (From Eggleton *et al.* 1996.)

Although ATBIs have a clear value, a complete inventory of all plants and animals is a time-consuming exercise. In most cases, however, the results of biodiversity assessments of given areas are needed rapidly as a basis for making decisions about, for example, where protected areas should be, how to manage particular (protected and other) areas and how to assess the success of management decisions. The often urgent need for biodiversity information has led to the development of biodiversity indicators and a range of rapid biodiversity assessment techniques. Biodiversity indicators may be defined as a genetic measurement, a species or species guild, a measurement of habitat structure or some other feature that provides a relative estimate of total biodiversity.

The extreme view of indicators is that a single species can indicate the presence of a certain type of biodiversity. This is unlikely to be the case for insects, but birds have often been proposed as indicator species in this, and other, contexts. For example, the resplendent quetzal (*Pharomachrus mocinno*), a large, frugivorous bird, has been used as an indicator of montane biodiversity in Central America (Powell & Bjork 1995). More generally, the species richness of an indicator group is thought to measure total biodiversity. Balmford and Long (1995), for example, suggested that at an international scale the number of bird species with breeding ranges of less than 50 000 km^2 was a useful indicator of endemism and, to a lesser extent, species richness in other animal and plant groups.

Among several insect groups recommended as indicators are the coleopteran families Scarabaeidae (Halffter & Favila 1993) and tiger beetles (Cicindelidae) (Pearson 1994). Pearson and Cassola (1992) considered tiger beetles (Fig. 8.6) to be particularly good for indicating regional patterns of biodiversity because of their stable taxonomy, well-understood biology and life history, ease of sampling, world-wide distribution, presence in a wide range of habitat types and specialization of individual species (within habitats). They also claimed that there is a close correla-

Fig. 8.6 Tiger beetle.

tion between the species richness of tiger beetles and the species richness of other vertebrate and invertebrate taxa, providing supporting evidence from tiger beetles, birds and butterflies across North America, the Indian subcontinent and Australia. The value of using tiger beetles, rather than other taxa, Pearson and Cassola argued, is that the number of tiger beetle species can be reliably estimated within 50 h in a single site. It would take much longer to do the same for other taxa, even relatively apparent ones such as birds or butterflies (although the time taken will, of course, depend upon the scale of the investigation).

Butterflies have also often been cited as a suitable indicator group. Beccaloni and Gaston (1995), however, have gone further and suggested that because the proportion of different butterfly families and subfamilies in the tropical forests of Latin America is relatively constant, it may be possible to use a single butterfly group as an indicator of overall butterfly diversity. They suggested the use of ithomiine butterflies (Nymphalidae: Ithomiinae), which make up an average of 4.6% of all butterflies.

There have been a few attempts at setting general criteria for indicators. Pearson (1994), for example, suggested the following criteria, merging practical and ecological considerations:

1 well-known and stable taxonomy;
2 well-known natural history;
3 readily surveyed and manipulated;
4 higher taxa broadly distributed geographically and over a breadth of habitat types;
5 lower taxa specialized and sensitive to habitat changes;
6 patterns of biodiversity reflected in other related and unrelated taxa;
7 potential economic importance.

Some of these criteria may be applied to certain insects but several of them exclude most insect groups in tropical forests, notably 'well-known natural history' and 'well-known and stable taxonomy'. And why, for example, choose taxa of 'potential economic importance' if the purpose is to measure overall biodiversity? Brown (1991) has provided the most complete list of criteria for insect indicator groups to date. Table 8.3 shows 15 insect groups ranked according to these criteria. Brown's analysis suggests that heliconid butterflies and ants are the best indicator groups. Termites, Collembola and selected Coleoptera and Hymenoptera families also rate highly in Brown's analysis.

Recent research has shown, however, that the use of a single indicator group is misguided. Prendergast (1997) (and Prendergast *et al.* 1993) showed that there is a poor correlation between the richness of different taxa at a regional level, and Lawton *et al.* (1998) showed that at a local level correlations between the diversity of different insect taxa is low and, therefore, reliance on a single indicator would give a poor measure of overall biodiversity.

Despite the problems with biodiversity indicators, potentially the easiest way to measure variation in total biodiversity, the fact remains that whether we want to study variation in biodiversity for its own sake, or to plan or monitor the success of conservation measures, we need ways of measuring biodiversity rapidly. Thus has come the development of a variety of techniques known as 'rapid biodiversity assessment'. Obviously, compromises are inevitable—any reduction in the number of plant and animal groups sampled and any deviation away from statistically robust sampling will lead to a degree of uncertainty about the reliability of the results as a measure of total biodiversity.

The first step in rapid biodiversity assessment is the choice of groups or taxa to be sampled. To overcome the limitations discussed above that are inherent in choosing one indicator group, entomologists and others concerned with biodiversity have suggested that a range of taxa should be sampled to adequately measure overall biodiversity—'predictor sets' (Kitching 1993) or 'shopping baskets' of taxa (Stork 1995). Many of the criteria mentioned above in relation to indicators also apply here; indeed, rapid biodiversity

Table 8.3 Evaluation of different groups of insects as indicators of biodiversity (scored in relation to twelve criteria; 0, + or ++). (From Brown 1991.)

	Insect groups														
										Lepidoptera					
Desirable quality for an indicator group in ecology and biogeography	Collembola	Odonata (dragonflies, damselflies)	Isoptera (termites)	Hemiptera: Corcidae, Pentatomidae, Cygaeidae, Tingidne, Myridae	Homoptera: Membracidae, Cercopidae	Coleoptera: Carabidae, Cicindellidae, Elateridae, Cerambycidae, Chrysomelidae, Curculionidae	Diptera: Asilidae, Tabanidae	Hymenoptera: Formicidae	Hym: Apoidea, Vespidae, Sphecidae	Sphingidae, Saturnoidea	Arctiidae	Papilionidae, Picridae	Morphinae, Satyrinae (*s.l.*)	Bait-attracted Nymphalinae	Heliconiini, Ithomiinae
Taxonomically and ecologically highly diversified	++	++	++	++	++	++	++	++	++	++	++	++	++	++	++
Species have high ecological fidelity	++	++	++	++	++	++	+	++	++	+	+	+	++	+	++
Relatively sedentary	++	++	++	+	+	++	+	++	+	+	+	+	+	+	++
Species narrowly endemic, or if widespread, well differentiated	+	+	+	+	+	++	+	++	+	+	+	+	++	+	++
Taxonomically well known, easy to identify	+	++	++	+	+	++	++	++	++	++	++	++	+	++	++
Well studied	+	+	++	+	+	++	+	++	++	++	++	++	+	+	++
Abundant, non-furtive, easy to find in field	+	++	+	+	+	+	++	+	+	+	+	++	++	+	++
Damped fluctuations (always present)	++	+	++	+	0	0	+	+	+	0	+	0	0	+	+
Easy to obtain large random samples of species and variation	+	+	+	0	0	+	+	++	+	0	+	0	0	+	++
Functionally important in ecosystem	++	+	++	+	+	++	+	++	++	+	+	+	+	+	+
Response to disturbance predictable, rapid, sensitive, analysable and linear	++	++	++	+	+	+	++	+	+	+	+	++	++	+	+
Associates closely with and indicates other species and specific resources	++	+	+	++	++	++	+	++	++	+	+	++	++	+	++
Total value as indicator (maximum score = 24)	19	18	20	14	13	19	16	21	18	13	15	16	16	14	21

assessment may be considered to be the use of sets of indicators to measure biodiversity. The key criterion for choosing a set of taxa is that together they ensure that the insects sampled represent as wide a range of taxonomic and functional groups as possible. In tropical forests, for example, we might choose butterflies, termites, ants and dung beetles. Together they represent four insect orders; butterflies are phytophagous, termites and dung beetles are decomposers, termites and ants are numerically the most dominant macroinvertebrate taxa in the soil and canopy, respectively (although neither group is restricted to these microhabitats), and ants are in themselves an extremely diverse group functionally including leaf-cutting species, wood-eating species, pollinators, Homoptera-tending species, predators and social parasites (Hölldobler & Wilson 1990).

The strategy of focusing on functionally diverse taxa is rather appealing but it depends upon being able to place individual species in different functional groups. With most insect species still unidentified let alone ecologically understood, it would appear to be a difficult task to categorize different species (in functionally diverse groups such as Hymenoptera and Coleoptera) according to their functional roles. However, although this may be true of taxa such as ants in most ecosystems, the functional role of individual species in, for example, the mega-diverse Coleoptera can be assessed. In many beetle families the functional role is well known and stable across the family, and in beetle families where there is species to species variation in functional roles, the role of individual species can be assessed by examination of the mouth parts. Table 8.4 shows an example of the classification of beetles into fungivores, xylophages, decomposers, herbivores and predators (Hammond *et al.* 1996).

Another important criterion for choosing taxa for biodiversity assessment is ease of identification. There are two aspects of this: availability of specialists and availability of user-friendly identification guides. The latter lag well behind the former and the only insects well served by identification guides for most of the world are butterflies (e.g. D'Abrera 1982, 1985, 1986; De Vries 1987, 1997). There is an urgent need to improve identification guides: 'No more turgid keys, please. Expert systems, picture keys . . . are a major step forward' (Janzen 1997). At present, most insect groups require specialist (and expensive) taxonomists for identification of species, and in many cases, family. Moreover, only a limited part of the world is reasonably well served by insect taxonomists (Gaston & May 1992).

It would be a mistake, however, to restrict studies on biodiversity to the most easily identifiable but least diverse groups, neglecting the difficult taxa that form most of insect, and total, biodiversity. There are two possible solutions to this problem: the use of higher-taxon measures of biodiversity as surrogates for species-based measures, and the use of morphospecies. Morphospecies are essentially individual specimens that can be placed in groups on the basis of very similar shapes, sizes, colours, etc. Although their taxonomic identities are unknown, it is often sufficient for biodiversity studies that they look the same.

Higher-taxon measures of biodiversity, such as the number of genera, rather than the number of species, have been applied in a few recent studies on plants, vertebrates and invertebrates. The use of higher-taxon richness as an estimate for species richness has been applied to plants with some success (Balmford *et al.* 1996a, b; but see Prance 1994). Williams and Gaston (1994) recommended this approach on the basis of analyses of plants, vertebrates and invertebrate taxa, specifically, in the latter case, British butterflies. The most relevant entomological study to date on the use of higher-taxon measures has been that on ants by Andersen (1995). He assessed the use

Table 8.4 The proportion of different feeding guilds (as defined by Hammond (1990)) of beetles found in forest canopies in Sulawesi, Brunei, Australia and the UK. (From Hammond *et al.* 1996.)

	Herbivores	Xylophages	Fungivores	Saprophages	Predators	No. of spp. total
Sulawesi	25.1	16.1	27.5	13.8	17.4	1355
Brunei	34.6	7.7	18.5	15.5	23.7	875
Australia	19.8	21.1	23.1	8.4	27.3	454
UK	23.5	10.0	26.5	11.0	29.0	200

of genus richness as a surrogate for species richness at 24 sites throughout Australia. He showed that, within a region, there was a strong relationship between the number of genera and the number of species, but overall the number of genera was a poor surrogate for the number of species. Andersen suggested, however, that the technique might be more successful for taxa in which a large proportion of species are found in a relatively small number of genera.

The second approach that has been developed specifically to overcome the problem of lack of specialist taxonomists and identification guides is the use of morphospecies, 'recognizable taxonomic units' (RTUs) or operational taxonomic units (OTUs). Oliver and Beattie (1993, 1996a, b) described the method in detail and presented a supportive case study, and Stork (1995) proposed a useful method of verification. Much of the criticism of this technique (Goldstein 1997) is valid only for countries where taxonomic knowledge of insects is good and there is a good supply of taxonomic specialists. In tropical forests, for example, there may be no alternative to using morphospecies for the rapid assessment of the diversity of most insect groups. Table 8.5 shows the proportion of insects in different groups that had to be sorted to morphospecies in a study in Cameroon (Lawton *et al.* 1998). Only 1% of the butterflies could not be assigned to known species but over 80% of the canopy beetles had to be assigned to morphospecies. The main problem with the use of morphospecies is that the identifications are, by definition, local. This is no obstacle to calculating species richness but it prevents comparison of the species composition of different sites (or the same site at different times) unless the same set of 'voucher specimens' or reference collection is used.

8.3 Patterns in insect diversity

8.3.1 How many species are there in the world?

In 1883, the British entomologist Ray estimated that there were 20 000 insect species in the world (Stork 1997). In 1982, Erwin suggested that there were 30×10^6 species of arthropods in the tropics alone. Although now discounted as an overestimate, Erwin's estimate provided an initial frame of reference for increasing concern about species extinctions (see also Chapter 1).

Erwin's calculation was appealingly simple. It was based on the sampling of the canopy of 19 trees of a single tree species (*Luehea seemannii*) in Peru. Erwin estimated that there were 1200 beetle species on these trees and that 20% of the herbivorous beetles, 5% of the predators, 10% of the fungivores and 5% of the scavengers were specific to that tree species. That is, 163 (13.5% overall) of the 1200 beetles on *L. seemannii* were estimated to be host-specific. Erwin also estimated that there are 70 tree species or genera in each hectare of tropical forest, that beetles made up 40% of all arthropod species and that there are twice as many arthropods in the canopy than the forest floor. Adding the non-specific beetles and other arthropods in the canopy and forest floor this gives an estimate of about 41 000 arthropod species per hectare of tropical forest. Erwin repeated his estimate for the tropics as a whole, based on 50 000 species of

Table 8.5 Number of insect and other species recorded and number of species that could not be assigned to known species in the Mbalmayo Forest Reserve, Cameroon. (From Lawton *et al.* 1998.)

Group surveyed	Total species recorded on study plots during survey	Morphospecies that cannot be assigned to known species (%)
Birds (Aves)	78	0
Butterflies (Lepidoptera)	132	1
Flying beetles (Coleoptera)	358 (Malaise) 467 (interception)	50–70
Canopy beetles (Coleoptera)	342	>80
Canopy ants (Hymenoptera: Formicidae)	96	40
Leaf-litter ants (Hymenoptera: Formicidae)	111	40
Termites (Isoptera)	114	30
Soil nematodes (Nematoda)	374	>90

tropical trees, and produced an estimate of almost 30 million tropical arthropods.

Several ecologists have considered the assumptions made by Erwin. The most critical assumption is the estimate of host specificity, which is now thought to be around 10 insects per plant species (Gaston 1994) (i.e. approximately four specific beetle species rather than the 163 beetles per plant species estimated by Erwin). Gaston's estimate is, however, based on temperate countries, which may be atypical compared with tropical areas. In Papua New Guinea, Basset *et al.* (1996) recorded 391 species of phytophagous beetles (from 4696 individuals) on 10 species of tree. They estimated that there are between 23 and 37 monophagous leaf-feeding species on each tree species, higher than Gaston's estimate but much lower than Erwin's estimate of 136 species.

Erwin's second assumption, that 40% of all canopy arthropods are beetles, is probably an over-estimate too (see below). Stork (1997) suggested a figure of about 20% may be more accurate. Research in Borneo and Sulawesi (see also below) suggests that Erwin's third assumption, the ratio of 2:1 arthropods in the canopy in comparison with those on the ground, underestimates the number of arthropods on the ground, which is twice the number found in the canopy (Hammond 1992).

Erwin prompted several other attempts to estimate the number of species in the world (Stork 1997). The simplest approach is to base estimates on the ratio between different groups of insects. For example, Stork and Gaston (1990) used the ratio of butterfly species (67) to all species of insects (22000) in the well-documented British fauna and an estimate of 15 000–20 000 butterfly species world-wide to give a global estimate of $(4.9–6.6)\times10^6$ insect species.

Hodkinson and Casson (1991) estimated global insect species richness by using the ratios (a) between the number of known and new Hemiptera in a sample of 1690 species collected in Sulawesi and (b) between the number of species of Hemiptera and other insects. They estimated that 62.5% of the Hemiptera they collected were new to science and that there are 184 000–193 000 described Hemiptera world-wide. Assuming that 7.5–10% of insect species are Hemiptera, they estimated that the number of insect species in the world is between 1.84×10^6 and 2.57×10^6. They also estimated global species richness by assuming that if the 500 tree species in their study area yielded 1056 new species of Hemiptera the 50 000 species of tree in the tropics (estimated by Erwin) means that there are 105 600 undescribed Hemiptera world-wide. Adding the 184 000–193 000 described species and using the ratio of Hemiptera to all insect species given above, Hodkinson and Casson again estimated that there are $(1.84–2.57)\times10^6$ insect species world-wide. Hodkinson and Casson's estimates are flawed in several ways (Stork 1997). It is unlikely that they sampled all the Hemiptera in their study area, their ratio of described to new species is probably not representative and the ratio of (described) Hemiptera to other insects is probably biased because of the large number of hemipteran pest species. Using data on beetles from the same study area, Stork (1997) produced an estimate of $(5.0–6.7)\times10^6$ insects world-wide but warned that such estimates are very sensitive to the intensity of sampling.

Other methods have been used to estimate global species richness, such as extrapolation based on body size distribution (May 1988) and analysis of species turnover (Stork 1997). Hammond (1992) reviewed the different approaches to estimating the number of species in the world and after consulting experts on different groups produced a conservative 'working estimate' of 12.5×10^6 species of which 8×10^6 are estimated to be insects.

8.3.2 Biogeographical patterns in biodiversity

Sixty-seven butterflies have been recorded in the UK, 321 in Europe, but over 700 have been recorded in one area (near Belem) in tropical Brazil (Robbins & Opler 1997). There are many similar examples of latitudinal variation in biodiversity, most demonstrating a general increase in biodiversity from polar to temperate to tropical latitudes (Gaston & Williams 1996). Latitudinal variation in biodiversity applies to vertebrates and plants as well as insects and other invertebrates. Many underlying mechanisms for this variation have been suggested, such as the effects of environmental stability, environmental predictability, productivity, area, number of habitats, evolutionary time and solar radiation (Rohde 1992). For example, after reviewing the various hypotheses, Rohde concluded that the lower degree of variation in climate in tropical areas has permitted a greater degree of specialization, and therefore speciation, than temperate areas.

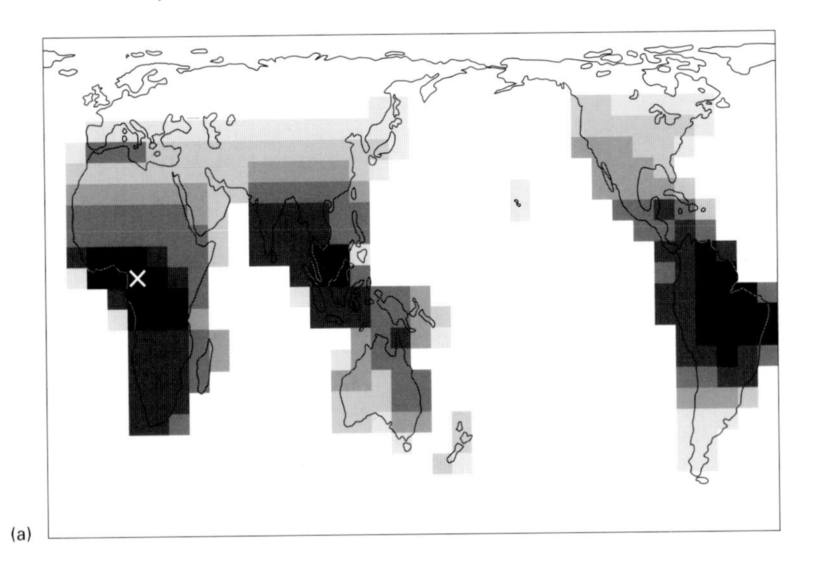

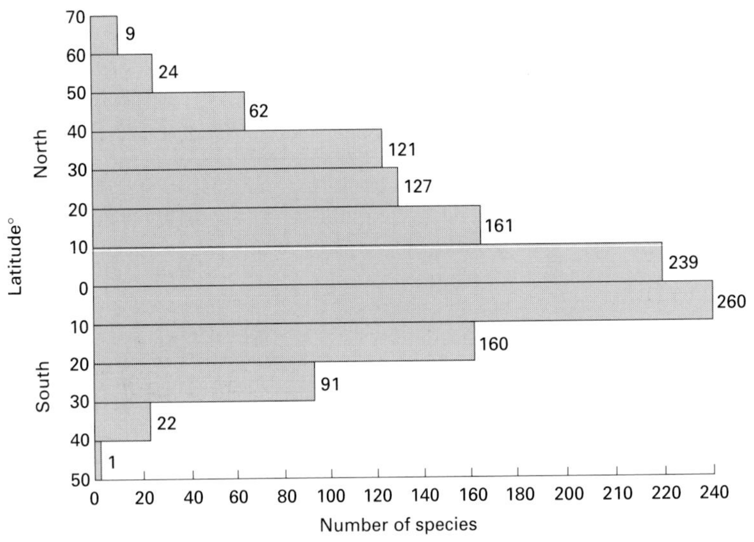

Fig. 8.7 (a) Regional variation in the number of termite genera (represented by a logarithmic grey scale from a minimum (light grey) to a maximum (black with a cross); white, no data). (From Gaston & Williams 1996; data from Eggleton *et al.* 1994.) (b) Latitudinal gradients in the number of swallowtail butterfly species. (From Sutton & Collins 1991.)

Although there is no consensus on what causes the latitudinal patterns in species richness, the fact remains that the diversity of most insects declines from tropical to temperate areas. Two well-documented examples are termites and swallowtail butterflies (Fig. 8.7) (Sutton & Collins 1991; Eggleton *et al.* 1994). However, there are some interesting exceptions. Most aphids, for example, are found only in temperate regions. It has been argued that aphids are particularly adapted to temperate conditions, but some aphids are endemic to the tropics and are well adapted to tropical conditions, as noted by Dixon *et al.* (1987). They argued that there are so few species of aphids, particularly in the tropics, because of (a) their high degree of host specificity; (b) their relatively poor efficiency in locating their host plants; and (c) their inability to survive for long without feeding. The problems aphids face are not limited to tropical regions—90% of plant species world-wide are not used by aphids. However, the situation is worse in the more plant-species rich tropical regions, where aphids' host plants are more difficult to locate. Thus as the number of plant species in a particular geographical region increases, the number of aphid

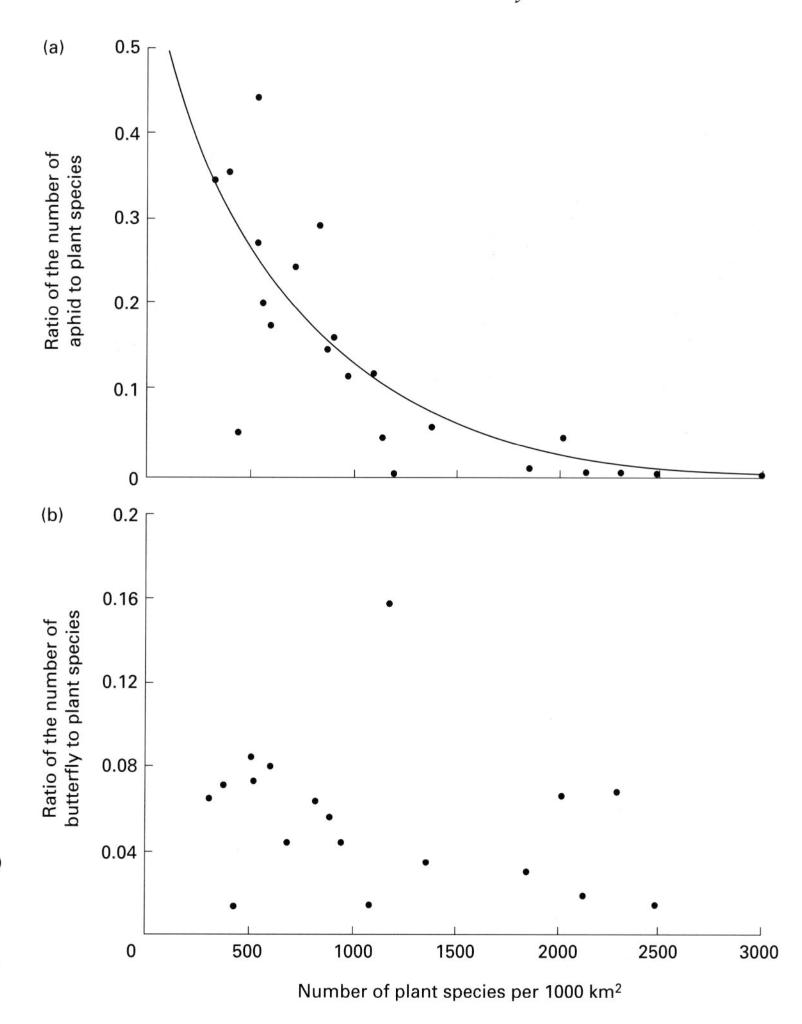

Fig. 8.8 The relationship between the ratio of the numbers of (a) aphid and (b) butterfly species to plant species and the number of plant species in a range of countries. (From Dixon *et al.* 1987, with permission from The University of Chicago Press.)

species actually declines (Fig. 8.8). In contrast, phytophagous insects that are more efficient at locating their host plants, such as butterflies, become more species rich as the number of plant species increases.

Sawfly (Hymenoptera, Symphyta) diversity also increases from tropical to polar regions, and does so even more sharply than aphid diversity (Kouki *et al.* 1994) (Fig. 8.9). Although sawflies, like aphids, are very host specific, they are much better at locating their host plants. Thus Kouki *et al.* concluded that the latitudinal trends in sawflies and aphids are caused by different factors. They suggested that sawfly diversity parallels the diversity of their most widely used plants, the willows (*Salix* spp.).

Altitudinal gradients in biodiversity have also been studied by many ecologists. Because environmental variation along an altitudinal transect is similar to variation along a latitudinal transect, we would expect insect diversity, in general, to decrease from low to high altitudes. The available data support this hypothesis. For example, Hanski and Niemelä (1990) studied the abundance and diversity of dung beetles along altitudinal transects in Sarawak, Borneo (from 300 to 1150 m above sea level). Beetle species richness declined steadily from 15 to 20 species at low altitudes to 5–10 species at around 800 m and to less than five species above 1150 m (Fig. 8.10).

As well as species richness, species composition

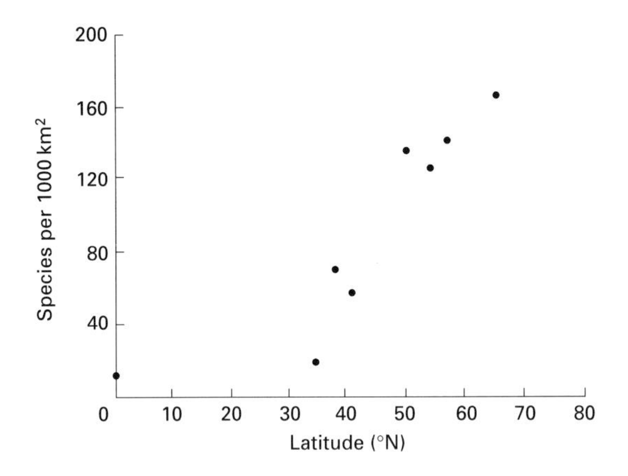

Fig. 8.9 Latitudinal variation in the number of species of sawflies. (From Kouki *et al.* 1994.)

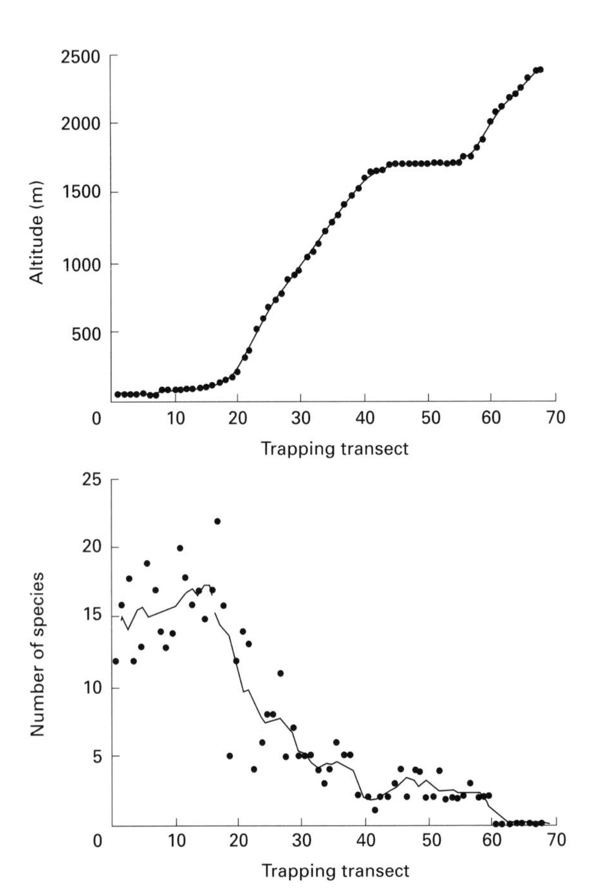

Fig. 8.10 Altitudinal variation in the number of species of dung beetles in the Gunung Mulu National Park, Sarawak, Borneo. (From Hanski & Niemelä 1990.) The upper figure shows the altitude of the trapping stations along the transect.

changes along altitudinal gradients. In the dung beetle study discussed above, DECORANA analysis (see Section 8.2.5) showed that altitude had a major effect on species composition. Table 8.6 shows a few examples of the abundance of different dung beetle species at different altitudes.

There are other large-scale patterns in biodiversity apart from the latitudinal variation discussed above. There is a surprising amount of variation, for example, from biogeographic region to region in

Table 8.6 Altitudinal distribution of six common dung and carrion beetles in the Dumoga-Bone Reserve and on Gunung Muajat, Sulawesi, Indonesia. (After Hanski & Niemelä 1990.)

	Number of individual beetles per trap					
Altitude (m a.s.l.)	*Onthophagus aereomaculatus*	*Copris macacus*	*Onthophagus aper*	*Onthophagus* sp.n. (1)	*Copris* sp.n.	*Onthophagus* sp.n. (2)
200	2	7	22			
300	5	8	40			
350	8	11	43			
400	3	22	37			
450	8	16	60			
500	3	38	26			
550	1	27	10			
600		19	27	1		
650		16	11	1		
700		18	20			
750		7	11			
800		6	12			
850			5			
900			5			
950		1	2	4	1	
1000			3	1		
1050						
1100		1	1	2	1	
1150				3	3	
1300			1		1	
1600				2		2
1700				19		13
1750				26		30

sp.n., new species; m a.s.l., metres above sea level.

addition to latitudinal variation. In general, biodiversity in tropical regions is greatest in the Neotropics, least in the Afrotropics and intermediate in the Indotropics (Gaston & Williams 1996). For example, the diversity of butterflies is particularly high in the Neotropics (Fig. 8.11). Not all butterfly families, however, show this pattern. Swallowtails, for example, are most species rich in East Asia and Australia (Fig. 8.12) (Sutton & Collins 1991).

Such patterns of biodiversity are interesting but too general to be useful in focusing efforts to conserve biodiversity. In contrast the 'hot spots' concept was developed to highlight the threatened nature of much of the Earth's biodiversity. Myers (1988, 1990) suggested that as many as 20% of plant species may be confined to 0.5% of the land surface of the Earth. He identified 18 biodiversity hot spots—concentrations of endemism as well as species richness, and called

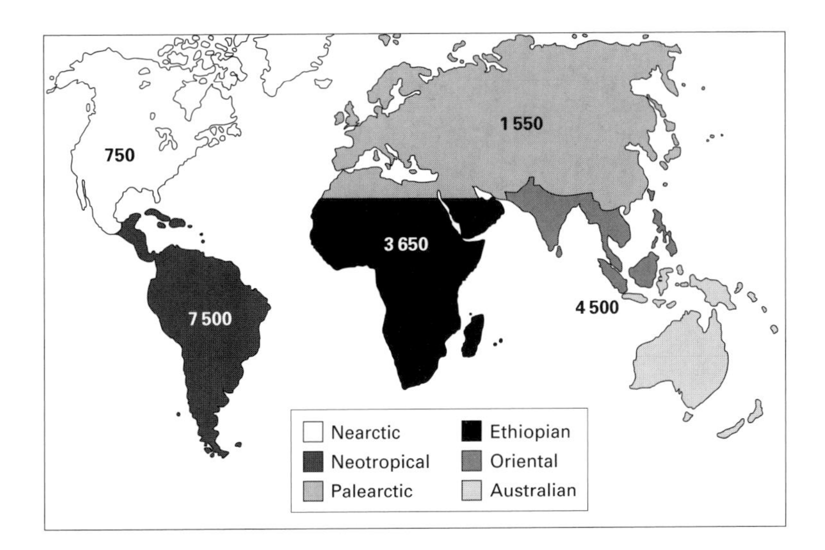

Fig. 8.11 The number of butterfly species in different biogeographical realms. (From Robbins & Opler 1997.)

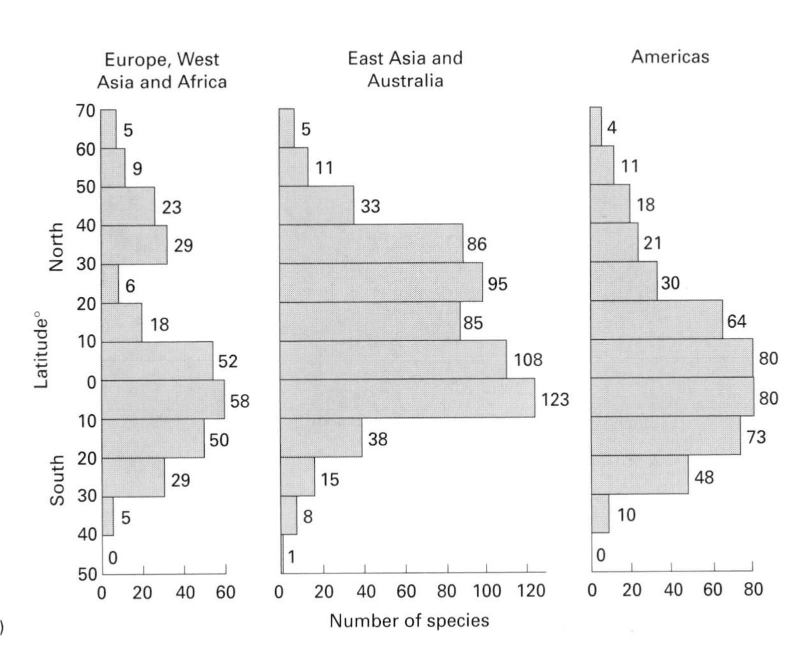

Fig. 8.12 Latitudinal variation in the number of species of swallowtail butterflies in: Europe, West Asia and Africa; East Asia and Australia; and the Americas. (From Sutton & Collins 1991.)

them 'hot spots' because of the threats implicit in biodiversity being concentrated in relatively small areas. These areas include Hawaii, Western Ecuador, the south-western Côte D'Ivoire, Madagascar, Sri Lanka and northern Borneo. Apart from plants, the best available data on 'hot spots' as originally conceived are for birds (Bibby *et al.* 1992), but among insects the best data are, not surprisingly, on butterflies. Figure 8.13 compares areas of endemism of butterflies, birds, reptiles and amphibians in Central America.

Several ecologists have considered the hypothesis that 'hot spots' hold concentrations of diversity of different taxa (Prendergast *et al.* 1993; Williams *et al.* 1994; Prendergast 1997). The conclusion of these studies is that species-rich areas of different taxa tend not to coincide. Perhaps, as Lovejoy *et al.* (1997) pointed out, this is not very surprising given the diverse ecological requirements of different taxa. However, it may be that the 'hot spot' hypothesis applies much better to the tropics and subtropics than to temperate and boreal areas (Myers 1997), and more rigorous analysis of the type performed by Prendergast for the British Isles needs to be carried out in the tropics. Unfortunately, only a very few countries, such as Costa Rica, are collecting data that might serve this purpose. Perhaps Myers' contribution has been misinterpreted—he highlighted the fact that biodiversity is vulnerable when it is concentrated. Other studies have shown that different taxa tend not to be concentrated in the same areas. Thus a much better understanding of the distribution of taxa other than plants and birds is clearly needed. This must focus on endemism, as implicit in the original 'hot spot' concept, and rather neglected in recent analyses that have focused on species richness (Gaston & Williams 1996).

8.3.3 Species–area relationships

One aspect of natural variation in biodiversity has long been of interest to ecologists: species–area relationships or 'island biogeography' (MacArthur & Wilson 1967). Many studies have shown that the number of species on islands increases as island size increases (Fig. 8.14) (see also Chapter 1).

There are two main theories for these species–area relationships: the equilibrium theory and the habitat diversity theory. The equilibrium theory (MacArthur & Wilson 1967) is that the number of species on an island is a consequence of the balance between extinction and colonization by new species. Extinction and colonization rates are affected by the number of species on the island, the size of the island and the distance of the island from other islands and the mainland. Although species are constantly colonizing the island and becoming extinct, an equilibrium is eventually reached that is directly related to island size and inversely related to the distance from other sources of species. The habitat diversity theory (Lack 1969, 1976) is that because larger islands tend to contain more habitats they will tend to contain more species. These theories are not mutually exclusive and there is evidence to support both the effect of habitat diversity and the colonization–extinction equilibrium. Wilson (1961), for example, showed that the number of ponerine ant species in the Moluccan and Melanesian islands increases as a function of island size (Fig. 8.14). In contrast, Torres and Snelling (1997), in a study on the numbers of ant species on Puerto Rico and 44 surrounding islands, found that island size had little influence on the numbers of ants.

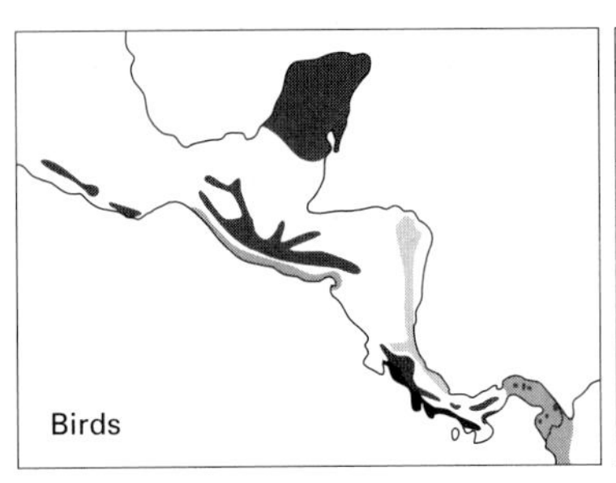

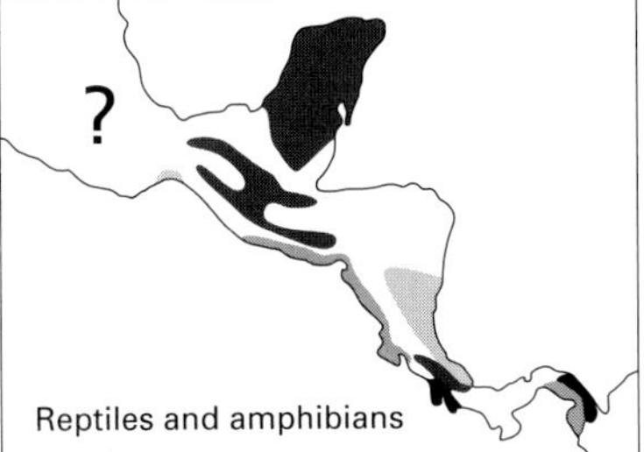

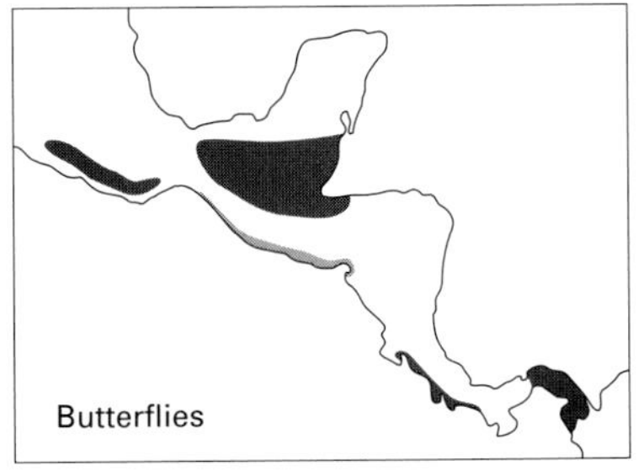

Fig. 8.13 Areas of high endemism of birds, reptiles and amphibians, and butterflies in Central America. (From Bibby *et al.* 1992.)

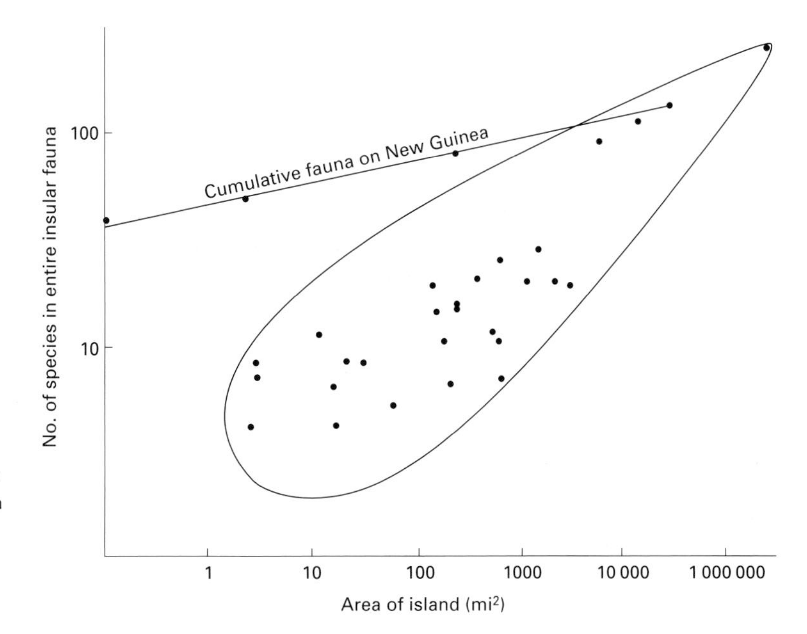

Fig. 8.14 The relationship between the number of (ponerine and cerapachyine) ant species found on different Moluccan and Melanesian islands and the area of these islands. (From Wilson 1961, with permission from The University of Chicago Press.)

They described the situation as a 'nonequilibrium case', with the number of colonizations exceeding the number of extinctions over a period of 18 years. They found that habitat diversity had the greatest effect of the factors studied (island size, isolation, etc.) on the number of ant species on an island. One other example where, unsurprisingly, the number of colonizations at present exceeds the number of extinctions is the number of butterfly species on Krakatau (Fig. 8.15). In the case of the ants of Puerto Rico, there is evidence that human occupation, disturbance and movement between islands has led to an increase in both habitat diversity and colonization by ants. Torres and Snelling also suggested that there has been too much emphasis in island biogeography on the numbers of species rather than their identity. They pointed out that colonizations of islands by several ant species, such as the fire ant *Solenopsis wagneri*, can have a disproportionate effect; these aggressive species cause the extinction of many other species.

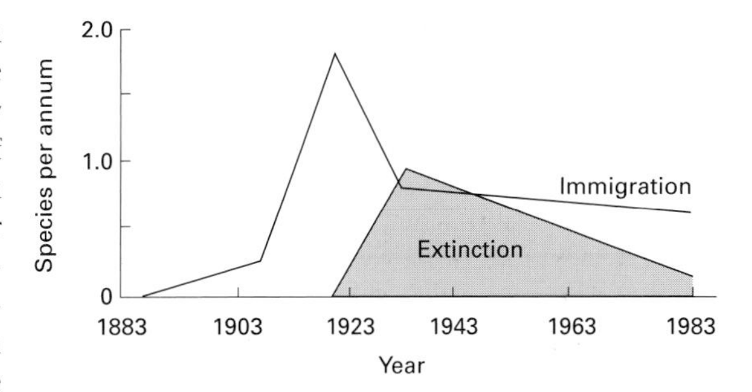

Fig. 8.15 Annual immigation and extinction curves for butterflies on Krakatau (Krakatoa) 1883–1989. (From Bush & Whittaker 1991.)

Species–area relationships have been widely applied to 'habitat islands', such as forest fragments, as well as 'true' islands. The most notable practical example of this application has been the 'Biological dynamics of forest fragments' project in Brazil (Lovejoy *et al.* 1986; Brown 1991). Figure 8.16 shows the numbers of species in different butterfly families in reserves ranging in size from 1 to 1000 ha: the numbers of species in some families behave as predicted by island biogeography, others do not.

For phytophagous insects, species–area relationships have been used to try to explain the number of insect species feeding on different plant species. Many studies have shown that the area occupied by a particular plant species is the most important determinant of the number of insect species that it supports (Janzen 1968; Strong *et al.* 1984) (Fig. 8.17). Several of the studies on the influence of area on phytophagous insect species richness have focused on insects feeding on British trees, using data on the number of 10 km × 10 km or 2 km × 2 km squares

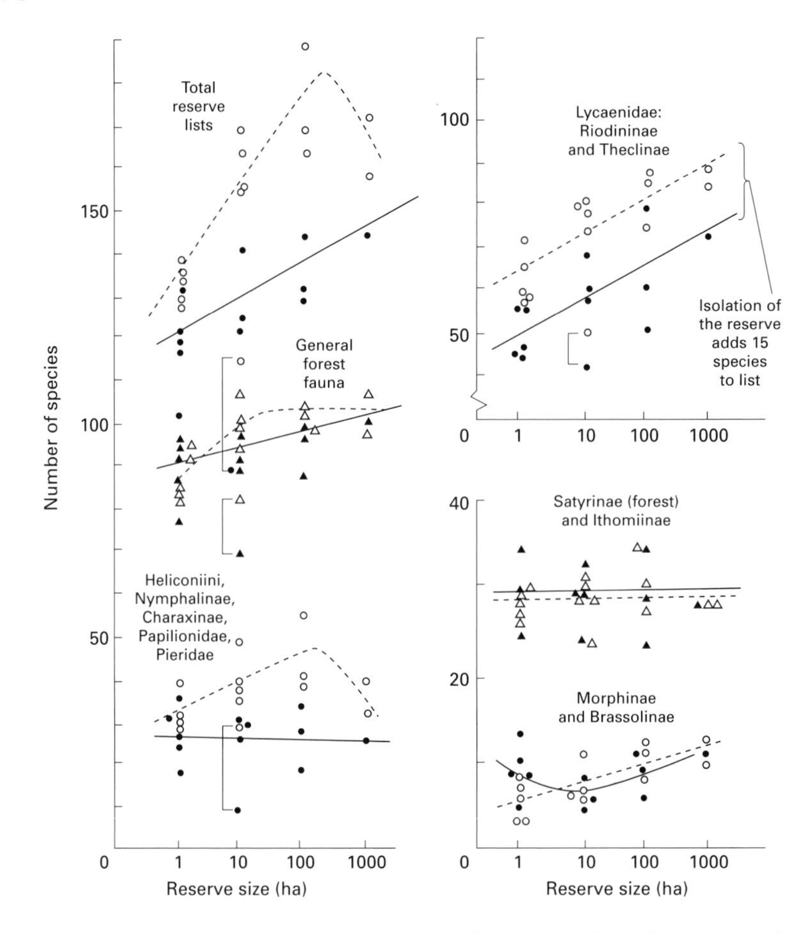

Fig. 8.16 Relationships between number of butterfly species and reserve (forest fragment) size in the 'Biological Dynamics of Forest Fragments' project near Manaus in the Brazilian Amazon: the total number of butterfly species is given in the top left and separate groups are shown elsewhere. Isolated sites are shown as open symbols and broken lines; sites surrounded by forest are shown as solid symbols and lines. (Linked symbols in most graphs show the numbers of species in a 10 ha reserve before and after isolation.) Note (i) that isolation from surrounding forest results in a marked loss of species in some groups (e.g. Lycaenidae, top right); (ii) the species richness of some groups increases with reserve area but other groups, notably the shade-loving Satyridae and Ithoniinae and general (widespread) species (centre left) are insensitive to reserve size. (From Brown 1991.)

occupied by each tree species. Analysis based on the 2 km × 2 km data explained 56% of the variation in the number of insect species on a tree (Kennedy & Southwood 1984). Interestingly, however, an analysis based on the most accurate data available on the area covered by each tree species, a thorough census (measuring the number of hectares occupied by each tree species) carried out by the British Forestry Commission, resulted in the explanation of only 17% of the variation in the number of insect species on a tree (Claridge & Evans 1990). This suggests that the number of insect species on a tree is influenced by the diversity of habitats occupied by, as well as total area covered by, a tree species—the 2 km × 2 km data will reflect habitat diversity better than the census (hectare) data.

Claridge and Evans also obtained a better explanation of the number of species on a tree by separately considering the mainly native broadleaved and the mainly non-native coniferous trees, particularly the latter (22% and 79%, respectively). It is worth recalling that Southwood (1961) concluded that the

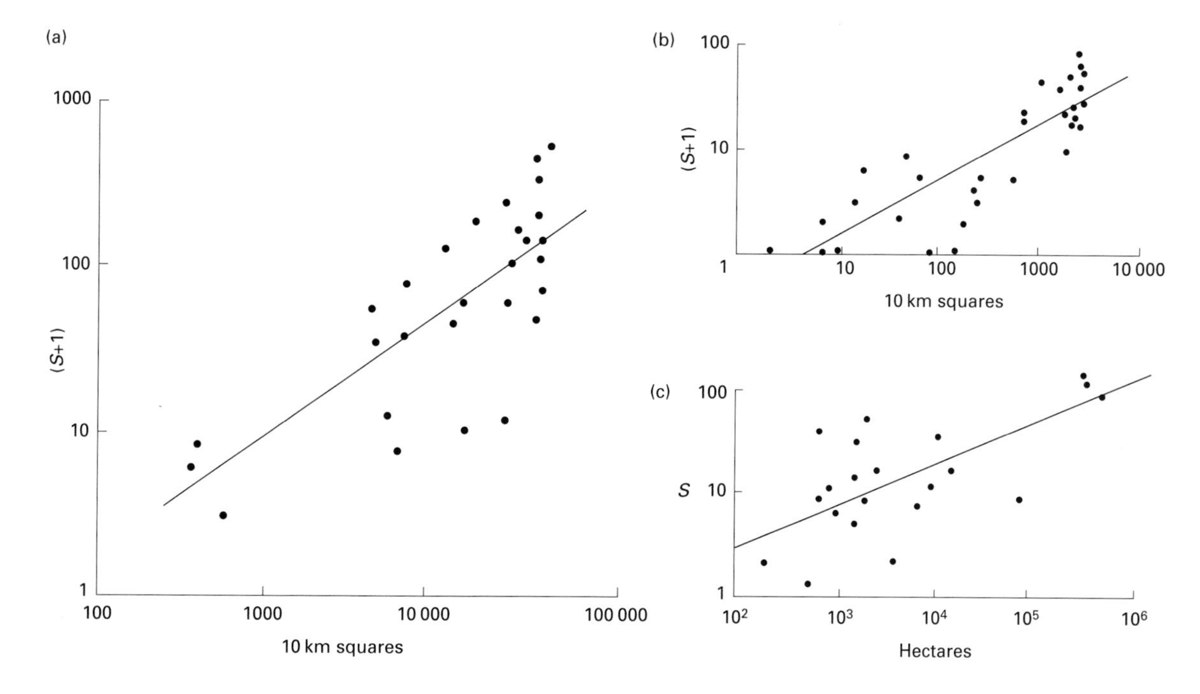

Fig. 8.17 Relationships between numbers of species and host plant range for: (a) phytophagous insects on genera of British trees; (b) phytophagous insects on British perennial herbs; (c) insect pests on cacao (each point is a different country). (From Strong 1974; Lawton and Schröder 1977; Strong *et al.* 1984.)

number of insects feeding on a tree species was determined by the present and past abundance of the tree. Data on the present abundance of trees may not accurately reflect the abundance of these species in the past, particularly in heavily deforested countries such as the UK. However, data on recently introduced species, such as most of the conifers in the British Isles, provide a better test of Southwood's hypothesis, with notable success, despite the small number of tree species.

There have been many other studies of the number of insect species on plants, including several that focused attention on particular feeding guilds such as leaf-miners (e.g. Claridge & Wilson 1982). In these cases, and in general, the area occupied by a host plant was found to be a better predictor of the number of insect species when the studies were restricted to groups of taxonomically related host plants, such as Californian oaks (Opler 1974) and British Rosaceae (Leather 1986), than when the studies included taxonomically diverse plant species.

The study on British Rosaceae mentioned above is a good example of the influence of plant architecture on the number of insect species feeding on a particular plant. Among the Rosaceae, trees (on average, 53 insect species on native host species) have more species than shrubs (25 species), which have more species than herbaceous plants (eight species), even when the area occupied by each species is taken into account (Leather 1986). Thus, the more complex the structure of the plant, the more insect species feed on it, presumably because of the increasing number of niches available on more architecturally complex plants (Fig. 8.17) (see also Chapter 1).

One practical application of species–area relationships has been to predict the effect of deforestation on species extinctions. This is considered below in section 8.4.2.

8.3.4 Insects and their habitats

In addition to the patterns of biodiversity discussed above there are, of course, marked differences in insect diversity between different habitats and ecosystems. In this section we will consider a few examples, concentrating on the insect communities of different habitats in the British Isles and in tropical forests.

Given the specialist nature of many insect herbivores, including Lepidoptera, it is not surprising that there is a difference in the species composition of moth communities in different ecosystems. However, there are even marked differences in the species composition of polyphagous moths in different ecosystems. Table 8.7 shows the abundance of the nine most common species of moth—all polyphagous—trapped in four land 'classes' in Britain. There are marked differences in the relative abundance of different species, particularly between the species found in lowland and upland habitats (Usher 1995).

Arable fields clearly lack the insect diversity of woodlands, certainly for insects associated with trees and the ground flora of woodlands. Some studies have also examined the diversity of ground- and soil-dwelling insects and other arthropods in woodlands and adjacent arable fields. In a study in the UK, Bedford and Usher (1994) found that there were more species of ground-dwelling ground beetles (Coleoptera: Carabidae) in arable fields than in adjacent woodlands. Peaks in species richness were found at the field-edge and the lowest number of species was recorded in the middle of the woodlands. However, Bedford and Usher (1994) also found that some carabid species were restricted to the woodlands and, more generally, Luff and coworkers (reviewed by Luff & Woiwod 1995) have identified the carabid species characteristic of different (agricultural and other) habitats and therefore those most and least vulnerable to land-use change. In fact, Sgardelis and Usher (1994) suggested that the frequent application of pesticides, repeated soil disturbance and the lack of organic matter in the arable fields had a detrimental effect on arthropod populations, and concluded that most species 'appear to be isolated in the woodland fragment and are surrounded by a hostile arable environment'.

The major difference between the insect communities of the temperate ecosystems discussed above and the insect communities of tropical forests is the greater species richness of the latter. Unsurprisingly, our knowledge of the ecology of most tropical forest insects is much poorer than that of temperate species. The ecology of a few tropical forest insect species is well known, notably that of ant species such as leaf-cutting ants (*Atta* species), army ants (e.g. *Eciton* spp.) and the weaver ants (*Oecophylla* spp.) (Hymenoptera: Formicidae) (Hölldobler & Wilson 1990). However, recent developments in forest canopy access methods have led to an increase in the study of tropical insect communities (Lowman & Nadkarni 1995; Stork *et al.* 1997).

For example, Stork (1991) surveyed the arthropod community of 10 forest trees in Borneo. Almost 24 000 individuals were sampled by insecticide knockdown fogging, a technique now widely used to sample the arthropod communities of the canopies of temperate and tropical forests (Stork *et al.* 1997). Stork (1991) found that the most abundant groups from the Borneo trees were Hymenoptera (27%), Diptera (22%), beetles (17%) and Hemiptera (11%) (Fig. 8.18). Of the Hymenoptera, two-thirds were ants (18% of the total sample). Studies elsewhere in the tropics have shown that ants can be even more common. In Cameroon, for example, Watt *et al.*

Table 8.7 Abundance of nine moth species in different land classes in Britain. (From Usher 1995.)

Species	Lowland cereal	Lowland grassland	Marginal upland	Upland
Luperina testacea (flounced rustic)	37	18	0	0
Agrotis exclamationis (heart and dart)	33	48	6	0
Spilosoma luteum (buff ermine)	10	20	0	0
Diarsia rubi (small square spot)	6	20	4	18
Spilosoma lubricipeda (white ermine)	16	31	15	6
Orthosia gothica (Hebrew character)	31	42	124	100
Xanthorhoe montanata (silver ground carpet)	19	21	69	96
Hydraecia micacea (rosy rustic)	6	5	105	25
Cerapteryx graminis (antler)	2	2	94	57

(1997c) found that ants made up 63% of the total canopy arthropods (Fig. 8.18). Studies in Brazil, Malaysia and elsewhere have shown similar levels of ant abundance (Stork *et al.* 1997; Chey *et al.* 1998).

The average number of arthropods in the Borneo study was 117 per m^2 (in a range of 51–218 per m^2). (That is, the average numbers of arthropods caught by knockdown fogging in each 1 m^2 tray was 117.) Other studies of tropical canopy arthropods have produced similar results (e.g. 214 per m^2 in Cameroon (Watt *et al.* 1997c)). However, results from studies of arthropods in temperate forests show even higher densities. Southwood *et al.* (1982) recorded 389 arthropods/m^2 on trees in England and South Africa, and Ozanne *et al.* (1997) recorded 1000–10 000 arthropods/m^2 in a transect across a Norway spruce forest.

What is most remarkable about tropical forest canopy arthropods is their species richness. Stork recorded over 3000 arthropod species in the canopy of 10 study trees in Borneo (excluding Psocoptera, Acari and some other groups that were not identified to species). Hymenoptera were the most speciose order (25.8% species), followed by Diptera (22%), beetles (21%) and Hemiptera (8%) (Fig. 8.19). In contrast to the abundance figures described above, ants made up only 3% of the total number of species. Studies elsewhere in the tropics also show that ants, in comparison with beetles for example, are numerically dominant but not proportionally rich in species. In contrast, the arthropod community of temperate trees is less species rich. For example, Moran and Southwood (1982) recorded 465 arthropod species on *Quercus robur* in southern England (and fewer on other tree species). Nevertheless, the guild structure of the arthropod community on different trees appears to be remarkably similar (see section 8.5).

Although the canopy of tropical forests is undoubtedly rich in insect species, other parts of tropical forests hold large numbers of species too. In a study of the Indonesian island of Seram, for example, Stork and Brendell (1993) measured the abundance of arthropods in the forest canopy, on tree trunks, on herb-layer vegetation, in the leaf litter and in the soil. They collected over 32 000 individual insects and other arthropods and expressed the numbers caught as numbers per m^2 (of ground). An average of 1201 arthropods/m^2 were recorded in the forest canopy, a much higher figure than recorded in the study discussed above on the nearby island of Borneo. However, almost twice that number of arthropods, 2372 per m^2, were recorded in the soil. Only 12 arthropods/m^2 were sampled from herb-layer vegetation and 52 arthropods/m^2 from the tree trunks, but 602 arthropods/m^2 were found in the litter layer. Thus over twice as many arthropods were found in the soil and litter (70%) than in the canopy (28%). The canopy arthropods were dominated by ants

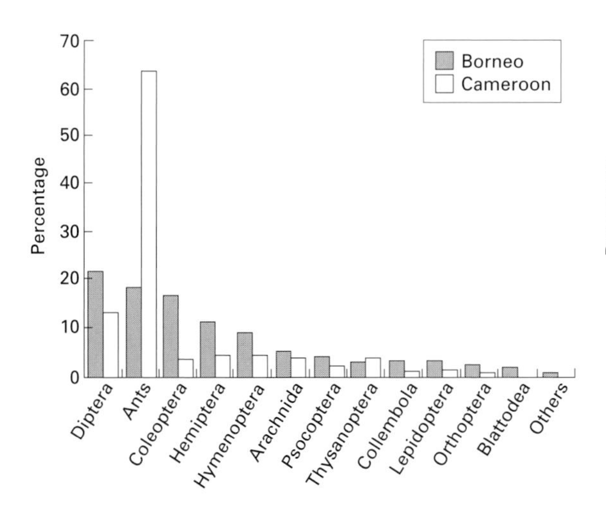

Fig. 8.18 Relative abundance of different groups of canopy-dwelling arthropods in Borneo and Cameroon. (Data from Stork 1991; Watt *et al.* 1997b.)

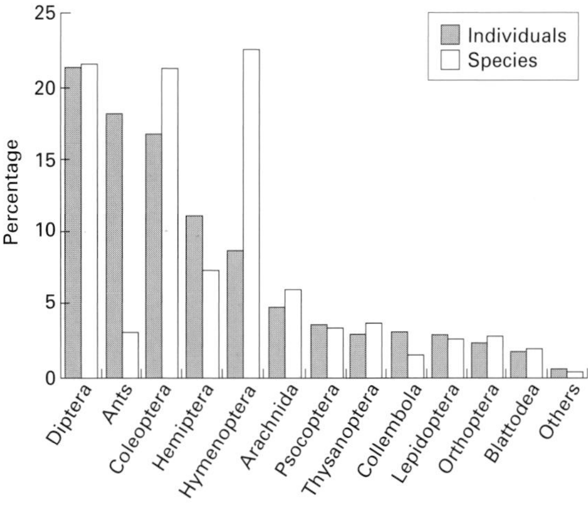

Fig. 8.19 Relative abundance and number of species in different arthropod groups collected from tree canopies in Borneo. (From Stork 1991.)

(48%) but the most abundant arthropods in the soil and litter were non-insect Collembola (60% and 70%, respectively) and Acari (25% and 16%, respectively). Other studies of the soil and litter communities of tropical forests have shown them to be numerically dominated by termites (Eggleton *et al.* 1996).

8.3.5 The guild structure of arthropods on trees

The biodiversity of tropical forests is remarkable, but what are all these species doing? This question was also posed by Moran and Southwood (1982) for the insect arthropod communities of British and South African trees. Moran and Southwood were the first to analyse the feeding guild structure of the arthropod fauna of trees. They classified the arthropods found in samples from six species of trees in each country into phytophages (herbivores), the fauna associated with epiphytes, scavengers (of dead wood, etc.), predators, parasitoids, ants and tourists. 'Tourists' were defined as 'non-predatory species which have no intimate or lasting association with the plant but which may be attracted to trees for shelter and sustenance (honeydew and other substances), or as a site for sunbasking and sexual display'. Moran and Southwood found that the ratio of different feeding guilds did not vary greatly either between tree species or between countries. Phytophages were the most abundant guild, making up about 68% of the individuals. They were also the most speciose guild—25% of all species were phytophagous arthropods. Parasitoids, which made up 3% of the total number of individuals, were the second most speciose guild (25%). The full guild structure is shown in Fig. 8.20b.

Stork (1987) also carried out an analysis of the guild structure of the 3000 species recorded from the canopy of trees in Borneo. As was found in the studies on temperate trees, the most abundant group were phytophages (26%), but they were slightly less species rich (22%) than parasitoids (25%). The abundance of parasitoids in the tropical forest study was similar to that of the temperate study. Indeed, overall, the relative abundance of different guilds in the temperate and tropical studies was surprisingly similar (Fig. 8.20a). There were relatively fewer phytophages, fewer insect predators and more tourists in the tropical study. In terms of species, there were fewer phytophagous arthropods, more scavengers and more ants in the tropical study. There were no other significant differences between the temperate and tropical studies (Stork 1987).

Unfortunately, but not surprisingly given the amount of work involved in sorting so many arthropods to species, there have been no other studies in tropical forests as complete as Stork's analysis of the arthropods on 10 trees in Borneo. As pointed out above, most other tropical canopy studies have recorded more ants than the Borneo study. However, until more studies are carried out, particularly in forests where more typical numbers of ants are found, we may conclude that there is a remarkable degree of uniformity in the guild structure of canopy arthropod communities.

8.3.6 Biodiversity and ecosystem processes

Before considering threats to biodiversity and its conservation, it is worth posing the question: does biodiversity matter? Several researchers have considered this question and listed a range of values that biodiversity may have (e.g. Kunin & Lawton 1996). Cultural, aesthetic and ethical values can be given to biodiversity, and, clearly, humans depend on many plant and animal species for their survival (see Chapter 1). Because of the last factor, biodiversity may therefore be considered to have a direct economic value. However, economic values may also be given to biodiversity generally because of the wealth of chemicals with medicinal and other valuable properties that are waiting to be discovered by 'bioprospecting', and the role that biodiversity plays in the healthy functioning, or 'sustainability' of ecosystems that humans directly or indirectly use.

Most bioprospecting focuses on plants, particularly where ethnobotanical information has identified candidate species (Martin 1995). The current bioprospecting programme of INBio (the Costa Rican Institute for Biodiversity), however, includes insects (R. Gamez, personal communication). Nevertheless, most interest in the value of insect diversity has been in the roles insects play in the functioning of ecosystems. The critical question is not whether insects play important roles in ecosystems; of course they do: as decomposers, pollinators of valuable plants and natural enemies of potential pests (see Chapter 7). Rather, the critical question is whether these vital ecosystem processes require so much diversity. Would ecosystems function adequately with fewer decomposers, pollinators, predators and parasitic insects?

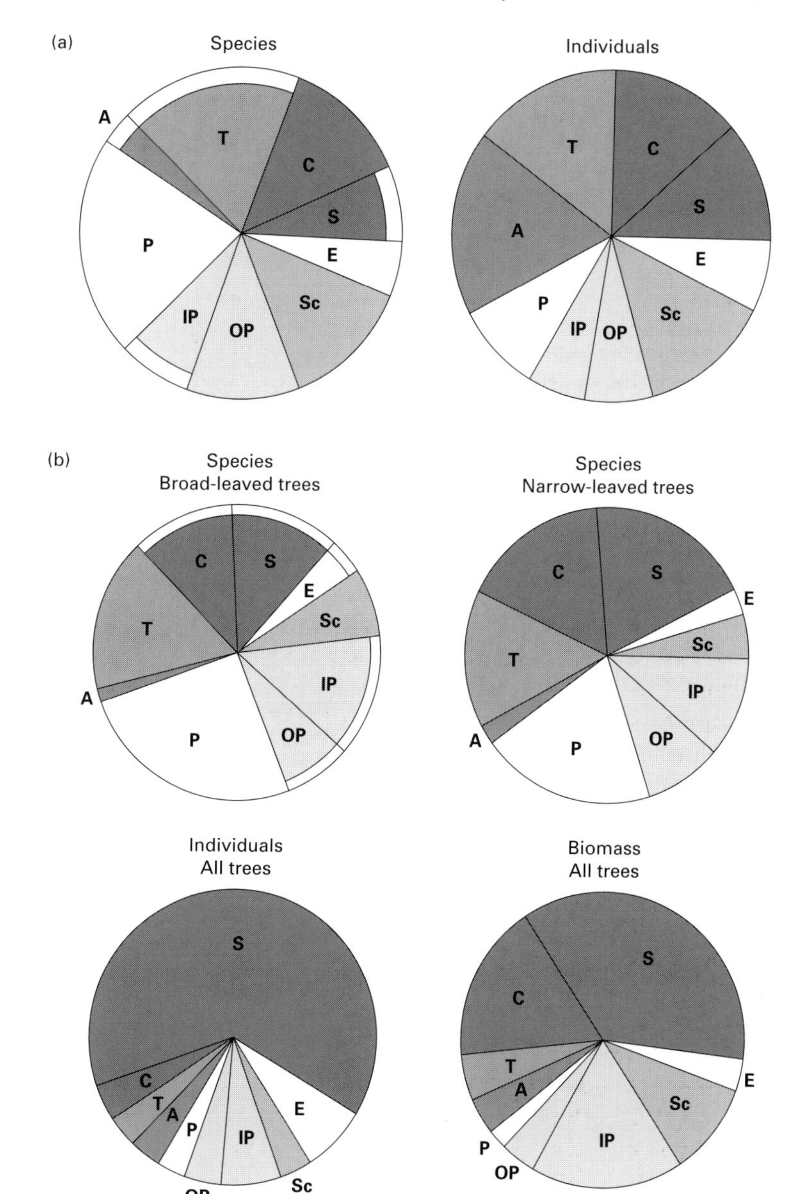

Fig. 8.20 (a) Guild structure (in terms of numbers of species and numbers of individuals) of the arthropod fauna of 10 Borneo trees: S, sap-suckers; C, chewers; T, tourists; A, ants; P, parasitoids; IP, insect predators; OP, other predators; Sc, scavengers and fungivores; E, epiphyte fauna. Segments with double external lines indicate guilds with constant proportions of species from tree to tree. (From Stork 1987.) (b) Guild structure (in terms of numbers of species on broad-leaved and narrow-leaved trees, individuals and biomass) of the arthropod fauna of South African and British trees: C, chewers; S, sap-suckers; E, epiphyte fauna; SC, scavengers; IP, insect predators; OP, other predators; P, parasitoids; A, ants; T, tourists. Segments with double external lines indicate guilds with constant proportions of species from tree to tree among broad-leaved species. (From Moran & Southwood 1982.)

There are several hypotheses to describe the possible roles of increasing biodiversity in ecosystem functioning (Lawton 1994; Johnson *et al.* 1996). Each hypothesis can be illustrated by showing the effect of increasing species richness on the rate of an ecosystem process (such as decomposition) (Fig. 8.21). First, the 'redundant species hypothesis' suggests that ecosystem processes benefit from an increase in biodiversity up to a threshold level after which there is no increase in the rate of ecosystem processes: further species are redundant (Fig. 8.21a). In contrast, the 'rivet hypothesis' suggests that each species plays a significant role in affecting the ecosystem process: even a small decrease in biodiversity (similar to the loss of a single 'rivet' from a complex machine (Lawton 1994)) will result in a decrease in the rate of an ecosystem process (Fig. 8.21b). Under this hypothesis, various forms of the function between

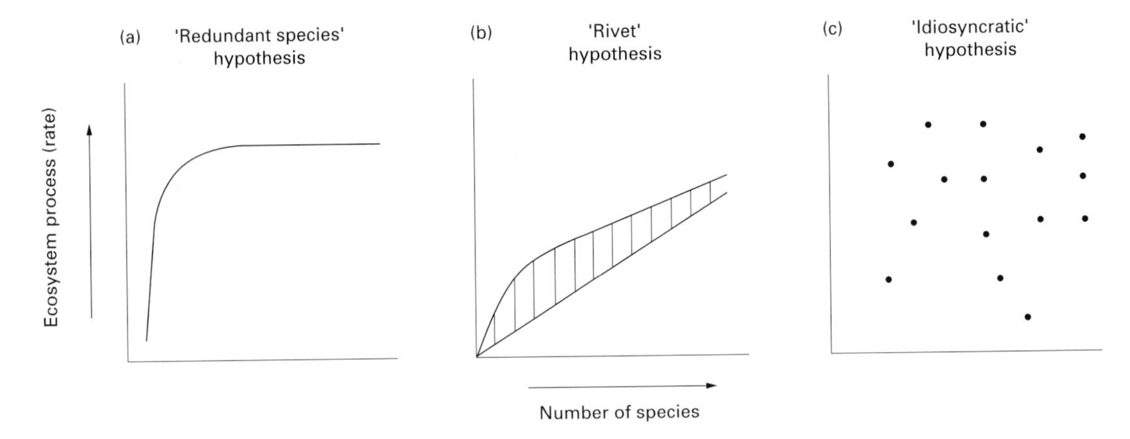

Fig. 8.21 Three hypotheses for the relationship between ecosystem processes and the number of species present in an ecosystem. (From Lawton 1994.)

the ecosystem process and biodiversity are possible but all assume that each species has a unique contribution to that process. Third, the 'idiosyncratic response hypothesis' suggests that increasing biodiversity affects ecosystem function in an unpredictable way because of the complex and varied roles of individual species (Fig. 8.21c).

There is very little evidence to support the different hypotheses for the role of increasing biodiversity in ecosystem functioning, and most of the available evidence concerns the role of increasing plant diversity in ecosystem functioning (e.g. Tilman & Downing 1994; Naeem *et al.* 1994, 1995). In one of the very few studies of insects, however, Klein (1989) examined the role of dung beetles (Coleoptera: Scarabaeinae) in the decomposition of dung in different habitats in central Amazonia—in continuous forest, forest fragments and cleared pasture. He found that the rate of dung decomposition declined by about 60% from continuous forest to cleared pasture (Fig. 8.22). The mean number of dung beetle species also declined about 80% from continuous forest to cleared pasture. However, the total abundance of dung beetles did not vary between different habitats. Taking this evidence together suggests that different dung beetle species

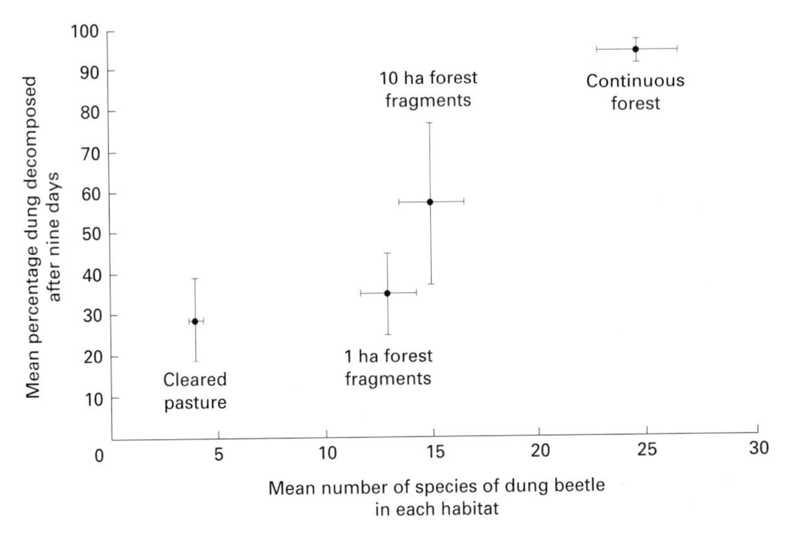

Fig. 8.22 Impact of the number of dung beetle species on dung decomposition in Central Amazonia. (From Klein 1989; Didham *et al.* 1996.)

play subtly different functional roles, and supports the rivet hypothesis (Fig. 8.21b).

Further research on this aspect of biodiversity is clearly needed, not only on decomposers, but also on pollinators and insect natural enemies. However, the history of biological control gives ample evidence for the importance of biodiversity. There are many examples where the successful control of non-native pests requires the introduction of particular natural enemy species from the native ranges of these pests (see Chapter 10). Clearly, biodiversity matters and is worth conserving. Rather worryingly, many of the species that are most worth conserving from an economic point of view, including many groups of insects, are least valued aesthetically. Such is the close linkage between the economic and cultural valuation of species that a wider appreciation of the economic importance of insects should lead to a greater cultural valuation of these species and a greater desire to conserve them. Meanwhile, however, insects and other species are being threatened with extinction as never before.

8.4 Threats to diversity

8.4.1 Recorded insect extinctions

Only 61 insect species are known to have become extinct globally since 1600 (Groombridge 1992), a small percentage (approximately 0.006–0.0006%) of the estimated total number of insect species. In the same time, 23 reptiles, 32 fish, 59 mammals, 116 birds and 596 higher plants have become extinct (Groombridge 1992), much higher percentages of the global species richness of these groups (Mawdsley & Stork 1995). In the UK, Hambler and Speight (1996) estimated that around 40 species of terrestrial invertebrates (mainly insects and spiders) have become extinct in the last 100 years.

Clearly, the figure for the number of insect species extinctions is an underestimate—for every plant, bird and mammal that have become extinct, insects solely dependent upon them as plant and animal hosts have become extinct too. For example, following the extinction of the passenger pigeon (*Ectopistes migratorius*), at least two insects, the lice *Columbicola extinctus* and *Campanulotes defectus*, became extinct too (Stork & Lyal 1993). Notably, the list of extinct insects given by Groombridge (1992) contains no ectoparasites. Of the recorded extinctions, 54% are Lepidoptera and 16% Coleoptera, probably because the former is the most well-known order and the latter the most species rich, not necessarily because most extinctions have taken place in these orders.

Most recorded insect extinctions (84%) have been amongst island species, 69% on Hawaii alone. These figures partly reflect the fact that insects in places such as Hawaii and mainland USA (where most of the continental extinctions have been recorded) are much better known than in most other parts of the world. However, the inevitably smaller ranges and smaller population sizes of insects on islands than on mainlands probably means that insects restricted to islands have been and will continue to be more prone to extinctions than insects with mainland distributions. In addition, as discussed above, the smaller the island, the greater the chance of extinction. Furthermore, the adaptations of insects on remote islands may be such as to make them prone to extinction. This is certainly true of many island birds and mammals, which evolved in the absence of predators. Tameness, flightlessness and reduced reproductive rates are characteristic of many island birds and have probably contributed towards the extinction of many of them by making them vulnerable to humans and introduced predators (Groombridge 1992). The evolution of flightlessness is also common among insect species endemic on islands. For example, 10 of the 11 orders of flighted insects that established in Hawaii have evolved flightless species (Howarth & Ramsay 1991). Flightless species have failed to evolve only in the Odonata in Hawaii. Insect species on islands may show adaptive radiation similar to the Galapagos finches studied by Darwin. The best-known example is probably the *Drosophila* (Diptera: Drosophilidae) of Hawaii (Carroll & Dingle 1996). Twenty per cent of the world's *Drosophila* species are found in the Hawaiian islands and most of these (484 of 509) are endemic.

8.4.2 Future insect extinctions

Is it possible to predict the future course of insect extinctions? As discussed above, what applies to 'real' islands often applies to habitat islands. Thus insects, and other animals and plants, are likely to become more and more vulnerable to extinction as their habitats become smaller. Not surprisingly, therefore, several ecologists have used island biogeography to attempt to predict the consequences of habitat loss, particularly deforestation, on global species richness (Reid 1995).

The relationship between the number of species (S) in an area and its size (A) (section 8.3.3) can be represented by the equation of a straight line:

$$\log S = \log c + z \log A$$

where c and z are constants.

Studies on islands, mainland areas and habitat islands have shown that the slope (z) generally lies in the range 0.1–0.7 (Begon *et al.* 1996). Reid and Miller (1989) based their predictions of species extinctions on a slope of 0.15–0.40. These slopes predict that a 90% reduction in island or habitat size will result in the loss of 30–60% of the species present. Reid and Miller then took a deforestation rate of 0.5–1% loss of forest area per year, 1–2 times the FAO (1988) estimate for 1980–85. The predicted rate of extinction based on these predictions was a loss of 2–5% of species per decade. Other predicted extinction rates range from 0.6–5% to 20–30% species per decade (Mawdsley & Stork 1995).

The main problem with predictions of species extinctions based on species–area relationships is that they assume that as habitats are removed (e.g. by deforestation), the species that formerly occupied them are unable to survive in the new habitat. This may be true in intensively farmed areas but not necessarily elsewhere, such as forest plantations, which may be rich in insects and other species (Lugo 1995) (see section 8.5.2).

8.4.3 Threats to insects

Many entomologists have assessed the various threats to insects. All conclude that the most serious class of threat to insects is habitat destruction and fragmentation (e.g. Hafernik 1992). More detailed lists of threats to insects are dominated by habitat-related threats. In Britain, for example, the major threats to butterflies were considered by Hyman & Parsons (1992) to be (in decreasing order of importance): agricultural expansion, deforestation, (habitat) management practices, afforestation and urbanization. The same workers concluded that the major threats to endangered beetles are: management practices, deforestation, agricultural expansion, natural succession and removal of dead wood. Thus threats to different insect groups show a strong degree of overlap, at least within a single country. However, threats to different insect groups may be greatest in different habitats (Hyman & Parsons 1992). In Britain, the habitats occupied by threatened butterflies are (in decreasing order of importance): chalk grassland, woodland edge, other grassland, hedgerow, heathland, and fen (measured by the number of threatened species found in each habitat). In contrast, the habitats occupied by threatened beetles are: deciduous woodland, other grassland, ancient woodland, coastal shingle, riparian habitats and sand dunes.

Threats to insects vary from region to region and are notably different between islands and continents. Gagne and Howarth (1985), for example, listed the following factors (in order of importance) for the extinction of 27 species of Macrolepidoptera in Hawaii: biological control introductions, habitat loss, alien mammals, host loss, alien arthropods and hybridization with a related invading alien species.

8.4.4 Biodiversity and climate change

There is increasing evidence that, as a result of human activity, mainly the burning of fossil fuels in the developed world, the global climate is changing (e.g. Bennetts 1995; UK Climate Change Impacts Review Group 1996). By 2050, the atmospheric CO_2 concentration will rise to almost double the preindustrial level; the climate is predicted to be about 1.6°C warmer. Hence, extremely warm seasons and years are expected to occur more frequently; the sea level will be about 37 cm higher; and other climatological changes are predicted to occur. Notably, most evidence suggests that these predicted changes imply a rate of climate change greater than ever experienced in the past, and it is possible that they will have a profound effect on biodiversity (see also Chapter 2).

Plants and animals will respond to climate change in one or more of four ways (Possingham 1993): by tolerating the change in climate (without adaptation), by adapting (genetically), by a changing distribution, or by becoming locally extinct. There has been little research on the way that species might respond to climate change by genetically based changes in tolerance or life history parameters. Partridge *et al.* (1994) showed that *Drosophila* adapts (in terms of survival, growth and development) to changes in temperature within 5 years (of continuous exposure). Species with much longer life cycles may, however, not be able to adapt genetically to climate change. A further problem with adaptation to climate change is that, in addition to global warming, other abiotic and biotic factors are likely to be changing rapidly, and plants and animals have a limited ability to adapt to multiple changes in their environments (Holt 1990).

The most likely outcome of future climate change is a change in distribution, as happened in the geological past (Coope 1995). The fossil record shows large-scale geographical changes in the distribution of Coleoptera and other insects in response to climate change during the latest glacial–interglacial cycle (Fig. 8.23). Some species of beetles now found in northern Scandinavia occurred throughout the UK during previous 'cold' episodes (Fig. 8.23b). Others, such as *Bembidion octomaculatum* (Coleoptera: Carabidae), were widespread in the UK during a brief period of high temperatures, but are now found much further south and rarely in the UK (although, interestingly, *B. octomaculatum* became re-established in Sussex in the late 1980s) (see also example in Fig. 8.23a). Overall, species extinctions of Coleoptera and other insects on a global scale during this cycle appear to have been uncommon despite the rapidity of climate change at the end of the glacial era (Coope 1995).

Some research is already producing evidence of the effects of current climate change. For example, monitoring in Mexico, the USA and Canada of the checkerspot butterfly, *Euphydryas* spp. (Lepidoptera: Nymphalidae) (Parmesan 1996) has shown that populations of this species more frequently go extinct at the southern end of its range—evidence for the first time across the entire range of a species that it is currently shifting in response to climate change. There is also a suggestion that spring events are occurring earlier; the appearance of some migrating animals and the leafing dates of common species of plants have been occurring earlier in the calendar year over the past 250 years, since the end of the mini ice age, in eastern England (Sparks & Carey 1995). The many studies of recent range changes of species in Britain are tentative in their implication of climate change as a cause, but some have shown a temperature-related response: the flight period of the hedge brown butterfly, *Pyronia tithonus* (Lepidoptera: Satyridae) is increasing as it spreads north (Pollard 1991); this species occurred in Scotland in the nineteenth century, and seems likely to return within the next two decades. Further examples are described in Chapter 2.

However, the view that insect and other species will respond to climate warming by changing distribution is based on the assumption that species can respond to climate change, and will have time to respond. It must be emphasized that species interact and that the net response to climate change will depend upon the interaction between species of the same and different trophic levels (Peters 1988). Randall (1982a,b) demonstrated that climate indirectly limits the distribution of insect species, through a study on the population dynamics of the moth *Coleophora alticolella* (Lepidoptera: Coleophoridae) along an altitudinal transect in northern England. At the highest altitude, the abundance of *C. alticolella* is limited by low availability of larval food, seeds of the rush *Juncus squarrosus*, and at the lowest altitude by parasitism, which becomes more intense with decreasing altitude. That is, the altitudinal difference in climate does not have a great influence directly on the moth but it has a profound influence through the insect–parasitoid relationship at low altitudes (and high temperatures) and the insect–plant relationship at high altitudes (and low temperatures).

Although the responses of particular species to changes in climate are unknown, it is probable that different species will respond in different ways and to different degrees. This will, in turn, lead to changes in the balance between competing species, plants and herbivores, and insects and their predators (Lawton 1995). This may result in local extinctions of species if currently benign insect herbivores become damaging 'pests' when they spread faster than their predators and parasitoids. Differences in the response of species to climate change are likely to result in the formation of different plant and animal communities with novel interactions (between competitors, plants and herbivores, herbivores and predators, etc.) (Peters 1988). Some of these interactions could threaten biodiversity, resulting in loss of mutualists (such as specialist pollinators) and the emergence of new diseases and increasing numbers of epidemics (Holmes 1996). Again, local extinctions are likely.

In the face of a direct or indirect decline in the suitability of the local environment as a result of climate change, the ability of species to disperse to new habitats is critical and is dependent on the presence of new habitats or environments, the distance to those new habitats or environments, and the dispersal ability of each species (Watt *et al.* 1997a). Thus the threat of climate change is made worse by the erosion

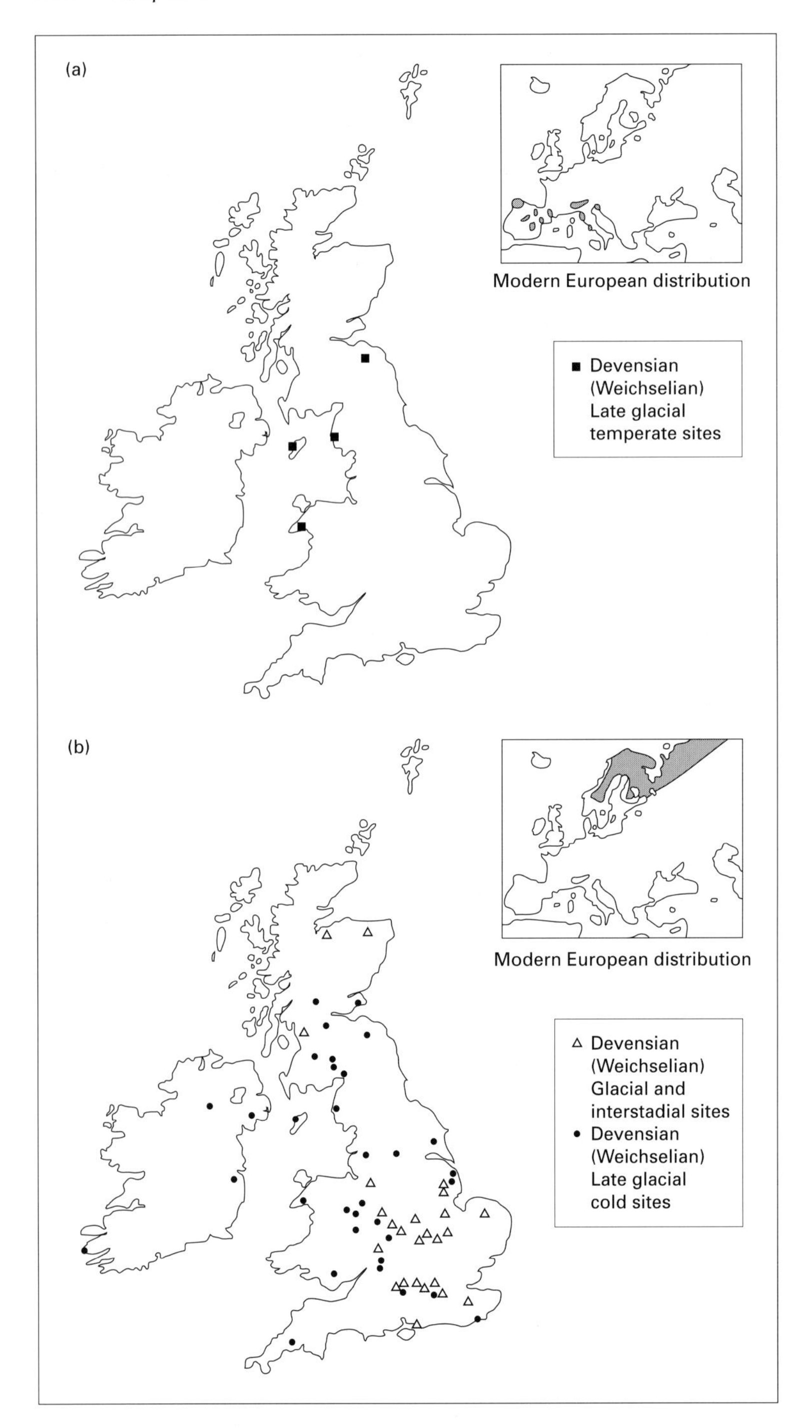

Fig. 8.23 Comparison between the fossil record and present day distribution of two beetle species in the British Isles: (a) *Asaphidion cyanicorne* and (b) *Boreaphilus henningianus*. (From Coope 1995.)

of natural habitats and the restriction of many species to protected areas or other refuges. Protected areas may cease to be suitable environments for the species that they exist to protect. Also, protected areas (or other suitable areas) with the right climatic conditions after climate change either may not exist or may be too far for the threatened species to disperse to readily (Peters & Darling 1985; Lawton 1995). In other words, past changes in climate occurred when the landscape was relatively intact. Movement of many species in response to future climate change will be severely restricted by the increasingly fragmented nature of our landscapes.

8.5 Insect conservation

8.5.1 Targeting conservation measures

The ideal way of deciding whether or not conservation measures for a particular insect species are needed (or to assess the success of conservation measures) is to monitor its abundance and distribution. For example, changes in the abundance and distribution of butterflies and day-flying moths are monitored in the British Isles through a variety of schemes, notably the Butterfly Walk Scheme (Pollard & Yates 1993) and the Biological Records Centre. The Butterfly Walk Scheme provides data on the abundance of British butterflies at a range of sites, potentially identifying species at risk locally and nationally (Fig. 8.24). The Biological Records Centre collates data on the distribution of plants and animals. Maps clearly show when species are becoming restricted in their distribution (Fig. 8.25).

However, accurate monitoring of threatened

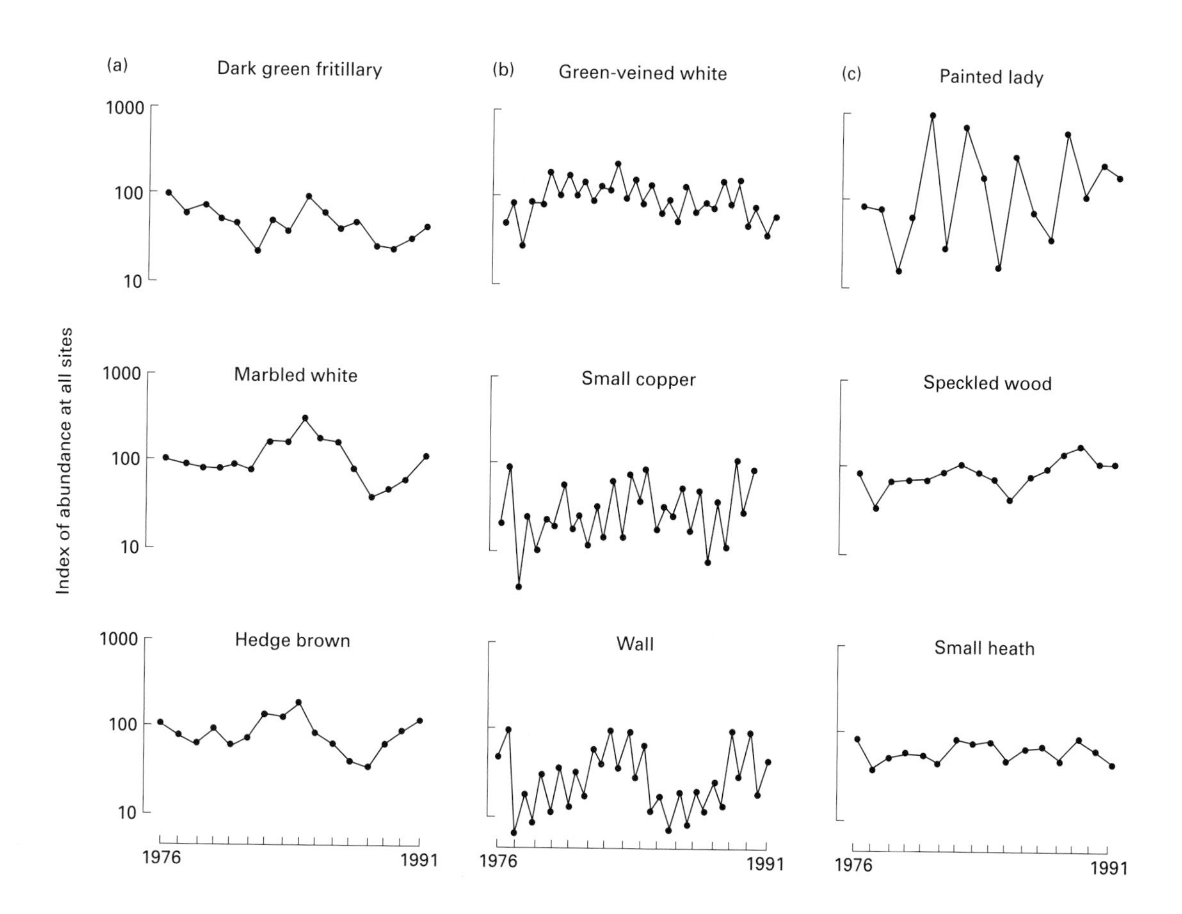

Fig. 8.24 Trend in abundance of nine butterfly species in Britain 1976–1991: (a) species with a single generation each year; (b) species with two distinct generations; and (c) species with overlapping generations, assessed by a single annual index. (From Pollard & Yates 1993.)

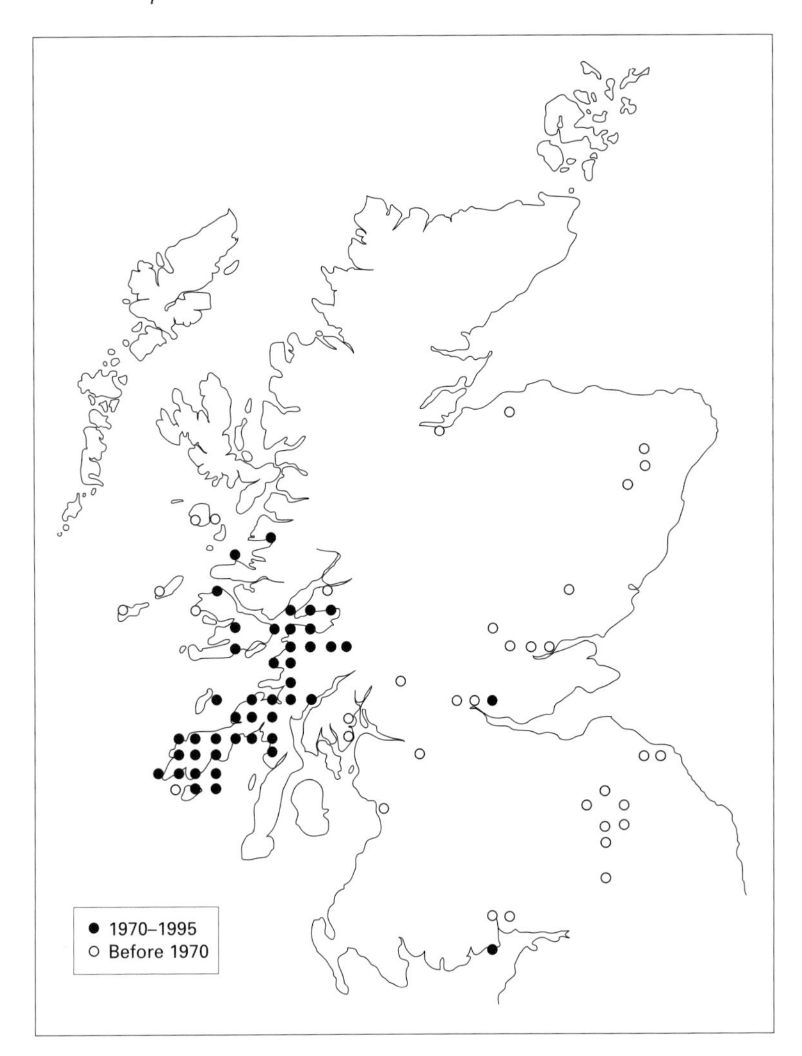

Fig. 8.25 Past and present distribution of the marsh fritillary butterfly in Scotland. (Data from the Biological Records Centre, Monks Wood.)

species is usually difficult and often undesirable—the rarity and inconspicuousness of threatened species usually prevent inexpensive, accurate assessment of abundance and distribution, and monitoring itself may be detrimental to the survival of insects. In these circumstances risk assessment may be based on 'expert' judgement. That is, entomologists make a judgement of the threatened status of a species based on knowledge of its current abundance and distribution, any known changes in its abundance and distribution, and an assessment of the vulnerability of its habitat. The best known classification system for the threatened status of plants and animals is the International Union for the Conservation of Nature and Natural Resources (IUCN) Red List classification. The Red List classifies threatened species as extinct, 'endangered', 'vulnerable' or 'rare' ('indeterminate' or 'insufficiently known') (Mace & Stuart 1994). Table 8.8 shows the numbers of species listed in each of these categories in 1995. The number of species in each category reflects the fact that some groups are better known than others: the threatened status of all known bird species has been assessed, in comparison with 50% of the world's mammals, 10% of the amphibians (Groombridge 1992) and an unknown but undoubtedly extremely small proportion of the world's insects. Red List coverage for insects is generally poor: only one group, swallowtail butterflies, has been adequately covered (Collins & Morris 1985).

The original IUCN Red List classification system

Table 8.8 Numbers of species considered to be 'endangered', 'vulnerable', 'rare' or threatened to an indeterminate degree by the World Conservation Monitoring Centre. (From Heywood 1995.)

Threatened	Endangered	Vulnerable	Rare	Indeterminate	Total
Mammals	177	199	89	68	533
Birds	188	241	257	176	862
Reptiles	47	88	79	43	257
Amphibians	32	32	55	14	133
Fishes	158	226	246	304	934
Invertebrates	582	702	422	941	2647
Plants	3632	5687	11485	5302	26106

operated for nearly 30 years and was clearly valuable in producing databases of threatened species, to provide a basis for setting conservation priorities and monitoring the success of conservation efforts (Miller *et al.* 1995). However, this system was frequently criticized for being subjective; classification of species by different authorities varied and did not always correspond to actual risks of extinction (Groombridge 1992). Therefore a new IUCN classification system was adopted in 1994 (IUCN 1994). The new system has 10 categories: 'extinct', 'extinct in the wild', 'critically endangered', 'endangered', 'vulnerable', 'conservation dependent', 'near threatened', 'least concern', 'data deficient' and 'not evaluated'.

In addition to the global Red List classification system, national Red Data Books have been produced for plants, birds, mammals, amphibians, reptiles and invertebrates, such as the Red Data Book for British insects (Shirt 1987). IUCN 'endangered' species were defined as those 'in danger of extinction and whose survival is unlikely if the causal factors continued to operate'. For the British Insect Red Data Book the criteria used for the endangered (RDB1) category were: species or subspecies known to occur as a single population in one 10 km square; species found in habitats known to be particularly vulnerable; species that are declining rapidly and continuously and are now found in fewer than five 10 km squares; or species, although considered extinct, that would require protection if rediscovered (Foster 1991). IUCN 'vulnerable' species were defined as species likely to become endangered unless the factors that caused their vulnerable status ceased to operate, and (IUCN) 'rare' species as those that are at risk but are not considered to be endangered or vulnerable. For the British Insect Red Data Book the criterion used for the rare (RDB3) category was that the species were known to exist in 15 or fewer 10 km squares (Foster 1991).

The problem with assessing the risk of extinction of insects is that we know relatively little about most species that may be considered to be at risk. An alternative approach to species-based risk assessment is therefore needed. The obvious alternative is to assess the risk to the habitats of insects. In fact, although the Red List and Red Data Book system is species based, as mentioned above, the actual classifications include assessments of risk to the habitats of species. Although globally the habitat requirements of most insect species are not known, nevertheless, information on the association between insect communities and particular habitats can help us to assess the risk posed by the loss of specific habitats to specific insect communities or insect species richness in general. For example, alpine meadows, pine woods and freshwater lakes all have specific insect species associated with them and therefore any risk to these habitats represents a risk to these insect species.

The main problem with purely habitat-based risk assessment is that particular species may be lost from habitats. This is clearly more frequent for vertebrates than invertebrates, and has been aptly named 'defaunation' (Phillips 1997). Nevertheless, there are various direct and indirect threats to insects and other species that may lead to the loss of insects from habitats. These include direct threats, e.g. butterfly collecting, and many indirect threats, e.g. pesticide application and other agricultural practices, pollution and climate change. Assessments of both the presence of these threats and their impact may be difficult. Although some potentially detrimental management practices, such as hedgerow removal, are obvious, others are not, such as the effects of pesticides on insects in terrestrial and aquatic habitats, but are, nevertheless, significant.

8.5.2 Applying conservation measures

There are two different strategies for conserving endangered species: '*in situ* conservation', the conservation of species within their natural or seminatural habitats; and '*ex situ* conservation', the conservation of species away from their natural habitats. *Ex situ* conservation is widely practised for plants and vertebrates, particularly mammals, but only a few insects are held in zoos as part of an *ex situ* conservation strategy. Exceptions include the field cricket in London Zoo (P. Pearce-Kelly, personal communication).

The most effective approach to conserving insect species is to conserve them in their natural habitats. The full complexity of habitat conservation for insects cannot be covered here but there are four major ecological aspects to consider:

1 all habitats important for insects should be retained;

2 habitat areas should be of sufficient size and shape;

3 there should be a sufficient number of areas of each habitat type, and these areas should be situated in such a way that the flow of individuals between different habitat areas ensures the long-term survival of the species unique to these habitats;

4 appropriate habitat management strategies should be adopted.

The question of what comprises an 'important' habitat is a crucial one. Because the ultimate aim of insect conservation is to ensure the survival of as many species as possible, the most important habitats are those containing species that are solely dependent on those habitats for their survival. In some cases, the association between insects and habitats is well known from direct observation, particularly the habitat requirements of phytophagous insects. In other cases, analyses of differences in insect species composition in different habitats are needed to identify the species associated with particular habitats.

The question of habitat size was discussed in section 8.3.3 and Chapter 1. The shape of habitat patches can be critical too. Usher (1995), for example, produced recommendations for the design of size and shape of 'farm' woodlands for Macrolepidoptera (Fig. 8.26).

There has been a great deal of theoretical research on the relevance of the spatial design of habitat patches for insects and other species. This includes the study of metapopulation dynamics (the study of subdivided and patchy populations, e.g. Hanski & Gilpin 1991). In parallel, there has been much practical consideration of the size, shape and arrangement of protected areas. Much of this relates to general conservation objectives, notably the (protected areas) 'many small or few large' debate (e.g. Margules *et al.* 1982) and the idea of 'conservation corridors' (e.g. Hobbs 1992) linking protected areas (which tend to include more than one habitat type).

Relatively few studies of the arrangement, and connections between, protected areas have explicitly considered insects. However, several studies have focused on the design of farm woodlands, already mentioned above in relation to habitat patch size. Figure 8.26 also summarizes the value for Macrolepi-

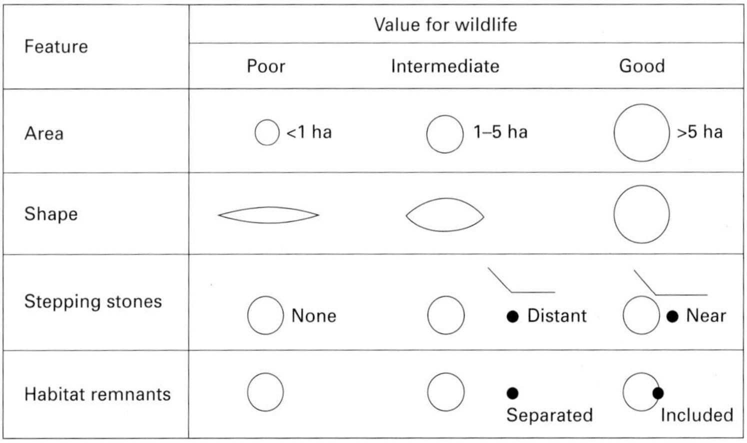

Feature	Value for wildlife		
	Poor	Intermediate	Good
Area	<1 ha	1–5 ha	>5 ha
Shape			
Stepping stones	None	Distant	Near
Habitat remnants		Separated	Included

Fig. 8.26 An assessment of the value of farm woodlands for Macrolepidopters according to their area, shape, isolation, the presence of stepping stones and habitat remnants. Woods are shown as open symbols, remnants as small black circles and hedgerows as straight lines. (From Usher 1995.)

doptera of placing farm woodlands near larger areas of woodland and of having other 'stepping stone' patches of woodland. Much of this is ecological 'common sense' but it is perhaps surprising how important habitat patch arrangement is for such mobile species as 'macro' moths.

Despite the importance of the biogeographical aspects of habitats discussed above, perhaps the most important aspect of conserving habitats for insects is habitat management. Returning to the example of farm woodland design, Dennis *et al.* (1995) considered a wide range of factors that might influence the number of arthropod species feeding on beech and other deciduous trees. He found that although factors such as woodland size and isolation had some effect, the most important factor was the amount of understorey vegetation. Dennis concluded that the absence of an understorey meant that the woods were more exposed and the overwintering stages of the insects associated with these tree species suffered greater mortality than insects in woodlands with understorey vegetation. In this case, therefore, habitat island biogeographical effects are insignificant compared with habitat management.

One of the most interesting aspects of Dennis's work on farm woodland insects is that by focusing on phytophagous insects he knew what species were potentially present in his study areas and therefore which tree-feeding species could benefit from management. Habitat management often focuses on individual species and there are many cases where ecological research has identified the habitat requirements of individual species. The habitat requirements of temperate butterfly species have been particularly well studied. The heath fritillary, *Mellicta athalia* (Lepidoptera: Nymphalidae), for example, requires sunny, sheltered woodland such as occurs in the first 3–4 years after coppicing, and has become rare following the marked decline of this practice in England (Warren *et al.* 1984). The reintroduction of coppicing (Fig. 8.27) has been shown to benefit species such as the high brown fritillary, *Argynnis adippe* (Lepidoptera: Nymphalidae) (Usher 1995) (Fig. 8.28). It has to be said, however, that not all researchers are in favour of coppicing as a practice to encourage insect conservation (Hambler & Speight 1996).

Most butterflies breeding in British woodland prefer very open, sunny rides or glades (i.e. less than 20% shade) (Fig. 8.29), but a few, such as the speckled wood butterfly, *Pararge aegeria* (Lepidoptera: Satyridae), prefer shaded (40–90%) rides or glades (Warren & Key 1991). Although the habitat requirements of moths are poorly known in comparison with those of butterflies, at least 60% of the 125 woodland Macrolepidoptera classified as 'Nationally Notable' are thought to be associated with open woodland, rides and clearings (Warren & Key 1991). The benefits of providing rides and glades in forests and woodland for Lepidoptera and other insects are such that precise recommendations for this form of management have been drawn up. For the least shade-tolerant species, Warren and Key cited minimum ride widths of 1–1.5 times the height of surrounding trees.

Fig. 8.27 Hazel coppice, New Forest, UK.

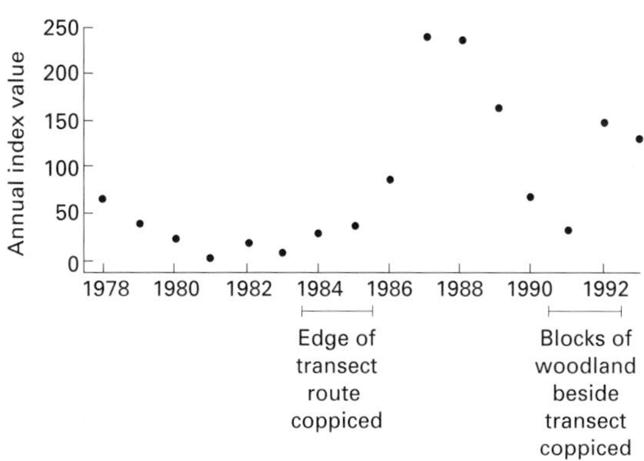

Fig. 8.28 The effect of coppice management on the abundance of the high brown fritillary butterfly (*Argynnis adippe*) in a National Nature Reserve in England. The woodland was coppiced in about 1970—other coppice dates marked. (From Usher 1995.)

Fig. 8.29 Woodland ride in the UK cleared and widened to reduce shade.

Apart from shade, the most important aspect of managing temperate woodland for insects is the provision of dead wood. Even modern conifer plantations can easily be managed to produce sunny conditions but the amount of dead wood in a primary forest takes many years to accumulate. The amount of dead wood in young plantations can be increased by cutting branches and felling trees but this tends to support only a limited number of dead wood species (Warren & Key 1991). Neither standing dead trees nor felled 'healthy' trees have the diversity of niches for dead wood species that very old naturally decaying trees have (Figs 8.30–31). Young felled trees rot from the outside (and rot develops inward) but old trees develop heart rot and rot holes. The size of decaying trees is important too: saproxylic insects tend to be sensitive to fluctuations in humidity but these are buffered in large trees and pieces of wood. In the absence of naturally decaying trees, there are a variety of ways of initiating premature decay: chainsaws to start the development of rot holes, inoculation of heart rot-inducing fungi, ring barking, fire, herbicides and explosives (Warren & Key 1991).

The challenge for managing woodland for insects is to create habitat heterogeneity: variation in shade, for example, to benefit a range of butterflies and dead wood in a range of conditions to provide habitats for a range of saproxylic insects. The latter include species with contrasting requirements, such as *Psilocephala melaleuca* (Diptera: Therevidae), which develops in very dry, red-rotted wood (Warren & Key 1991), to the many Diptera, Ephemeroptera, Trichoptera and Coleoptera that live in dead wood in water (Dudley & Anderson 1982).

Fig. 8.30 Decaying stump of a rainforest tree.

The above examples focused on insects in temperate woodlands. However, insect conservation is perhaps most urgently needed in tropical forests. These forests, despite covering only 8% of the land surface, probably contain 50–80% of the world's species (Stork 1988; Myers 1990). Notably few extinctions have been recorded in continental tropical forests, but this is more a reflection of our ignorance of tropical insects than any true measure of their susceptibility to extinction. The average annual rate of tropical deforestation stands at about 1% overall (Whitmore & Sayer 1992) but varies from region to region and country to country. The highest regional rate of deforestation is in West Africa: 2% overall (and 4% in closed forest). Within this region, some countries are losing their forests faster than others. For example, 5.2% of the total forest area was lost annually in the 1980s in Côte d'Ivoire (6.5% in closed forest) but deforestation rates in Cameroon were about 0.4–0.6% per annum. Although these estimates are based on data collected in the 1980s, recent figures for tropical log production show that the pressure on tropical forests for timber remains high, particularly in countries where deforestation was not as marked as elsewhere in the 1980s, such as Cameroon (Johnson 1995).

As discussed in the previous section, theoretical considerations of the impact of deforestation are no substitute for direct measurement of its impact on

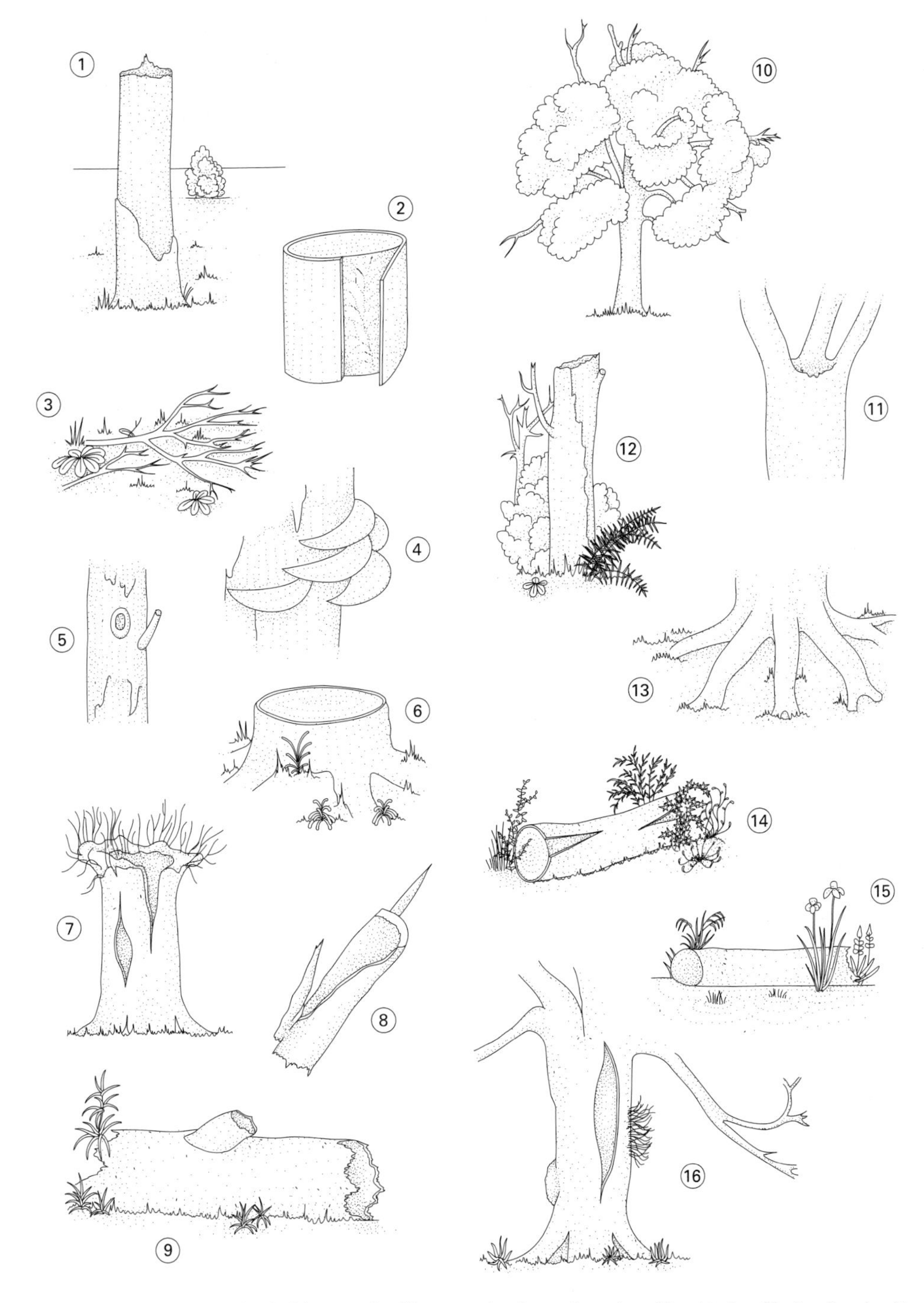

Fig. 8.31 The variety of dead wood microhabitats, each with a distinctive fauna: 1, sun-baked wood; 2, fungus-infected bark; 3, fine branches and twigs; 4, bracket fungi; 5, birds' nests; 6, stumps; 7, hollow trees; 8, burnt wood; 9, large fallen timber; 10, dead outer branches; 11, rot-holes; 12, standing dead trees; 13, roots; 14, well-rotted timber; 15, wet fallen wood; 16, red-rotten heartwood. (From Kirby 1992.)

biodiversity. Accordingly, several entomologists have tried to quantify the effects of deforestation by sampling insects in uncleared forest and areas cleared of forest for a range of purposes such as the establishment of forest plantations. In Cameroon, for example, several entomologists measured the effect of deforestation on ants, beetles, butterflies and termites (Watt *et al.* 1997c; Lawton *et al.* 1998). The species richness of each group of insects was assessed in: (a) forest with no evidence of previous clearance (primary forest); (b) forest known to have been logged in the past (old secondary forest); (c) plantations partially cleared and replanted with a West African tree species *Terminalia ivorensis*; (d) planta-

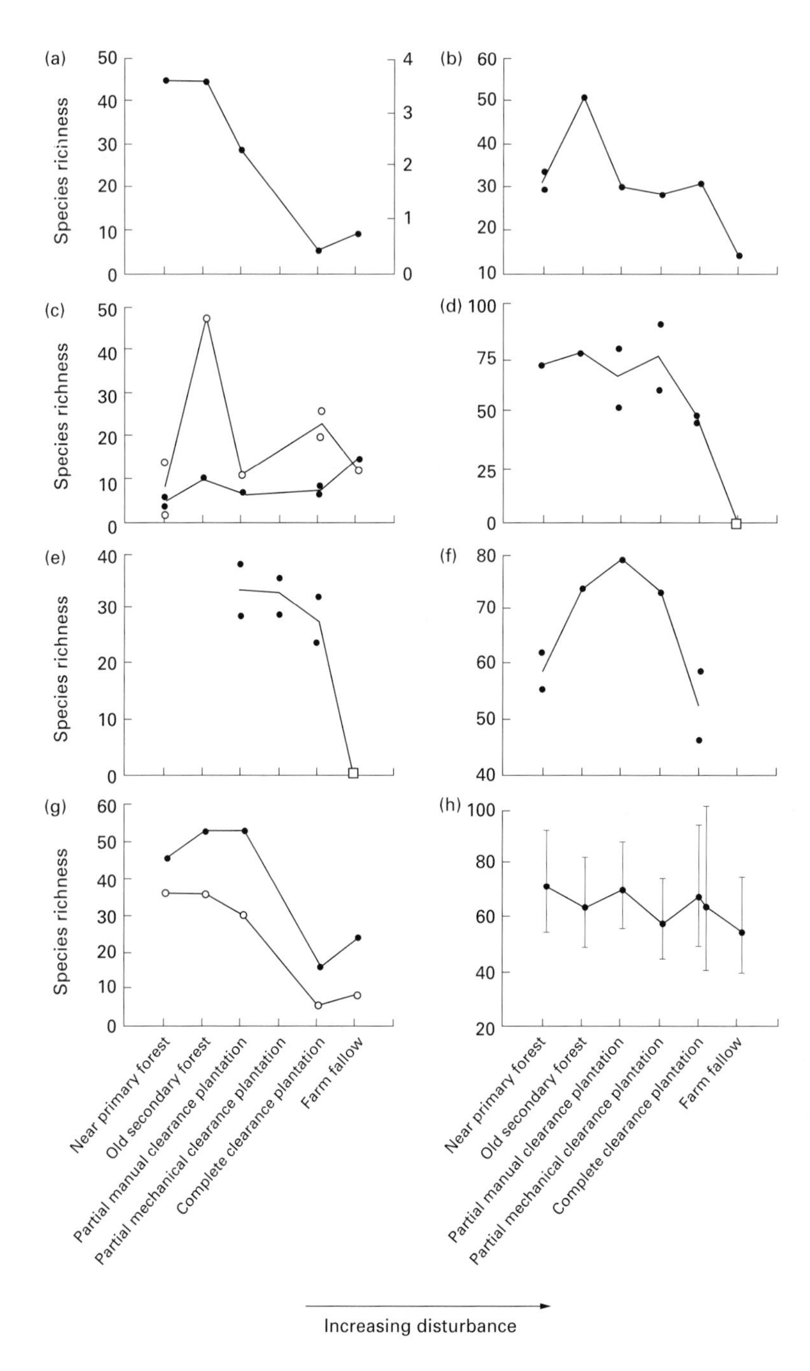

Fig. 8.32 Trends in the number of insect and other species along a disturbance gradient in the Mbalmayo Forest Reserve in southern Cameroon: (a) birds; (b) butterflies; (c) flying beetles (solid circles, malaise traps; open circles, flight-interception traps); (d) canopy-dwelling beetles; (e) canopy-dwelling ants; (f) leaf-litter ants; (g) termites (two surveys shown); (h) nematodes (with 95% confidence limits). (Note open symbols in (d) and (e) where the numbers of canopy-dwelling species are assumed to be zero in the absence of a canopy.) (From Lawton *et al.* 1998.) Reprinted by permission from *Nature*, ©1998 Macmillan Magazines Ltd.

tions completely cleared and replanted with *T. ivorensis*; (e) fallow farmland. As might be expected, the numbers of species of each group of insects are lowest in the fallow farmland (Fig. 8.32). Although each group of insects shows a different response to forest disturbance, it is notable that, in terms of species richness, forest plantations can be comparable with uncleared forest for most of the insect groups studied. However, the type of plantation is crucial: plantations established after complete forest clearance are much poorer in insect diversity than plantations established after partial clearing of existing forest. In this study, the plantation tree was a native species: the contrast between the ecologically sensitive establishment of plantations of native trees and plantations of alien tree species established after complete forest clearance is likely to be even more marked than that observed in the Cameroon study. Sadly, reforestation projects are small in scale in comparison with deforestation. For example, during the 1980s, only 360 km^2 of plantation forests were established each year in West Africa (Lawson 1994) in comparison with a rate of deforestation of an estimated 12 000 km^2 annually (FAO 1991). In addition, these plantations often involved the use of alien tree species, have had a poor record of maintenance and survival, and were usually established in areas of existing forest (Lawson 1994).

Perhaps the best strategy for conserving tropical insects is to ensure that there is an adequate coverage of protected areas (nature reserves, national parks, etc. (Groombridge 1992)) in tropical regions. Approximately 5% of the tropical humid forests are protected to some degree. The value of protected areas for vertebrates, particularly birds, is well understood. Round (1985), for example, found that 88% of Thailand's birds occurred in protected areas, and Sayer and Stuart (1988) concluded that 90% of vertebrate species associated with tropical moist forests in Africa would be protected with a slight increase in the 10% of tropical moist forests currently under some form of protection. Other regions are less adequately protected and the presence of a species within a protected area does not guarantee its long-term survival (Groombridge 1992). Moreover, we do not know how many insect species these areas actually protect.

8.6 Conclusion

Clearly, habitat loss is the most serious threat to insect conservation. Ways of combating other threats, such as the introduction of alien species, need to be found. In addition, we need to plan for future threats to insects, such as man-made climate change (section 8.4.4). We need to know where plants and animals would move, if free to do so in a future climate, so that species conservation efforts can be concentrated where they will be effective (Watt *et al.* 1997a). There is little point in trying to conserve species where their environment is no longer suitable and, more importantly, there is a need to identify areas where new protected areas (or conservation efforts in general) should be focused. From the point of view of conservation, the most worrying aspect of climate change is that the current distribution of protected areas may be inappropriate in the future. We are not currently in a position to identify accurately existing or new areas that will serve as future protected areas. However, we can take practical steps by increasing the number of protected areas, particularly in parts of the country where they do not currently exist, and focusing on areas with altitudinal variation, linking protected areas as far as possible with wildlife 'corridors' or 'stepping stones' (for all major habitats), developing protocols to translocate species that cannot disperse fast enough, and promoting conservation outside protected areas.

In conclusion, our knowledge of many endangered temperate insect species is remarkably good. Some of these insects are covered by protective legislation, and conservation action plans have been prepared for a few of them. In some cases, we know enough to be able to reintroduce species that have gone locally extinct. For example, when the large blue butterfly, *Maculinea arion* (Lepidoptera: Lycaenidae), went extinct in England, knowledge of its ecology was good enough to allow a successful reintroduction programme (Thomas & Lecoington 1991). Successful conservation of insects generally, particularly in the tropics and under the threat of increasing habitat loss and climate change, will require a much better knowledge of insect ecology.

Chapter 9: Insects and Diseases

9.1 Introduction

The ecology of insects that feed on plants (herbivores) has been discussed in Chapter 3, and that of insects feeding on other insects (predators and parasitoids) in Chapter 4. Some insects such as mosquitoes and bedbugs also feed on vertebrates. In fact, we are able to draw parallels between this latest habit and that of sap feeding on plants, as both systems rely predominantly (though by no means exclusively) on piercing mouthparts, and utilize the circulatory fluids of the host, either sap or blood, as the primary food supply. During these feeding bouts, both plant- and animal-feeding insects are potentially able to come into contact with pathogenic organisms present in the fluids of the host, and many have evolved some sort of relationship with these organisms, which promotes their spread and proliferation. The insect, known as a vector, may or may not benefit from this association, but the host, be it a plant or animal, certainly does not. It becomes infected with pathogens, which cause disease.

The dictionary definition of the term 'disease' is loose and woolly; many forms of deterioration in the health of a plant or animal might be thought of as a disease, from heart problems in humans to drought stress in trees. For the purposes of this chapter, we will be considering only those animal and plant diseases that have their basis in the actions on the host of pathogenic organisms with which insects have some sort of association. Even the term 'pathogen' seems hard to define satisfactorily. Clearly, a pathogen is a type of parasite, but not all parasites are pathogens. Thus, fleas are well-known parasites of mammals and birds, whereas the disease-causing organisms (bacteria in this case) that they transmit to cause bubonic plague are pathogens. Perhaps the best way of defining 'pathogen' is by description, without worrying too much about the reasons for the classification. In this context, a variety of organisms commonly have associations with insects, including bacteria, fungi, nematodes, protozoans, spirochaetes and viruses.

Both plants and animals become infected with disease agents attributable to the various groups of pathogens, and the characteristics of the insect, host and pathogen are often similar ecologically. In most cases, though by no means all, the main role for insects in host–pathogen interactions is as a vector of the pathogen from one isolated host to another. Here, the insect transports pathogenic organisms either on the outside of its body, or in its gut, salivary glands or another internal location. Some plant and animal diseases, however, do not involve insects as vectors, and the ecology of these associations will be considered first.

9.2 Non-vectors

On occasion, by virtue of insects feeding on plants or animals, they may predispose the host to pathogen infection. In the absence of insect attack, the host would remain unavailable or resistant to infection. An excellent example of this is beech bark disease. This disease infects European beech, *Fagus sylvatica*, in Europe and North America, and begins when the beech scale insect, *Cryptococcus fagisuga* (Homoptera: Eriococcidae) (Fig. 9.1), uses its long piercing mouthparts to penetrate beech bark from the outside to feed on the living bark cambium inside (Houston 1994). This feeding predisposes the bark to infection by airborne fungi, *Nectria coccinea* var. *coccinea* in Europe, and *N. coccinea* var. *faginata* and *N. galligena* in North America. Heavy infestations of the scale insect allow the only weakly pathogenic *Nectria* spp. to spread rapidly within tree bark, unrestricted by defence reactions such as wound periderm (where special tissues develop around wounds and act as a barrier to infection) or callous formation. Without the scale insect, the fungus is usually unable to infect healthy trees on suitable soils.

First instar scale 'crawlers' move over the bark of host trees to establish their own feeding sites, and only a relatively small proportion move to adjacent trees on air currents within a stand. Only around 1%

Fig. 9.1 Beech scale of tree trunk.

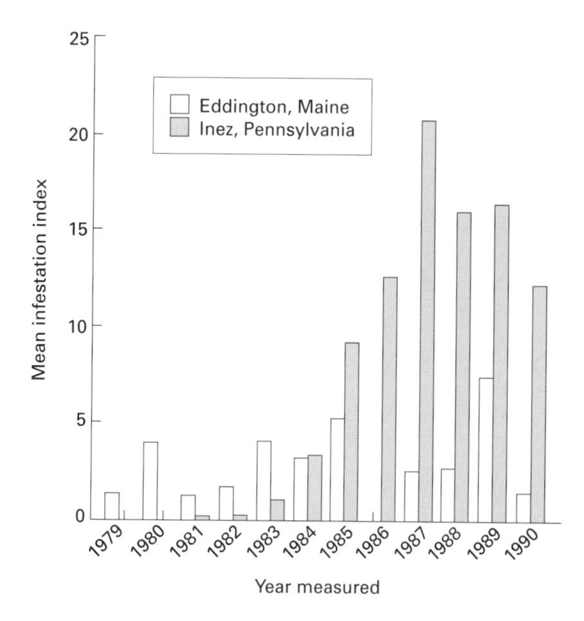

Fig. 9.2 Annual population levels of *C. fagisuga* from 1979 to 1990 in the first decade of infestation (Inez, Pennsylvania) and in the sixth decade after initial infestation (Eddington, Maine). The infestation index was calculated as a weighted average of infestation scores for approximately 200 trees per plot. Trace populations were scored as 1; very heavy populations as 40. (From Houston 1994. With permission, from the *Annual Review of Phytopathology*, Volume 32, ©1994, by Annual Reviews.)

of crawlers spread greater distances facilitated by wind above beech canopies (Wainhouse 1980). New outbreaks are characterized by rapid population build-up on individual trees; established ones, on the other hand, tend to fluctuate around lower infestation levels (Fig. 9.2). Many factors are thought to contribute to these observed variations in density, including host plant resistance, weather conditions, and natural enemies. As the disease relies predominantly on the presence of the insect, however, the epidemiology of *Cryptococcus* is closely mimicked by that of *Nectria*. Beeches free of the insect are not able to be infected and subsequently killed by the fungus, so that though no pathogen transmission occurs, the insect is vital in the life of the fungus.

9.3 Insects as vectors of disease-causing pathogens

A great number of different groups of insects have been shown to act as vectors for an equally wide range of animal and plant disease-causing pathogens. Table 9.1 summarizes some of these vectors and their associated diseases. Some of the most serious will be considered individually below, but some unifying features can be observed. Taxonomically, pathogen transmission is fairly widespread within the Insecta. The two major orders, the Hemiptera and the Diptera, one from the Exopterygota and the other from the Endopterygota, are somewhat analogous as vectors of animal and plant pathogens, respectively. Both have mouthparts derived from generalist mandibles, maxillae and labiae, now adapted for the highly specialized function of piercing the epidermis of the host and transferring liquid contents to the insect's buccal cavity, or mouth. Both transmit a whole range of pathogens, though, with notable exceptions such as malaria and onchocerciasis, animal and plant viruses form the major group. It should be noted, however, that there is one huge difference between the two orders. Juvenile stages of the Hemiptera (nymphs) tend to exhibit much the same lifestyle as their parents, whereas the larvae of the Diptera have extremely different ecologies from the adults. Thus, only adult tsetse flies (and, in the case of mosquitoes, adult females only) can act as vectors, whereas all stages of a plant-hopper or aphid can, in theory, transmit a pathogen, including the males if they occur.

Certain morphological and behavioural traits are required for an insect to be a good pathogen vector. A

Table 9.1 Examples of insects that can act as vectors of pathogens, and the diseases with which they are associated (note that most genera consist of several species world-wide).

Insect genus or species	Order: family	Common name	Pathogen	Disease or syndrome
Animal diseases				
Aedes	Diptera: Culicidae	Mosquito	Virus	Rift Valley fever
Aedes	Diptera: Culicidae	Mosquito	Virus	Equine encephalitis
Aedes	Diptera: Culicidae	Mosquito	Virus	Yellow fever
Aedes	Diptera: Culicidae	Mosquito	Virus	Dengue fever
Anopheles	Diptera: Culicidae	Mosquito	Protozoan	Malaria
Chrysops	Diptera: Tabanidae	Horse fly	Nematode (filarial)	Loa loa (loiasis)
Culex	Diptera: Culicidae	Mosquito	Virus	Japanese encephalitis
Glossina	Diptera: Glossinidae	Tsetse fly	Protozoan (flagellate)	Sleeping sickness, nagana
Lutzomyia	Diptera: Psychodidae	Sandfly	Protozoan	Leishmaniasis
Simulium	Diptera: Simuliidae	Blackfly	Nematode (filarial)	Onchocerciasis
Triatoma	Hemiptera: Triatomidae	Kissing bug	Protozoan	Chagas disease
Xenopsylla cheopis	Siphonaptera: Pulicidae	Flea	Bacterium	Bubonic plague
Plant diseases				
Delphacodes	Hemiptera: Delphacidae	Plant-hopper	Virus	Maize dwarf virus
Monochamus alternatus	Coleoptera: Cerambycidae	Longhorn beetle	Nematode	Pine wilt
Myzus persicae	Hemiptera: Aphididae	Aphid	Virus	Sugar beet yellows
Scaphoideus titanus	Hemiptera: Cicadellidae	Leaf-hopper	Phytoplasma	Asters yellows
Scolytus	Coleoptera: Scolytidae	Bark beetle	Fungus	Dutch elm disease
Sirex/Urocerus	Hymenoptera: Siricidae	Wood wasp	Fungus	Wet rot
Sitobion avenae	Hemiptera: Aphididae	Aphid	Virus	Barley yellow dwarf virus
Thrips	Thysanoptera: Thripidae	Thrips	Virus	Tomato spotted wilt

completely immobile insect might be thought to have little chance of carrying any pathogen from one host to another (but see section 9.4.3), so that the ability to move, preferably over long distances by flying and short distances by hopping or crawling, both within and between a crop or population of hosts, is fundamental. Most disease-causing pathogens are located for most if not all of their life cycles inside the host organism (fungal fruiting bodies being one obvious exception), so that a means of penetrating the host's epidermis to collect pathogens from its circulatory or transport system is also essential. This normally demands a piercing, chewing, or in the case of sandflies for example, rasping, mouthpart system, as mentioned above.

The majority of pathogens are fairly host-specific. Thus for example, barley yellow dwarf virus attacks only certain cereal species, whereas malaria is essentially a human disease (though non-crossinfective forms are also found in birds and monkeys). Moreover, some pathogens have alternative hosts, with varying degrees of obligateness, so that an efficient pathogen vector should restrict its own feeding habits to a narrow host range too, to avoid depositing pathogens in hosts to which they are not suited. When these requirements for an efficient pathogen vector are summed, it is not at all surprising that insects are so well known in this capacity, with their dispersal abilities, mouthpart structures and complex host associations.

The role of insect vectors in the epidemiology of pathogens ranges from entirely accidental or facultative, to complete obligateness. Reference to a few examples will illustrate this point. Enterobacterial infections such as *Shigella* (Pellegrini *et al.* 1992) can be carried passively and mechanically by cockroaches or houseflies walking over contaminated food and transporting it on their bodies, legs and mouthparts to other locations including human food (Cohen *et al.* 1991). This sort of passive transmission can also be seen in certain plant diseases. Pine pitch canker is a common disease of conifer cones, caused by the fungus *Fusarium* spp. In California, research has shown that it appears to be transmitted on the bodies of various species of small beetle that tunnel into cones and twigs of both native and exotic pines

(Hoover *et al.* 1996). Simple artificial wounding of cones did not produce canker, showing that the beetles were required to vector the fungus into the cones, though this was entirely accidental on the part of the insect. In complete contrast, however, on the same genus of tree, the woodwasp, *Sirex gigas* (Hymenoptera: Siricidae), is entirely dependent on the fungus *Amylosteruem* spp., which it vectors from one stressed pine tree to another to break down the host's heartwood so that woodwasp larvae can feed on otherwise unusable timber. The fungus, in return, is transported between ephemeral hosts, and in fact cannot live without its wasp symbiont (see also Chapter 6).

9.4 Vector ecology

Insect vectors are influenced by many abiotic and biotic factors just as are other insects. For example, weather conditions may affect growth, reproduction and dispersal (Chapter 2), natural enemies and competition may limit their abundance (Chapters 4 and 5), and the availability of both major and alternative hosts may influence their persistence in a habitat and spread between habitats. General principles of vector ecology will be considered here, but reference should also be made to the detailed examples of specific major insect-vectored diseases discussed later.

9.4.1 Epidemiological theory

We feel that it is helpful at this stage to consider briefly one or two aspects of the theory of epidemiology as related to insect ecology, as some basic parameters of vector–host–pathogen interactions can be explained and then applied by the reader to case studies later in the chapter. Fundamental to epidemiology is the basic case reproduction number (labelled R_0), which is defined as the average number of secondary cases of a disease arising from each primary one in a particular population of susceptible hosts (Dye 1992). Another equally important parameter is the vectorial capacity (labelled C), defined as the daily rate at which future inoculations arise from a currently infective situation. Vector survival is most critical here; if, for example, a pathogen has to develop inside the body of the vector before its being deposited in a new host (the time taken for this development is known as the extrinsic incubation period, which can be considered to be a latent period), then it is clearly vital that the vector itself has a life span longer than this period. If the vector dies before its 'on-board' pathogens are ready for inoculation, then of course transmission will never occur!

Two related equations, first derived in studies on malaria, describe the derivation of R_0 and C:

$$R_0 = \frac{ma^2bp^n}{-r \ln p}$$

$$C = \frac{ma^2bp^n}{-\ln p}$$

where m is the number of vectors per host, a is the number of times the vector bites the host per day, p is the daily survival rate of the vector, n is the extrinsic incubation period mentioned above, b is the proportion of vectors that actually generate a new infection in the host when it is bitten, and r is the daily rate of recovery of the infected host.

These somewhat cryptic equations are simply split into ma, the biting rate, and $p^n/-\ln p$, the so-called longevity factor, and R_0 can be estimated as C/r (D.W. Kelly, unpublished data). So what does all this mean for vector ecology? Simply put, if R_0 is greater than unity (i.e. the recovery rate is less than the vectorial capacity), then the disease will spread through a host population, whereas if R_0 is less than unity (i.e. the recovery rate is greater than the vectorial capacity), the disease will become eradicated over time.

9.4.2 Host range

Some vector species may be very specific to the hosts on which they feed, and hence are able to vector pathogens between only a narrow range of plants or animals. Elm bark beetles such as *Scolytus scolytus* and *S. multistriatus* (Coleoptera: Scolytidae) feed exclusively on trees in the genus *Ulmus*, and hence the fungi for which they act as vector (see below) can never venture beyond this narrow host range. The transmission of mutualist fungi by bark beetles is discussed in more detail in Chapter 6.

However, not all potential vectors have the ability to transmit pathogens between hosts, and pathogens in turn may be specific to certain species of vector. This is especially the case when viruses, for example, replicate inside the vector that has acquired them, before infecting a new host. The mosquitoes *Culex quinquefasciatus* and *Aedes aegypti* (Diptera: Culicidae) are both renowned as vectors of a range of arboviruses (arthropod-borne viruses), though they do not perform this task equally well. Dengue fever is widespread over SE Asia, and although rarely fatal, causes serious aches, high fevers, extreme fatigue and rashes (Schwartz *et al.* 1996). Dengue virus replication in the mosquito vector, however, does not occur in *Culex*, only in *Aedes*, so that only species in the latter genus can serve as vectors (Huang *et al.* 1992).

More often though, insect vectors have a fairly wide range of hosts, many of which may act as reservoirs of pathogen infection, thus maintaining a disease within a locality, even when their host is absent. Blackflies (Diptera: Simuliidae), in the genus *Simulium*, are vectors of onchocerciasis (river-blindness) in humans. However, no species of blackfly feeds exclusively on humans (Service 1996), and some of those that do bite people may do so only when other large mammals such as donkeys or cattle are unavailable. Clearly, this is no comfort to men, women and children who have contracted the disease, but as the filiarial nematode, *Onchocerca volvulus*, that causes river-blindness has no other animal reservoirs, village livestock may be considered to be advantageous in keeping the vectors away from humans and reducing the incidence of the pathogen.

Bubonic plague (the Black Death of medieval Europe) is vectored by fleas (Siphonaptera), and in urban areas epidemics are normally thought of as having a very restricted host range, that is, rats and humans. However, outside urban situations, plague, caused by the bacterium *Yersinia pestis*, is primarily a disease of wild rodents and other mammals (Service 1996), and such hosts of the plague bacilli range from gerbils and voles to chipmunks and ground squirrels. This is a good example of how human activities can localize and focus a naturally occurring epizootic by forcing an otherwise wide host range into a narrow one, with extreme consequences. Human diseases with animal reservoirs such as in this example are termed zoonoses, as opposed to anthroponoses where only humans are involved.

Somewhat unlikely mammals may act as alternative hosts for insects that vector human diseases. One of the vectors of Chagas disease in South America, the bug *Triatoma sordida* (Heteroptera: Triatomidae), utilizes various food sources as well as humans. Bugs in this genus occur naturally in various habitat types, including hardwood forests, where they can be found inhabiting areas dominated by trees, logs, cacti or even bromeliads (Wisnivesky-Colli *et al.* 1997). With all these ecosystems in which to find hosts, it is to be expected that many other species of mammal are regularly encountered by the triatomids, and in fact opossums provide an important alternative to humans, and a significant source of pathogen infection (Diotaiuti *et al.* 1993).

Many mosquitoes show wide host ranges too. *Aedes albopictus* is a vector of dengue fever in SE Asia, which was accidentally introduced into North America in the 1980s. Precipitin (antibody reaction) and ELISA (enzyme-linked immunosorbent assay) tests were carried out on wild-caught adult female mosquitoes to investigate the animals on which they had been feeding (Savage *et al.* 1993). Some of the results of these tests are shown in Fig. 9.3. In fact, the mosquitoes not only fed on a wide range of mammalian species, but 17% of those tested had also fed on various birds. *Ae. albopictus* is clearly an opportunistic feeder with a wide range of hosts, and hence has the potential to become a vector of indigenous arboviruses (arthropod-borne viruses).

Plant disease vectors may also show wide host ranges and hence are able to vector plant pathogens to and from large numbers of host species. The peach–potato aphid, *Myzus persicae* (Homoptera: Aphididae) is much less host-specific than its common name suggests. Studies in the UK showed that this aphid was able to transmit beet yellows virus (BYV) and/or beet mild yellowing virus (BMYV), both serious diseases of sugarbeet, to nine species of common arable weed (Stevens *et al.* 1994). These included *Stellaria media* (chickweed), *Papaver rhoeas* (poppy) and *Capsella bursa-pastoris* (shepherd's purse). The weeds are likely to play an important role

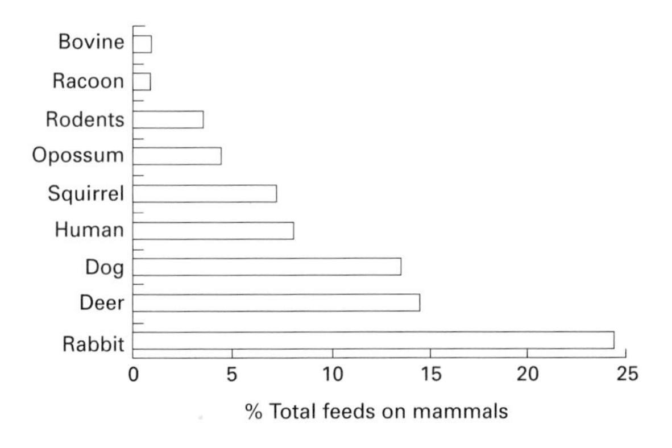

Fig. 9.3 Mammalian host feeding patterns of *Aedes albopictus* in North America as percentages of positive reactions in precipitin and ELISA tests. (From Savage *et al.* 1993.)

in the persistence of disease-causing pathogens in arable crops (see section 9.5.4). Whitefly (Homoptera: Aleyrodidae) are also important vectors of plant viruses. In various parts of India, adults of *Bemisia tabaci* act as vectors of tobacco leaf curl virus (TbLCV) in 10 tobacco-growing areas. This crop, though possibly not popular everywhere in the world, is a vital source of revenue, and the virus causes a variety of debilitating symptoms from severe leaf curling and paling to irregular swellings and thickening of the veins (Valand & Muniyappa 1992). However, the whitefly does not confine its attentions to tobacco, and is able to transmit TbLCV to 35 other plant species, some of which are shown in Table 9.2. Once more, the role of non-crop reservoirs in the epidemiology and management of insect-vectored diseases needs very careful appraisal.

9.4.3 Vector dispersal

As mentioned above, the prime importance of an insect vector of a plant or animal pathogen is its ability to transport the pathogen from one host to another, and the amount of vector movement can have a great influence on disease dynamics and spatial distribution (Ferriss & Berger 1993). Vectors may travel only short distances, from one plant in a crop to an adjacent one, or over relatively long ones from one crop to another remote from the first. Short distances may be covered by flying, jumping or simply crawling, activities to which insects are supremely well adapted. Even the least mobile of insects seem capable of some vector transmission. Mealy bugs (Hemiptera: Pseudococcidae) are a very sedentary group of sap-feeders; neither the nymphs or the adults move very much, and the only really mobile stage is the first instar nymph, known as a crawler because of the way in which it moves around! The citrus mealy bug, *Planococcus citri*, despite its name, is a polyphagous species found over most of Europe and North America, and one of its claims to fame is as a vector of grapevine leafroll virus (GLR) in the Galician wine-growing region of north-western Spain (Cabaleiro & Segura 1997), at least over short distances within a vineyard. This disease is found from Europe to America, and Australia to South Africa, and it causes various symptoms on commercial grape vines including uneven and retarded ripening of fruit, and small, pale-coloured grapes that are difficult or impossible to export (Saayman & Lambrechts 1993). Scale insects (Hemiptera: Coccoidea) are even less mobile than mealy bugs, though one species, *Pulvinaria vitis*, is also said to be able to vector GLR at least in experimental settings (Belli *et al.* 1994).

For these small-scale movements, the type of crop and its planting system may help to dictate the success of the dispersal and hence spread of the pathogen. Leaf-hoppers (Homoptera: Cicadellidae) transmit a wide range of plant viruses and, as the name suggests, are able to move by jumping. In the case of the corn leaf-hopper, *Dalbulus maidis*, which spreads maize rayado fino virus in maize (*Zea mays*), plant-to-plant movement appears to depend on various planting characteristics, with the basic rule that the shorter the distance to the next adjacent plant, the more likely a leaf-hopper is to move between them (Power 1992). However, things are never quite as simple as they seem, and several more complex observations are available (Table 9.3). Thus, part of a vector–virus management system has to take

Table 9.2 Indication of the wide host plant range to which the whitefly *Bemisia tabaci* transmits tobacco leaf curl virus. (From Valand & Muniyappa 1992.)

Plant species	Common name
Beta vulgaris	Beetroot
Capsicum annuum	Sweet pepper
Carica papaya	Papaya
Cyamopsis tetragonoloba	Cluster bean or guar
Lycopersicon esculentum	Tomato
Phaseolus vulgaris	Bean
Petunia hybrida	Petunia
Sesamum indicum	Sesame

Table 9.3 Patterns of dispersal of the leaf-hopper *Dalbulus maidis* within a maize crop. (From Power 1992.)

(A) Leaf-hoppers less abundant and spread of maize rayado fino virus slower in dense stands of maize than in sparse stands
(B) With constant plant density, leaf-hoppers more abundant in stands with uniform spacing than in those with densely sown or clumped rows, though virus incidence did not differ
(C) Leaf-hoppers were less likely to move to adjacent plants in uniform spacing patterns than in linear or clumped ones
(D) Item C may explain the lack of higher virus incidence in uniform stands despite higher leaf-hopper abundance

into account the dispersal behaviour of the insect, and alter the crop husbandry systems accordingly.

Although dipteran vectors of animal pathogens can travel relatively long distances in the order of kilometres (see below), evidence shows that short-range dispersal (displacement) is also important. This can occur, for example, in the transmission of malaria between members of a sleeping group of people, all of whom may be infected with a single strain of the parasite having been bitten by the same, single mosquito (D.W. Kelly, unpublished observations). Displacement can be even more important for vectors such as bedbugs (Hemiptera: Cimicidae), which transmit pathogens mechanically (Jupp & Williamson 1990) and have to rely on the blood from one host being fresh.

Medium-range dispersal of vectors usually relies on the powers of flight of the insect in question. An example concerns phlebotomine sandflies (Diptera: Psychodidae), particularly notorious as vectors of leishmaniasis in both the Old and New Worlds. The leishmaniases are a spectrum of diseases, caused by parasitic protozoans in the genus *Leishmania*, each with its own characteristic association of vectors and reservoir hosts. The parasites multiply in the gut of the sandfly once they have been acquired via a blood meal, and people may be affected in various ways. Skin infections can cause seriously debilitating sores and disfigurements, and visceral infections (known as kala-azar) are widespread through parts of Africa, Asia and South America (Kettle 1995). Depending on the species, the parasite occurs in a variety of animal reservoirs such as dogs, foxes and other primates apart from humans. Clearly, the dispersal of sandfly vectors from host to host will dictate the efficiency and spread of the disease. In a study in Kenya, Mutinga *et al.* (1992) used mark–recapture techniques to assess the flight range of 11 sandfly species, and found that 54% of their flies were recaptured within a radius of a mere 10 m from the release site, and 95% within 50 m. Only two flies of a single species were found 1 km from the release point. In this example, it is apparent that hosts separated by some distance are not so likely to be infected from a reservoir as those close to each other, i.e. in high-density populations.

Long distance (or passive) dispersal of vectors is often more dependent on wind than on the flying abilities of the insects themselves. Biting midges such as *Culicoides brevitarsus* (Diptera: Ceratopogonidae) are mainly a physical nuisance—to quote Service (1996), one midge is an entomological curiosity, a thousand are sheer hell! However, one or two arboviruses of livestock are also transmitted, and for such a small insect, it is not surprising that their long distance dispersal by wind is commonplace (Murray 1995). The whitefly *Bemisia tabaci* has already been mentioned as a vector of tobacco leaf curl virus (TbLCV), but it is also responsible for the transmission of viral diseases of, for instance, cassava and okra (Fargette *et al.* 1993). Spatial spread of the insect and hence its associated plant viruses is characterized by strong border or field-edge effects. Field edges are influenced by prevailing wind strength and direction, which results in pronounced gradients of insects and disease through crops.

Finally, for vectors with limited dispersal abilities of their own, it is highly advantageous to hitch a ride with a more mobile animal. Hence plague fleas are well known to use their rat hosts not only as a source of food, but also as a long-range dispersal mechanism, and woodwasps in the genus *Sirex* travelled from Europe to Australia and New Zealand during the early 1950s on ships carrying logs, thus carrying the rot fungus *Amylostereum areolatum* to the colonies. The latter's vector–disease association has caused much more severe problems for foresters than occur in Europe, because of the lack of natural biological control agents and the abundance of susceptible, particularly drought-stressed, trees (Speight & Wainhouse 1989).

9.4.4 Pathogen transmission

Insect vectors transfer pathogens to new hosts in various ways, and the mechanism of transmission frequently dictates not only the type of pathogen to be vectored, but also its location within the body of the host. As most pathogens are relatively immobile in their own right, part of the vector's role in disease transmission is to deliver the pathogen to the appropriate part of the host's tissue from where the pathogen can continue its life cycle without further dispersal. The host may even assist in this process; cockroaches provide an excellent example. The genus *Blatta* (Dictyoptera: Blattaria) is now established globally, and is known to frequent all manner of human habitation, including the supposed most hygienic hotels, restaurants and even hospitals. The pathogens with which cockroaches are associated are wide-ranging; viruses, bacteria, protozoans and nematodes are all thought to be transmitted by them (Service 1996), mainly by simple contamination of food over which they forage, disgorging their own partially digested meal, or depositing their excreta. All that is required for host infection beyond this point is that we eat this contaminated material, thus completing the job begun by the cockroach. It is therefore not surprising that most cockroach vectored diseases of humans are intestinal. Many of the enterobacteria vectored by cockroaches are themselves considered to be opportunistic (Pellegrini *et al.* 1992), and the vector merely provides a fortuitous means of mechanical dispersal.

A somewhat analogous system can be seen in plant-feeding insect vectors. In the case of Dutch elm disease, a vascular-wilt fungus, *Ophiostoma* (= *Ceratocystis*) *ulmi*, is carried from an infected tree to a healthy one by bark beetles predominantly in the genus *Scolytus* (Coleoptera: Scolytidae) (Plate 9.1, opposite p. 158). Fruiting bodies of the fungus proliferate in the larval galleries of the beetle under the bark of infested trees. When the young adults leave their pupal chambers in the bark to seek out healthy trees for maturation feeding (a prereproduction method for enhancing food reserves) fungal spores are carried on their bodies and mouthparts (Bevan 1987). There is no suggestion that this transmission is anything but mechanical. Maturation feeding occurs at the twig crotches of healthy trees, which causes exposure of sapwood tissue by the chewing action of the beetles. Fungal spores are then introduced into these wounds, from where they are able to germinate and infect the tree's vascular system, eventually rendering the new tree susceptible to attack by a new generation of beetles. Figure 9.4 depicts this admirable cycle of infestation and infection. Undoubtedly, the association between beetle and fungus is symbiotic (mutualistic), and though vectoring of the pathogen is simply via mechanical contamination of the vector's external surfaces, Dutch elm disease has proved to be one of the most serious causes of tree decline and death in Western Europe and parts of North America (Houston 1991) (see also below).

Sap-feeding insects are able to transmit plant viruses in a variety of ways (Table 9.4). Simple mechanical transmission is again apparent, with

Table 9.4 Modes of transmission of plant viruses by sap-feeding Hemiptera (partly from Agrios 1988).

Mode	Description
Non-persistent (stylet-borne) viruses	Mechanical transmission from sap or epidermal cell contents of infected host to a healthy one, e.g. turnip mosaic virus (TuMV)
Persistent (circulative) viruses	Vectors accumulate the virus internally, which passes through the tissues of the vector and is introducted to the new host again via the mouthparts, e.g. potato leaf roll virus (PLRV)
Propagative viruses	Persistent viruses, which multiply in the vector before transmission to the new host plant, e.g. tomato yellow leaf curl virus (TYLCV)

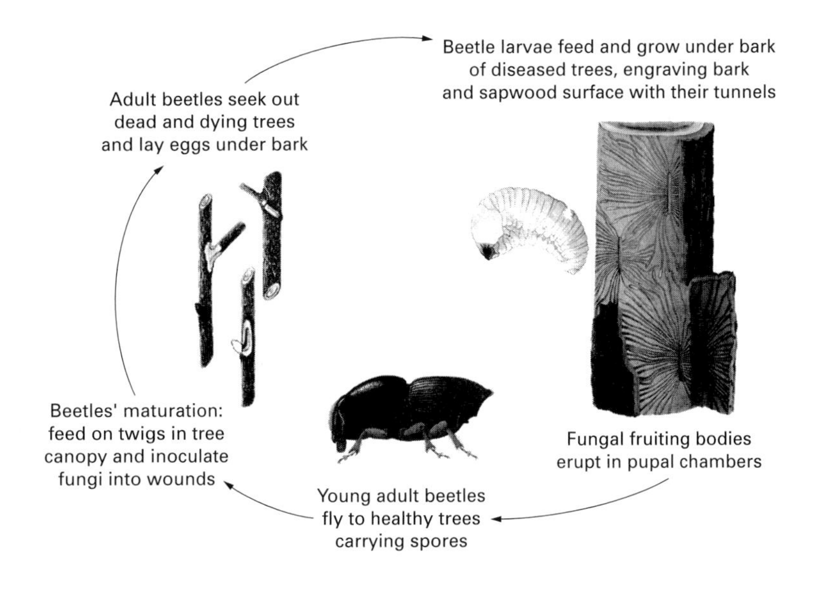

Fig. 9.4 Life cycle of scolytid bark beetles in association with elm disease fungi.

viruses from the sap of an infected plant adhering to the mouthparts of the aphid or leaf-hopper, in a manner analogous to that of a contaminated hypodermic syringe. This is a similar system to that of biting and displacement seen in blood suckers. Such viruses are readily transmitted to new hosts during initial probings by sap-feeders, carried out in attempts to identify suitable host plants. In fact, the use of plant species or varieties that are resistant to sap-feeders may encourage the transmission of non-persistent plant viruses (Table 9.4), as probings on a number of plants become more frequent if the insect cannot find suitable hosts. An example of this problem is shown in Fig. 10.4, where aphids change their behaviour when confronted by resistant varieties of cowpeas, leading to a proliferation of plant virus incidence (Mesfin *et al.* 1992).

9.4.5 Pathogen multiplication

So far in this section we have considered simple mechanical transmission of pathogenic organisms by insect vectors, but many associations of disease-causing agents and their vectors involve more complex interactions, where the pathogen not only relies on the vector for dispersal, but is also able to replicate or undergo part of its own life cycle within the insect. Pathogen transmission may involve the passage of the disease-causing organism through the body of the vector, with the pathogen emerging at the other end of the alimentary tract in an infectious state, during which time the vector has moved on from one host to another. An example of this involves triatomid bugs (Heteroptera: Triatomidae) that vector Chagas disease. The protozoan *Trypanosoma cruzi* is responsible for this human disease, which is widespread in Central and South America. Parasites are ingested by the bugs via a blood meal on an infected host (humans or other mammals) and undergo their entire development within the gut of the vector (Service 1996). After a week or two, infective stages of the protozoan are present in the bug's excreta, which can enter a person's body via wounds, scratches, or even eyes and other mucous membranes. Transmission is therefore not through the bite of the insect, merely through its faecal deposits. Entry into the host's body may be enhanced when the person scratches his or her irritating bites. This type of transmission is termed stercorarian, as opposed to salivarian, which we will now discuss.

Vector–pathogen associations can become even more complex, wherein the parasite invades the vector's tissues and replicates before inoculation into a new host. Often, the vector's salivary glands are the site of this multiplication; ecologically, this must be considered to be an optimal site, as the pathogen can build up large numbers immediately before the vector deposits it into the tissues of a new plant or animal host. Saliva is employed by insects feeding on animal hosts as an anticoagulant and vasodilator, whereas in plant feeders it may be used to increase sap pressure within host tissues to 'force-feed' the insect. In other words, the injection of saliva into the host is an extremely common part of feeding, and hence a very successful route for a pathogen to enter a new host. Figure 9.5 illustrates a diagrammatic cycle of plant pathogens, in this case viruses, through the tissues of an aphid (Schuman 1991), typical of a persistent, circulatory virus such as barley yellow dwarf virus (Irwin & Thresh 1990). Because virus multiplication takes some time to complete, a latent period (see section 9.4.1) is often observed between the vector first acquiring the pathogen and its becoming infective for new hosts. This latent period may last from at least 12 h for aphids to several weeks for leaf-hoppers (Schuman 1991). Once established, this infectivity may last for some time, possibly even up to 100 days (Chiykowski 1981).

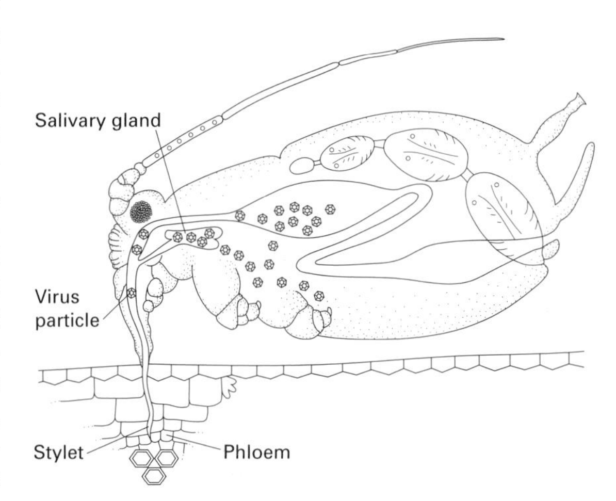

Fig. 9.5 Virus transmission by an aphid. In persistent transmission, the virus enters the digestive tract of the aphid and eventually arrives in the salivary glands. Aphids and other sucking insects frequently probe deep into plant tissue and feed in phloem. (From Schuman 1991.)

A similar multiplication of pathogens occurs in insect vectors of animal diseases. The parasitic protozoans that are responsible for human malaria, *Plasmodium* spp., undergo several stages of their life cycle in the gut and haemocoel of the mosquito vector (Fig. 9.6; Kettle 1995), and these processes may take up to 12 days depending on temperature and species of *Plasmodium* before the infective stage, the sporozoite, is abundant in the vector's salivary glands and hence ready to be inoculated into a fresh host at the next blood meal (Service 1996). Once infective, however, female mosquitoes usually remain so for the rest of their lives.

9.4.6 Effects on the vector

Insect-transmitted pathogens can have positive, negative or neutral effects on the vector insect (Vega *et al.* 1995), and these effects may be direct, caused by the pathogen, or indirect via the host organism. Where simple mechanical transmission of a pathogen occurs, as in the case of cockroaches or houseflies walking over sources of contamination and infecting new sites, there will be no effect on the vector at all; the insect is merely a fortuitous carrier. However, as pathogen–vector associations become more intimate, the pathogen has to be able to survive and frequently replicate itself in the body of the insect vector. Anatomical and biochemical barriers to these pathogens do exist in the insect body; key resistance factors include the toxicity of digestive fluids, impermeability of the peritrophic membrane, which surrounds the blood meal as it progresses down the vector's gut, and adverse intracellular environments of insect cells, which can be hostile to the multiplication of pathogens such as viruses (Glinski & Jarosz 1996).

Nevertheless, some pathogens can survive in insect bodies, and a trend may be observed in the benefits or disadvantages of a vector's actions. Certain of these interactions are decidedly unwelcome by the vector, a prime example involving fleas, which carry bubonic plague between infected rats and humans. Bubonic plague was certainly well known by the 6th century AD (Bari & Qazilbash 1995). The Black Death, as it was known in the Middle Ages, killed somewhere in the region of 130 million people (or very approximately, 40% of humans in Europe) from the 14th right through until the 17th century. Based on demographic models, it has been estimated that if such enormous mortality had not occurred, thus causing a severe setback in population growth, the world population might now exceed 9.8 billion, instead of the 'mere' 5.8 billion we actually have. Plague is still very much with us today. World Health Organization statistics (WHO 1994) show, for example, that during

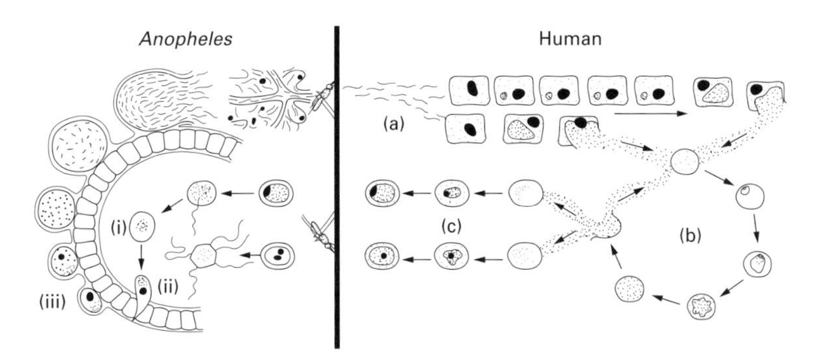

Fig. 9.6 The life cycle of malaria parasites in the *Anopheles* mosquito and the human host, according to present views on exo-erythrocytic schizogony. (a) Sporozoites injected into a human by an *Anopheles* female develop in the liver either into latent hypnozoites (above), which sometime later undergo schizogony to cause relapses, or undergo immediate schizogony (below). Both release merozoites, which enter red blood cells. (b) Erythrocytic schizogony involving release of merozoites. (c) Some merozoites develop into male or female gametocytes, which develop further only when ingested by *Anopheles*. (d) Male gametocytes undergo exflagellation to produce male gametes, one of which will fuse with a female gamete to form a zygote (i). The zygote becomes a motile ookinete (ii), which passes between the cells of the mid-gut to form an oocyst (iii). (e) The oocyst enlarges; there is much nuclear division, ending in the formation of motile sporozoites, which invade the haemocoel and penetrate the salivary glands, from which they are passed into the host with the saliva when *Anopheles* next feeds. (From Kettle 1995.)

the 15 years from 1978 to 1992, 14856 cases were reported from 21 countries in South America, Africa and Asia. Of these, 1451 involved fatalities.

Fleas in the genus *Xenopsylla* (Siphonaptera) are the most important vectors of so-called urban plague, and they acquire plague bacilli whilst feeding on the blood of diseased rats. These bacilli multiply enormously in the vector's stomach, and move forward into anterior sections of the gut, where they frequently block the alimentary tract. When a flea so affected feeds again, with increasing urgency because starvation will set in if the gut is blocked, some bacilli are regurgitated into the new, human, host. Eventually, the vector may starve to death, but not before bacilli have been transmitted onwards. As the original source of infection, a rat, is also killed by the bacilli, all three organisms associated with infection are disadvantaged, with the obvious exception of the bacillus itself. It may be argued that the pathogen can get away with killing rats, fleas and humans as it is also able to spread through the air in high-density host populations (hence the traditional nursery rhyme), and thus does not rely completely on a vector to move from one person to the next.

The above example is an extreme case of where a pathogen has adverse effects on its vector. In many other situations, the abundance of the pathogenic organism in the vector insect may cause various debilitating effects, in a density-dependent fashion, i.e. the higher the density of pathogens, the more seriously is the vector affected. Human onchocerciasis, which is discussed in full in section 9.5.1, is widespread in parts of Africa and South America, and is caused by microfilarial nematodes in the genus *Onchocerca*, which are vectored between human hosts by blackfly, *Simulium* spp. (Diptera: Simuliidae). In the laboratory, three species of blackfly showed decreasing survival times as the density of microfilariae in their blood meals increased (Basanez *et al.* 1996). Because there is a latent period between the acquisition of nematodes by the insect and its ability to inoculate a new host, during which time the pathogen is developing in the vector's body, it is possible that flies will not live long enough to pass the pathogen on if they are seriously debilitated by its presence. In fact, life expectancy at high pathogen loads is species dependent, and only certain simuliids can tolerate these loads for long enough periods of time (i.e. long enough to become infective).

Pathogens do not have things all their own way, and some vectors make an effort to restrict the presence of pathogens in their bodies, and to prevent these pathogens multiplying within their tissues. This may be seen as evidence of the basically disadvantageous nature of infection from the vector's point of view. Filariasis is a disease of humans and other animals that can take various forms, and is the responsibility of various nematode worms, including *Brugia malayi*. Mosquitoes are the vectors in this system, and one species, *Armigeres subalbatus*, has been shown to react to the presence of filarial nematodes in its tissues by encapsulating and killing up to 80% of the parasites within 36 h of acquiring them through a blood meal on an infected host (Ferdig *et al.* 1993). This defence system is not without cost to the insect, however, as mosquitoes performing this feat seem to show longer periods of time before they can lay eggs, and the normal processes of egg development, including ovary size and protein content, are significantly reduced.

So far, we have considered the down-side of being a pathogen vector, but in many cases living on a diseased population of hosts may be beneficial to the vector. It may be argued, for instance, that a blood-feeding mosquito or tsetse fly is less likely to be disturbed during feeding if its host is debilitated by parasites within, which cause disease symptoms in the host of fever or lassitude. It is possible that fevers open up the peripheral blood vessels in host skin, thus providing extra food supplies for the insect. In some cases, it may even be that pathogens in the vector's tissues promote the insect's survival, especially in adverse conditions.

Rift Valley fever (RVF) is a type of encephalitis that has been found in goats, cattle, horses and camels as well as humans (Olaleye *et al.* 1996) in at least a dozen African countries from Mauritania in the west to Zambia in the south and Egypt in the north-east. It can be carried by a broad range of arthropods. Even ticks and sandflies are implicated (Hogg & Hurd 1997), as well as many mosquitoes including *Anopheles* spp., *Aedes* spp. and *Culex* spp. In the case of *Culex pipiens* (Diptera: Culicidae), insects infected with RVF virus had a higher mean survival time under conditions of carbohydrate deprivation than did their uninfected counterparts (Dohm *et al.* 1991). However, the reasons for this phenomenon are rather unclear.

A somewhat similar mutualistic relationship between plant pathogens and their vectors can frequently be observed. Barley yellow dwarf virus (BYDV) is the most widespread and economically important disease of cereals world-wide (Irwin &

Thresh 1990) (see also section 9.5.4). Many species of aphids (Homoptera: Aphididae) routinely act as vectors for BYDV, and Table 9.5 summarizes some of the benefits that accrue to the insects by acting as vectors. It has to be said that not all species of aphid respond in the same fashion to BYDV infections in their host plants, but one basic mechanism likely to fuel the benefits shown in the table may involve the fact that BYDV infection seems to increase the total amino acid (e.g. alanine and glutamine) content of cereal leaves, a commodity well known to favour aphid success (see Chapter 3).

Mutualistic associations between insect vectors and pathogenic fungi have already been described in Chapter 6. Perhaps the quintessence of obligate mutualism is seen in the so-called ambrosia beetles (Coleoptera: Platypodidae). Beetles such as *Trypodendron* (*Xyloterus*) *lineatum* can be serious pests of young trees in Western Europe, especially when the trees are temporarily stressed via heavy defoliation by various lepidopteran species. Defoliation interrupts the normal transpiration stream of the tree, and hence sap pressure, a simple but usually effective physical defence mechanism, is reduced, allowing adult female beetles to bore through the bark and into the wood. So-called blue-stain fungi are intimately associated with ambrosia beetles, and wood tissues are inoculated with the fungi as the beetle tunnels along (see also Chapter 6). Eggs are laid in chambers in the wood, the walls of which are infected with the fungi. The developing beetle larvae are in fact mycetophilic (fungus feeding), rather than direct wood-feeders, and harvest the fungal fruiting bodies that proliferate in their chambers. Without the fungus, beetle larvae cannot survive, and though the tree is not killed, severe wood degradation by the joint presence of beetle galleries and fungal-stained timber is the final result.

Table 9.5 Some of the advantages reported for aphids when vectoring barley yellow dwarf virus (BYDV). (From Irwin & Thresh 1990.)

Variable	Effect
Development rate	+
Longevity	+
Reproductive period	+
Number of offspring	+
Number of alatae (winged adults)	+

+, enhancement of the parameter in the presence of virus.

9.4.7 Epidemiology; the spread of disease

The ecology of disease spread encompasses both theoretical concepts and field observations. First, it must be remembered that though we have concentrated on insect–pathogen associations, the role of insect transmission in virus ecology, for example, is only one of several major routes a virus may take to spread from one host to the next. Figure 9.7 presents these various routes (Garnett & Antia 1994), which

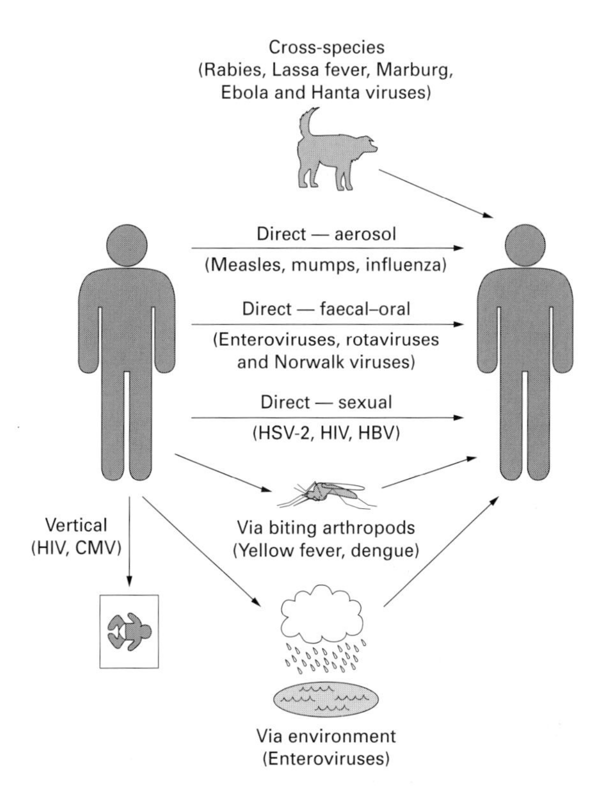

Fig. 9.7 A schematic representation of the transmission routes of viral infections in humans, with some examples. Direct transmission can include sexual or faecal–oral contact. The sexually transmitted viruses, human immunodeficiency virus (HIV) and hepatitis B virus (HBV), may also be transmitted by other processes during which blood is exchanged. Some viruses that are transmitted by faecal–oral contact may also be released into the environment, which provides a source of infection. The cross-species virus transmission can be from rodents (the Hanta viruses and the arenaviruses such as Lassa fever), from monkeys (African haemorrhagic fever—the Marburg and Ebola viruses), or from domestic animals infected by wild animals (rabies). There is evidence of some transmission between humans for these infections, but the extent of this interperson transmission is limited. (From Garnett & Antia 1994.)

illustrate the diversity of dispersal mechanisms evolved by these pathogens. It might be considered, in fact, that the employment of a mobile vector organism, especially one in which the pathogen undergoes an essential part of its development, is a highly specialized form of dispersal, evolutionarily and ecologically complex. A case in point concerns the alphaviruses, which are typical of RNA viruses that readily undergo mutations and are hence quick to exploit changes in their environment or the behaviour of their hosts or vectors (Scott *et al.* 1994). Major alphavirus diseases include Ross River virus, and three equine encephalitis viruses, Eastern (EEE), Western (WEE) and Venezuelan equine encephalitis (VEE). Ross River (RR) virus, far from being confined to the developing world, is present in urban areas such as Brisbane, Australia. It causes painful joints, rashes and fevers, which can persist for several months, and it occurs in cattle, horses and kangaroos as well as in humans. It is carried by various *Aedes* and *Culex* species that breed in fresh- and brackish-water locations in suburban areas (Ritchie *et al.* 1997). The equine encephalitis viruses, as the name suggests, are predominantly problematic in horses and donkeys, but humans do suffer as well. In 1995 a VEE outbreak in Colombia caused an estimated 75 000 human cases, of which 300 or so proved fatal (Rivas *et al.* 1997). Much further north and a year later, people living in Rhode Island, New York, Long Island, and Connecticut were forced to remain indoors in the evenings to prevent being bitten by mosquitoes carrying EEE; 1% of mosquitoes in the area were found to carry the potentially fatal virus (Anon. 1996). It is thought that all three alphaviruses evolved from the same ancestor, but these days, a number of different forms with different vectors and host mammals can be recognized (Table 9.6; Scott *et al.* 1994). This diversity may be due in part to the restricted dispersal of mammalian hosts, especially in South America, and hence the chance establishment of isolated, founder populations of virus, which then adaptively radiate under local selection pressures. Vector diversity in

Table 9.6 Classification, distribution and transmission characteristics of New World equine encephalomyelitis viruses. (From Scott *et al.* 1994.)

Virus	Subtype	Variety	Distribution	Primary vector	Primary vertebrate host
EEE		North America	Eastern North America	*Cs. melanura*	Songbirds
			Caribbean	?*Cx. taeniopus*	?Birds
		South America	Central and South America	*Cx. (Mel.)* sp.	Small mammals and birds
WEE			Western North America	*Cx. tarsalis*	Songbirds and small mammals
	R-43738		South Dakota	?	?
	Ag80-646		Argentina	?*Cx. ocossa*	?
	BeAr 102091		Brazil	?*Cx. portesi*	?
VEE	I	AB	South, Central, and North America	Numerous mosquitoes	Equines, other mammals and birds
	I	C	South, Central, and North America	Numerous mosquitoes	Equines, other mammals and birds
	I	D	Ecuador, Panama, Colombia, Venezuela	*Cx. ocossa* *Cx. panocossa*	Rodents and birds
	I	E	Central America	*Cx. teeniopus*	Rodents and birds
	I	F	Brazil	?	?
	II	(Everglades)	Southern Florida	*Cx. cedecei*	Rodents
	III	A (Mucambo)	South America	*Cx. portesi*	Rodents and birds
	III	B (Tonate)	South America	?	Birds
	III	C	Peru	?	?
	IV	(Pixuna)	Brazil	?	?
	V	(Cabassou)	French Guiana	?	?
	VI		Argentina	?	?

Cs., Culiseta; Cx., Culex.

these restricted areas will also have a role to play in the direction and result of this radiation.

Disease-causing pathogens that are carried by insects rely, of course, on the dispersal activities of the vector (see section 9.4), and, in this context, one fundamental aspect involves the life span of the adult vector in relation to that of the pathogenic agent in its tissues. Let us take, for example, the case of mosquitoes taking a blood meal that contains the malarial parasite *Plasmodium*. If the time between initial infection and the parasites appearing in human-infectious form in the vectors' salivary glands (the latent period) is close to or even exceeds the normal life span of the fly, then clearly there is little chance of the vector infecting humans (Anderson & May 1991). This returns us to the concept of extrinsic incubation period described in section 9.4.1. In fact, adult *Anopheles* mosquitoes have an expected life span in the field of only some 6–14 days, with latency periods before the parasite becomes infectious of up to 12 days at 25–27°C for *P. falciparum*, for example (Anderson & May 1991). In the case of sandflies that transmit leishmania, the vector life span is particularly short. After the latent period, the average female might be able to bite only once or twice more before dying. To ameliorate this problem, the parasite multiplies in the mid-gut, migrates forward and blocks the pyloris (part of the fore-gut), thus forcing the fly to bite repeatedly and to regurgitate parasites as it does so (D.W. Kelly, unpublished observations).

The population dynamics including mortality factors in the life of vectors that influence life expectancy are clearly of vital importance in disease epidemiology, and in evolutionary terms, it might be expected that it would be in the interests of a parasite to extend the life span of its vector to increase the likelihood of the former reaching a new host. This is certainly the case for aphids that vector barley yellow dwarf virus (see Table 9.5).

9.4.8 Human activities and vectors

A large number of animal and plant diseases occur in natural communities, and usually tick along at endemic levels in wild plant and animal populations without affecting humans at all. Yellow fever provides a good example. Yellow fever and its mosquito vector *Aedes* spp. are said to have spread through the African slave trade (Morse 1994). The zoonotic, so-called sylvatic phase of African yellow fever is maintained in monkey populations in the natural forest. Monkeys seem to be little affected by the virus, with only a few individuals actually dying from yellow fever (Service 1996). Only when humans enter the jungle, or convert it to agricultural land, enticing monkeys into closer proximity to villages, does the virus transfer via mosquitoes into human populations, where it is carried by other *Aedes* species, *Ae. aegypti* in particular. Figure 9.8 provides a diagrammatic representation of African yellow fever, illustrating the complex interactions between natural and human host populations. It should be noted that several vector species are involved, as different species of mosquito frequent certain habitats. It is also important to realize that mammals including humans can act as pathogen vectors as well as insects; a villager bitten by a virus-infected mosquito whilst working on his farm may travel to the market in town to sell his produce, and is there bitten by another mosquito, which acquires the pathogen from its blood meal, and is then ready to infect other people who have never left their urban environment.

It is thus easy to see that human activity, for instance agriculture and urbanization, can readily interfere with natural disease epidemiology, resulting in serious consequences for people, their crops and their livestock. Many human activities may be implicated in the enhancement of insect-borne disease. Triatomid bugs that transmit the trypanosome responsible for Chagas disease live in the walls and rafters of village houses, and their presence in these dwellings can be directly linked with the incidence of the pathogen in the blood of village children (Andrade *et al.* 1995a). In the case of malaria, natural or seminatural habitats such as transitional swamps or unmanaged pasture have been shown by geographical information system (GIS) analysis to be important elements in predicting the abundance of mosquito vectors in Mexican villages, especially in the mid-to-late wet season (Beck *et al.* 1994), but as mosquitoes are necessarily aquatic as larvae, any increase in the abundance of slow-moving or still water is likely to enhance the vector population in the locality. Irrigation ditches, water supplies, paddy-fields and water butts provide excellent sources of vectors (Sadanandane *et al.* 1991; Service 1991; Reisen *et al.* 1992). In fact, there is even strong evidence that the mosquito *Aedes aegypti* has spread throughout the tropics by virtue of the international trade in second-hand truck and car tyres! Larvae

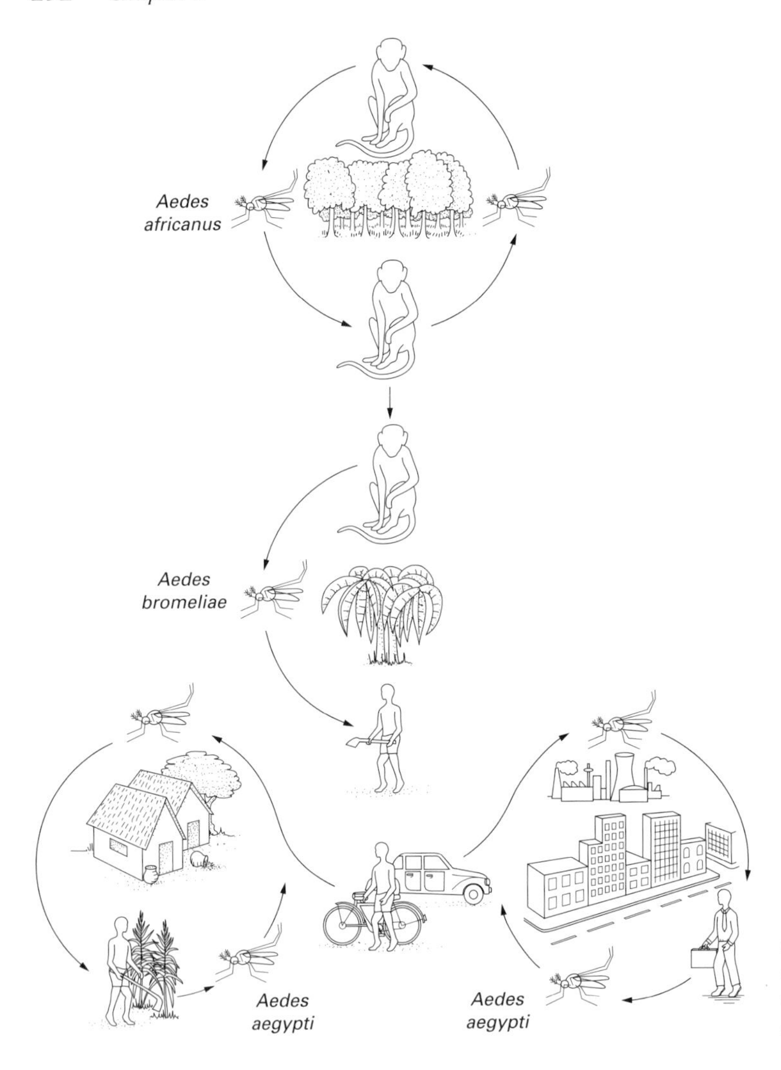

Fig. 9.8 Diagrammatic representation of the sylvatic, rural and urban transmission cycles of yellow fever in Africa. (From Service 1996.)

will happily live in small pools of water conveniently provided inside the rims of these artificial habitats.

The relative abundance of certain vector species may also be altered by human activities. In Sri Lanka, for example, irrigation development for rice culture, which provided newly abundant breeding sites for mosquitoes such as canals and reservoirs as well as numerous paddy-fields, resulted in long-term changes in the species composition of vector populations, which was characterized by the increasing dominance of mosquito species with high potential to transmit human pathogens (Amerasinghe & Indrajith 1994).

Similar principles apply for plant pathogens. In many examples of crop diseases transmitted by insects, wild vegetation either at field boundaries or further away acts as reservoirs for plant viruses whilst the crop is absent or in a non-susceptible stage. In the case of barley yellow dwarf virus, for instance, grasses are known to be a perennial source of infection from where aphids carry the virus to annual crops (Irwin & Thresh 1990); these wild plants are relatively resistant to the virus, whereas our high yielding, interbred domestic varieties of cereals certainly are not. Another example concerns beet curly top virus (BCTV) in California. This virus was found to naturally infect a wide range of host plants, including weed species from 14 different families (Creamer *et al.* 1996). These naturally infected weeds were generally

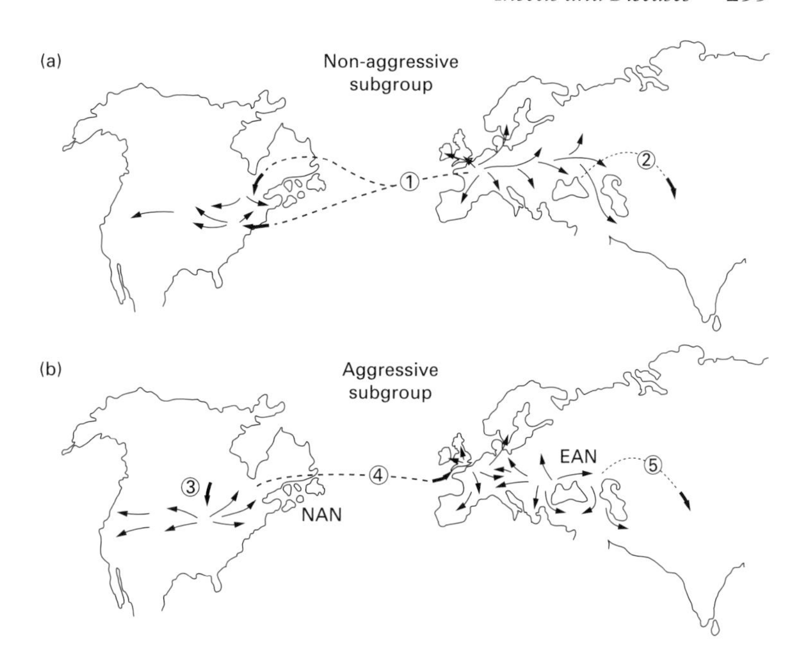

Fig. 9.9 Proposed pattern of spread of (a) the non-aggressive, and (b) the EAN and NAN aggressive subgroups of *O. ulmi* during the first and second epidemics of Dutch elm disease. Small arrows, overland spread; large arrows, major introductory events as follows. (1) Introduction of the non-aggressive subgroup from NW Europe to North America, *c.* 1920s. (2) Introduction of the non-aggressive subgroup from Krasnodor to Tashkent, *c.* late 1930s. (3) Introduction of a form close to the EAN aggressive subgroup into North America (Illinois area), *c.* 1940s, and its subsequent evolution into the NAN subgroup. (4) Introduction of the NAN subgroup from the Toronto area into the UK, *c.* 1960. (5) Introduction of the EAN subgroup into the Tashkent area, *c.* mid-1970s. (From Brasier 1990.)

asymptomatic (i.e. did not show morphological evidence of infection), and frequency rates of BCTV infection ranged from 2 to 11%. Clearly, the pathogen is easily able to persist all year round in the crop area, posing a permanent threat to commercial crops.

Complex interactions between vectors, pathogens, hosts and human activity in the epidemiology of plant disease are well exemplified by the examination of Dutch elm disease (DED) in Europe and North America during this century (Brasier 1990). As mentioned above, DED is caused by the wilt fungus *Ophiostoma (Ceratocystis) ulmi*, which is carried from diseased to healthy trees by elm bark beetles, predominantly *Scolytus* species (Coleoptera: Scolytidae). Trees infected with the fungus during maturation feeding by the beetle die rapidly, within 1 or 2 years of infection, thus providing highly suitable breeding sites for new generations of the vector. There have been two epidemics of DED this century. The first appeared in various parts of NW Europe around 1920 and then spread eastwards across central Europe in the 1930s. It also spread in a westerly direction and reached Britain and North America during the 1920s. This first epidemic appears to have been caused by a weakly pathogenic, so-called non-aggressive, subgroup of *Ophiostoma*. The second epidemic, or more exactly, a set of epidemics, is still in progress, and involves two races of a much more aggressive subgroup of the pathogen. One race, the Eurasian or EAN race, has been spreading across Europe, probably since the 1940s. The North American or NAN race has also been spreading across that continent for the same period of time, and has crossed the Atlantic, presumably on log and other wood-product imports, to Britain and neighbouring parts of Europe, during the late 1960s. Figure 9.9 illustrates the progress of these two epidemics of DED (Brasier 1990).

It may be argued that the scolytid beetle vectors of DED were 'innocent parties' in the tussle between pathogen and host tree. Both continents have indigenous elm bark beetle species, which can persist in the absence of any fungal pathogen, by utilizing dead or dying trees resulting from wind damage or old age. Beetles can only realistically transmit the fungi over relatively short distances, say from one tree to another in rather close proximity, but when people start to move logs infested with beetle larvae or young adults either nationally by road transport or internationally by ship, then the insect vector receives a considerable helping hand. A somewhat anecdotal example of this comes from the first evidence of the second epidemic in central England in the 1970s, when dying trees were observed alongside major roads where trucks carrying elm logs would routinely stop. Beetles contaminated with fungal spores could easily hop off their truck and 'high-tail it' to the nearest healthy elm tree growing by the roadside!

9.5 Case studies

Many of the ecological principles discussed in the above section need to be combined to explain the complexities of major insect-vectored disease syndromes. In this section we present details of a series of major animal and plant diseases that illustrate this; examples of control procedures are also presented. It should be noted that we have already discussed some aspects of these cases earlier in the chapter. The reader should pay particular attention to the significance of epidemiological theory (see section 9.4.1). Specifically, the reader should remember the importance of maintaining R_0 values for vectors below unity to achieve eradication, and the overriding significance of the vector's longevity as the most important parameter of vectorial capacity.

9.5.1 Onchocerciasis

Onchocerciasis, or river-blindness, is a human disease caused by the filarial nematode parasite, *Onchocerca volvulus*. Though, strictly speaking, it is a non-fatal disease, the life expectancy of sufferers following the onset of blindness is much reduced. Adult filariae live in various human tissues, especially the skin, where they may produce unsightly nodules. The dispersive or migratory stage of the parasite, microfilariae, cause much more serious skin irritations, and large populations may be found in the eyes of infected humans. Here, lesions are caused when the microfilariae die, which lead to increasing and finally complete opacity of the cornea.

Fig. 9.10 Blackfly feeding on human.

Over the years, many millions of people in Africa, and a somewhat lesser figure in Central America, have been infected. Statistics from the World Health Organization (WHO 1997) describe 120 million people at risk from onchocerciasis, and around 18 million actually infected. Of these, 99% are from 28 endemic countries in sub-Saharan Africa. Some idea of the scale of the problem is exemplified by the situation in West Africa in the late 1980s. Here, a survey of over 600 villages in Guinea, Guinea-Bissau, western Mali, Senegal and Sierra Leone found that 1475 367 out of a rural population of 4464 183 people (33%) were infected by the parasite, and of those, 23 728 were blinded (de Sole *et al.* 1991). The only vectors of the parasite are blackflies in the genus *Simulium* (Diptera: Simuliidae) (Fig. 9.10); the main species (or complex of species) in Africa is *S. damnosum*, and in Central and South America, *S. ochraceum* is the major vector.

When an adult fly cuts a hole or lesion in human skin, providing access to superficial blood capillaries, secretions in the fly's saliva are inserted into the wound, which not only inhibit the aggregation of blood platelets and slow down coagulation, but also increase vasodilation so that the wound bleeds more freely (Cupp & Cupp 1997). Some species of *Simulium* seem to have better vasodilatory abilities than others, and it is those that excel at this, such as *S. ochraceum*, that also happen to be amongst the most efficient vectors of the nematode. The saliva of one species of blackfly, *S. vittatum*, is even reported to affect immune cell response in humans, indicating just how complex insect–host relationships can become. Of course, this system has probably evolved to benefit the insect alone, but the prolonged bouts of undisturbed feeding will also benefit pathogen transmission. Blackfly use their rasping mouthparts to feed on blood, and the microfilariae contained within it then spend a developmental latent period inside the thoracic muscles of the vector. Nematodes then become concentrated again in the insect's proboscis before being introduced into a new host. There are no animal hosts except humans, but as *O. volvulus* can remain active in humans for 10–15 years, a high incidence of infection amongst people in rural communities maintains the parasite for a very long time, including during periods of low vector abundance.

In the late 1980s, a filaricidal drug, ivermectin, became available, which drastically reduces the numbers of microfilariae in the skin, such that relatively few flies are able to pick up an infection during a blood meal. In 1997, 8 million people living in Africa received ivermectin treatment (WHO 1997), though this is a relatively small percentage of those actually at risk. However, the mainstay of the control of onchocerciasis has since the 1930s involved vector control based on insecticidal treatment (Davies 1994). The major target of these control programmes has been the larval stage of blackfly, which is spent in fresh, flowing water such as streams or rivers. The larvae anchor themselves to stream beds, but many can be swept downstream for long distances. Thus when adult flies emerge, these potentially infective vectors may have been dispersed over considerable distances. This phenomenon, coupled with the ability of adult blackflies to disperse on the wind, has resulted in Ecuador in the formation of new foci of onchocerciasis in areas previously free of the disease (Guderian & Shelley 1992). However, various species of *Simulium* prefer different types of running water, in a variety of environmental conditions. In Mexico and Guatemala, the main vector species of onchocerciasis predominates in shallow water at altitudes of between 800 and 1100 m above sea level (Ortega & Oliver 1990). As might be expected, such specific observations are appropriate for only fairly local areas, but onchocerciasis infections in humans do seem in general to be positively correlated with altitude, and, not surprisingly, negatively correlated with distance from blackfly breeding sites (the nearest river) (Mendoza-Aldana *et al.* 1997). With this sort of ecological information, it is possible to restrict control operations to a subset of the total range of all *Simulium* species.

Undoubtedly, onchocerciasis vector control provides a very good example of where cheap, persistent and relatively safe insecticides such as the otherwise abhorred DDT are vitally important. Early attempts at larval control in the late 1940s and early 1950s, for instance, using DDT in Kenya, Zaire and Uganda resulted in virtually complete eradication of blackfly vectors, success stories that have remained so up to the present day (Davies 1994). Admittedly, copious quantities of DDT were released into the environment (10 applications of the chemical were applied at 10 day intervals in Kenya in 1952 and 1953), but this is arguably a small price to pay for the cessation of a seriously debilitating human disease. DDT has now been replaced as the insecticide of choice by pyrethroids.

More recently, the WHO has embarked on the Onchocerciasis Control Programme (OCP), a huge and far-ranging undertaking employing aerial spraying of various larvicidal insecticides, covering, for example, 1300 000 km^2 of land and 50 000 km of rivers in West Africa (Davies 1994). This scheme began in 1974, and continues today, with spectacular results. Figure 9.11 shows the extent of the project (WHO 1997). In the original central area of West Africa, in Burkina Faso and the Ivory Coast, transmission of the disease has been reduced to almost zero in 90% of the area, with the prevalence of onchocerciasis falling from 70% to a mere 3%. In central Sierra Leone, biting rates of potentially infective blackfly were reduced to 2% of their pretreatment levels after 4 years of efficient larviciding (Bissan *et al.* 1995). In many areas, it has been possible to abandon treatment, and the niches of the blackfly vectors have been taken over, seemingly permanently, by other *Simulium* species that are unable to carry the parasite. Of course, it is dangerous to be complacent about the long-term effects of such a strategy, and though the OCP began the cessation of larviciding as early as 1989 in certain West African river basins, back-up studies were required to ensure that vector flies did not re-establish themselves (Agoua *et al.* 1995). In one example, the precontrol infectivity levels of blackflies ranged from 2.5 to 8.9%, unacceptably high values. After the Programme, back-up surveys revealed that these levels have been reduced to less than 1% in the main. In a few sites where infectivity remains significant, larviciding has had to be resumed. On the whole though, the OCP has been very successful, and over 90% of rivers in the original Programme areas are no longer in need of vector control.

The OCP in West Africa has now been extended into a new initiative, the African Programme for Onchocerciasis Control (APOC) (WHO 1997). Countries in Central and East Africa are now being targeted, including Burundi, Chad, Kenya, Malawi, Nigeria, Sudan, Tanzania and Uganda. Amongst the standard techniques of drug distribution and vector control, APOC is also supporting the acquisition and use of Rapid Epidemiological Mapping of Onchocerciasis (REMO), a mapping system that delineates high-risk communities. In Cameroon, the levels of onchocerciasis endemicity were evaluated in 349 villages by examining the skin nodules caused by the

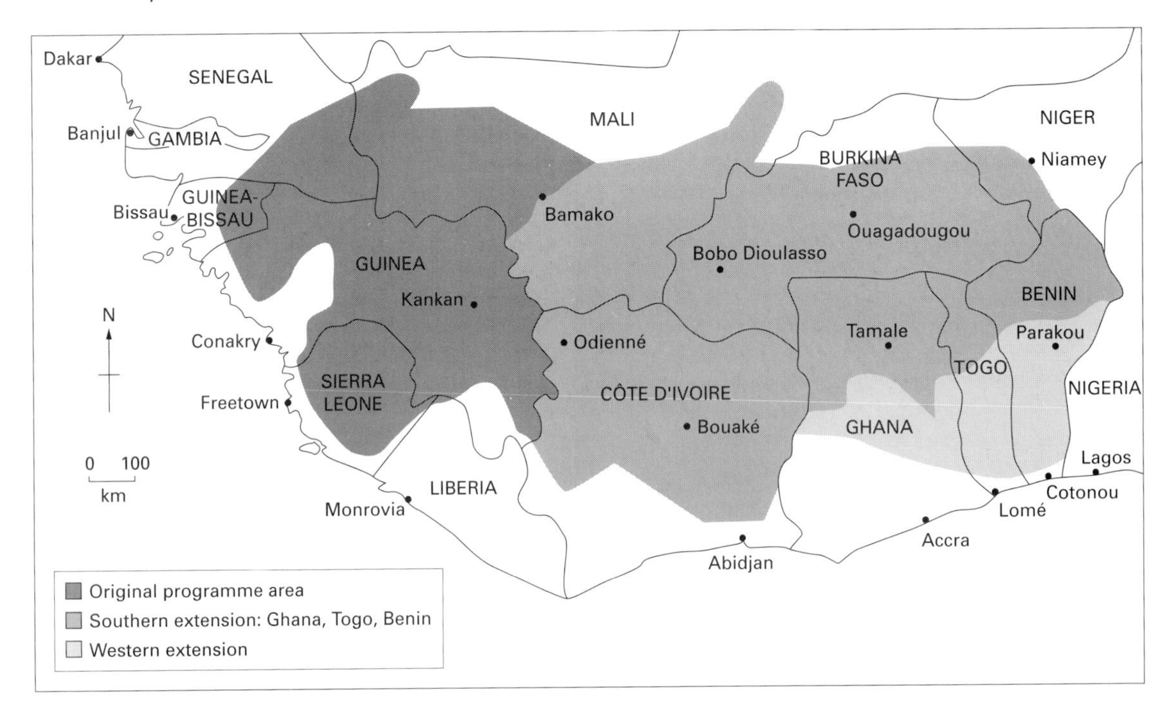

Fig. 9.11 The extent of the Onchocerciasis Control Programme in West Africa (WHO 1997).

parasite in men. REMO maps were then constructed, which identified high-risk localities where ivermectin treatment was urgently required (Mace *et al.* 1997).

In the final analysis, the resounding success of the control of the insect vectors of onchocerciasis is at least partially attributable to the rather specific and identifiable ecology of blackfly. Insects that transmit other serious disease-causing pathogens whose ecology is more complex and disparate are much harder to deal with.

9.5.2 Trypanosomiasis

Protozoan trypanosomes in sub-Saharan Africa cause two related diseases, sleeping sickness in humans and nagana in livestock, both of which have enormous consequences for human health and livestock production. The vectors of trypanosomiasis are tsetse flies in the genus *Glossina* (Diptera: Glossinidae), and these flies are a major constraint to animal production in approximately 10×10^6 km^2 of Africa (Rogers & Randolph 1991). Annual livestock losses are estimated to amount to US$5 billion (Kettle 1995). In the same vast area, some 50 million people are also at risk to the human form of trypanosomiasis (Chadenga 1994), and according to WHO reports, men, women and children in 36 African countries are at risk from the disease. Despite enormous research effort and investment by aid-agencies in vector control for more than 70 years, the disease still persists in vast areas of scrub savanna, forest edges and in the vegetation along watercourses and lakes (Dransfield *et al.* 1991; WHO 1997). The distribution of tsetses is roughly between the 15th parallel north and south of the equator, and is shown in Fig. 9.12 (FAO 1998b).

Tsetse flies (Fig. 9.13) are extremely unusual insects, as adult females do not lay eggs, but instead deposit single late-stage larvae (one larva roughly every 10 days), which have developed from the egg inside their mother's body; such larvae immediately pupate after being released in soil or litter. Hence, the insect's reproductive rate is very low, and tsetse flies are classical low-density pests (see Chapter 10). However, because of long life spans and their efficiency at transmitting trypanosomes to mammalian hosts, notwithstanding their low population densities, they are exceedingly difficult to control completely.

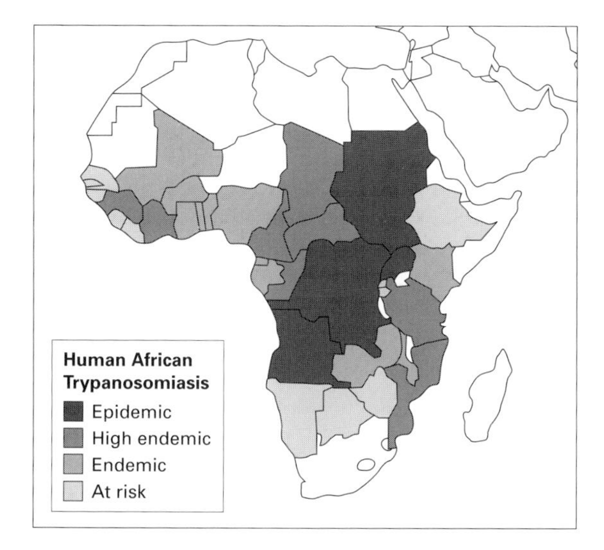

Fig. 9.12 Distributions of sleeping sickness in Africa (FAO 1998b).

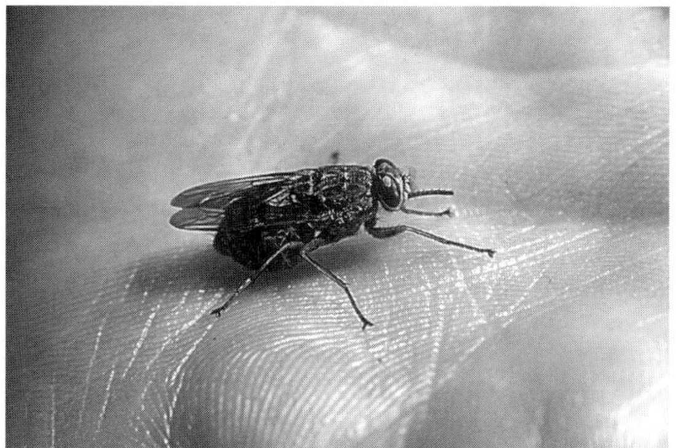

Fig. 9.13 Adult tsetse fly, *Glossina morsitans*, after a blood meal in Zambia. (Courtesy of D.J. Rogers.)

In the case of sleeping sickness, two subspecies of the pathogen exist. *Trypanosoma brucei gambiense* occurs in western and central Africa, whereas *T. brucei rhodesiense* is found in eastern and southern Africa. The latter tends to cause a more virulent and acute form of the disease than the former, although individual virulence can vary considerably (Dumas & Bouteille 1996). Once the pathogen has been transferred to a new host, he or she begins to exhibit fever, weakness and headache, joint pains and itching. As the disease progresses, anaemia, abortion and kidney, cardiovascular and endocrine disorders occur. In the advanced stages, the victim becomes indifferent to events, exhibits lethargy and aggression in cycles, followed by extreme torpor and exhaustion. Death is preceded by a deep coma (WHO 1997). This final stage may only occur several years postinfection. With nagana, infected cattle, for example, may die within a few weeks; those that survive may persist with chronic infections for years, acting as reservoirs of the parasite within husbandry systems. Trypanosomes also circulate in the host's bloodstream, where they can be acquired by both male and female tsetse flies whilst pool-feeding (lapping blood oozing from wounds cut by the insect's blade-like mouthparts) on the host. The protozoans multiply in the vector's gut and eventually migrate to its salivary glands, ready to be inoculated into a new host.

The taxonomy, ecology and regional distributions of *Glossina* species are complex; some species feed only on game animals or livestock, others prefer humans, and certain species or subspecies are able to move between both groups of hosts. Table 9.7 presents a summary for the most important species of tsetse fly. Tsetse flies appear to be rather dependent on moist habitats. For example, they are particularly

Table 9.7 Feeding habits and host ranges of various species of tsetse fly (*Glossina*) that vector sleeping sickness to humans. (From Service 1996.)

Species	Location	Main hosts	Main habitat type
G. morsitans	East Africa	Wild pigs, wild and domesticated cattle	Savannah
G. morsitans	West Africa	Warthogs	Savannah
G. pallipedes	East Africa	Wild cattle	Wooded savannah
G. palpalis	West Africa	Reptiles, humans	Riverine, mangroves
G. swynnertoni	East Africa	Wild pigs	Savannah, dry thicket
G. tachinoides	S. Nigeria	Domestic pigs	Riverine
G. tachinoides	West Africa	Cattle, humans	Riverine, coastal

abundant in shady, humid riverine vegetation; fortuitously for the vector and pathogen, habitats frequented by game, livestock and people alike, especially during the heat of the African day. However, because of these humid habitat requirements characteristic of lush vegetation, the distributions of tsetse fly populations, and hence presumably the likelihood of epidemics of sleeping sickness or nagana, can be plotted and predicted according to the development of vegetation types surveyed by remote imagery (Rogers & Williams 1994). Normalized Difference Vegetation Indices (NDVIs) can be obtained from data derived from orbiting meteorological satellites, which integrate a variety of environmental factors, such as temperature and rainfall, that are known to be important for tsetse fly survival. These NDVIs can then be correlated statistically with known fly distributions, using discriminant analysis (a multivariate statistical technique), along with more straightforward variables such as temperature. Figure 9.14 shows one example from this analysis (Rogers & Williams 1994), where the distribution of *Glossina morsitans* in Kenya is predicted with more than 80% accuracy. In this example, from a central region for the insect, vegetation index was the major predictive variable, whereas in Zimbabwe, which is on the edge of the tsetse fly range, temperature was the most important variable. Hence, in situations where detailed and labour-intensive ground surveys are problematic, remotely sensed data may provide useful insights into the ecology and distributions of vector populations. In addition, it should be possible to predict high-risk areas of disease transmission based on satellite surveys (Rogers & Randolph 1991).

Though drugs may be employed to treat humans and livestock infected with trypanosomes, especially in the early stages of the disease, vector control is the only reasonable way to prevent the disease, and many techniques have been tried to this end. An ideal solution is to site villages and farms outside tsetse belts, but this of course prevents the inhabitation of large areas of Africa. The widespread use of insecticides is no longer possible or desirable; DDT was a rather useful tool for fly control in the old days, but its

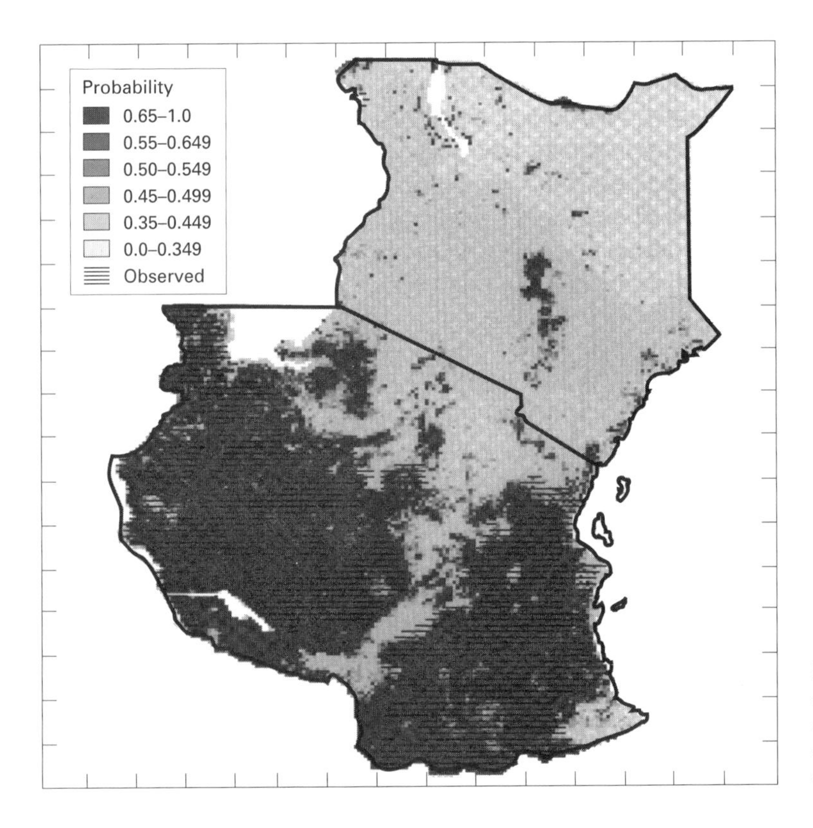

Fig. 9.14 Probability of tsetse fly occurrence in Kenya predicted by discriminant analysis using NDVI data, compared with observed distributions. (From Rogers & Williams 1994.)

banning in the developed world has meant that DDT and related compounds are no longer an option for Africa (Langley 1994). However, if it is possible to predict where large numbers of flies are to be found at different times, then chemical control could be localized. In the Ivory Coast, for example, *Glossina palpalis* rests during the day on lianas, coffee bushes and *Eupatorium* foliage (Seketeli & Kuzoe 1994), mostly between 10 cm and 2.5 m above the ground. Such sites are readily spot-treated with insecticides, thus minimizing wastage and pollution, and greatly reducing the cost of control. In Uganda, such bush resting sites were sprayed with the pyrethroid insecticide lambda-cyhalothrin, which resulted in the numbers of human sleeping sickness cases dropping from between 4 and 12 per month to no more than one per month (Okoth *et al.* 1991).

In some countries, specially designed tsetse fly traps impregnated with pyrethroid insecticides have been used. Trials, again in Uganda, placed a mere 10 of these traps per square kilometre, reducing fly populations by more than 95% (Lancien 1991). Trials in the 1980s particularly have studied the prospects of sterile male release, where artificially sterilized male tsetse flies are released into an area to out-compete natural, fertile ones (Moloo *et al.* 1988). A combination of traps and insecticide-impregnated targets with sterile male release did lead to the final eradication of *Glossina palpalis palpalis* in Nigeria in the 1980s. The traps were used to reduce fly populations along rivers and streams, with sterile males being released when the vector population was low enough to achieve a ratio of 10 sterile to one wild male (Myers *et al.* 1998).

In general, however, it is dangerous to be too optimistic about tsetse fly control. Rogers and Randolph (1991) have pointed out that eradication successes obtained in some countries are insignificant when compared with the continental scale of the problem; it is certainly more than likely that areas in which intensive management has resulted in huge reductions in fly populations may easily be reinvaded by flies from adjacent areas where control has been ineffective, or simply lacking.

Finally, it is somewhat thought-provoking to speculate about the effects of a really successful eradication programme for nagana and sleeping sickness. If the non-human trypanosome can be controlled effectively, it is thought that a great deal of already degraded land may become overstocked with grazing animals (FAO 1998b). Indeed, testes flies have even been considered as the 'Guardians of Africa'! (D.J. Thompson, unpublished observations). Human exploitation of dwindling resources may also increase dramatically if people are able to inhabitat and develop previously dangerous places. Neither problem, however, can really be considered a good enough reason not to view the control of trypanosomiasis as an essential component of sustainable rural development in Africa (FAO 1998b).

9.5.3 Malaria

Malaria is undoubtedly the most serious and widespread human disease transmitted by insects in the world. Somewhere around 300 million people are infected (Collins & Paskewitz 1995), and 120 million clinical cases are estimated globally per year (Coluzzi 1994). Nearly 40% of the world's population live in regions where malaria is endemic (Collins & Paskewitz 1995), but sub-Saharan Africa accounts for a large percentage (maybe up to 80%) of the reported cases. In this region, more than 1 million children below the age of five are thought to die of the disease each year (Sexton 1994). Figure 9.15 shows the distribution of malaria in the world, comparing the situation in the mid-19th century with that of the 1990s, roughly 150 years later (Kettle 1995). The disease is no longer endemic in large areas of Europe, Asia, North America and Australia, owing in part at least to the drainage of swamps for farming, higher living standards and improved education, and especially the widespread use of insecticides such as DDT, starting in the 1940s (Collins & Paskewitz 1995). Malaria is still rife in most of Central and tropical South America, Equatorial Africa, India and the Far East.

Human malaria is thought to have originated in tropical areas of the Old World, but Pleistocene glaciations delayed its spread in the Northern Hemisphere (de Zulueta 1994). During the last ice age, low temperatures in southern Europe, for example, precluded the activity of vector mosquitoes, and hence malaria was unable to spread. However, as temperatures rose postglacially about 10 000 years ago, some vector species were able to move northwards. The decline of malaria in Europe really began only during the 19th century, mainly because of new agricultural practices and changed social conditions such as better housing, drainage and health care. Because, in this example, climate change appears to be one of the

Fig. 9.15 Comparison of the global extent of malaria in the mid-19th century (dotted area) with that in the 1990s (shaded area). (From Kettle 1995.)

driving forces in natural disease epidemiology it is possible that predicted rises in global temperatures may once again change the geographical distribution of insect vectors of malaria (and many other insect-vectored diseases such as yellow fever, dengue and encephalitis) (McMichael & Beers 1994).

Far from declining in the late 20th century, malaria appears to be increasing rapidly on a global scale, with the alarming appearance of vectors resistant to insecticides, and parasites resistant to drugs (Sharma 1996). A shift in the dominance of species of *Plasmodium* has also been observed in many countries, with an increase in the most dangerous and potentially fatal *P. falciparum*. The most acute and frequently lethal form of this disease, cerebral malaria, is becoming commonplace, especially in parts of Africa and Oceania. It is thought to be responsible for 2 million deaths a year on its own (Reeder & Brown 1996), and very disturbing figures from Malawi, for example, show that a staggering 39% of pregnant women where found to be infected with *P. falciparum* at their first antenatal visit (Brabin *et al.* 1997). To make matters even worse, many species of malarial parasite are now showing resistance to classic antimalarial drugs. Work in India in 1994, for instance, found that 95% of *P. falciparum* isolates from people in an epidemic in Rajasthan were showing resistance to chloroquine (Sharma *et al.* 1996).

The life cycle of the protozoan parasites that cause malaria is illustrated in Fig. 9.6 (Kettle 1995). As can be seen, the infective stage for humans of *Plasmodium* is the sporozoite, and an important concept in malaria epidemiology and control is the sporozoite rate, i.e. the percentage of female mosquitoes with sporozoites in their salivary glands. The sporozoite rate is an indication of the likelihood of malaria being transmitted to a human after being bitten by one of the 40 or more species of *Anopheles* mosquito known to be important vectors of the disease world-wide. Table 9.8 shows some quoted sporozoite rates from around the world, which indicate a high degree of variability caused by many ecological factors such as availability of breeding sites, the incidence of wet and dry seasons, and the species of vector and parasite involved. In theory, the average sporozoite rate in a mosquito species during a transmission season would reflect its transmission efficiency (Mendis *et al.* 1992), but it must also be pointed out that the detection of a

Table 9.8 Examples of sporozoite rates detected in anopheline mosquito vectors of malaria.

Country	Mosquito species	*Plasmodium* species	Sporozoite rate (%)	Reference
Cameroon	*An. gambiae*	*P. falciparum*	5.70	Robert *et al.* (1995)
China	*An. anthropophagus*	*P. falciparum*	10.90	Liu (1990)
Gambia	*An. gambiae*	n.a.	6.1–7.7	Thomson *et al.* (1995)
Kenya	*An. gambiae*	*P. falciparum*	5.4–13.6	Beier *et al.* (1990)
Madagascar	*An. gambiae*	n.a	1.7–3.2	Fontenille *et al.* (1992)
Papua New Guinea	*An. punctulatus*	*P. falciparum* or *P. vivax*	0–3.3	Burkot *et al.* (1988)
Sierra Leone	*An. gambiae*	n.a.	3.90	Bockarie *et al.* (1993)

n.a., not available.

high sporozoite rate does not necessarily point to high risk; if the population density of vectors is very low, the biting rate may be also very low, even if each bite has a fairly high probability of being an infective one. The combination of sporozoite rate and bite intensity is known as the entomological inoculation rate (EIR). One of the most important problems in malaria control is the fact that, in an increasing number of studies, a high incidence of the disease occurs even under conditions of very low levels of EIR. The consequence is that decreases in transmission by even the most efficient control systems may still not yield corresponding long-term reductions in the incidence of severe malaria (Mbogo *et al.* 1995).

The control, and, ideally, eradication, of malaria is based on two concepts that can be combined, at least in theory. One tactic involves the management of the parasitic protozoan in humans, and the other, control of vector insects. The pharmacology and medical application of malarial prophylactic drugs such as chloroquine and maloprim are beyond the scope of this book, as are exciting but controversial developments in the immunization of people against the pathogen. Suffice it to say that the use of drugs for malaria prevention is limited in scope, expensive, and has undesirable side-effects in the medium to long term, even if the majority of people in the world prone to malarial infections had access to them, which in fact they do not. Malarial vaccines are taking a long time to develop, and it is unlikely that the first-generation ones will help travellers to malarial areas very much (Greenwood 1997). Indeed, it is possible that these only partially effective new vaccines may actually upset the complex balance that exists between parasite and host in local populations, and make matters worse rather than better! Indeed, simply having a new antimalarial drug, an effective malaria vaccine, or a new way to kill mosquitoes, is still a long way and many years from achieving the effective and dependable control of malaria (Beier 1998). None the less, for most countries, where humans live in permanent potential contact with vector mosquitoes, perhaps the only realistic management system is that of vector control.

Mosquitoes in the genus *Anopheles* can be subdivided into two groups, depending on where they feed or rest between meals. Species such as *An. albimanus* are termed exophagic, because they bite people outdoors, during agricultural activities or hunting. Endophagic species such as *An. gambiae* feed mainly indoors, and rest inside buildings or dwellings, feeding on the blood of people at night whilst their hosts are sleeping. Differences in ecology indicate that different control measures need to be employed for these two groups, though if the larvae can be eradicated, the behaviour of the adults is of little significance. The larvae (similar to that shown in Fig. 9.16) may be found in all sorts of locations, as described earlier in this chapter, and in fact any standing water, whether it be aquaculture ponds, irrigation ditches, paddy-fields, water butts or old car tyres, may provide excellent breeding sites for one mosquito species or another. Thus, there are two basic targets in malaria vector control: the removal of larvae from breeding sites (or, in fact, the destruction of these very sites), and the eradication of adult females (only female mosquitoes are blood feeders; males utilize sugar sources) from sites frequented by humans in their various day-to-day activities.

The control of mosquito larvae is based on either the removal of their habitat, which is often impossible—people, their crops, and their livestock need water—or the destruction of larvae while in their aquatic habitat. As with a myriad other insect

Fig. 9.16 Larva of *Culex* mosquito.

pests (see Chapter 10), the latter control systems include two basic strategies, chemical and biological control. The inundation of water courses and irrigation ditches with insecticides such as DDT and its more recent (but more expensive) replacements might work to some extent, but because of the frequent multipurpose use of such habitats for everything from growing crops and fish, to washing and drinking, this tactic may be undesirable. Simple sanitation and other types of removal of unnecessary excess standing water in a community may help, such as the covering up of cess-pools, and the emptying of redundant water containers, but it is thought that if larval control has any real future, it is likely to involve biological control.

In general, it has been suggested that the control of mosquito larvae should be based on sound appreciation of natural control components (Legner 1995). Many species of freshwater fish, for example, are able to eat large numbers of mosquito larvae under laboratory conditions (Fletcher *et al.* 1993), but it is questionable whether such predation could really make a great impact on vector populations in the field. New strains of the bacterial toxin derived from *Bacillus thuringiensis* (Bt) (see Chapter 10) are available to those who can afford it, but, realistically, this can only be of very limited application. If vector control is to work efficiently, the target has to be adult mosquitoes.

No-one would be concerned about even large populations of vector mosquitoes if they could be prevented from biting people, and perhaps the most important advancement in this area is the widespread encouragement of the use of mosquito-proof bed-nets that have been impregnated with insecticides. Such systems are likely to be particularly effective against endophilic species of *Anopheles*. Emphasis on malaria control is now centred around sustainable programmes that can be implemented by local communities, with only limited assistance from primary health-care services (Sexton 1994). To this end, bednets impregnated with pyrethroid insecticides seem to work well. One study has described trials in various countries, where the incidence rate of malaria has been reduced by as much as 50% (Choi *et al.* 1995). In the Sichuan province of China, 2.42 million bed-nets have been impregnated annually with deltamethrin, with marked reductions in malaria incidence, and, in some cases, 100% mosquito mortality (Cheng *et al.* 1995). Simple precautions are still required; holes in nets must be avoided, the nets must not be washed, and people have to be strongly encouraged to use them on a routine basis.

Much research is now being carried out on the genetic manipulation of insect vector populations. It is highly unlikely that systems such as those for screw worm (Diptera: Calliphoridae) would work for mosquitoes. In the former case, artificially sterilized males are released in huge numbers to out-compete natural, fertile ones and hence bring about enormous reductions in population densities; mosquitoes are far too numerous and dispersed, and most countries where malaria is a severe problem could not afford the technology. Instead, genetic manipulation of mosquitoes may be successfully targeted towards reducing the capacity of the vectors to support parasite development within their tissues (Collins & Paskewitz 1995). Anti-parasite genes are being sought, and the mosquitoes' own defence mechanisms may be enhanced at some stage in the future. For now, however, malaria must be considered to be the most serious vector-borne disease threat to human life, not only for the millions of people living in endemic regions, but also for the growing number of visitors to these regions from areas where malaria was eradicated many years ago, or never existed at all (see section 9.6).

9.5.4 Barley yellow dwarf virus

Barley yellow dwarf virus (BYDV) (Plate 9.2, opposite p. 158) is the most economically important and widespread disease of cereals in the world (Irwin &

Thresh 1990). It affects over 100 species of plants, including barley, wheat, oats, sorghum, rye, maize, rice and many wild grasses. Wild grasses play an important role in the epidemiology of BYDV, acting as alternative hosts and vector–pathogen reservoirs within and around arable land. Over 20 species of aphid vector the virus, in a circulative, persistent manner. These include *Metopolophium dirhodum*, *Rhopalosiphum padi* and *Sitobion avenae* (Hemiptera: Aphididae) (Irwin & Thresh 1990), all of them serious pests of cereals in the UK, Continental Europe and North America in their own right. Each aphid species tends to transmit rather specific strains of BYDV, which are designated under their own acronyms according to the vector species; the strain of BYDV vectored by *R. padi*, for example, is RPV. The various strains of pathogen are to some extent regionally distributed, so that RPV is more common in the south of the UK, whereas others predominate further north (Kendall *et al.* 1996). We have suggested earlier (section 9.4.5) that BYDV has some sort of mutualistic association with its aphid vectors. Little is known about these mechanisms at the cellular level, though it seems that there may be an involvement of the aphid's own endosymbionts in 'chaperoning' BYDV through the vector (Filichkin *et al.* 1997). In any event, the effects of BYDV on the host plant are definitely debilitating. Infection may contribute to crop death over the winter in colder regions, induce plant stunting, inhibit root growth, reduce or prevent flower production, and generally weaken the plant, rendering it more susceptible to other pathogens and environmental stresses (Irwin & Thresh 1990). Roots are particularly affected, with severely reduced lengths and general stunting being observed in trials only 4 days after inoculation with the virus (Hoffman & Kolb 1997).

Some of the complexities of BYDV epidemiology are illustrated in Fig. 9.17, and these, plus the dispersive nature and host range of the aphid vectors, mean that the disease is not an easy one to manage or control. Three main areas of control can be identified: (a) host plant resistance to virus and vector; (b) the use of insecticides to control the vector; (c) manipulation of the crop environment or agricultural practices to minimize epidemics. Reliable host plant resistance appears difficult to find; the effectiveness of resistance varies with type of cereal, species of vector and even regional locality (Irwin & Thresh 1990), so the production of a coverall resistance mechanism is unlikely. However, some wild perennial grass species do exhibit real resistance, or at least, a low rate of virus multiplication, and molecular techniques may

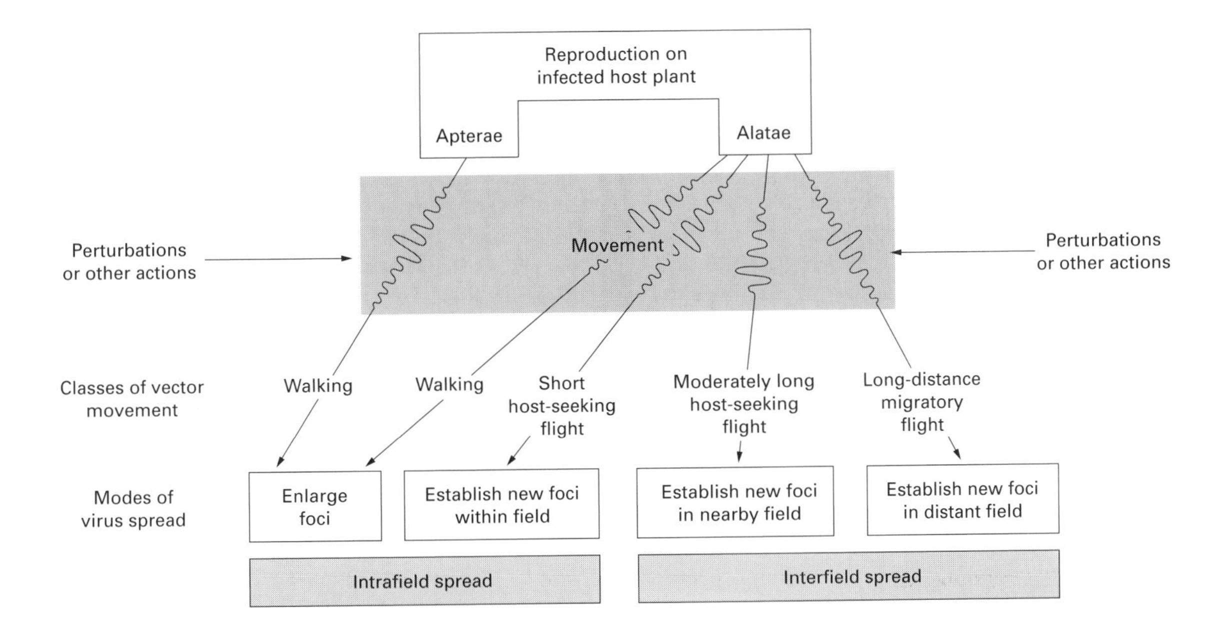

Fig. 9.17 Conceptual diagram of aphid movement resulting from various perturbations to the system. These movements can result in spreading BYDV to other fields, or establishing new or enlarging existing barley yellow dwarf virus foci within the field. (From Irwin & Thresh 1990. With permission, from the *Annual Review of Phytopathology*, ©1990 by Annual Reviews.)

allow such characteristics to be transferred to crop cereals. If this resistance to the pathogen can be achieved, then additional, heritable, characteristics of cereals that deter aphid feeding may provide a dual-acting package (Gray *et al.* 1993).

The crop environment may be manipulated in the management of BYDV epidemics. Most of these strategies target the aphid vectors rather than the pathogen directly, though the elimination of perennial weeds and volunteer cereals adjacent to cereal crops may remove local reservoirs of BYDV infection. Strategies to deter aphids cover a variety of measures. Many attempt to reduce the crops' attractiveness to aphids, using tactics such as crop spacing and planting density. Others try to grow cereals such that they are less available to peak aphid populations, by using late sowing in autumn or spring, or the removal of cereals from arable land for a time to break the otherwise permanent cycle of aphids and virus, using crop rotations. Sowing date has proved to be particularly successful, in fact. By delaying the sowing of wheat to late autumn or early winter, workers in Australia were able to reduce BYDV levels and hence increase grain yield significantly (McKirdy & Jones 1997). The vectors were presumably unable to find the new plants effectively because of poor weather, though it must be said that these particular results were significant only in areas of high rainfall.

Cereal aphids that damage crops without vectoring viruses are routinely controlled with timely doses of chemical insecticides, based on monitoring and prediction strategies (see Chapter 10). When the pests are infected with BYDV, the situation becomes more complicated, as virus spread is due to vector movement between and within crops (Fig. 9.17). Insecticidal compounds may influence aphid movement so that they become more mobile and hence transmit the pathogen more effectively. Clearly, aphidicides that kill rapidly before the vector has time to move are most desirable, but as only a few aphids left alive after a spray programme can still carry on the spread of BYDV, very high percentage kills must be achieved. Another problem is the ability of aphids to move considerable distances from alternative wild host plants, or to move into a recently sprayed field from another, non-treated one. If chemicals are to be used at all, appropriate timing of treatments is the key to success, so that the incidence of aphid migrations can be matched with susceptible stages in crop development (Mann *et al.* 1997). In general, Irwin and Thresh (1990) considered that pesticides must be used carefully and wisely (whatever that concept implies), and as a last resort for the control of barley yellow dwarf virus, rather than a front-line defence.

Some, if not all of these strategies may be commercially unattractive, especially in areas of intensive, monocultural cereal production, such as occurs in many regions of the world. In essence, the associations of cereal aphids with barley yellow dwarf virus strains, wild plants and commercial crops are exceedingly complex, and even now not completely understood. Ecologically, these associations are highly successful; the result of coevolution between the major players has to be seen as most admirable. This is another example of insect ecology and evolution defeating human aspirations and economic targets!

9.5.5 Pine wilt

So far, we have mainly discussed the ability of insects to carry nematodes responsible for human diseases, such as onchocerciasis. There are also some examples of insects that transmit plant pathogenic nematodes, and the now globally worrying pine wilt story is one of these.

This disease (Plate 9.3, opposite p. 158) is a major killer of trees in the Far East, especially in China and Japan, where it was first described in 1905 (Fielding & Evans 1996). The pine wood nematode (PWN), *Bursaphelencus xylophilus*, is the causative pathogen, which is transmitted by beetles in the genus *Monochamus* (Coleoptera: Cerambycidae) (Fig. 9.18). Cerambycids, of which there are well over 20 000 species so far described in the world, are notorious pests of a huge range of trees. The life cycle of *Monochamus* is fairly typical of the family. Adult females lay eggs in slits cut in the bark of trees that are usually lacking in vigour for one reason or another. Healthy trees are not suitable for oviposition because of high resin flows, which prevent larval development. Stressing agents may involve unsuitable soils, drought stress, or attacks by more primary pests such as defoliators. The hatching larvae tunnel under the bark, feeding between the inner bark cambium and the sapwood, where they excavate broad, flattened tunnels filled with coarse frass consisting of wood fibres. This activity may kill trees by girdling them. Towards the end of the larval stage, cerambycid larvae 'duck-dive' into the sapwood, where they excavate U-shaped galleries. They finally pupate, sometimes after several

Fig. 9.18 Adult, larva and pupa of pine longhorn beetle, *Monochamus alternatus*, from Japan. (Courtesy of H.F. Evans.)

years as larvae, in chambers near the wood surface. Young adults then re-emerge through the bark of infested trees, leaving characteristic oval exit holes. Most will then seek out new trees on which to lay eggs, but some species, including *Monochamus*, undergo an extra stage called maturation feeding, where freshly emerged adults of both sexes fly to the tops of healthy trees, where they chew away at nutrient-rich tissues, before they are able to mature their eggs and sperm.

Nematodes live predominantly in the xylem tissues of host trees, where they feed on live wood cells, or on a variety of fungi that occur on moribund trees. This mycetophilic (fungus feeding) habit can support PWN for a considerable time. The large number of nematodes plus the phytotoxic effects of their feeding activities cause blockages of the tree's transport vessels, inducing local drought effects in the plant's tissues, a wilt. Seriously infected trees rapidly die from the top down. During winter and spring, dispersive PWN larvae move to the beetle's pupal chambers, where they moult to produce specialized forms (dauer larvae), which enter the body of the new adult beetle through its spiracles. Up to 200 000 PWN may then accompany a single beetle on the next stage of its life cycle. There would appear to be a strong, coevolved association between nematode and vector, especially in terms of the dauer stage (Fielding & Evans 1996). Transmission of PWN to healthy trees then occurs via the wounds caused by maturation feeding, thus initiating the decline of new trees, which subsequently become suitable oviposition sites for a new generation of beetles. PWN can also be inoculated into trees at oviposition, but as these host trees are already likely to be infected in epidemic conditions, this form of transmission is considered secondary to the feeding process. Figure 9.19 summarizes the combined life cycles of vector and pathogen.

The problem does not seem to be completely restricted to pines. Although, in Germany, experiments showed that spruces (*Abies* spp.) largely tolerated PWN infestations without developing any wilt symptoms (Braasch 1996), Fielding and Evans (1996) listed 54 species of conifer susceptible to PWN, including firs, cedars and larches, as well as pines. The extent of the damage in Japan has been very considerable indeed, with many millions of *Pinus* spp. being killed over the years; $2.4 \times 10^6\,m^3$ of timber were reported lost to pine wilt in a single year (Mamiya 1984). Trees in forest stands are often observed to die back in clumps rather than as singleton trees (Togashi 1991), possibly indicating stress-related site interactions, or the presence of vector and pathogen in one locality exacerbating the syndrome in adjacent trees. As with so many similar situations in the modern world (see section 9.4.7), human activities in recent years have made the situation worse by artificially spreading the nematode in timber along railways and roads (Xu *et al.* 1996). Beyond Japan, pine wilt disease has spread to many countries, including the USA, where it is associated with susceptible tree species, suitable vector species, and high summer temperatures (Fielding & Evans 1996). The nematode has also been isolated from imported wood material in Europe, though the disease itself has yet to be noticed there. It is highly likely that the nematode would survive in Europe, and that tree mortality could be severe in warmer southern countries. To avoid further introductions of PWN or its vector into Europe, legislation and quarantine practices have been set in motion, whose requirements include the heat treatment of coniferous material from outside the European Community, such as chips, packaging and pallets; round wood or sawn timber must be stripped of all bark, and if holes exist in the wood over 3 mm in diameter (indicating the likely presence of *Monochamus*), then this material must be kiln-dried at the point of origin.

9.6 Conclusion

The associations between insects and pathogens of animals and plants are extremely complex and

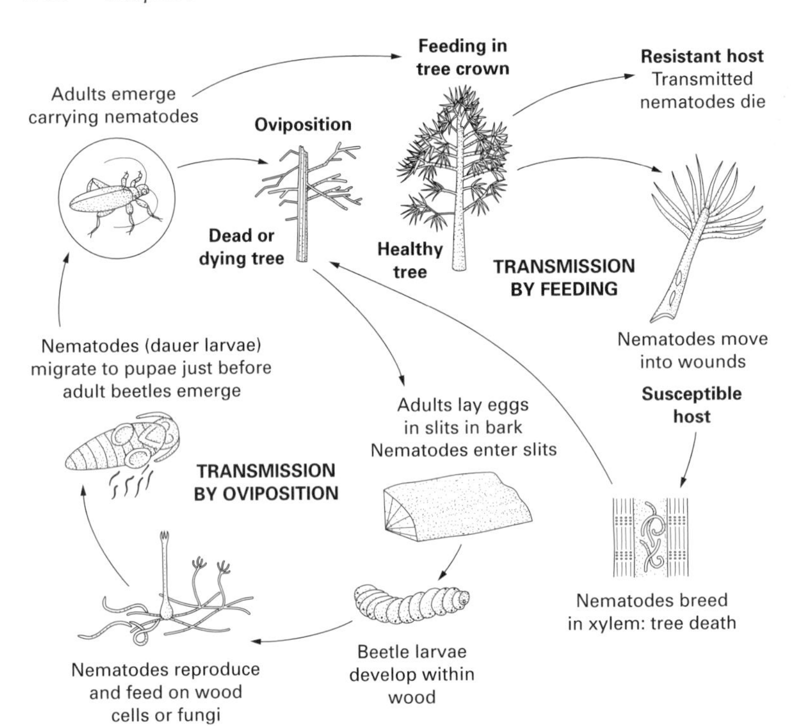

Fig. 9.19 Life cycle of pine wilt nematode and its cerambycid beetle vector. (From Fielding & Evans 1996.)

common. In many cases, the ecology of the associations shows a high degree of mutualism, where the vector and pathogen benefit, and only the host suffers. There are, of course, some pathogens that do not appear to be carried by insects (or any other animal), and their future may be watched with interest. What, for example, is the likelihood of hepatitis or human immunodeficiency virus (HIV) being vectored by insects? This topic was vigorously debated in the 1980s, when it was concluded that potential vectors such as bedbugs (Hemiptera: Cimicidae) and horse flies (Diptera: Tabandidae), were unlikely to act as mechanical transmitters of HIV from one human to another (Jupp & Lyons 1987). However, by 1990, the same first author had concluded that there were indications that the bed-bug, *Cimex lectularius* (Hemiptera: Cimicidae), might be able to transmit hepatitis B virus (HBV) mechanically, by multiple feeding on human hosts. There is also some evidence that it may be possible, at least in theory, for horse flies to transmit hepatitis C virus (HCV) (Silverman *et al.* 1996) in a similar fashion. With pathogens as mutable as RNA viruses, of which HIV is one, might it not be considered that in large human population densities, with a high incidence of the infection, blood feeding insects could begin to play a role in transmission? If this were to occur, the ecology of insect vectors could take on a whole new magnitude of importance!

Meanwhile, major killers such as malaria seem to continue virtually unchecked. Australia, for example, has in theory been free of the disease for many years, but as Brookes *et al.* (1997) reported, in 1996 a man from the far north of Queensland was diagnosed as having malaria, though he had not visited a malarious area for 19 years! It is suspected that the infection was transmitted by local mosquitoes from a neighbour who had been infected during a visit to Papua New Guinea. The future of malaria and other supposed tropical diseases in temperate climates remains hard to predict. Certainly, many cases of malaria, some of which are fatal, occur in people in the UK who have returned from overseas. The role of wetter, warmer climates predicted by global warming in insect vector–pathogen interactions will be of great interest.

Chapter 10: Insect Pest Management

10.1 Introduction

Insects compete at many levels with humans for the crops they grow and the livelihood they try to make from all forms of production, including agriculture, horticulture and forestry. It is clear that without some form of control or management of these insects, enormous and unacceptable losses will regularly occur in all parts of the world. We have shown in many previous chapters just how complex insect ecology can be. We are now at the stage where interactions such as insect–plant relationships, population dynamics and abiotic influences can be brought together in a synthesis that leads to the modern concept of insect pest management. No longer do we think of 'pest control' as a strategy to reduce crop damage. This black and white approach has been the cause of many basic problems and upsets, and in many cases has simply been ineffective. Management these days is more concerned with the prevention of pest outbreaks whenever possible, so that more curative systems such as the use of chemical methods come into play only when prevention fails. The key to modern pest management is the striving for the goal of integrated pest management (IPM) in all appropriate cropping systems, and this chapter presents various important components that might function in complementary or additive ways to achieve this goal.

10.2 The concept of 'pest'

The term 'pest' is of course highly anthropocentric. Large numbers of insects in natural habitats are usually a source of wonder. Let us consider, for example, the enormous population densities of monarch butterfly, *Danaus plexippus* (Lepidoptera: Danaidae), which overwinter in Mexico. Adults from eastern North America migrate each autumn to avoid climatic extremes (Mastrers *et al.* 1988), returning in the following spring when the larvae feed on milkweeds, *Asclepias* spp. (Malcolm *et al.* 1989). As milkweeds are not exploited as a source of food or revenue by man, there is no problem. Only if *Danaus* in some way swapped to eating a commercially important plant would it then earn the label 'pest'.

However, it is not just insects in large numbers that can be pests; in some cases, even one insect on its own can be too many to tolerate. These low-density pests include those species that act as vectors of many diseases of crop plants, domestic animals, and humans themselves (see Chapter 9). Clearly, strategies for the management of low-density pests must differ markedly from those appropriate for insects that become pests only by virtue of their high numbers (high-density pests). In the former case, all insects must be prevented from attacking the host plant or animal, and prophylactic (insurance) measures are usually employed. In the latter case, however, it is desirable to wait until some threshold density is exceeded before any further action is taken. For example, not even one small larva of the codling moth, *Cydia pomonella* (Lepidoptera: Tortricidae), is tolerable in an apple to be sold in the supermarket, whereas fairly large numbers of defoliating Lepidoptera such as winter moth and green oak roller moth can be tolerated on oak trees. This tolerance of insect numbers usually relates to cost–benefit considerations, where the impact of the pest's damage in yield or monetary terms is weighed against the cost of the management tactic to reduce it. Figure 10.1 shows a conceptual framework for this type of analysis. The acquisition of the economic threshold based on sound and dependable impact data is a vital component of most IPM systems. Monitoring programmes review the densities of pests on the crop, on which decisions about control tactics are made.

10.3 Why pest outbreaks occur

To prevent the numbers of insects on crops reaching levels where significant damage takes place, it is of fundamental importance to consider why such outbreaks occur, on the understanding that if we know

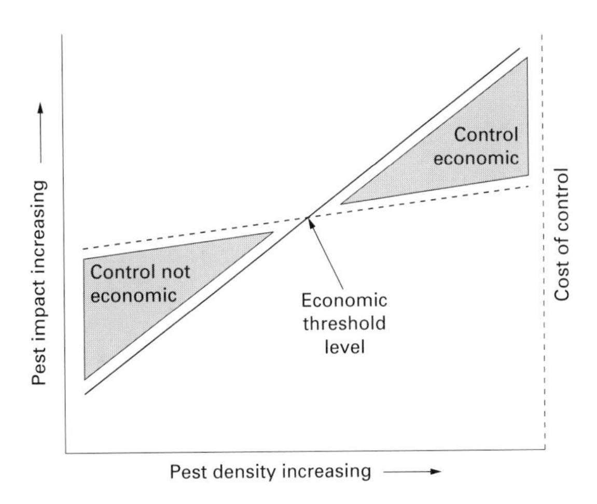

Fig. 10.1 Concept of economic threshold in insect pest management.

Table 10.1 Some major reasons for insect pest outbreaks. (From Speight 1997b.)

Natural disasters
Fire, windthrow, drought, etc. providing food and/or breeding sites
Crop management: avoidance of stress and susceptibility
Site choice and crop species matching in the planning stage
Genetics and variable crop susceptibility
Nursery management and the production of healthy transplants
Choices at planting—monocultures
Crop management—high densities may promote suppression
Post-harvest manipulation
Pest invasions
International—accidental imports from foreign countries
National or regional—invasions from infested crops
Misuse of control systems
Removal of natural enemies
Stress caused by phytotoxicity

what has caused an upsurge in pest numbers, then it may be possible to avoid these, or at least reduce their intensity, in the future. Some of these factors are predominantly unavoidable. For example, it is well known that insect population densities can fluctuate tremendously over time (see Chapter 1). Fairly predictable cycles may be observed, caused in the main by delayed density-dependent processes (see Chapters 1, 4 and 5), but on occasion, stochastic events such as weather conditions may drive numbers higher and less predictably than in stable cycles (Cavalieri & Kocak 1994). Even changes in the landscape or patterns of agriculture may lead to higher pest densities. Lack of habitat connectivity because of fragmentation is thought to release some insect herbivores from regulation by their natural enemies (Kreuss & Tscharntke 1994), but, as with weather variations, these factors are beyond the influence of the pest manager.

Table 10.1 summarizes some of the important factors that contribute to pest outbreaks (Speight 1997b), only some of which may be manipulated in a preventive pest management strategy. Many of them relate directly or indirectly to the provision of a large quantity of high-quality food and/or breeding sites for herbivorous insects. Intensive crop production systems, whether they be from agriculture, horticulture or forestry, almost invariably represent gross departures from natural habitats within which these same insects have coevolved with their enemies, host plants, competitors, etc., so it is hardly surprising that the ecology of insects in these novel situations also changes. Various important differences between natural habitats and cropland are summarized in Table 10.2; these, in most cases, are situations with which pest managers have to live. Most crop production in the world necessitates the modification and usually simplification of natural ecosystems to produce high yielding, specific crops. Even the plants themselves are frequently 'unnatural', having been specially bred for certain desirable characteristics. Tongue somewhat in cheek, it may be of interest to consider most examples of attack by insects as mainly caused by the crop producer and not the sole responsibility of insects, who are in fact merely responding to an extra provision of resource! In reality, insect pest management cannot normally try to change crop husbandry techniques radically; rather it has to deal with intensive production systems in most of the world, and attempts to manipulate them rather subtly to reduce the advantages provided by these systems to herbivorous insects. We will now consider some of the major topics relating to the causes of insect pest outbreaks summarized in Table 10.1 in more detail.

10.3.1 Host plant susceptibility

Whether or not a plant species, genotype or individ-

Table 10.2 Summary of differences between natural habitats and managed cropland, in relation to the increased likelihood of pest outbreaks.

	Natural habitats	Cropland
Alternative hosts for enemies	Many	Few
Artificial inputs such as agrochemicals	None	Many
Ease of finding specialized host plant	Low	High
Habitat disturbance	Rare	Common, often annual
Numbers of natural enemies	High	Low
Plant genetic diversity	High	Low
Plant species exotic or domesticated	No	Yes
Plant species richness	High	Very low
Plants chosen for high yields	No	Yes
Population limitation by intra-specific competition	Less important	Highly important
Resistance mechanisms in plants	Common	Rare

ual is susceptible or resistant to an insect herbivore concerns the provision of food or breeding sites, or both, by the plant for the insect, and the ability of the insect to utilize these resources. Two distinct ecological features of susceptibility are important in pest management: the first involves environmentally derived changes in the host plant, which are not heritable, and the other involves genetically based mechanisms, which can be passed on in selection and breeding programmes to subsequent generations.

Environmental influences

The conditions in which a plant grows are fundamental to determining its vigour. Plants that are in some way stressed do, on many occasions, provide enhanced levels of organic nitrogen to insects feeding upon them, whilst showing reduced levels of physical and chemical defences (White 1993). Deterioration in the health of plants tends to render them more susceptible to attack by certain groups of insects. Boring and sucking insects seem to perform better on stressed plants, whereas gall-formers and chewers are adversely affected (Koricheva *et al.* 1998). Although the links between environmentally derived plant stress and insect attack are not universally accepted (Watt 1994), it is vital, whenever possible, to grow plants under conditions that promote their health and reduce the likelihood of their becoming stressed if pests such as aphids, scales and boring beetles are to be avoided. One example is the occurrence of the horse-chestnut scale, *Pulvinaria regalis* (Hemiptera: Coccidae), on urban trees in the UK (Speight *et al.* 1998) (Plate 10.1, opposite p. 158). This insect feeds as a nymph on leaves of limes, sycamores and horse-chestnut trees in urban areas in parts of Western Europe, and adult females lay large, obvious egg-batches on the main stems. In studies in Oxford, the density of the insect was found to rise dramatically on trees that were growing alongside streets or in car parks, where the permeability of the soil was seriously impaired (Fig. 10.2). Furthermore, trees with morphological indications of low vigour such as dieback, rot or wounds almost always showed higher

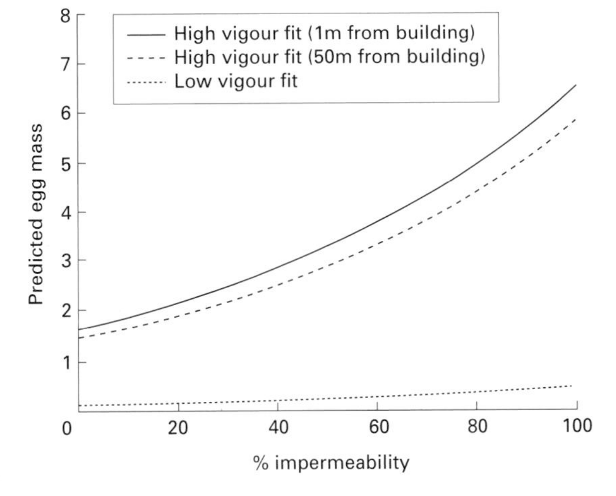

Fig. 10.2 Lines fitted from regression equations describing the influence of soil permeability and building distance on the numbers of adult *Pulvinaria regalis* laying eggs on urban trees. (From Speight *et al.* 1998.)

levels of infestation no matter where they were growing.

Even manipulations of host plants during routine management practices can promote pest attack, presumably via a form of environmental stress. In Kenya, pruning of *Cassia samia*, a leguminous fodder tree, to provide extra graze for livestock, invokes significant increases in attacks by stem boring moth larvae (Lepidoptera: Hepialidae) (Opondo-Mbai 1995), with a resultant decline and eventual death of the plant. It is likely that environmentally derived losses of vigour in host plants leading to the increase in pest attack are more a feature of forestry than agriculture, though even arable crops can be influenced. Mexican bean beetle, *Epilachna varivestis* (Coleoptera: Coccinellidae), is an unusual ladybird in that its larvae are herbivores rather than predators, and can be serious pests of crops such as soybean in North America. Jenkins *et al.* (1997) found that both total mortality and larval development time increased as soil moisture decreased, indicating that the pest performs better on plants growing in wetter areas. This environmental influence in fact also influenced soybean plants selected for resistance to the beetle, so that decreased expression of this resistance was observed under conditions of high soil moisture. In other words, the plant was such a suitable host for the pest under wet conditions as to override any inherent resistance mechanisms.

Genetic resistance

Van Lenteren *et al.* (1995) considered that host plant resistance is one of the cornerstones of integrated pest management. Some of the background to this topic is considered in more detail in Chapter 3, but, in essence, certain genotype-based characteristics of a plant are more able to reduce or tolerate the feeding impacts of insects than others, and crop breeding programmes attempt to isolate the germplasm containing these characters, which can then be incorporated into new resistant strains (Abel *et al.* 1995). These heritable resistance mechanisms are normally considered under three separate headings, i.e. non-preference, antibiosis (for instance, lower fecundity) and tolerance (Wiseman 1994). The first two categories at least can involve physical and/or chemical mechanisms for resistance. Resistance can reduce the impact of insects feeding on plants as a direct effect, or it may enhance the ability of natural enemies to regulate the pest density, again resulting in a reduced pest problem.

One problem in detecting plant resistance focuses on distinguishing between environmentally induced changes in susceptibility to insects (see above) and those that are genetically based. For example, antibiosis in the form of reduced weights of larvae and increased larval development periods were observed for the soybean looper, *Pseudoplusia includens* (Lepidoptera: Noctuidae), feeding on fields of soybean in the USA (Lambert & Heatherly 1991). However, it transpired that these observations, which in theory pointed towards a plant resistance mechanism, were again in fact a result of changes in local soil water potential caused by local variations in soil type, and hence not heritable. This example compares well with that of the Mexican bean beetle described in the previous section.

The various types of resistance mechanisms will be examined in turn, bearing in mind that they probably operate most of the time as complementary systems within the host plant–herbivore–enemy complex.

Non-preference

Non-preference, or antixenosis, can be observed when insects tend to avoid or reduce their feeding rate on one species or genotype of plant when compared with another, on which they feed more readily. Most herbivorous insects tend towards host specialization, so that in a gross example, elm bark beetles (Coleoptera: Scolytidae) will not feed or breed in ash, nor will ash bark beetles utilize elm. There are also marked differences to be observed within a plant genus, so that, for example, the pine bark beetle, *Ips grandicollis*, imported from Europe into Australia some years ago, prefers to oviposit in logs of radiata pine (*Pinus radiata*), rather than in those of pinaster pine (*P. pinaster*) (Abbott 1993). Table 10.3 shows how the green spruce aphid, *Elatobium abietinum* (Plate 10.2, opposite p. 158), responds to different species of spruce in selection trials (Carter *et al.* 1985). Clearly, those species that originate in North America are much preferred to those of European provenance. It is important to note that the most preferred species, *Picea sitchensis* (Sitka spruce), is the most widely planted of all spruce species in the UK at least! One obvious strategy for reducing aphid problems would be to change the crop species to a less suscepti-

Table 10.3 Differences in resistance of spruce species planted on the same site to the green spruce aphid, *Elatobium abietinum*. (From Carter *et al.* 1985.)

Species of *Picea*	Natural infestation levels (aphids/8.5 cm shoot)	Aphid growth rate (μg/μg per day)
American species		
P. engelmanni	36	135
P. glauca	97	104
P. mexicana	n.a.	129
P. sitchensis	113	123
European and Asian species		
P. abies	n.a.	133
P. brachytyla	0	99
P. glehnii	0	82
P. omorika	0	96
P. orientalis	0	96

n.a., not available.

ble one, but forestry practices and market forces must prevail.

More closely related crop varieties also show preference effects amongst their herbivores. The Russian wheat aphid, *Diuraphis noxia* (Hemiptera: Aphididae), seems to be able to discriminate wheat from oats, though this may also be related in part at least to which host they were reared on (Worrall & Scott 1991). This is another example of the problems of distinguishing absolute, inheritable, resistance from other effects.

The most common way to investigate non-preference is to allow insects to choose between different crop genotypes under controlled conditions, so that if one genotype is repeatedly ignored or rejected by the pest, then this is a good indication of antixenosis. In the case of the sorghum midge, *Contarinia sorghicola* (Diptera: Cecidomyiidae), a serious pest of grain sorghum, adult flies were shown to prefer certain crop genotypes on which to lay eggs (Sharma & Vidyasagar 1994). Always supposing that the crop husbandry system can utilize these resistant genotypes economically, then this is a simple but effective method for managing the pest. Finally, a common resistance system is based on the hirsuteness of leaves of crop plants. Hairs or trichomes on leaves (Fig. 10.3) make it more difficult for sap feeders to reach the leaf surface to penetrate its cuticle, or to move around on it. Clearly, the relative size of the hairs vs. the insect are crucial. For example, cultivars of maize with numerous trichomes on both upper and lower leaf surfaces are much less preferred as oviposition sites by the stem borer, *Chilo partellus* (Lepidoptera: Pyralidae) (Kumar 1992).

Fig. 10.3 Whitefly surrounded by trichomes on leaf surface.

Antibiosis

Antibiosis (not to be confused with antibiotic) mechanisms in plants can take various forms, such as variations in either physical deterrents to insect feeding or nutritional compounds, or chemical defences. Effects on insects are many and varied, and Table 10.4 provides some examples. In all cases, antibiosis acts on the performance of the insect rather than its preference between host plants. As can be seen, both sap feeders and leaf chewers are variously affected, exhibiting higher mortalities, longer development times and reduced final weights when feeding on

Table 10.4 Examples of the effects of antibiosis plant resistance to insects.

Plant	Insect	Effect of resistance	Reference
Alfalfa	*Spissistilus festinus* (Homoptera: Membracidae)	Reduction in male hopper weights	Moellenbeck and Quisenberry (1992)
Chrysanthemum	*Spodoptera exigua* (Lepidoptera: Noctuidae)	Longer development time, reduced pupal weight	Yoshida and Parrella (1992)
Cowpea	*Clavigralla tomentosicollis* (Hemiptera: Coreidae)	Increased nymph development time, high nymph mortality	Olatunde and Odebiyi (1991)
Rice	*Nilaparvata lugens* (Homoptera: Delphacidae)	Decreased feeding rate, high nymphal duration and mortality	Senguttuvan *et al.* (1991)
Sorghum	*Contarinia sorghicola* (Diptera: Cecidomyiidae)	Adult emergence significantly lower	Sharma *et al.* (1993)
Soybean	*Spilarctia casignata* (Lepidoptera: Arctiidae)	High larval and pupal mortalities, reduced pupal weight	Neupane (1991)

resistant crops. Figure 10.4 illustrates how behavioural changes can be observed on resistant plants, where the aphid *Aphis craccivora* (Hemiptera: Aphididae) reduces the time spent feeding on resistant varieties of cowpea, but increases the time spent probing (Mesfin *et al.* 1992). This system is a dubious advantage from the point of view of the transmission of plant pathogenic viruses by the pest, as increased probing may also increase the efficiency of the transmission of stylet-borne viruses (see Chapter 9).

Physical defence mechanisms can take the form of tough leaves, hairy leaves or more complex mechanical systems within the host plant, but most resistance based on antibiosis is a function of host plant chemistry, either nutrients or defences, or both. We have already seen how host plant chemistry can fundamentally influence the ecology of herbivorous insects (Chapter 3). Variations in primary metabolites (such as nutrient compounds) may be particularly effective as defences against highly oligophagous herbivores (Berenbaum 1995). Studies on black bean aphid, *Aphis fabae*, and peach–potato aphid, *Myzus persicae* (Hemiptera: Aphididae), in Poland found that resistant varieties of sugar beet had reduced concentrations of proteins when compared with susceptible ones, and that the number of aphids per cultivar was directly proportional to the total content of free amino acids in the host plant (Bennewicz 1995).

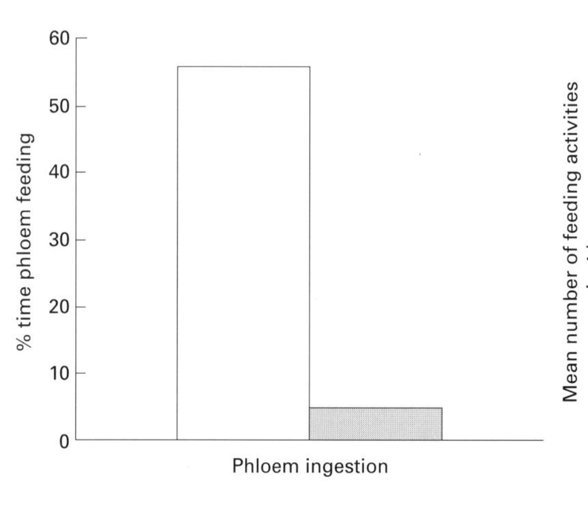

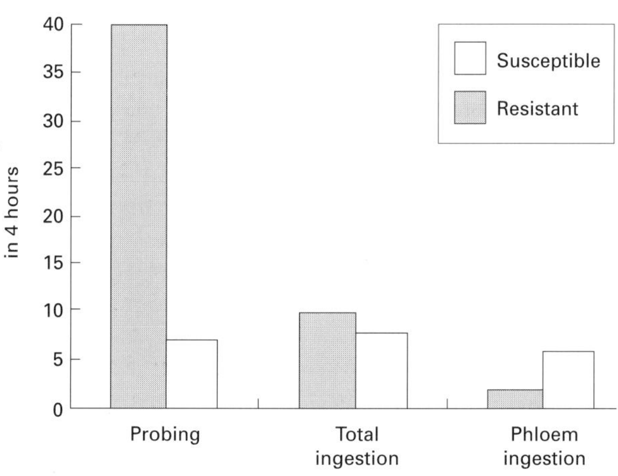

Fig. 10.4 Variations in feeding activity of the aphid *Aphis craccivora* on resistant and susceptible cowpea cultivars in Nigeria. (From Mesfin *et al.* 1992.)

Rather similar studies in China also found that resistance by sorghum to the sorghum aphid, *Melanaphis sacchari*, was negatively correlated with plant soluble nitrogen, sugars and free amino acids (He *et al.* 1991). However, other investigations have not been able to demonstrate such clear-cut effects. Although plant nutrients may vary between resistant and susceptible plants, simple nutritional effects may not be sufficient to explain observed antibiosis effects, as in the case of pea aphid, *Acyrthosiphon pisum*, feeding on lucerne (Febvay *et al.* 1988). The balance of amino acids in the host plant may be more important than the simple gross quantity.

Plant secondary compounds (allelochemicals) such as toxins vary considerably in their concentrations between different plant genotypes, and because many of these systems may have evolved in part to deter insect herbivores, it is not surprising that we find them at the basis of many resistance systems. Domestic potatoes (*Solanum tuberosum*) are derived from wild relatives and contain powerful glycoalkaloid antiherbivore compounds such as solanine and tomatine. Some genetic lines of potato are known to have antibiosis-like effects on Colorado potato beetle, *Leptinotarsa decemlineata* (Coleoptera: Chrysomelidae), as measured by the weight gains of fourth instar larvae fed on the different lines (Horton *et al.* 1997). These effects resulted in very low defoliation levels in the field. There are many chemicals in plants that may produce antibiosis to insect pests. Some cotton varieties, for example, are known to produce higher concentrations of a terpenoid aldehyde (known as gossypol) than others, and these chemicals have important antibiosis activity against one of the most serious pests of cotton in Australia, the cotton bollworm *Helicoverpa armigera* (Lepidoptera: Noctuidae) (McColl & Noble 1992).

Often, chemical and physical defences combine to produce resistance. The trichomes on the leaves and stems of tomatoes not only make it more difficult for small insects such as whitefly (Hemiptera: Aleyrodidae) or even defoliating Coleoptera to reach the leaf surface, but they also produce poisonous exudates, which can kill pests (Erb *et al.* 1994). Finally, it may be the case that insect herbivores feeding on toxic plants have managed to adapt to the chemicals from the host, and these allelochemicals sequestered by the pest may adversely affect their natural enemies, thus reducing the efficacy of biological control (Van Emden 1995) (see Chapter 3).

In the final analysis, it may be rather difficult to pinpoint the causes of antibiosis, as physical and chemical features vary so much between plant genotypes, and some investigations may yield less than convincing results. For example, in experiments to evaluate rice cultivars for resistance to the gall midge, *Orseolia oryzae* (Diptera: Cecidomyiidae), eight resistant varieties had considerably higher concentrations of polyphenolic compounds in the basal stem where the insect larvae attack (Sain & Kalode 1994). Unfortunately, in the same study, seven other resistant cultivars showed significantly lower levels of the same compounds when compared with even the susceptible varieties. Clearly, the full story has yet to be revealed.

Tolerance

Some plants are able to grow well and provide high yields despite being attacked by insect pests. In the case of various genotypes of chickpea attacked by the pod borer, *Helicoverpa armigera* (Lepidoptera: Noctuidae), the highest seed yield was obtained with fairly minimal damage (7%) to pods by the pest. However, another genotype, which suffered over three times this amount of pod damage, still yielded over 60% of the highest yield, indicating a degree of tolerance to the borer (Chauhan & Dahiya 1994). Tolerance is considered in more detail in Chapter 3.

The use of genetic resistance

In general terms, of course, the mechanisms behind an observed plant resistance system are less important to the pest manager than the dependability of its action to deter pests, and there is no doubt that plant resistance selection and breeding will increasingly be a crucial part of integrated insect pest management. Of particular importance in tropical countries especially is the development of crop cultivars that are resistant to several pest insects, pathogens and even nematodes, all in the same genotype, as a whole battery of pests and diseases may attack a crop at the same time (Heinrichs 1994). One of the best examples of multiple resistance of this type in a crop is rice, where millions of hectares of multiple resistance varieties are now planted all over the world (see section 10.7.2).

However, it is one thing to detect resistance, but it is quite another to decide what to do with it. It is unlikely that resistance on its own will be sufficiently effective to reduce and maintain a pest population below economic thresholds all the time, but it may be that other control systems may be enhanced by the use of resistant crop cultivars. The actions of natural enemies may be a case in point. Figure 10.5 shows the effect that a resistance-based drop in fecundity can have on the equilibrium density reached by a pest population. In this hypothetical example, density-dependent (see Chapter 5) predation rate increases with pest density, and the lower fecundity of the pest on the resistant crop results in a lower equilibrium density. It should be noted that equilibrium occurs when death rate equals birth rate (the lines cross).

Gowling and van Emden (1994) set up cage experiments wherein the cereal aphid, *Metopolophium dirhodum* (Hemiptera: Aphididae), was grown on two wheat cultivars, one susceptible to aphids and the other partially resistant. The parasitic wasp *Aphidius rhopalosiphi* (Hymenoptera: Aphidiidae) was then introduced to both systems and the reductions in peak aphid numbers caused by parasitism were recorded. In the presence of this parasitoid, aphid populations were reduced by 30% on the susceptible host plant, but by 57% on the partially resistant one. Furthermore, the numbers of aphids that left the plants in the presence of the parasitoid nearly doubled on the resistant host. In other examples, this time from cucumbers (Van Lenteren *et al.* 1995) and soybeans (McAuslane *et al.* 1995), less hairy plants supported fewer white fly (Homoptera: Aleyrodidae) than did hairy ones, because of the improved parasitism by the parasitic wasp, *Encarsia* spp. (Hymenoptera: Aphelinidae) on the non-hairy varieties. The enemy was better able to forage for hosts when its progress

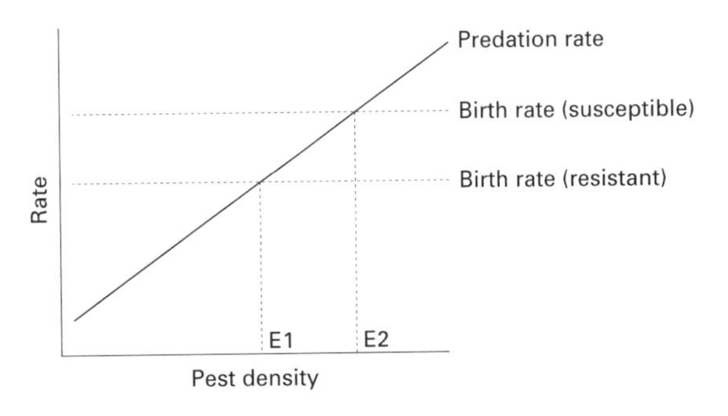

Fig. 10.5 Hypothetical example of the effect of predation on an insect pest population feeding on either a susceptible or a resistant crop.

and searching were not inhibited by hairs or trichomes.

Finally, combinations of host plant resistance and predation by the mirid bug, *Cyrtorhinus lividipennis* (Hemiptera: Miridae), have been found to have a cumulative effect on population densities of the green leafhopper, *Nephotettix virescens* (Hemiptera: Cicadellidae), a serious pest of rice (see section 10.7.2). Again in cage studies, on a strain of rice (IR29) resistant to leafhoppers, the number of leafhoppers reached only six in the presence of the predator, and 31 in its absence. On a leafhopper-susceptible strain (IR22), however, there were 91 and 220 leafhoppers, respectively (Heinrichs 1996).

However, host plant resistance does not always positively contribute to the suppressive effect of natural enemies (Van Lenteren *et al.* 1995). If a resistant plant causes reductions in fitness and/or abundance of pests, then there may be a knock-on effect on the pest's enemies because their prey or hosts are not so suitable for their own development (Reed *et al.* 1992). A whole range of plant characteristics may influence insect predators and parasitoids, often adversely. These include plant size, shape, hairs, waxes and colours, chemicals (both attractants and inhibitors), plant density and vegetation diversity (Bottrell *et al.* 1998). All of these factors impinge on the likely success of biological control programmes (see section 10.5).

A final problem remains. Unfortunately, insect populations have a wide range of genetic variability, and because of the widespread use of just one variety of crop in monocultures, some pests have been able to overcome the resistance of certain crop varieties. These new types of pest are known as biotypes. Heinrichs (1996) provided an example of biotypes of brown plant-hopper *Nilaparvata lugens* (Hemiptera: Delphacidae) that have been a very severe problem in rice cultivation in South and South-east Asia (see section 10.7.2). Rice variety IR26 was the first brown plant-hopper resistant cultivar to be released by the International Rice Research Institute (IRRI) in 1974, but within the Philippines, brown plant-hopper outbreaks were observed in IR26 after only 2 or 3 years (roughly six crops) of commercial cultivation, because of the selection of a plant-hopper strain that could utilize the previously resistant rice. Other resistant varieties have gone the same way. To cope with this biotype problem, so-called gene deployment strategies have been proposed, which include (a) sequential release, where a rice variety with a single major resistant gene replaces a variety whose resistance has been overcome by a new pest biotype, (b) gene pyramiding, where two or more major resistant genes are incorporated into the same rice variety to provide resistance to two or more biotypes of insect, (c) horizontal resistance, where a type of resistance is expressed equally against all pest biotypes, (d) gene rotation, where varieties with different major resistant genes are used in different cropping seasons to minimize selection pressure, and (e) geographical deployment, where rice varieties with different resistant genes are planted in adjacent cropping areas. Clearly, the battle will never be won completely, and continued research and development of plant resistance is vital to produce commercially viable crops using IPM (see section 10.7.2).

10.4 Ecological pest management

Once the reasons for pest outbreaks have been reviewed, the crop grower then has to consider that many practices carried out throughout the life of the growing crop may also influence the potential for pest problems. From planting to harvest, and beyond, the ecology of the crop and its surroundings can be fundamental to the promotion of pest numbers by enhancing both biotic and abiotic factors under which the pests can thrive. As mentioned above, every effort must be made to ensure that the crop plants are optimally healthy, and that the ecosystem provides the best conditions for them to flourish. The same habitat should also be optimized for the proliferation of natural enemies if the cropping system allows. All stages in crop husbandry may provide such enhancements, and each of these stages will be considered in turn.

10.4.1 Cultivation

Modern farming systems have a variety of cultivational or tillage tactics available to them. Traditional deep ploughing may give way to minimum tillage, where seeds are directly drilled into the stubble or other remains of the previous crop. This system reduces the time and energy invested in reseeding, such that machinery is moved over the soil less frequently, with consequent savings on fuel, compaction, soil erosion, manpower and so on. Table 10.5 provides some examples of the effects of cultivation on insect pests, most of which indicate that minimum tillage tends to enhance the success of insect pests. In economic terms, there has to be a cost–benefit assessment to compare the advantages of minimum tillage with the promotion of insect (and, indeed, weed) problems. Thus, pest population decreases brought about by soil surface disturbance regimes such as cultivation may be desirable as a form of pest management (McLaughlin & Mineau 1995), though the reasons for this effect may be variable. Some pests in the soil, such as cutworms (Lepidoptera: Noctuidae) or wire worms (Coleoptera: Elateridae), may be exposed to desiccation or bird predation by ploughing (Speight 1983), whereas pests that feed on stubble after harvest may starve if the ground is tilled (Robertson 1993).

10.4.2 Nursery management

For crops that are not sown directly into their final

Table 10.5 The effects of minimal tillage on insect pests.

Crop	Insect	Common name	Effect of minimal tillage	Reference
Arable	Lepidoptera: Noctuidae	Cutworms	+	Turnock *et al.* (1993)
Corn	*Papaipema nebris* (Lepidoptera: Noctuidae)	Corn stalk borer	+	Levine (1993)
Cotton	*Heliothis* spp.	Bollworms	+	Roach (1981)
Oats	*Oulema melanopus*	Cereal leaf beetle	+	Leibee and Horn (1979)
Soybean	Hemiptera: Pentatomidae	Stinkbugs	+	Buntin *et al.* (1995)
Soybean	*Anticarsia gemmatalis* (Lepidoptera: Noctuidae)	Velvetbean caterpillar	+	De Bortoli *et al.* (1994)
Wheat	*Cephus cinctus* (Hymenoptera: Cephidae)	Wheat stem sawfly	+	Morrillo *et al.* (1993)

+, enhancement of pest problem.

growing site, such as many horticultural and almost all forest species, the careful management of the young plants in the nursery is crucial in giving them a vigorous start in life, which in turn enables the transplants to withstand pest attack later on. In tropical forestry, it is all too easy to damage little trees in the nursery by rough handling at the pricking out stage. In Sabah (NE Borneo), a large number of nearly mature *Acacia mangium* were killed by secondary pests such as longhorn and roundhead borers (Coleoptera: Cerambycidae (Fig. 10.6) and Buprestidae), which were able to attack the sickly trees. These trees had reduced vigour caused by deformed root systems. This damage was acquired in the nursery, 8 years previously, because of rough handling during potting, which produced severely coiled roots (Fig. 10.7) (Speight 1997a). Other forms of nursery management can take the place of conventional chemical control. Again in the tropics, nursery plants of hot pepper (*Capsicum* spp.) that were protected from the hot sun by screens covering the beds had no problems with aphids, or the viruses that the aphids transmitted, when compared with unshaded plants. In addition, these shaded plants established much better in the field later on (Vos & Nurtika 1995).

10.4.3 Planting regime

Planting regimes for crops can take many forms, including planting date, plant spacing, crop rotations, mixed or monoculture crops, intercropping, trap cropping, and so on, and most of these tactics can influence the potential for pest problems once the crop has established. Obviously, many of these tactics are inappropriate for intensive crop husbandry where high yielding, mono-specific or even monogenetic crops are grown year after year in the same place for sound and unavoidable economic reasons. However, in some situations, growers may be able to modify their tactics to incorporate some of these ecologically sensible strategies.

Varying the time of planting may reduce the likelihood of pest problems. Where climates vary seasonally, many examples illustrate the fact that crops planted early in the season are less likely to suffer from pest outbreaks than those planted later. In crops such as cotton (Slosser *et al.* 1994), tobacco (McPherson *et al.* 1993), and rice (Thompson *et al.* 1994), insect pest problems were reduced or absent in early planted crops. One of the reasons for this may be the fact that early planted crops have a chance to become better established before insects appear. In this way such plants may be less palatable to herbivores, or simply be able to tolerate higher pest densities before succumbing. Certainly, in the example of rice attacked by the water weevil, *Lissorhopterus oryzophilus* (Coleoptera: Curculionidae), the early planted crop did not escape being attacked, but was able to tolerate infestations without loss of yield (Thompson *et al.* 1994). General rules rarely if ever hold true for all ecological situations, so that though work in South Korea on various other pests of rice showed agreement with the suggestion that early planting reduced pest problems, some insects such as striped rice borer, *Chilo suppressalis* (Lepidoptera: Pyralidae), actually had higher population densities

Fig. 10.6 Longhorn beetle larva in gallery under Acacia bark.

Fig. 10.7 Seedlings of *Acacia mangium* in forest nursery in NE Borneo showing root coiling.

in early transplanted paddy fields (Ma & Lee 1996). The example of barley yellow dwarf virus discussed in detail in Chapter 9 is another situation where early planting exacerbates the problem.

The effects of plant spacing on insect attack are complex, and need to be considered for each individual insect–crop interaction. In some cases, planting individual crop plants wide apart provides a lot of bare soil, which is preferred by insects for oviposition, whereas in other situations, the reverse occurs. Cabbage root fly, *Delia radicum* (Diptera: Anthomyiidae), adult females landed about twice as often on brassica plants growing in bare soil when compared with plants surrounded by other vegetation (Kostal & Finch 1994). In contrast, the southern corn rootworm, *Diabrotica undecimpunctata howardi* (Coleoptera: Chrysomelidae), laid more eggs in the soil around corn seedlings when broadleaved weeds were present rather than bare soil (Brust & House 1990). It is rather hard therefore to make broad generalizations about plant spacing and its ability to reduce the likelihood of pest problems. The best strategy would probably be to identify the most serious pest species in a crop or region of planting and follow specific planting guidelines.

The use, or some would say, misuse, of crop monocultures has been at the forefront of debate in pest management for a very long time. There can be problems with crop mixtures. They may produce decreased yields when compared with monocultures, because of competition for resources between the two plant types (Hernandez-Romero *et al.* 1984), and because of different ripening times, or the need for different harvesting technologies, the use of mixtures may be commercially impractical. The links between ecosystem diversity and population stability are tenuous, but suffice it to say that in terms of probability of pest attack, extreme monocultures must be considered to be more of a risk.

The mechanisms behind this observation vary from situation to situation. Host plant location may be enhanced in a monoculture via either physical or chemical means. Early work illustrated in Fig. 10.8 demonstrates how the damage caused to Douglas fir by the spruce budworm, *Choristoneura fumiferana* (Lepidoptera: Tortricidae), rises dramatically as this preferred host constitutes more and more of the stand (Fauss & Pierce 1969). In this case, it seems that the larvae, which disperse from high densities of their siblings to neighbouring trees, are more likely to find an equally suitable host, free from competition from other larvae, in single species stands. Host location by pests using chemical means may be reduced in mixtures. In line with the conventional wisdom of generations of gardeners, mixing carrots with onions reduces attacks by carrot fly, *Psila rosae* (Diptera: Psilidae), compared with monoculture carrots, probably because of the deterrent effect of onion volatile chemicals masking the more subtle odour of carrots (Uvah & Coaker 1984). Finally, pests may stay on mixed crops for less time than in monocultures. One of the most serious pests of broccoli and other brassicas in many countries are flea beetles in the genus *Phyllotreta* (Coleoptera: Chrysomelidae), whose larvae and adults can seriously defoliate the crop. Garcia and Altieri (1992) found substantially lower flea beetle densities on broccoli crops mixed with beans (which *Pyllotreta* does not eat), when compared with the numbers on broccoli monocultures. In this case, it seems that beetles tended to remain for less time in the mixed crop habitat, and more adults flew away from it.

Two or more entirely different crop plants intercropped together may reduce the damage to one or both species. In Kenya, for example, Skovgard and Pats (1997) intercropped maize with cowpea (a cereal with a legume), and were able to reduce the number of maize stem-boring Lepidoptera by 15–25%, with an increase in maize yield of 27–57%. Such a tactic will, of course, succeed only if the grower is able to harvest both crops easily and independently—it is unlikely to work in large-scale arable systems. The

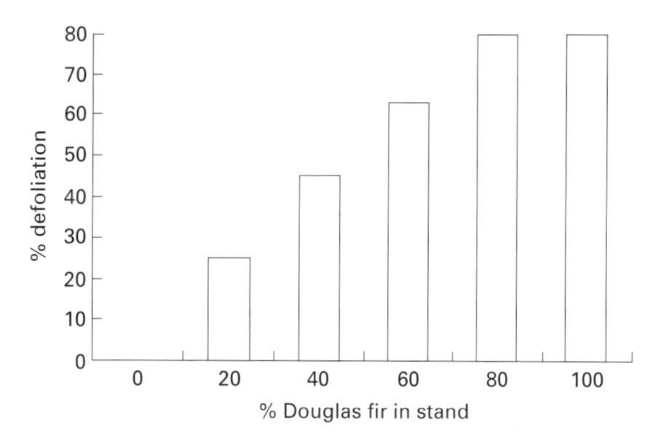

Fig. 10.8 Defoliation of Douglas fir by the spruce budworm *Choristoneura fumiferana* in mixed stands of trees, according to the percentage of Douglas fir in the mixture. (From Fauss & Pierce 1969.)

second crop species in a mixture may not in itself be of great commercial value. The under-sowing of crops such as leeks with clover can produce drastic reductions in pests such as onion thrips, *Thrips tabaci* (Thysanoptera) (Theunisse & Schelling 1996), and trap crops can be used to attract insect pests away from the more important target species. For example, alfalfa strips in cotton fields retain lygus bugs, *Lygus hesperus* (Hemiptera: Miridae), keeping them out of the cotton crop (Godfrey & Leigh 1994). If the alfalfa is cut on a 28 day schedule, sufficient bugs are removed from the cotton fields to substantially reduce subsequent damage. In fact, the trap crop need not be a different species from the commercial one. Other trials in cotton crops have shown that early planting of cotton around later sown crop cotton attracts many thousands of cotton boll weevil, *Anthonomus grandis* (Coleoptera: Curculionidae) to the trap, which can then be destroyed, thus removing a great number of pests from the fields before the main crop becomes susceptible (Moore & Watson 1987).

Another type of crop mixture involves just one commercial plant species, but this time, associated with non-crop, wild vegetation. In many instances, natural enemies such as parasitic wasps (see below) rely on sources of pollen and nectar to provide energy for the adults. The establishment or maintenance of wild flowers (so-called 'companion' plants) in and especially around, crops may enhance biological pest control (Bowie *et al.* 1995). It is unlikely that most growers will readily tolerate weeds and flowers within their intensively grown crops, but field boundaries can be suitable. So, sowing certain, but not all, species of wild flower in cabbage fields and adjacent habitats increased the fecundity and survival of parasitoids of the diamondback moth, *Plutella xylostella* (Lepidoptera: Yponomeutidae), in North America (Idris & Grafius 1995), and the numbers of adult hover flies (Diptera: Syrphidae), in New Zealand (White *et al.* 1995), the larvae of which prey on aphids. It must be borne in mind, however, that it is one thing to augment the numbers of natural enemies around a crop, but quite another to actually achieve significant reductions in crop losses caused by insect attack (see section 10.5).

10.4.4 Growing crops

Once an annual crop has established, there is little opportunity for ecological manipulation, and further pest management takes the form of biological or chemical control, or a combination of both. However, in a perennial system, such as plantation forestry, further manipulations can be carried out, such as stand thinning. This is a common enough silvicultural practice, where a number of trees are removed from a growing stand so that those that remain are subject to significantly less competition for light and other resources, and hence grow more vigorously. Once again, the concept of vigorous trees being less susceptible to insect attack has to be invoked; many, though not all, judicious thinning programmes reduce the problems from forest pests such as bark beetles or ambrosia beetles (Coleoptera: Scolytidae & Platypodidae) (Speight & Wainhouse 1989). There is no doubt that thinning can dramatically enhance the performance of trees. In ponderosa pine stands in California, the growth of trees in a thinned stand 5 years after thinning was five times greater than in a control plot where no thinning had taken place (Fiddler *et al.* 1989). This increase in vigour would certainly be expected to reduce the susceptibility of trees to insect attack. In loblolly pine in the USA, southern pine beetle, *Dendroctonus ponderosae* (Coleoptera: Scolytidae), attacks were significantly lower in thinned stands (Brown *et al.* 1987), probably because of the thinning effects on the biochemical and physical characteristics of the trees (Matson *et al.* 1987).

10.4.5 Senescing crops

Senescing foliage provides increased quantities of organic nitrogen, which some insects are able to utilize, thus enhancing their performance. This is rarely a problem that reaches economic proportions, but as trees become older, for example, their growth rate declines, until a stage called over-maturity is reached. Once more, low vigour now renders them susceptible to secondary insect attack, and silvicultural practices need to ensure that commercial crops are harvested before this stage ensues. A good example comes from Sabah in NE Borneo, where *Acacia mangium* trees were found to be severely attacked by various borers such as longhorn beetles (Coleoptera: Cerambycidae) and ambrosia beetles (Coleoptera: Platypodidae) (M.R. Speight, unpublished observations). The age at which such trees should be harvested to avoid their becoming overmature and susceptible to attack is as little as 8 years;

the trees in question were found to be at least 14 years old, and should have been felled and processed years earlier.

10.4.6 Post-harvest

Insect pest problems do not necessarily cease once the crop has been harvested. For example, felled trees must be removed from commercial forest stands within a few weeks of spring cutting in the UK to prevent bark beetle populations escalating in the plantations, and treatments such as bark peeling or constant water spraying may have to be employed to protect logs from beetle attack before they are processed. This last strategy may have to continue for several years if large quantities of timber are suddenly made available by large-scale sanitation felling after disasters such as forest fire (F.R. Wylie, unpublished observations). In agriculture, storing products such as grain can be very problematic. A variety of pests are notorious destroyers of stored products all over the world. In the UK, the grain beetle, *Oryzaephilus surinamensis* (Coleoptera: Cucujidae) (Fig. 10.9), can proliferate very quickly indeed in grain silos where inadequate drying of the wheat or barley has been carried out before storage. At moisture contents of over 30% or so, the beetles aggregate and breed in 'hot spots' in the silos, where their larvae tunnel into the grain, destroying it. Usually, some form of insecticidal control has to be administered, though controlled atmospheres such as elevated carbon dioxide levels may be established in the storage containers (Alagusundaram *et al.* 1995).

Fig. 10.9 Grain beetle in silo of barley.

10.5 Biological control

To distinguish it from the multifarious techniques of crop ecosystem management, we define biological control as the use of natural enemies of insect pests to reduce the latter's population densities. In most cases, it is hoped that these reduced densities will be maintained (i.e. regulated) through time without further manipulation (see Chapter 5). Therefore, it may be considered, perhaps pedantically, that the term biological 'control' is a misnomer, as what we are in fact looking for in most cases is biological 'regulation', a fundamentally density-dependent process. Over the last century or so, biological control has had some rousing successes, but it has also suffered many failures. It is vital to appreciate the limitations of the various types of biological control, to place them in the perspective of other pest management strategies, in particular, as part of IPM.

10.5.1 Limitations of biological control

Many types of natural enemies of insect pests exist, each with their own advantages and disadvantages; Table 10.6 summarizes the most important ones. It must be pointed out that the third column in the table, which indicates the likely success of the various enemy types, is a very general view, and that for nearly all the categories there are examples where biological control has been successful (see below). However, one fundamental facet of biological control is the ability of a particular natural enemy, or group of enemies, to respond to differing densities of the pest's population. Specifically, the rate (or per cent) mortality caused by a natural enemy has to increase as pest density increases if the biological control agent is to regulate the pest population.

A parasitoid (Fig. 10.10) is a special type of parasitic insect where the larva of a wasp or fly consumes the body of its host so that the host dies within one generation of the parasitoid. We have discussed the population dynamics of predator–prey and parasitoid–host interactions in detail in Chapter 5, and reference to this will help to explain why parasitoids and particularly predators have a tendency to be unable to regulate pest populations when the latter reach high, epidemic levels. In the absence of any other mortality-causing factor, food limitation usually brings about pest population decline; such pests are known as resource limited. Only pathogens

Table 10.6 Major types of natural enemies of insect pests that may have a role in biological control.

Type of enemy	Commonly attacked pests	Likely efficiency in biological control
Vertebrates		
Mammals (mice, voles)	Soil larvae and pupae of defoliators	Low
Birds	Defoliating larvae of moths and sawflies	Low
Amphibia (frogs)	Soil larvae	Low
Fish	Aquatic insect larvae	Low
Invertebrates		
Spiders	Small flying or crawling pests, e.g. aphids	Low to medium
Predatory insects (beetles, bugs, lacewings, hoverflies, ants)	Aphids, lepidopterans, dipterans and sawflies, soil larvae and pupae	Low to medium
Parasitic insects (wasps and flies)	All pest types, all life stages	Medium to high
Pathogens		
Bacteria Fungi Nematodes Protozoa Viruses	Most pest types, except those permanently concealed within the host plant	Medium to very high

Fig. 10.10 Adult female parasitoid (*Rhyssa* spp.) ovipositing on woodwasp larva in tree trunk.

such as viruses seem to have no upper limit to their ability to reduce very high insect pest numbers (see Fig. 1.15, Chapter 1). A further problem is the fact that many predators, parasitoids and even pathogens have their own natural enemies, so-called higher-order predators or hyper-parasitoids. Natural enemies of herbivorous insects are often not top of their food chain, and it is likely that in some cases, their ability to regulate pest densities effectively is constrained by the action of trophic levels above them (Rosenheim 1998).

As food limitation is so important in influencing pest population densities, any consideration of a biological control programme must also include a careful appraisal of the ways in which the pest responds to its host plant. This is especially important when that plant is presented in an intensively grown crop system, or where the environment of the host tends to promote the herbivore above the carnivore, such as when the host is non-vigorous. The failure of many biological control attempts may at least in part be due to a lack of understanding of these basic relationships.

10.5.2 Biological control manipulations

Biological control involves various basic manipulatory techniques, which can be employed to maximize the efficiencies of natural enemies. These techniques vary according to the type of pest, its origin, and the nature of the crop ecosystem or husbandry system.

The control of an introduced pest by its introduced natural enemy is known as *classical* biological control, mainly because of early success stories such as the cottony cushion scale (see below). Natural enemies may be introduced from elsewhere to control native pests whose own complex of predators, parasitoids

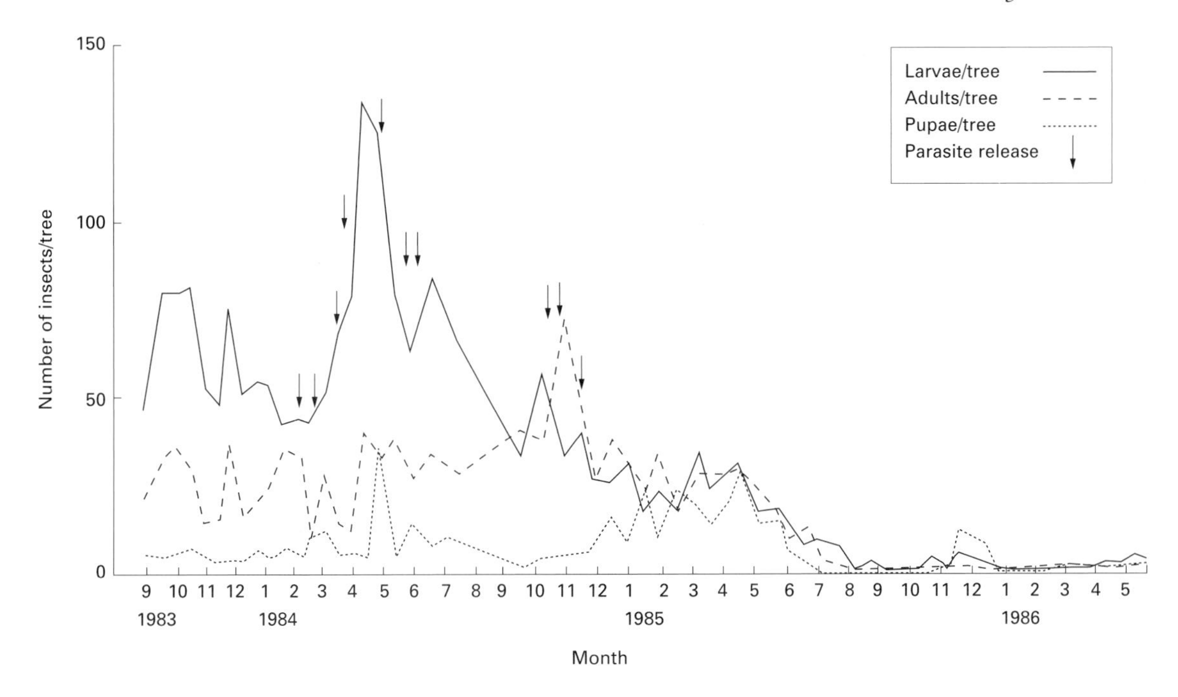

Fig. 10.11 Population densities of coconut leaf beetle in Taiwan, before and after release of the parasitoid *Tetrastichus brontispae*. (From Chang 1991.)

and pathogens fail to do the job properly. This may occur by chance, and is called *fortuitous* control, or by design where, for example, a virus disease is moved from one region or country to another to establish it in an area where the pest is active but the pathogen is absent (see below). Native or exotic natural enemies may be encouraged in crop systems by the provision of suitable habitats or alternative hosts. This is *conservation*, and the background populations of enemies may be topped up by limited introductions of more of the same species in *augmentation*. Finally, large numbers of usually exotic enemies may be released into a pest-ridden crop; this is *inundation*, a good example of which takes place routinely in some glasshouse systems.

10.5.3 Biological control using predators and parasitoids

Predators and parasitoids certainly can exert heavy impacts on pest insects, and, as mentioned above, classical biological control involves large-scale releases of enemies into crops to regulate pest populations in the future. A good example of this comes from Taiwan, where the coconut leaf beetle, *Brontispa longissima* (Coleoptera: Hispidae), causes severe yield losses to coconuts (Chang 1991). The larvae and adults of the beetle feed on developing leaflets in the central leaf bud of coconut palms, gnawing long incisions in the tissues, thus causing severe dieback when the leaves expand. In 1983, a parasitic wasp, *Tetrastichus brontispae* (Hymenoptera: Eulophidae), was introduced into Taiwan from Guam, and Fig. 10.11 shows the suppression of coconut leaf beetle populations following successive releases of the parasitoid. By 1991, about 980 000 wasps had been released, preventing the potential destruction of around 80 000 coconut palms.

One of the first really good examples of a successful biological control programme, and a blueprint for many others to follow, is the oft-quoted story of the cottony cushion scale, *Icerya purchasi* (Homoptera: Margarodidae) (Plate 10.3, opposite p. 158), in citrus orchards in California in the 19th century. The pest was accidentally introduced into California in 1868 from Australia on acacia plants, and within 20 years, the scale was seriously threatening the citrus industry, with no obvious way of controlling it. So, a search was made for natural enemies in the pest's native home, where eventually two species were discovered that seemed to hold the scale in check. The vedalia ladybird, *Rodolia cardinalis* (Coleoptera: Coccinellidae), and a parasitic fly, *Cryptochaetum iceryae* (Diptera: Tachinidae) were released into Californian citrus groves. Eighteen months later, the cottony cushion scale was reduced to a non-economically threatening level all over the area (Huffaker 1971). In this case, the pest's biology acted in favour of a successful programme. Scale insects such as *Icerya*, although having serious impacts on their host plant, do not reproduce quickly and tend to disperse from tree to tree rather poorly, giving their predators opportunities to dominate.

Since those days, biological control programmes have been set up all over the temperate and tropical world. One modern example mimics that of the cottony cushion scale from a century earlier rather well, and involves a pest called the cassava mealy bug, *Phenacoccus manihoti* (Homoptera: Pseudococcidae) (Fig. 10.12). Table 10.7 describes the programme. The native home of the mealy bug is South America, and like so many other exotic pests, it was introduced accidentally into Africa in the late 1960s or early 1970s (Herren 1990). The sap-feeding mealy bug causes stunting of the growing shoots of cassava. Peak densities of the pest vary a great deal, from 600 to 37 000 bugs per plant (Schulthess *et al.* 1991), an enormous infestation at the upper end of the scale. Yield losses to the crop itself, the cassava tubers, caused by the mealy bug were of the order of 52–58% when compared with non-infested plants (Sculthess *et al.* 1991). In 1981, the South American parasitic wasp, *Epidinocarsis lopezi* (Hymenoptera: Encyrtidae), was imported into Nigeria, where it was reared in an insectary to bulk up numbers before being released into cassava crops. Further releases were carried out, so that by 1985, the parasitoid was established over 420 000 km² in West Africa and 210 000 km² in Central Africa (Herren *et al.* 1987). After this establishment, densities of cassava mealy bug were very much reduced. In Chad, for example, average pest numbers were around 1.6 per shoot (Neuenschwander *et al.* 1990), and the biological control programme resulted in yield increases of around 2.5 tonnes/ha in savanna regions of West Africa (Neuenschwander *et al.* 1989). It has to be said that some workers feel that the introduction of a single exotic parasitoid species into a huge region of ecological diversity typified by the African cassava growing region may not solve the pest problem entirely (Fabres & Nenon 1997), and 20 years or so after the first parasitoid introductions, this biological control is not as efficient as had been expected.

Another successful introduction of an exotic parasitoid is the biological control programme of the California red scale, *Aonidiella aurantii* (Hemiptera: Diaspididae). This sap feeder is a world-wide pest of citrus (Murdoch *et al.* 1996), including oranges, lemons and grapefruits.The scale insect attacks all aerial parts of the tree, including twigs, leaves, branches and fruit (Plate 10.4, opposite p. 158).

Fig. 10.12 Cassava mealy bug.

Table 10.7 Biological control of the cassava mealy bug, *Phenacoccus manihoti* (Homoptera: Pseudococcidae) by *Epidinocarsis lopezi*. (Hymenoptera: Encrytidae) in Africa. (From Herren *et al.* 1987; Herren 1990; Dent 1991.)

Pest spread	
1973	Congo–Zaire river
1976	Gambia
1979	Nigeria and Benin
1985	Sierra Leone and Malawi
1986	25 countries (70% of African cassava belt)
Parasitoid spread	
1981	Introduced to Nigeria from Paraguay
1983	Recovered from all samples within 100 km of release site
1985	Over 50 releases in 12 Africa countries
1986	Established in 16 countries, covering 750 000 km²
Result	
Pest numbers reduced by 20–30 times, saving around 25 tonnes/ha in savannah region	

Severe infestations cause leaf yellowing and twig dieback, and infested fruit becomes seriously devalued (University of California 1998). One parasitoid wasp in particular has successfully suppressed red scale in many regions including California for years. *Aphytis melinus* (Hymenoptera: Aphelinidae) was originally found in India and Pakistan (De Bach 1964), but is now firmly established in the USA. The adult female parasitoid lays eggs under the armoured cover of the scale insect, where the resulting larvae eat the pest. In addition, *Aphytis* frequently kills scales without parasitizing them, by probing their bodies with its ovipositor. This, coupled with the fact that the parasitoid has about three generations for every one of the scale (Murdoch *et al.* 1995) makes it a particularly efficient biological control agent.

Curiously, though the interaction between parasitoid and host appears to be stable, in that populations persist over long periods of time, fluctuations in densities are limited, and no trends in mean population density can be seen (Murdoch *et al.* 1995), there is no concrete evidence of a density-dependent response to changing pest densities in the parasitoid (Reeve & Murdoch 1986). Clearly, the suppression of the pest is efficient and dependable, as Table 10.8 shows. When the scale is unable to hide from attacks by the parasitoid by sheltering in bark crevice 'refuges', the mean percentage parasitism reaches over 19% (though it can peak at over 30% on second instar male scales). Pest densities are drastically reduced, and the numbers of first instar nymphs that mature to reproduce (the recruitment rate) is reduced by more than four times in regions of the tree exposed to parasitism (Murdoch *et al.* 1995). *Aphytis* is available commercially from specialist suppliers in the USA for around US$30 per 10 000 insects (1998 figures). They can be released into citrus groves in an inundative fashion at a rate of between 2500 and 5000 per ha, preferably just before peak male scale flight so that virgin female pests are parasitized. Growers can then monitor levels of parasitism in their crops as the season progresses, according to prescribed guidelines (University of California 1998).

Table 10.8 Numbers of California red scale, recruitment rate per female (measured as (number of first instar)/(number of crawler-producing female adults)), and per cent parasitism by *Aphytis melinus*, in a grapefruit grove in California, within and outside bark refuges. (From Murdoch *et al.* 1995.)

	Exterior	Refuge
Number of scale/100 cm^2	1.9±0.3 SE	189.6±19.31 SE
Recruitment/female	9.56±1.36 SE	2.22±0.21 SE
% Parasitism	19.2±1.0 SE	2.3±0.3 SE

An example of successful biological control in modern times, which employs predators rather than parasitoids, concerns the great spruce bark beetle, *Dendroctonus micans* (Coleoptera: Scolytidae), in the UK (Figs 10.13 & 10.14). Unlike most species of bark beetle, whose larvae excavate solitary tunnels away from their maternal gallery in the bark cambium of

Fig. 10.13 Spruce trunk attacked by *Dendroctonus micans*, showing copious resin exudations. (©UK Forestry Commission.)

Fig. 10.14 Part of larval brood of *Dendroctonus micans* under spruce bark. (©UK Forestry Commission.)

infested trees, and who are almost always only able to attack non-vigorous or moribund trees and logs, *D. micans* larvae exhibit aggregative behaviour. All the larvae from one brood live together in a chamber under the bark of spruce trees. They take part in communal feeding, which enables them to overcome the tree's resin defences, and hence fairly vigorous trees can be attacked and killed (Evans & Fielding 1994). In continental Europe, *D. micans* has been an extremely serious pest of spruce for many years (Legrand 1993), and it was first discovered in the UK in 1982 (Evans & Fielding 1994), having been accidentally introduced on imported timber, it is assumed, some years previously. In 1983, a predatory beetle, *Rhizophagus grandis* (Coleoptera: Rhizophagidae), was introduced into infested spruce stands in Wales and the west of England from Belgium (Fielding *et al.* 1991; King *et al.* 1991). The predators are extremely efficient at locating prey eggs and larvae under the bark of attacked trees, using the volatile compounds in the pests' frass as an attractant (Grégoire *et al.* 1991). Because *D. micans* larvae aggregate in one brood chamber, once a female *R. grandis* has located one prey larva, she is able to lay sufficient eggs to produce larvae to destroy the whole pest brood. *D. micans* has a relatively long generation time when compared with *R. grandis*, and the pest disperses very slowly by virtue of the adults' high flight threshold temperature (see Chapter 2). There is now strong evidence that *R. grandis* has established well in the UK, and is an important factor in the regulation of *D. micans* (Fig. 10.15a,b) (Evans & Fielding 1994), and now performs a key role in the integrated pest management of the pest.

Unfortunately, not all biological programmes are as successful as these examples. Indeed, it seems in some cases that the actions of predators and parasitoids make little or no difference to the course of insect pest outbreaks. Some insect pest populations simply do not respond to the actions of their natural enemies in a manner useful for biological control. We discussed the green spruce aphid, *Elatobium abietinum* (Hemiptera: Aphididae), earlier in this chapter. This species can cause serious defoliation and increment loss to spruce trees in Western Europe, and as Chapter 2 describes, one of the most important mortality factors is late winter temperatures. The decline in spring populations appears to be a function of the production of alatae (winged adults) in response to changes in the host quality and to photoperiod (Leather & Owuor 1996). Voracious predators do abound, however. The two-spot ladybird, *Adalia bipunctata* (Coleoptera: Coccinellidae), has both larvae and adults that consume spruce aphids, but studies have shown that they fail to make an appreciable impact on the pest population density because of their slow numerical response (see Chapter 5) (Leather & Owuor 1996).

Even when high rates of parasitism are recorded in a pest population, reports of which can be found throughout the scientific literature, the pest may still be a serious problem if those individuals that survive parasitism are capable of causing serious damage. As

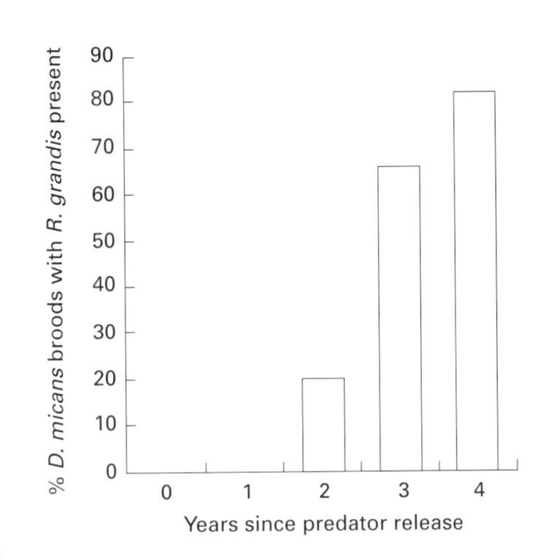

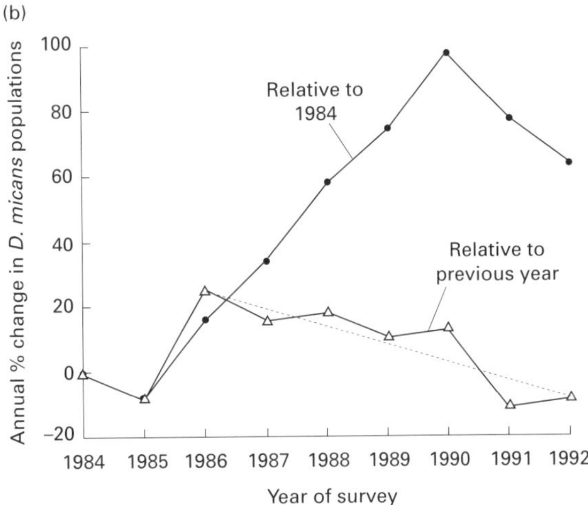

Fig. 10.15 Biological control of the spruce beetle *Dendroctonus micans* showing (a) an increase in the population density of the predatory beetle *Rhizophagus grandis*, and (b) the decline in pest numbers over 5 years after the predator was first released. (From Evans & Fielding 1994.)

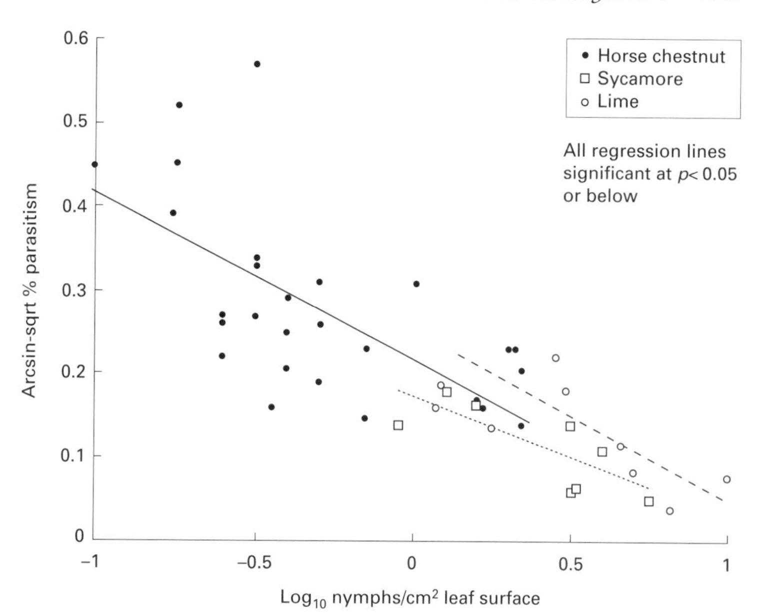

Fig. 10.16 Parasitism of horse-chestnut scale *Pulvinaria regalis* by *Coccophagus obscurus*, showing inverse density dependence. (M.R. Speight, unpublished data.)

mentioned above, the system hinges on the pests' biology as well as that of the enemy, and if there are other mortality factors that dominate the life of the pest, such as food plant quantity or quality, or abiotic mortality factors such as climate or dispersal losses, then the enemies may merely track the population fluctuations of the pest, and not regulate it. Let us take, for example, the case of the horse-chestnut scale, *Pulvinaria regalis* (Homoptera: Coccidae), another relatively new species to the UK (and the rest of Europe too, in fact). As described in section 10.3.1, this species feeds as a young nymph on leaves of a variety of urban trees, causing considerable growth losses (Speight 1992). Though *P. regalis* arrived in the UK only in the 1970s, at least one parasitoid species, *Coccophagus obscurus* (Hymenoptera: Chalcidae), has swapped from a native host, in this case an indigenous scale insect feeding on ivy. The parasitoid is now frequently found attacking young *P. regalis* nymphs, but its response to the density of the herbivore is inversely density dependent, and thus non-regulatory (Fig. 10.16). Rates of parasitism actually decline as pest density increases. It seems that the most important mortality factor in the life of the pest is the often huge losses of first instar crawlers as they migrate on the wind from tree to tree down town streets.

Even when inundative releases are carried out, biological control may not work adequately. One example of this is the attempts to control the sap-feeding bug, *Lygus hesperus* (Hemiptera: Miridae), a pest in commercial strawberry fields in the USA, using repeated releases of the egg parasitoid *Anaphes iole* (Hymenoptera: Mymaridae). Here, following the release of a staggering 37 000 adult parasitoids per hectare of strawberries, 50% of *Lygus* eggs were found to be parasitized. This resulted in a 43% reduction in the numbers of nymphs of the pest, and a 22% reduction in the amount of fruit damaged (Norton & Welter 1996). Whether or not the huge inundation programme for parasitoid release merits the rather small reduction in damage in the final analysis will only be revealed by detailed cost–benefit analysis.

Natural enemies may be abundant in a crop ecosystem, but unfortunately this does not mean that they will automatically have a role to play in efficient biological control. Polyphagous predators such as beetles and spiders abound in British cereal fields in summer, but their influence on cereal aphids seems from some work at least to be negligible. Holland *et al.* (1996) excluded predators from plots in wheat fields, and recorded a reduction in their activity of up to 85% when compared with control areas. However, no difference was found in the number of grain aphids, *Sitobion avenae* (Hemiptera: Aphididae) between the two sites. Aphid pests of wheat such as *S. avenae* and *Rhopalosiphum padi* have an enormous potential for increase, and can reach peak populations in a very short time (Ohnesorge 1994). Many polyphagous predators such as ladybirds can certainly build up to large numbers, but this usually occurs after aphids have peaked and crashed again (Pankanin-Franczyk & Ceryngier 1995). The overall results are serious yield losses to wheat, and epidemics of coccinellids in late summer who are so hungry that they are anecdotally reported to bite anything including babies in prams!

The habitat of the pest as well as its relationships with its host plant may influence the potential success of a biocontrol programme. Many pests are in some way concealed, by living in shoots, stems, galls, mines, roots or bark (see Table 10.19, p. 278), and though there are examples of such pests being regulated successfully by natural enemies (see *Dendroctonus micans* above), there are many occasions where parasitoids and predators simply are unable to attack enough pests in a given time to affect the latter's density significantly. In general terms, it is thought that the success of biological control is inversely related to the proportion of insect pests that are in some way protected from enemy attack, i.e. they inhabit some sort of refuge (Hawkins *et al.* 1993). The more concealed a pest, the less likely it might be thought that biological control will work satisfactorily. Once again, this is not to say that borers can never be controlled biologically, but there is a conventional wisdom suggesting that when ecological factors such as food, climate, refuges and so on are able to promote the proliferation of the pest, then we should perhaps look elsewhere for primary pest management systems. We can then retain biological control by predators and parasitoids as a back-up measure within an IPM system (see below).

One final consideration hinges on whether or not it is better to introduce more than one natural enemy at a time to improve the chances of efficient biological control. Specialist parasitoids, such as an egg parasitoid and a later larval parasitoid might well be expected to augment the action of each other, but a problem certainly arises when predators such as ladybirds are introduced into a system already utilizing parasitoids. Experiments have shown that this type of predator will readily prey on parasitized aphids as well as healthy ones (in fact, the former may be easy to catch), and the effects of multiple natural enemies in this scenario are clearly non-additive (Fergusen & Stiling 1996).

10.5.4 Biological control using pathogens

For a good many years now, pathogenic organisms have been growing in popularity as alternative biocontrol agents to parasitoids and predators, and also as alternatives to synthetic chemical insecticides. There are a variety of organisms that fall under the general heading of 'pathogens', summarized by Table 10.9. The use of these so-called microbiological insecticides, both natural and genetically engineered forms, is at the forefront of insect pest management in many parts of the world. Several excellent textbooks are devoted to these topics, such as those by Entwistle *et al.* (1993) and Miller (1997). For the sake of brevity, we consider only two major groups of pathogens here, the bacteria and the viruses.

Table 10.9 Summary of types of pathogenic organism used in the biological control of insect pests.

	Commercial or experimental example				
Pathogen type	Pathogen	Insect	Common name	System	Reference
Bacterium	*Bacillus thuringiensis*	*Plutella xylostella*	Diamondback moth	Arable–brassicas	McGaughey (1994)
Fungus	*Verticillium lecanii*	*Myzus persicae*	Peach–potato aphid	Glasshouses	Milner and Lutton (1986)
Nematode	*Nosema locustae*	*Locusta migratoria*	Locust	Grassland	Li *et al.* (1994)
Protozoan–Microspora	*Vairimorpha necatrix*	Lepidoptera: Noctuidae	Leatherjackets	Various	Maddox (1987)
Virus	*Ns* NPV	*Neodiprion sertifer*	European pine sawfly	Pine plantations	Doyle and Entwistle (1988)

Bacteria

Bacteria of use in pest management belong to a mere three species, *Bacillus popilliae*, *Bacillus thuringiensis* and *Bacillus sphaericus* (see Table 10.10, below) (Payne 1988). *B. popilliae* is a bacterium that causes milky disease in numerous species of scarab beetles all over the world (Cherry & Klein 1997), and is used, not always successfully, to control turf pests such as Japanese chafer beetle larvae (Coleoptera: Scarabaeidae) (Redmond & Potter 1995). Unlike *B. popilliae*, *B. thuringiensis* and *B. sphaericus* produce toxins that have insecticidal activity. *B. sphaericus* is able to grow saprophytically in polluted water (Payne 1988), and is showing promise as a self-replicating microbial control for disease vectors such as mosquitoes (Becker *et al.* 1995). *B. thuringiensis* is, however, the most widespread bacterial pesticide, and it is available in a whole variety of strains and commercial formulations all over the world. *Bacillus thuringiensis*, or *B.t.* for short, is a spore-forming pathogen, which is easy to grow on many media, including cheap waste products of the fish and food processing industries (Payne 1988). It was discovered as far back as 1911, and it was later found that the bacterial spores were accompanied by a proteinaceous toxic crystal, which had strong insecticidal properties (Burges 1993). This so-called δ-endotoxin is the product of a single bacterial gene, which has great potential for breeding and selection for specific targets, and for genetic engineering. Normally, it is the δ-endotoxin itself that forms the active ingredient in commercial formulations of *B.t.*, so that the pathogen is unable to replicate in the environment and is instead a true biological insecticide. In all cases, the toxin has to be ingested by the host. Unlike fungi, *B.t.* cannot enter the body through the cuticle or spiracles. So, any insect that feeds in a concealed manner, even if only its mouthparts are concealed as in the case of sap feeders, will be unable to pick up *B.t.*, and is hence effectively immune to its effects. For example, the only way to control a lepidopteran larva such as a shoot or fruit borer with *B.t.* would be to ensure that the very young larvae hatching from eggs on the outside of the plant consumed the pathogen before entering the plant tissues. Timing of application for this type of pest is therefore critical (see section 10.6.1). Nowadays, several distinct varieties or subspecies of *B.t.* are available (and even different varieties have different serotypes!), with different properties and target insect groups (Table 10.10). *B.t. kurstaki* is still the most widely used (Table 10.11), though *B.t. israelensis* is being employed increasingly in vector control. Even then, the use of *B.t.* cannot begin to rival that of synthetic chemical insecticides, though in some major cases, it has rather taken over from more traditional compounds. For example, the spruce budworm, *Choristoneura fumiferana* (Lepidoptera: Tortricidae), is one of the most serious forest defoliators of firs and other conifers in the USA and Canada. Aerial spraying is a routine annual treatment for the pest over very wide areas; in fact, it used to be said that a larger area of forest was sprayed against spruce budworm every year in North America than the entire area of forest in all of the UK! With the changing attitudes to insecticides and the environment, *B.t.* has replaced synthetic compounds in the fight against budworm in many areas (Van Frankenhuyzen 1993) (Fig. 10.17).

One problem with *B.t.* is its rather broad spectrum of activity. Clearly, growers would prefer to use a chemical that will kill a whole range of pests, rather than having to select a specific compound for each insect species they wish to control. For the grower, this wide host range is good news, though for the conservationist, this may be less attractive. Furthermore, it would indeed be folly to use *B.t.* against crop pests in close proximity to a silk farm; silkworm larvae (Lepidoptera: Saturnidae) are eminently sus-

Table 10.10 Types of bacterial pesticides available. (From Payne 1988.)

Name	Target
Bacillus popilliae	Chafer grubs
Bacillus sphaericus	Mosquito larvae
Bacillus thuringiensis var. *kurstaki*	Lepidopteran larvae
Bacillus thuringiensis var. *israelensis*	Mosquito larvae
Bacillus thuringiensis var. *tenebrionis*	Beetle larvae

Table 10.11 Major global uses of *Bacillus thuringiensis* in 1990. (From Van Frankenhuyzen 1993.)

Global sales worth US$50 to US$80 million
60% subspecies *kurstaki* in agriculture (mainly North American vegetables)
20% subspecies *kurstaki* in forestry (North America and Europe)
20% subspecies *israelensis* in vector control (West Africa, especially in Onchocerciasis Control Programme; Germany and USA)

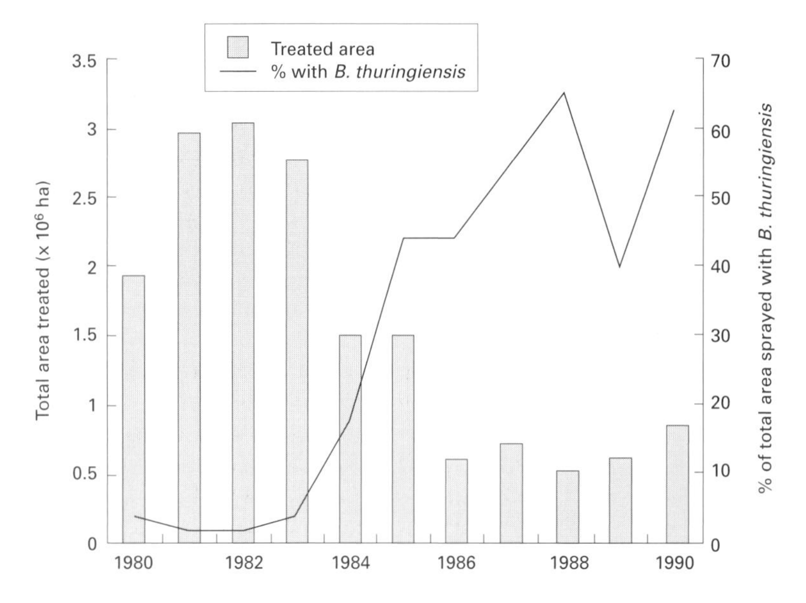

Fig. 10.17 Control of spruce budworm *Choristoneura fumiferana* in eastern Canada with *Bacillus thuringiensis*. (From Van Frankenhuyzen 1993).

ceptible to *B.t. kurstaki*, so the pathogen has to be used with great care and foresight. However, all strains of *B.t.* are entirely specific to arthropods, and hence have zero environmental impact on vertebrates, except perhaps indirectly by removing the food of certain insectivorous birds and mammals.

Summing up these pros and cons, *B.t.* is used in many parts of the world, and is even available in some countries, including the UK and Australia, as a garden insecticide for the domestic market. Perhaps its most elaborate use is in the production of transgenic insecticidal cultivars (TICs) of crop plants. This system addresses the problem mentioned above of controlling concealed pests, or pests feeding on external plant tissues that may for many reasons be out of reach of the spray. It is now routine to transform many species of plant so that foreign genes, such as the *B.t.* gene coding for the δ-endotoxin, are stably introduced into the plants' genome (Ely 1993). In 1996, for example, nearly 1×10^6 ha of transgenic cotton were planted in the USA (Gould 1998), and Table 10.12 lists some of the other common crops that are now available, either experimentally or even commercially, with these characteristics. The advantages of such systems are various. Even internal plant tissues are protected from insect damage, weather conditions such as rain do not effect the persistence of the toxin in the environment, timing of applications are no longer critical, and environmental impact to non-target but susceptible insects is much reduced, as only insects actually feeding on the genetically engineered plant will ever come into contact with the toxin (Ely 1993). Transgenic plants certainly influence the feeding and survival of insect herbivores. Poplar clones expressing a *B.t.*-δ-endotoxin gene significantly reduced the feeding and weight gain by gypsy moth, *Lymantria dispar* (Lepidoptera: Lymantriidae), larvae in experiments, and the forest tent caterpillar, *Malacosoma disstria* (Lepidoptera: Lasiocampidae), showed significantly increased mortality after feeding on the engineered plants (Robison *et al.* 1994). In trials with transgenic cotton expressing δ-endotoxin, between 70 and 87% protection was achieved against the cotton bollworm, *Helicoverpa zea* (Lepidoptera: Noctuidae) (Perlak *et al.* 1990). Not all results are so encouraging though. Field trials with transgenic corn have shown that some moth species, such as the cutworm *Agrotis ipsilon* (Lepidoptera: Noctuidae) showed no reduction in crop damage on the engineered corn when compared with normal corn (Pilcher *et al.* 1997).

In general then, the future for *B.t.* looks rather promising, but one major problem has already begun to arise, that of resistance in the insect pests to the bacterium. The potential for resistance to *B.t.* has been shown in at least five insect species (McGaughey 1994), with two moth species, the flour moth, *Plodia interpuntella* (Lepidoptera: Pyralidae), and the dia-

Table 10.12 Examples of crop species that have been stably transformed to express *Bacillus thuringiensis* (*B.t.*) δ-endotoxin. (From Ely 1993.)

Roots and tubers	*Crucifers*
Carrot	Lettuce
Potato	Kale
Sugarbeet	Cauliflower
	Celery
Fruiting vegetables	*Legume vegetables*
Aubergine	Pea
Cucumber	Soybean
Pepper	Bean
Tomato	
	Other crops
Fruit	Asparagus
Apple	Cotton
Papaya	Rape
Raspberry	Sunflower
Walnut	Tobacco
Plum	Mint
Currant	Alfalfa
Strawberry	
Orange	*Trees*
	Poplar
Cereals	
Maize	
Rice	

mondback moth, *Plutella xylostella* (Lepidoptera: Yponomeutidae), exhibiting widespread resistance. The use of *B.t.* in transgenic plants (TICs) may significantly increase the risk of resistance (Sanchis *et al.* 1995), because they are likely to select intensively for insect adaptation to the toxin involved (Tabashnik *et al.* 1991). At least four basic strategies have been suggested to delay the adaptation of pests to TICs. These are (a) the mixture of toxic and non-toxic cultivars in a field (the so-called refuge approach), (b) the use (or stacking) of two or more distinctly different toxins in one cultivar, (c) the use of low doses of toxins that act in conjunction with natural enemies, and (d) the insertion of toxins in plants so that only certain parts of the plant (such as buds or developing leaves) are toxic, or so that the toxin is expressed only at certain times of year (Gould 1998). One example of the use of multiple forms of the toxin, rather than the single δ-endotoxin (called CrylVD) normally employed comes from trials using *B.t. israelensis* against the mosquito *Culex quinquefasciatus* (Diptera: Culicidae) (Georghiou & Wirth 1997). It was found that resistance in the insect populations was much lower when *B.t.* lines were used that were selected for different forms of the toxin. Obviously, careful management using *B.t.* has to be designed, integrating it with other tactics, to prolong the useful life of *B.t.* as an environmentally benign bio-insecticide. Much development and investment is clearly required.

Viruses

Insects, like humans, suffer from a wide variety of virus diseases. By the end of the 1980s over 1600 virus isolates had been discovered attacking insects (Payne 1988). However, only one group, the Baculoviridae, are exclusively found in the arthropods, with no known vertebrate associations whatsoever (Plate 10.5, opposite p. 158). Baculoviruses are large rod-shaped DNA viruses and, though modern nomenclature changes rapidly, they have been split in the past into three basic types (Table 10.13). Both the nuclear polyhedrosis viruses (nucleopolyhedroviruses or NPVs) and the granulosis viruses (granuloviruses or GVs) have their virus rods, called virions, embedded in a proteinaceous sheath, the polyhedral inclusion body (PIB) (Fig. 10.18). In the NPVs, each PIB is relatively large, certainly visible under a compound microscope, and contains many virions, whereas GV PIBs are much smaller, and usually contain but one virion. Oryctes-type viruses (recently renamed 'unclassified') are a special type of baculovirus, which have so far only been found commonly in beetles such as the palm rhinoceros beetle, *Oryctes* spp. (Coleoptera: Scarabaeidae), hence the name.

The development of baculoviruses for the biological control of insect pests is still progressing, but it is hampered not so much by biology, ecology or technology, but more by public and even scientific distrust of the imagined danger of 'germ warfare' in our crops. None the less, insect pest populations can be regulated very efficiently by baculoviruses, and Table 10.14 provides some examples. Though these results look encouraging, merely to quote mortality figures does not necessarily imply satisfactory pest suppression. Indeed, though Podgwaite *et al.* (1992) found up to 98% reduction in gypsy moth egg masses after NPV was applied, differences in defoliation caused by the hatching generation of larvae between sprayed and control woodland areas were not significant. In

Table 10.13 Types of baculovirus.

Subgroup	Polyhedral inclusion bodies	
	Size	Number of virions
Nuclear polyhedrosis virus (NPV)	0.8–15 μm	Many
Granulosis virus (GV)	0.3–0.5 μm	Rarely more than one
Oryctes type	None	

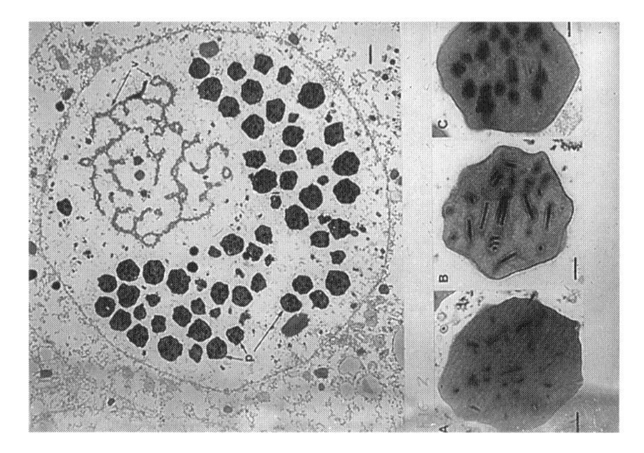

Fig. 10.18 Electron micrograph of nucleus of midgut cells of moth larva showing polyhedral inclusion bodies of NPV (also magnified PIBs with virions). (Courtesy of IEVM.) NPV, nuclear polyhedrosis virus; PIB, polyhedral inclusion body.

the case of pine sawfly in Canada, 78% mortality caused by NPV was not enough to control the pest outbreak (Olofsson 1988).

All baculoviruses so far employed are essentially natural entities, which occur in nature as a result of insect–pathogen coevolution. Pest managers are merely in the business of moving the viruses from one place to another, where for some reason the pathogen has not established itself. Natural epizootics of baculoviruses are regularly recorded, and various pest species seem to suffer natural declines as a result. For example, the European spruce sawfly, *Gilpinia hercyniae* (Hymentoptera: Diprionidae), in Canada; the teak defoliator, *Hyblaea puera* (Lepidoptera: Hyblaeidae), in India; and the yellow butterfly, *Eurema blanda* (Lepidoptera: Pieridae), in Borneo, are all known to commonly exhibit spectacular population declines because of NPV epizootics (Entwistle *et al.* 1983; Ahmed 1995; M.R. Speight, unpublished observations). Whether naturally erupting, or purposely applied, NPVs have a variety of characteristics that render them very suitable indeed for certain types of biological control of insect pests. Table 10.15 presents these characteristics. Their major advantages over other types of natural enemy hinge upon their ability to remain potentially infective for considerable periods of time outside the hosts' body, and their enormous replication rate once ingested by the usually very species-specific target. This replication rate is well exemplified by a control programme carried out in Colombia to manage defoliating Lepidoptera on cultivated palms. The moth *Sibine fusca* (Lepidoptera: Limacodidae) was successfully eradicated by aerial spraying of a viral solution containing the equivalent of a mere 10 dead caterpillars per hectare (Philippe *et al.* 1997)!

In Fig. 10.19 we illustrate the various pathways by which baculoviruses, NPVs in particular, are able to spread rapidly and independently from an epicentre of infection into the surrounding crops or habitat, where new hosts may encounter them, either immediately, or during subsequent generations of the pest. NPV spread can be via various routes (Richards *et al.* 1998). First, infected larvae themselves may transmit NPV to clean substrates such as leaves and stems as they move around. Larvae of the cabbage moth, *Mamestra brassicae* (Lepidoptera: Noctuidae), even when infected with a sublethal dose of NPV, live longer (Goulson & Cory 1995) and hence move further than non-infected ones, and their faeces and other body secretions may contaminate new leaf surfaces where more susceptible young larvae can pick up a fatal dose. Other moth larvae spread over much greater distances. Gypsy moth, *Lymantria dispar* (Lepidoptera: Lymantriidae), larvae balloon away from high-density patches of their peers. If these emigrants are dispersing from populations infected with NPV, then at least in the first few weeks of spread, this dispersal has been shown to be a good predictor of virus spread through a forest (Dwyer & Elkinton 1995). Other organisms may also help to disperse NPVs. During an outbreak of soybean looper, *Anticarsia gemmatalis* (Lepidoptera: Noctuidae), in southern USA into which a pest-specific NPV had been introduced, 80%, albeit from a small sample in this case, of bird droppings in the area contained viable NPV (Fuxa & Richter 1994). Because of zero vertebrate

Table 10.14 Examples of baculoviruses in the control of insect pests.

Insect	Common name	Crop	Country	Virus	Effect	Reference
Oryctes rhinocerus (Coleoptera: Scarabaeidae)	Coconut rhinoceros beetle	Coconut palms	Maldives, Sulawesi	Oryctes	Dramatically reduces adult life span	Zelazny *et al.* (1992)
Orcytes rhinocerus (Coleoptera: Scarabaeidae)	Coconut rhinoceros beetle	Coconut palms	India	Oryctes	Population suppression to low levels for at least 3 years	Mohan and Pillai (1993)
Pieris rapae (Lepidoptera: Pieridae)	Cabbage white butterfly	Cole	NE USA	GV	Approx. 30× fewer large larvae on treated crops; damage reduced by around 60%	Webb and Shelton (1991)
Chilo infuscatellus (Lepidoptera: Pyralidae)	Sugarcane shoot borer	Sugarcane	India	GV	Cost–benefit ratio of 1 : 2.5 against no control	Parameswaran *et al.* (1993)
Cydia pomonella (Lepidoptera: Tortricidae)	Codling moth	Apples	Europe	GV	Winter diapausing larvae significantly reduced after 2 years	Guillon and Biache (1995)
Euproctis chrysorrhoea (Lepidoptera: Lymantriidae)	Browntail moth	Hawthorn and bramble (public health pest)	UK	NPV	Peak larval mortalities over 75%	Speight *et al.* (1992)
Lymantria dispar (Lepidoptera: Lymantriidae)	Gypsy moth	Broadleaf woodlots	USA	NPV	Number of egg masses reduced by 80–98%	Podgwaite *et al.* (1992)
Neodiprion sertifer (Hymenoptera: Diprionidae)	European pine sawfly	Lodgepole pine	Canada	NPV	Larval mortality 78% after 2 years	Olofsson (1988)

Table 10.15 Characteristics of nuclear polyhedrosis viruses. (From Ivory & Speight 1994.)

Specificity	Complete within the Phylum Arthropoda; usually within families, genera, even species; no effects on vertebrates ever detected
Occurrence	Natural within insect populations, especially in indigenous areas (but wait for genetic engineering)
Life stage	Infect larvae; eggs, pupae and adults may rarely carry inactive virus
Mode of action	Infection from contaminated substrate by ingestion only; virus attacks nuclei of gut cells in sawfly larvae and of other cells (e.g. fat body) in Lepidoptera
Transfer between hosts	In faeces of infected but living larvae, from disintegrating dead bodies, in guts of predators (e.g. birds and beetles), maybe on parasitoid ovipositors
Persistence	Environmentally very stable except in UV light, persist outside host on leaves or bark from generation to generation of pest, or in soil for years; secondary epizootics may negate need for repeated applications
Efficiency	Very high in pest epidemics (ease of transfer of virus between hosts greatest in high pest densities); enormous replication potential
Cost	Cheap to produce and bulk-up by rearing host insect, or from field collections; complex technology not usually required beyond semi-purification
Drawbacks	Must be ingested; not plant systemic, limited or no use for sap feeders or borers High host specificity; often not cross-infective from one insect species to another May not be available from nature Do not kill immediately; may be weeks until infected pests die Must be applied in the same way (timing and technology) as an insecticide Public and government distrust; unfounded worries about safety; overtight rules

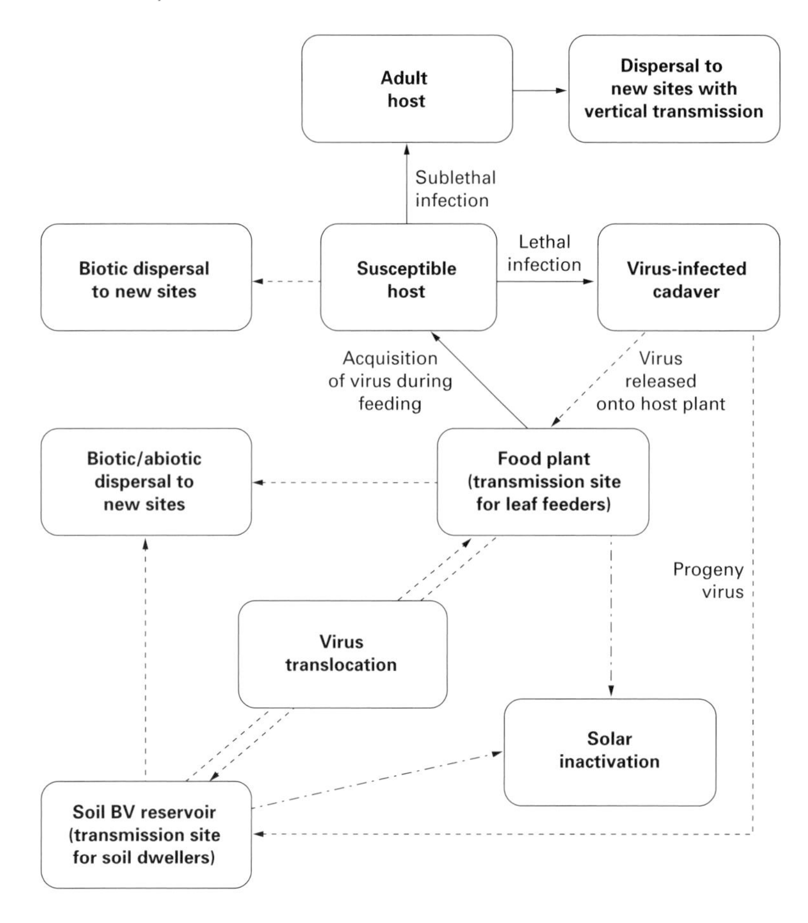

Fig. 10.19 The fate of lepidopteran baculoviruses in the environment. Continuous arrows, BV transmission routes; dashed arrows, BV dispersal routes; dashed and dotted arrows, BV inactivation. (From Richards *et al.* 1998. With permission, from the *Annual Review of Entomology*, ©1998, by Annual Reviews.)

toxicity, NPVs are able to pass right through a predator's body, causing no ill-effect to the animal, emerging in the faeces still effective, but some distance from where the infected larva was eaten. Even invertebrate enemies have been shown to be effective vectors of NPVs. Again in the case of *A. gemmatalis*, NPV from the pest population was found in at least six species of predatory hemipteran, one coleopteran, nine species of spider, plus one hymenopteran and two dipteran parasitoids (Fuxa *et al.* 1993). Finally, rain is well known to redistribute NPVs from soil reservoirs where they are protected from UV radiation, which damages them, back onto plants where insect pests may encounter them, through a phenomenon called rain-splash (D'Amico & Elkinton 1995) (see Fig. 10.19).

Though public and professional resistance to the commercial use of insect viruses exists in some regions, notably Europe and America, NPVs are used regularly, commercially and over large areas of cropland in other parts of the world. One of the best examples has already been alluded to briefly above, that of the soybean looper, *Anticarsia gemmatalis*. Figure 10.20 illustrates the magnitude of the project to control this extremely serious pest in Brazil using a species-specific NPV. Clearly, the system works most efficiently, and should be considered as a model for many other crop pest management scenarios in many parts of the world, including Europe, where environmentally benign control systems are required as alternatives to the large-scale use of chemical insecticides. However, ground swells of nervousness and distrust abound in governmental institutions, environmental movements and even the scientific community when it comes to the commercialization of baculoviruses. Like any other pesticide, legislation is very tight, and stringent safety tests are rightly demanded. Tests on mammals should include acute

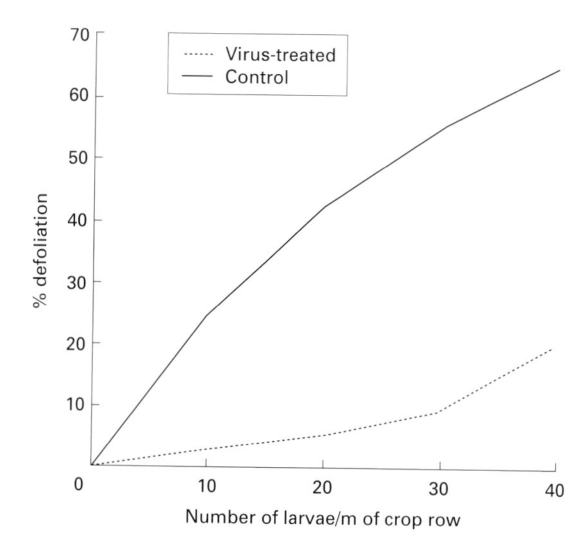

Fig. 10.20 Defoliation of soybean by velvet bean caterpillar *Anticarsia gemmatalis*, at varying population densities, with and without an application of NPV. (From Moscardi & Sosa-Gomez 1996.)

oral toxicity, dermal toxicity, eye irritation, and carcinogenicity. In practice, even detailed tests with NPVs such as the reaction of vertebrate embryos or the mutagenicity of human body cells have shown no responses (Liu *et al.* 1992). A final concern then must be with the vanishingly small probability that an NPV will mutate to become infective to vertebrates, including humans. DNA viruses, unlike those based on RNA such as influenza or HIV, are well known not to mutate at all easily, and hence the cause for concern can be considered negligible.

The next stage in the development of baculoviruses centres around the genetic engineering of natural NPVs, to diminish some of their less desirable properties, in particular, the time it takes for an infected insect pest to die from the disease. Unlike a chemical insecticide, which can normally be expected to kill the pest immediately on application, NPVs may take many days or even several weeks before the host dies and its body breaks down, releasing large numbers of PIBs back onto the crop. In that time, some feeding damage may still occur, so that in a valuable crop, this time delay may be unacceptable. In an effort to speed up the time to death after an NPV treatment, work has been carried out to genetically improve the NPV of the alfalfa looper, *Autographa californica* (Lepidoptera: Noctuidae), so that it is able to express an insect-selective toxic gene derived from scorpion venom (Cory *et al.* 1994). In field trials with the engineered (recombinant) NPV against the cabbage looper, *Trichoplusia ni* (Lepidoptera: Noctuidae), a relative of *A. californica*, larvae of the pest died significantly sooner than when treated with the natural (wild-type) virus (Fig. 10.21). Not only that, the virus yield from larval cadavers infected with the recombinant virus was significantly less than for the wild-type, showing its reduced virulence and hence relatively low potential for transmission to other hosts and general environmental impact. Similarly promising results have also been achieved using genetically enhanced isolates of *A. californica* NPV that express insect-specific neurotoxin genes from spiders rather than from the more sinister scorpion (Hughes *et al.* 1997). One species of spider utilized in this way belongs in fact to the same genus, *Tegenaria*, as the benign European house spider. So, although this technology has a long way to go for it to be accepted as a commercial pest management tool, if it ever is, the system has been shown to work, and must be considered for now to be one of the potential goals of biological control. To this end we need to improve our knowledge of baculovirus genetics, molecular biology and ecology so that we may be able to design effective genetically engineered viruses with minimized (or zero) perceived and actual environmental impacts (Richards *et al.* 1998).

10.6 Chemical control

The use of chemicals in insect pest management has come a long way since the indiscriminate use of organochlorine insecticides in the 1940s and 1950s. It can be considered to comprise three distinct sets of tactics, one involving the use of (mainly) synthetic toxins, another that of insect growth regulatory compounds, and a third that of insect pheromones.

10.6.1 Insecticides

Huge reductions in crop yield caused by the depredations of insect pests, in terms of both quantity and quality, are clearly intolerable, especially when gross yield is the overriding priority of third world subsistence farmers. In the developed world, in contrast, net yield and hence profit are probably more important than gross figures, but most important for many of us is the quality of the produce in the marketplace or back garden. The vast majority of people simply

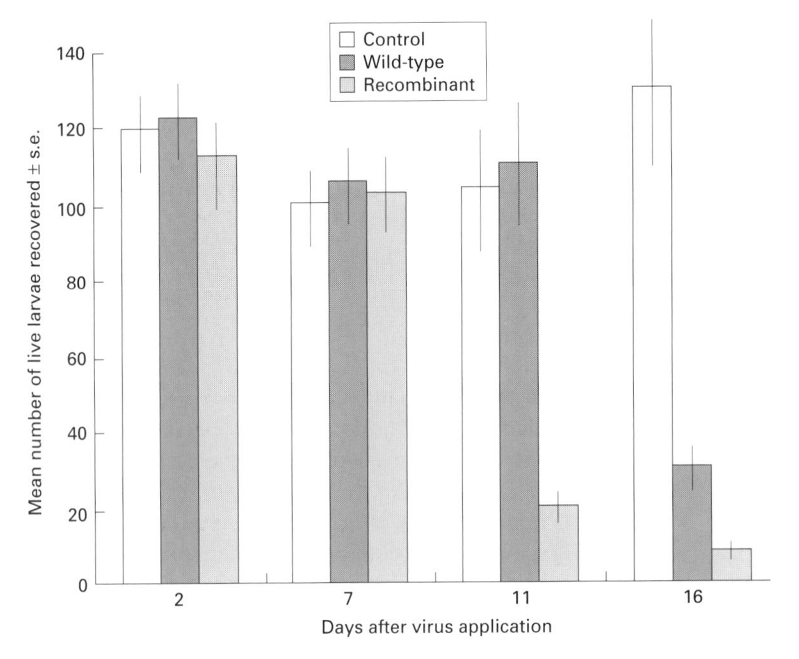

Fig. 10.21 Numbers of live *Trichoplusia* larvae recovered from three field treatments, illustrating the efficacy of genetically modified (recombinant) NPV. (From Cory *et al.* 1994.)

will not tolerate apples with grubs in them, cut-flowers with leaf-miners, or strawberries full of holes. Organic farming strives to produce high-quality crops without resorting to chemical inputs, and it is certainly the case that organic fields have been shown to support significantly higher numbers of beneficial arthropods when compared with intensively managed ones (Berry *et al.* 1996). But with the best will in the world, the pragmatist must accept that farmers will continue to need to apply pesticides of various types to meet global requirements of food supply and commerce (Matthews 1992). Figure 10.22 shows the magnitude of pesticide use in the world in 1989. Insecticides made up nearly 30% of the total market value of around US$22 billion, though the distribution of this enormous quantity is far from even geographically, with the developed world using the lion's share.

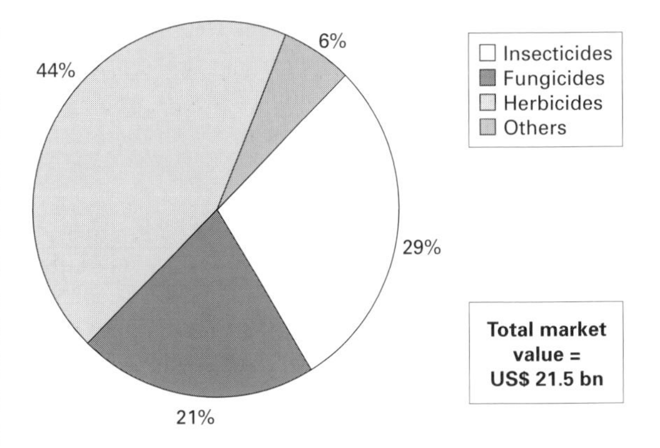

Fig. 10.22 Global agrochemical market by pesticide type in 1989. (From Matthews 1992.)

The fact that we still use so much pesticide may come as a shock when it is considered how long ago Rachel Carson wrote the book *Silent Spring* (Carson 1962). It is of some significance that this work is still often quoted, though so seldom actually read, and has for all intents and purposes entered the realms of mythology and race-memory. The basic arguments are plain enough. In 1944, *Scientific American* published a glowing testimonial to the new insecticide, dichloro-diphenyl-trichloroethane, otherwise known as DDT (BICONET 1996), saying that 'painstaking investigations have shown it [DDT] to be signally effective against many of the most destructive insects that feed on our crops'. Indeed, DDT appeared to be a godsend. Figure 10.23 shows how

Table 10.16 Published half-life of various organochlorine insecticides.

Compound	Country or region	Climatic regime	Half-life	Reference
BHC (HCH)	USA (assumed)	Temperate	5 or 6 years	Carson (1962)
Heptachlor	USA (assumed)	Temperate	4 to 5 years	Carson (1962)
DDT	Norway	Temperate (marine)	5 years ± 2.3 years	Skare *et al.* (1985)
DDT and HCH	Not specified	Temperate	6 months to 3 years	Rajukkannu *et al.* (1985)
DDT	Highland Indonesia	Tropical	159 days	Sjoeib *et al.* (1994)
DDT	Lowland Indonesia	Tropical	236 days	Sjoeib *et al.* (1994)
DDT and HCH	Southern India	Tropical	35–45 days	Rajukkannu *et al.* (1985)

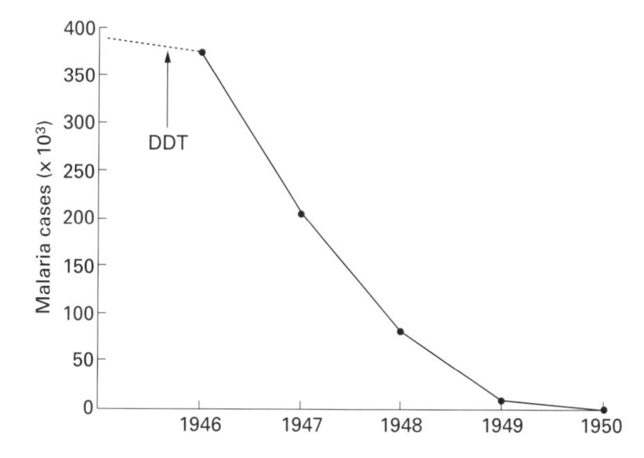

Fig. 10.23 Impact of DDT on malaria cases in Italy. (From Casida & Quistad 1998. With permission, from the *Annual Review of Entomology*, ©1998, by Annual Reviews.)

effective it was in virtually wiping out malaria in Italy (Casida & Quistand 1998).

However, by the late 1950s, Carson and others were strenuously pointing out that DDT and its relatives belonging to the cyclodeine group of compounds within the organochlorines were destroying a large variety of insects, fish, birds and mammals by virtue of their broad-range toxicity and above all, their long-term persistence in the environment and their concentration up food chains. Certainly, these compounds can be discovered in a whole range of localities, including human adipose tissue (Barquero & Constenla 1986), the milk of nursing human mothers (Umana & Constenla 1984), dolphins (Tanabe *et al.* 1993), both scavenging and fish-eating birds (Ramesh *et al.* 1992), and in mayflies, bark and leaves in streams flowing from tropical forest watersheds (Standley & Sweeney 1995). DDT is highly toxic to a myriad of organisms, from estuarine bacteria (Rajendran *et al.* 1990) to peregrine falcons (Thomas *et al.* 1992), but, perhaps surprisingly, it is much less toxic to humans in the short term than many of the newer and supposedly more desirable insecticides on the market today.

It is clear that the now legendary side-effects of DDT and its ilk need to be examined rather more critically. The half-life of a compound describes the time taken for 50% of it to disappear, and this measure is one of the most fundamental criticisms of all the organochlorine insecticides. Depending on the region of the world and its climatic regime, insecticides such as the organochlorines disappear at very variable rates (Table 10.16) and there can be little doubt that the turnover of DDT in the tropics is significantly faster than in temperate areas (Berg 1995), with a half-life measured in a few months rather than years. Therefore, in wet tropical countries, DDT becomes a completely different chemical from the notorious destroyer of ecosystems we hear about in northern temperate regions. Reasons for this much more rapid breakdown in humid tropical regions involve increased volatilization and inundation (Samuel & Pillai 1989), such that it may be that these chemicals are perfectly reasonable for certain types of pest management in the tropics where cheapness and human safety are paramount, such as for vector control (Mnzava *et al.* 1993) (see Chapter 9). Whether or not the old organochlorine insecticides are of some practical use in the late 20th century, the fact remains that most countries in the world have or are in the process of banning their use.

On a world-wide scale, insecticides are undoubtedly dangerous in the hands of untrained personnel, and it is suggested that about 25 million people are poisoned annually by them (Rengam 1992). Most of these people do not die, but sublethal effects can be serious and long-lasting. Exposure to these compounds can result in skin irritations from pyrethroid insecticides, or complex systemic illness resulting from cholinesterase inhibition by organophosphates (O'Malley 1997). People who apply the chemicals tend to suffer most, because of their use of inappropriate (and often unknown) compounds, by the use of the wrong or outdated technology, by not using protective clothing, and by general inefficiency and wastage. These problems are worst in developing countries, where advisory services and access to safer compounds and technologies are often unavailable. It is also suggested that pesticide companies in some developing countries exert an unethical but tight hold on growers, where aggressive marketing and unscrupulous dealers are commonplace (Rengam 1992).

Irrespective of the hazards to people, the insect pests themselves are becoming resistant to many insecticides, especially where large numbers of applications are required during one generation of a particular crop, such as cotton. For some pest species, insecticides have simply ceased to be a viable control tactic because of resistance. For example, the tobacco budworm, *Heliothis virescens* (Lepidoptera: Noctuidae), a defoliator of numerous crops in the USA, is now partially or totally resistant to compounds in all the major groups of insecticides, including the organophosphates, the carbamates and the pyrethroids (Elzen 1997). A notorious pest of brassica crops in both temperate and tropical regions is the diamondback moth, *Plutella xyllostella* (Lepidoptera: Yponomeutidae). Table 10.17 summarizes the colossal incidence of resistance now found in this single species of defoliator (Lim 1992). A further problem is the fact that resistance to chemicals induced by the use of large quantities on crops has the side-effect of also rendering disease vectors in the vicinity such as mosquitoes resistant. The only light on the horizon in this context may be that not only are many pests becoming resistant to insecticides, but there is also now some evidence that certain parasitic wasps are too (Baker *et al.* 1997).

We should therefore be looking to more modern groups of insect-toxic compounds, to reduce the problems discussed above. Ideally, novel toxins should be under continual development and commercialization, but this is no longer a viable expectation. Gratz and Jany (1994) pointed out some of the problems, with particular reference to the control of insect vectors of diseases, but their comments have more general relevance as well. Because of the development of insecticide resistance, and toxicological and environmental considerations, coupled with the cost of development and registration, the number of insecticide compounds available for use has declined, and the number of new insecticides submitted for laboratory and field trials has dwindled even more. Essentially, although legislation and safety testing are indispensable when new chemicals are to be released into the environment (we do not want *another* 'Silent Spring'), these mean that the cost of producing desirable but safe compounds becomes prohibitive, and agrochemical companies see little profit in the venture. Some new compounds do appear from time to time none the less. A relatively recent addition to the insecticide armoury comes from the chloronicotinyl group of compounds, and goes by the name of imidacloprid (Boiteau *et al.* 1997). This is one of several new-generation systemics available to crop producers (Casida & Quistad 1998), but there are some signs that pests such as aphids may already be developing tolerance or resistance to it (Nauen &

Table 10.17 Insecticidal resistance in the diamondback moth in South-east Asia (from Lim 1992).

	Number of compounds within a chemical group in which resistance has been found				
Country	Organochlorines	Organophosphates	Carbamates	Pyrethroids	Growth regulators
Indonesia	3	15	3	5	1
Malaysia	4	9	4	4	2
Philippines	1	9		2	
Thailand		12	3	8	5
Vietnam	4	10	1		

Elbert 1997). Overall, we have to make do with the relatively few insecticides that we are still content to use, on the understanding that they remain our most powerful tool in pest management (Matthews 1992). It is therefore vital that we use them carefully and efficiently, to extend their useful lives for as long as possible.

The efficiency of insecticide use

Historically, the efficiency of insecticide use, measured as the percentage of total chemical applied to a crop that actually kills insects, has been shown to be woefully poor (Table 10.18). The incredible waste is plain to see, and this is compounded by the certain knowledge that the vast percentage of insecticide that does not kill the target will go somewhere else in runoff and pollution. It is obvious that modern pest management must address the problem of maximizing insecticide efficiency while minimizing waste. This ideal may be achieved by integrating several tactics, namely, choice of compound and its concentration, application timing, and application technology.

Choice of compound

Insecticides are pre-eminently nerve poisons, and they gain access to the nervous tissues of pests in various ways. Contact chemicals may be applied directly to the insect's body, or the target may acquire them as it roams over the plant surface. The poison is then absorbed through the cuticle. Fumigants may penetrate the pest's body through the spiracles, and compounds that are not able to penetrate the epidermis can be ingested when the insect eats contaminated foliage or drinks the droplets from leaf surfaces. All these types of compound are effective only on pests that easily come into contact with externally applied insecticides. To kill insects feeding on internal tissues of plants, such as sap feeders, or pests contained within the plant such as borers, a systemic insecticide is required, whereby the chemical is absorbed by the roots, stems or foliage of the plant and translocated in the sap vessels to wherever the insect is feeding. Because different insecticides have different characteristics in this context, it is vital to choose one that is appropriate for the target pests' ecology and habitat. Using a contact-only insecticide against a boring beetle or moth larva will be doomed to failure.

Concentration of compound

It might be thought that a modern, environmentally aware, pest control programme would seek to use as little insecticide as possible, thus introducing minimum amounts of toxic compounds to the ecosystem. Although this is indeed an ideal, there is a flaw in the argument, in that if the insecticide is in too low a concentration, then not enough pests may be killed. Let us take, for example, a population of aphids feeding on a cereal crop at the height of the plants' nutrient availability (the so-called milk-ripe stage), and during optimally warm, settled weather. Such an aphid, for example, might be expected to have a generation time of around 12–30 days (Zhou & Carter 1992), and an average fecundity per female (i.e. the entire population), of approximately 40–80 nymphs (Kocourek *et al.* 1994). If a chemical spray kills 90% of the pest population, then the 10% that survive will have been freed of most potentially limiting intraspecific competition, and hence within 2 weeks or so, the population might easily be expected to climb back to its pre-spray density. An added problem is the high likelihood that any incipient resistance mechanism in the pest population would be concentrated in these survivors, so that the resulting generations would quickly and efficiently build up resistance to the insecticide. It is, in fact, more

Table 10.18 Examples of the efficiency of utilization of insecticides. (From Graham-Bryce 1977.)

Insecticide	Method	Pest and crop	% Efficiency
Dimethoate (organophosphate)	Foliar spray	Aphids on field beans	0.03
Lindane (organochlorine)	Foliar spray	Capsid bugs on cocoa	0.02
Disulfoton (organophosphate)	Soil incorporation	Aphids on wheat	2.9
Lindane, dieldrin (organochlorines)	Aerial spray	Locust swarms	6

than likely that the many examples of insecticide resistance in pests are due in part at least to the failure to kill a large enough percentage of a pest population at an early stage in the selection for this resistance. It is clear that in the absence of any expected back-up in pest control from natural enemies, a very high percentage kill should be striven for, up to and beyond 99%.

Timing of application

To kill as many pests as possible, and to reduce the damage done by them to a crop, it is critical to apply an insecticide, whether it be a synthetic chemical or indeed a pathogenic organism, at the right time. Furthermore, in basic toxicological terms, young insects are much more susceptible to poisons of a given concentration than are older ones. One example that combines several of these benefits involves the southern corn rootworm, *Diabrotica undecimpunctata howardi* (Coleoptera: Chrysomelidae). The adult of this pest is also known as the spotted cucumber beetle (Davidson & Lyon 1979), but it is the larvae that are the real problem for a variety of crops from maize to peanuts. The tiny white larvae burrow into the roots and lower stems of crop plants, frequently causing the crop to lodge (fall over). Injury can be particularly serious to young plants, which cannot withstand much damage. In peanut culture in the USA, there is a choice of times to apply insecticides against this pest (Brandenburg & Herbert 1991): (a) a preplanting treatment, (b) during the flowering stage, or (c) later as the crop matures. The actual yield of peanuts is not strikingly different between the timings, but the advantages of an earlier-season treatment include (a) less damage to the crop, because it is smaller and thus less affected by machinery, (b) early season control of other insect pests, not just rootworm, and (c) fewer problems with secondary pests, as small plants allow soil incorporation of granular insecticides, so that predatory birds are less likely to pick up the poisons.

It is obviously important to target pests when they are most exposed to sprays, and as many insect species spend most of their lives feeding in concealed locations, it is useless to try to control them when the likelihood of their coming into contact with poisons or pathogens is remote. Table 10.19 shows some examples of pests that are normally difficult to reach with chemicals, and hence for which application timing is critical. The key to obtaining the correct timing of an application lies, in many cases, with accurate monitoring and prediction tactics, which tell the grower when the vulnerable life stage of the pest is about to appear, or is at peak. Many commercial

Table 10.19 Examples of insect pests concealed for most of their active lives, for whom the accurate timing of spray application is very important.

Insect	Common name	Habit	Crop	Vulnerable life stage
Hylobius abietis	Great spruce weevil	Larvae under bark of stump roots; adults girdle young transplants	Pine, spruce	Adult on young trees
Cydia pomonella	Codling moth	Larvae in fruit	Apples	Young larvae before bud entry
Rhyacionia buoliana	European pine shoot moth	Larvae in leading shoots	Pine	Young larvae before shoot boring
Euproctis chrysorrhoea	Browntail moth	Larvae partially in silk tents	Many shrubs and trees; public health pest	Young larvae on foliage
Laspeyresia nigricana	Pea moth	Larvae in pea pods	Peas	Very young larvae on outside of plant
Phyllonorycter crataegella	Apple blotch leafminer moth	Leafmining larvae	Apple foliage	Adult before oviposition
Panolis flammea	Pine beauty moth	Young larvae feed head-down in needle bases	Pine	Older larvae feeding externally (note: applies to *B.t.* control only)

monitoring systems employ pheromones to predict the best time to apply an insecticide (see below), but simple counts of appropriate life stages may also be employed. The tomato fruit worm, *Helicoverpa zea* (Lepidoptera: Noctuidae), feeds on the foliage and especially green fruits of tomatoes, rendering them unmarketable. In this case, the eggs of the pest are fairly easily seen on crops, and a system called egg-scouting is employed whereby insecticide treatment is carried out only when fruit worm eggs are found on the foliage (Zehnder *et al.* 1995). An application just as these eggs hatch ensures that the very young larvae are killed before any damage is done. This system is much preferable to routine spraying, and in this example, around 50% fewer insecticide treatments were required, saving around US$100 per ha.

Application technology

Having chosen the appropriate insecticidal compound and the time at which to apply it, the third vital decision concerns the method by which it is applied to the crop. Nowhere else in ecology does the need to combine widely differing sciences come so much to the fore: physics, engineering, chemistry and biology all need to be integrated into a package that optimizes the delivery of the insecticide to the target while minimizing waste and off-target pollution. There are many ways of delivering a pesticide to a target, including the use of dusts, smokes, granules and liquid-based sprays. Perhaps the most widespread method is the last system, where the insecticide itself, the active ingredient, is dissolved or suspended in an oil- or water-based formulation, and delivered to the target insects through some sort of machine.

The type of machine is critical to the success of the operation, and old-style technologies that are still universally used are based on a hydraulic nozzle principle, where liquid is forced under pressure through a constriction, producing a supposedly fine spray or mist of droplets. There is an intimate relationship between the size of droplets in a spray cloud, the volume of spray used per unit area of target, and the efficiency of the operation (Matthews 1992). The size of drops is of paramount importance, as Table 10.20 shows. For a variety of reasons, which include the fluid dynamics of differing particle sizes in a turbulent or linear airflow, only a small size range of insecticidal droplets actually does the job of coming into contact with, and, it is hoped, killing, insect pests. The optimal size for insects on foliage is between 30 and 50 μm. Anything much larger does not penetrate crop canopies well, and if the drops do manage to contact foliage, they do not readily adhere because of their size. Very tiny drops, on the other hand, will tend to stay in air suspension even at very low velocities, and also, they contain such small amounts of active ingredient because of their small volume that there may well be insufficient to kill adequately. The largest percentage of drops from a normal hydraulic nozzle are well outside the size range required for killing insects efficiently (Matthews 1992), and it may be the case that as little as 5% or even less of a spray cloud from a backpack or tractor-mounted boom sprayer may be in this range. Hence, most of a conventional application of insecticide is very unlikely to kill pests. On the assumption that an adequate control is achieved, however, by this sort of technology, then it must be inferred that the same result could be achieved by reducing the spray volume very considerably indeed, as long as technology is available to produce droplets only in the size range of 30–50 μm or so. This technology is called controlled droplet application (CDA), and its development allows the use of ultra-low volumes (ULV) of pesticide, often as little as 1 L/ha in many cases.

CDA technology is based on two types of machine, one of which relies on electrostatic charging of spray droplets, and the other on rapidly spinning discs, the so-called centrifugal nozzle (Fig. 10.24). The latter system is much more widely used than the former. Oil formulations of the active ingredient are gravity or pump fed through flow-rate constrictors onto metal or plastic discs spinning at high velocity. The liquid is flung to the edges of the discs by centrifugal force, where it streams off the discs in filaments that fragment into droplets. The size of these droplets depends mainly on the flow rate of the liquid, and the velocity

Table 10.20 Optimal droplet sizes for pesticides used on a range of targets. (From Matthews 1992.)

Target	Droplet diameter (μm)
Flying insects	10–50
Insects on foliage	30–50
Foliage	40–100
Soil (avoiding drift)	>200

Fig. 10.24 CDA machine (Micron ULVA). (Courtesy of J.S. Cory.)

of the spinning disc. In reality, such machines are not actually sprayers at all; they are merely droplet producers. Some rely on air currents to disperse the drops amongst the target crop, whereas others provide a petrol- or battery-driven fan to blow the drops onto the target. These air-assisted ULV machines have been tested in field conditions and permit application of insecticides at considerably lower volumes than conventional technologies (Mulrooney *et al.* 1997). ULV–CDA machines are now available from simple and cheap hand-held devices to complex and powerful fixed-wing- or helicopter-mounted rigs for large-scale applications.

It is not to be said that the old technologies never work properly. The same degree of control may be achieved, but often at much greater cost in terms of chemicals, time, labour, and environmental impact. As Fig. 10.25 shows, the use of a CDA spinning disc machine for the control of various groundnut pests in India does not produce strikingly different results when compared with a conventional knapsack sprayer, but the crucial point is that there is a very much reduced demand on labour and costs. These systems are particularly appropriate when large volumes of water are in short supply, or the crop to be treated is isolated. Clearly, if 1 L/ha will do the same job as 100, then the former system has to be preferable, everything else being equal.

As with all systems, CDA–ULV does have its drawbacks. For example, the insecticides used through the newer machinery are more complex and expensive, being based usually on special oil-based formulations, which may not be easily available in many countries. They are an alien technology to many growers all over the world, from British farmers to subsistence growers in the third world, who may well be happier relying on their well-tried (though probably much less efficient) traditional machines. Furthermore, the use of CDA technology does involve more care and thought. Wind speed and direction have to be monitored carefully, and as the droplets are so small, they can frequently be hard to see, thus making attempts to cover all the target crop uni-

Fig. 10.25 Density of insect pests per groundnut plant in India 36 h after spraying with insecticide using two types of spray machine.

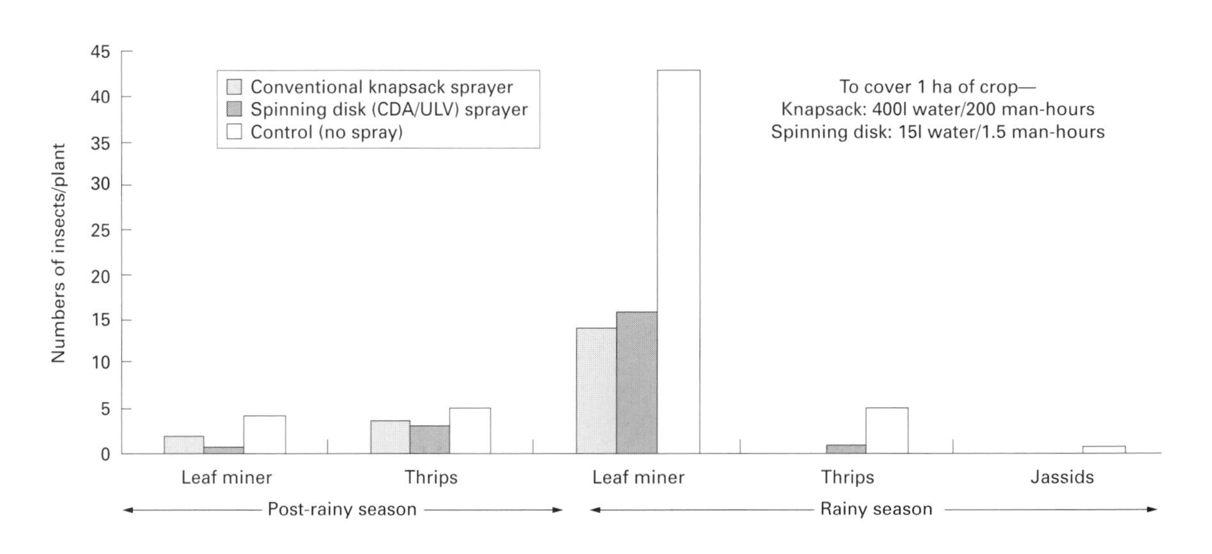

formly rather haphazard. On the plus side, most application systems involving pathogenic organisms such as bacteria, and especially viruses, now also involve the use of CDA spinning disc technology.

10.6.2 Insect growth regulators

In recent years, insecticides that are merely poisons have begun to be replaced, in some situations at least, by much more subtle compounds, which in various ways are able to interfere with the growth and development of insects, especially the young stages. Juvenile hormones are involved with the control of larval development and pupation, whereas the moulting hormones (ecdysones) control skin shedding in nymphs and larvae. Synthetic analogues of some of these hormones are now available, which, although expensive, tend to have a much narrower host spectrum than conventional insecticides, and are hence much safer for the environment. One of the most widely used compounds is diflubenzuron (trade name Dimilin), an inhibitor of chitin production, which prevents larval and nymphal ecdysis, as well as having ovicidal properties, where treated eggs fail to hatch (Saraswathi & Ranganathan 1994). Dimilin has to be ingested by the target insect, but it has a fairly high residuality on foliage, for example, where it may persist for most of a growing season (Wimmer *et al.* 1993). It is used on a wide range of pests and cropping systems, from orchards to forests, and mushroom growing to vector control. Even with Dimilin, prolonged exposure of insect populations to the chemical has resulted in some species such as the Australian sheep blowfly, *Lucilia cuprina* (Diptera: Calliphoridae), becoming tolerant to diflubenzuron (Kotze *et al.* 1997).

10.6.3 Natural insecticides

All the compounds considered so far have been synthetic, though some, such as the pyrethroids and ecdysones, mimic the basic chemical behaviour of naturally occurring substances. It is well known that plants produce a huge range of so-called secondary chemicals that act as defences against herbivory (see Chapter 3), and one or two of these compounds are being considered as commercial insecticides. The best known is neem-oil, a chemical produced from seed kernels of the neem tree *Azadirachta indica*. The compound is available commercially under the trade name Margosan-O. This plant extract has been shown to have significant effects on a variety of pest insects. Larvae of the gypsy moth, *Lymantria dispar* (Lepidoptera: Lymantriidae), grow much more slowly when exposed to neem-oil than non-treated controls, so that when most untreated larvae were in their last instar, ready to pupate, the treated ones were still only in the second and third instar (Shapiro *et al.* 1994). In addition, combining neem treatment with nuclear polyhedrosis virus infection resulted in a faster virus-caused mortality. Trials of neem have also proved encouraging in the control of mosquitoes and brassica pests, for example (Rao *et al.* 1995; Klemm & Schmutterer 1993), but it seems rather unlikely that the chemical will replace the more toxic and effective synthetics in the near future.

10.6.4 Pheromones

A very large number of animal phyla, both terrestrial and aquatic, communicate with members of their own species using complex chemical signals, known as pheromones. This system probably reaches its quintessence in the insects, who use pheromones to find mates, to signal aggression or overcrowding, or to aggregate in a particular part of the habitat where mates or breeding sites are abundant. In most cases, the system works by air currents delivering pheromone 'plumes' downwind from a point source, which is very often the female, or less often the male, who is ready to mate. These sex-attractant pheromones are widely used in the management of insect pests, either by luring adults to artificial pheromone sources, or to broadcast sex-attractants through a crop so that the point source system is overwhelmed. Aggregation pheromones may also be employed to concentrate pests in areas where they can be dealt with by other control methods.

A very large number of insect species have now been investigated in this context, and many sex-attractant chemicals identified. A publication on the Internet, 'Pherolist' (Arn *et al.* 1996) cites, for example, over 670 genera from nearly 50 families of Lepidoptera for whom the pheromones are known. Most pheromone compounds are of a rather low molecular weight, highly volatile and species specific. They may convey complex signals about the location of a mate or host, but in general are easy to synthesize in a laboratory, and an enormous range of synthetic compounds are now available on the market for a whole host of insect pests. Concomitant with the chemistry, and equally important, is the design of the lure or trap to which insects are attracted, and many designs of pheromone trap are also available. The chemicals themselves are usually placed in slow-release formulations, such as plastic or polythene capsules, microfibres, or are impregnated in rubber bungs, so that the active ingredient is slowly released into the habitat. The traps in which these lures are located have to be of a design whereby air currents flow readily through them to take the attractants away, and many also have some form of capture device such as a sticky insert or bottle, to retain the insects that are lured to the point source (Fig. 10.26a,b). Attractant pheromones can be used in several ways in pest management, including aggregation, mass trapping, monitoring and mating disruption.

Aggregation and anti-aggregation

Some pheromones attract insects to localized areas within a habitat, where, for example, a suitable host tree has been located by foraging young adult beetles. Others work in the reverse way, causing adults to disperse. A combination of aggregation and anti-aggregation pheromones can be used to reduce the probability of pest infestation in small stands of trees. For example, the Douglas-fir bark beetle, *Dendroctonus pseudotsugae* (Coleoptera: Scolytidae), is an extremely serious pest in North America. Adults lay eggs under the bark of host trees, and the larvae tunnel between the sapwood surface and the bark cambium, eventually ring barking the tree when present in high population densities. Non-vigorous or moribund trees are particularly susceptible, even those that are only temporarily stressed. Ross and Daterman (1994) placed capsules containing anti-aggregation pheromones around the perimeter of 1 ha circular plots. In addition, funnel traps baited with an aggregation pheromone for *D. pseudotsugae* were placed outside this perimeter, thus producing a dual protection for the inner plot of trees. Results showed that mass attacks on trees inside the plot by the beetle were reduced by 80% compared with untreated controls, but there was an eight-fold increase in attacks outside the plot in the vicinity of the funnel traps. Clearly, these types of pheromone can be used successfully to draw pests into certain areas, and to exclude them from others, though in practice this system would be economical only in the protection of small, high-value stands. Using aggregation pheromones, it is possible to lure insects to their death. In Belgium, where natural disasters such as windblow create enormous quantities of breeding material for the spruce bark beetle, *Ips typographus* (Coleoptera: Scolytidae), trap trees have been coated with pyrethroid insecticides, and adult beetles

(a)

(b)

Fig. 10.26 (a) Delta pheromone trap. (b) Boregaard pheromone trap.

attracted to these trees by pheromone dispensers attached to each (Drumont *et al.* 1992). Beetles are drawn away from susceptible timber and attracted instead to the trap trees, where they are killed.

Mass trapping

The concept of mass trapping using sex-attractant pheromones is simple enough. Sufficient adult insects (normally male) are lured into a trap where they are collected and killed, so that the likelihood of females in the crop mating to produce the next generation of pests is very much reduced, or ideally, removed altogether. A fundamental problem with this idea is that most male insects are polygamous. Each adult male of the maize stem borer, *Busseloa fusca* (Lepidoptera: Noctuidae), for example, will, given the opportunity, mate with an average of five or so females (Unnithan & Paye 1990). So, in general, the population density of males has to be reduced to extremely low levels before there are not enough left to ensure successful matings. An allied problem would be the immigration of new males from adjacent habitats to fill the vacuum left by the mass trapping. For instance, even after 5 years of large and continuous trials of mass trapping with pheromones, it was proven to be impossible to substantially influence the population densities of *Ips typographus* (Dimitri *et al.* 1992) in European plantations.

None the less, some successes have been achieved with mass trapping. In oil palm plantations in Costa Rica, nearly 250 pheromone traps were located in a 30 ha area, and the numbers of the weevil pest, *Rhynchophorus palmarum* (Coleoptera: Curculionidae), were counted in the traps for 17 consecutive months (Oehlschlager *et al.* 1995). The larvae of the pest burrow in the crowns of growing palms, feeding on the young tissue and causing the destruction of the growing point (Hill 1983). By the end of the mass trapping period, over 60 000 adult weevils had been removed from the crop, and expected outbreaks of the pest during the following dry season did not occur. For mass trapping to have some chance of success, the number of traps per unit area must be considerable. The citrus flower moth, *Prays citri* (Lepidoptera: Ypnomeutidae), infests the flowers of lemon trees, and as long as a minimum of 120 pheromone traps per hectare of orchard are employed, then mass trapping has been found to be significantly cheaper and rather more efficient than spraying 3–6 times a year with insecticides (Sternlicht *et al.* 1990).

Monitoring

The most widespread use of insect sex-pheromones and aggregation pheromones in the crop protection world is as monitoring tools, whereby estimates of adult insects in a crop can be made on a day-by-day or week-by-week basis. These estimates can then be related to dependable economic threshold figures (see below), and decisions made about whether or not to treat (usually with insecticide), and if so, when. Of crucial importance to the success of a monitoring programme is establishing the link between the numbers of adult insects caught in a trap and the density of the pest population to be expected in the crop in the forthcoming generation. Figure 10.27 gives one example of this relationship, for the bollworm, *Helicoverpa armigera* (Lepidoptera: Noctuidae), in southern Spain, where it is a serious pest of tomatoes and carnations (Izquierdo 1996). The relationship between mean weekly captures of males in pheromone traps and the numbers of eggs laid in carnation fields is significant, though the data are rather variable. Plant phenological and weather data can also be added to the prediction system to improve the regression fit shown in the figure, but it is clear that there is certainly a potential here for placing pheromone traps in a crop and counting the number of adult moths caught per week. From the figure, it can be inferred that, for example, an average of 300 males per trap per week would indicate that between 50 and 100 eggs could be expected per 100 sampling units later in the season. The next step, of course, is to relate such a density to a prediction of economic damage.

The practice of relying on pheromone trap captures to predict pest problems to come is commercially viable in a number of other crops too, from timber yards to apple orchards and pea fields. Timber storage yards, by their very nature, provide massive quantities of highly suitable breeding material for bark and wood boring beetles, in the form of logs waiting to be processed. Although it may be the case that felled logs can no longer be damaged by bark beetles (Coleoptera: Scolytidae), they can produce enormous numbers of these pests, which may then form a mass outbreak that certainly could cause much damage to standing trees in the vicinity of the timber yard. Fur-

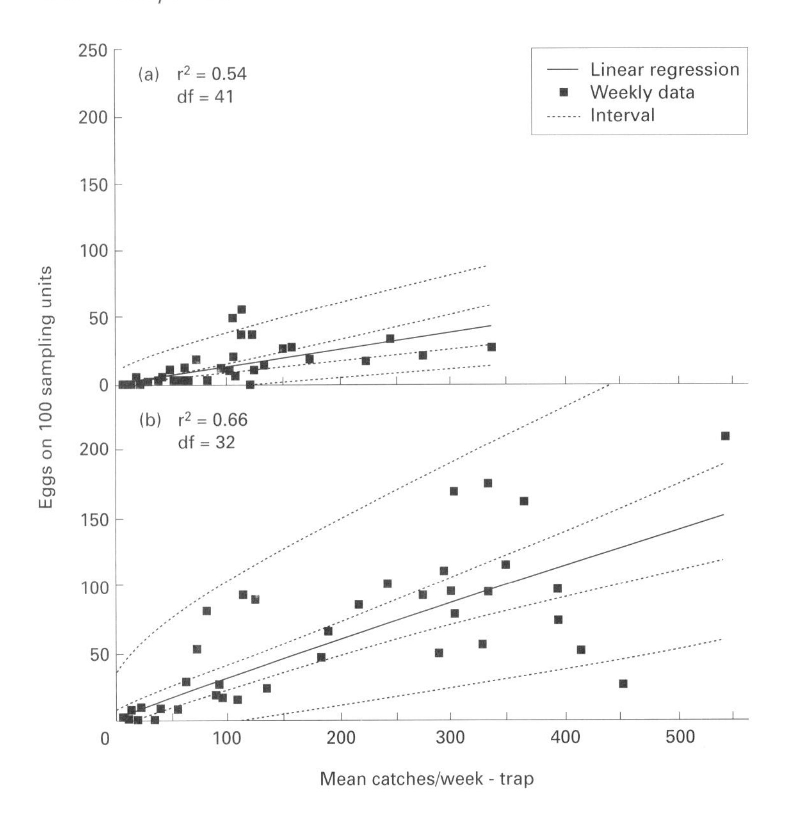

Fig. 10.27 Relationship between mean weekly catches of *H. armigera* in pheromone traps and egg density in carnation fields. (a) Early period (before mid-August), (b) late period (after mid-August). The dotted lines closer to the linear regression show the confidence interval ($P = 0.05$) and the farther ones the prediction interval ($P = 0.05$). (From Izquierdo 1996.)

thermore, wood borers such as some of the ambrosia beetles, *Trypodendron* spp. (Coleoptera: Platypodidae), can and do cause extensive degradation of stored logs, not only by their tunnelling in the timber itself, but also by their inoculations of blue-stain fungi into the wood (see Chapter 6). Pheromone traps baited with synthetic aggregation pheromones for *Trypodendron lineatum* and *Ips typographus* are used routinely in wood yards all over Europe to monitor densities of these pests in the locality (Fig. 10.28). One example from Slovenia is illustrated in Fig. 10.29, where the large numbers of captures of both species are depicted, along with the seasonal nature of their flight periods (Babuder *et al.* 1996). If large numbers of beetles are detected during this monitoring procedure, action must be taken to protect the timber in the yard, such as debarking, water storage or insecticidal sprays.

Fig. 10.28 Wood yard in France with pheromone trap to monitor *Ips typographus*.

Mating disruption

As male moths usually find their mates by following the female's pheromone plume upwind to its source, it is possible to reduce the likelihood of matings by broadcasting synthetic sex-pheromone through a crop. Disruptants are likely to work in various ways (Cardé & Minks 1995). Males may become habituated to the sex-attractant, which then loses its 'appeal', synthetic lures may 'compete' with calling

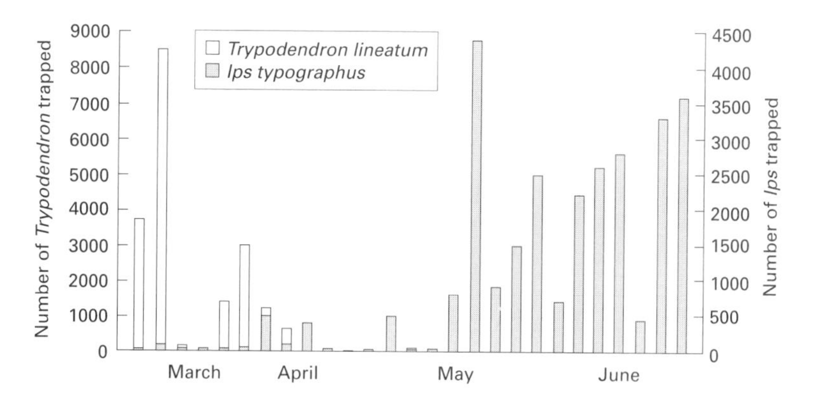

Fig. 10.29 Captures of two beetle species in aggregation pheromone traps in a wood yard in Slovenia. (From Babuder *et al.* 1996.)

females, so that males are attracted to the wrong place, or the synthetic may simply smother and camouflage the females' less powerful scent. Many different formulations of synthetic sex-pheromone are now available for disruption. These include polyvinyl chloride beads saturated with the sex-pheromone sprayed with conventional aerial or ground-spraying equipment (Hulme & Gray 1994), twisted-tie or wire 'ropes' (Kehat *et al.* 1995), or flakes, wafers or polythene microtubules. Such tubules can either be applied from the air to broadcast the synthetic, or attached to trees or stakes within a crop to provide point sources (Speight & Wainhouse 1989). Puffer machines are now available to release small amounts of pheromone blends into crops at predetermined intervals (Shorey & Gerber 1996).

As with most pest management systems, both failures and successes can be expected. In the case of codling moth, *Cydia pomonella* (Lepidoptera: Tortricidae), attacking organically grown apples in Ontario, Canada, mating disruption techniques using pheromones reduced the number of adult male moths found in the crop by between 75 and 96% when compared with control plots (Trimble 1995). However, this treatment with pheromone did not prevent an increase in codling moth damage, and in fact pest densities rose considerably. In stark contrast to this, in a very similar trial in British Columbia, Canada, pheromone mating disruption systems were employed to dispense the synthetic pheromone of codling moth, codlemone, at varying concentrations into orchards over a 6 month period. Fruit damage ranged from 0 to 1.5% overall in the treated orchards, but reached as high as 43.5% in untreated controls (Judd *et al.* 1996). Why the results of these two trials on the same insect using very similar tactics produced conflicting results is rather difficult to explain. It is, of course, important to remember the pest's ecology: a low-density pest such as codling moth can still cause serious economic damage by the tunnelling of its larvae in apples and other pome fruits even when the adults are at fairly low densities. Furthermore, mating disruption may not work when pests are present in a crop at very high densities (Kehat *et al.* 1995). Moths, for example, use other sensory systems to locate members of their own species, such as visual and tactile signals, and at very high densities day-flying adults may be able to see each other and hence do not require a sex-attractant system when they are very close together.

Some examples of the use of pheromone mating disruption show more dependable results. One of the most notorious cotton pests is the pink bollworm, *Pectinophera gossypiella* (Lepidoptera: Gelechiidae). The larvae of the moth are well protected from conventional insecticides because they feed within flower buds and cotton bolls, and thus the pest is particularly difficult to control (Cardé & Minks 1995). Table 10.21 presents some statistics on the development of mating disruption systems against pink bollworm in the USA, showing that disruption really can work on a large commercial scale. An extra benefit of these systems is, of course, the very significant reduction in the use of conventional insecticides in cotton crops, with the added advantages of a reduced incidence of resistance in the pest, less harm to neutral and beneficial insects, and even a decrease in direct phytotoxic effects of the insecticides on the crop itself.

Table 10.21 Examples of control of the pink bollworm, *Pectinophera gossypiella*, by mating disruption in cotton. (From Cardé & Minks 1995.)

Region	Year	Type of treatment	Area treated	Result of control
S. California and Arizona	1981	Aerial application of hollow fibres filled with disruptant	40 000 ha	Yields enhanced, 5% of crop damaged compared with 30% in conventionally treated fields
S. California and Arizona	1985	Wire-based, sealed polyethylene tube filled with disruptant (twisted-rope) placed by hand at base of cotton plants	1000 dispensers/ha	Reduced damage; no insecticide needed until pheromone in the dispensers had evaporated (2–3 months)
Arizona	1990–1993	Twisted rope or fibres	11 000 ha	Conventional insecticidal treatment (no pheromone) gave 23.4% boll infestations in 1989; in pheromone-treated crops, infestation level was 9.9% in 1990, 1.42% in 1991, 0.086% in 1992, and no larvae in 25 200 sampled bolls in 1993

10.7 Integrated pest management

We are now at the stage where all the individual techniques and tactics for insect pest management can be drawn together and used in a system that combines the most appropriate characteristics of each to produce an ecologically sound, economically viable package, called integrated pest management (IPM). Various definitions are available. One of the most recent suggests 'IPM is a decision support system for the selection and use of pest control tactics, singly or harmoniously coordinated into a management strategy, based on cost/benefit analyses that take into account the interests of and impacts on producers, society and the environment' (Kogan 1998). Most definitions stress the key points that, for IPM to succeed, it is crucial to have detailed information about the pest's biology, and how it interacts with the environment, natural enemies and the crop itself. All these factors must also be judged in the context of economic injury levels.

Figure 10.30 shows a framework for IPM in tropical forestry (Speight 1997b), which illustrates the various components of such a system, and their relationships to each other. Clearly, not all of these components would be available or even required for one IPM programme, but elements of these may be detected in the examples we discuss later in this chapter. It is of fundamental importance to consider IPM as first and foremost a preventive strategy, which relies on control measures only if and when prevention fails for some reason. As the figure suggests, if the tactics in stages A and B are adopted efficiently, then stage D may never be required in an ideal world. Complacency has of course to be avoided, however, so that stage C links the two parts of the strategy via monitoring and predictive systems that depend on the sound knowledge of threshold levels (see below)

Undoubtedly, IPM is more complex and sophisticated than any 'spray by date' pest control system, and as such cannot be considered to be an easy option for growers. Figure 10.31 shows how it becomes more complex in stages, and how there exists a continuum of adoption (Kogan 1998). As illustrated by the figure, as IPM programmes become more complex, their adoption by growers decreases very markedly. Pragmatically, IPM is likely to be adopted with enthusiasm only when other, essentially easier tactics have failed, or proved to be too expensive. IPM will be particularly attractive if it can be shown to save money. An example of the last situation was illustrated by Trumble *et al.* (1997), who analysed the profits to be made using an IPM programme in celery crops in California. They compared the standard systems of

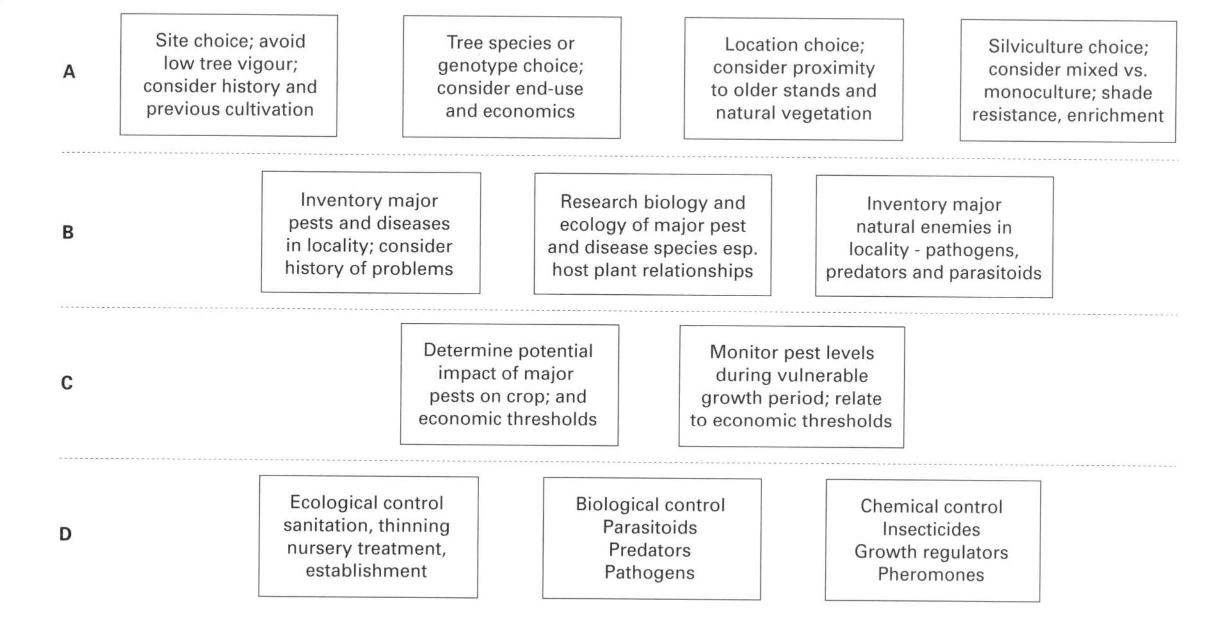

Fig. 10.30 Theoretical components of a generalized IPM system. Stages A and B are entirely preventive, stage C involves monitoring and prediction, and stage D covers control strategies that are available should prevention fail, or monitoring reveals high risk. (From Speight 1997b.)

chemical pesticide use, which consisted of nine applications through the crop's life, with an IPM strategy based on sampling and pest threshold assessments (see next section) followed by three or four treatments with *Bacillus thuringienis* if required. In a commercial operation, the IPM system generated a net profit of over US$410 per ha higher than that of the standard chemical use, with the added bonus of approximately 40% reduction in pesticides.

It is very unlikely that efficient IPM will be achievable without the recourse of growers to advice from well-informed experts, via advisory or extension schemes. The absence of such 'luxuries' in many developing countries, especially for subsistence crops, is a very serious drawback to the uptake of much needed IPM. Even if an IPM programme has reached the stage where it should be extended to commercial growers in the thick of pest problems, there is some doubt as to whether merely sending students on IPM training courses equips them with the ability to set up and execute an IPM system for real (Liau 1992); a more complex and sophisticated infrastructure may be required. Figure 10.32 provides a conceptual layout of various components and prerequisites for an ideal IPM system, whereby basic research provides the vital ecological and economic data concerning a particular pest–crop interaction. Support services including database reference, extension services and funding bodies then take this information and pass it on to the industry, wherein pest management becomes a crucial and integral part of crop husbandry (Speight 1997b). Ideals are infrequently met, but some examples of IPM systems from different crop systems should illustrate how some of these ideals may become reality.

10.7.1 Monitoring and economic thresholds

The impact of insect pest damage has been mentioned above, but it is of great importance that we are able to call on dependable data to tell us when the density of pests in our crops exceeds the level where crop damage is worth more than the cost of control (see Fig. 10.1). To this end, economic thresholds have been ascertained for a whole range of crop–pest sce-

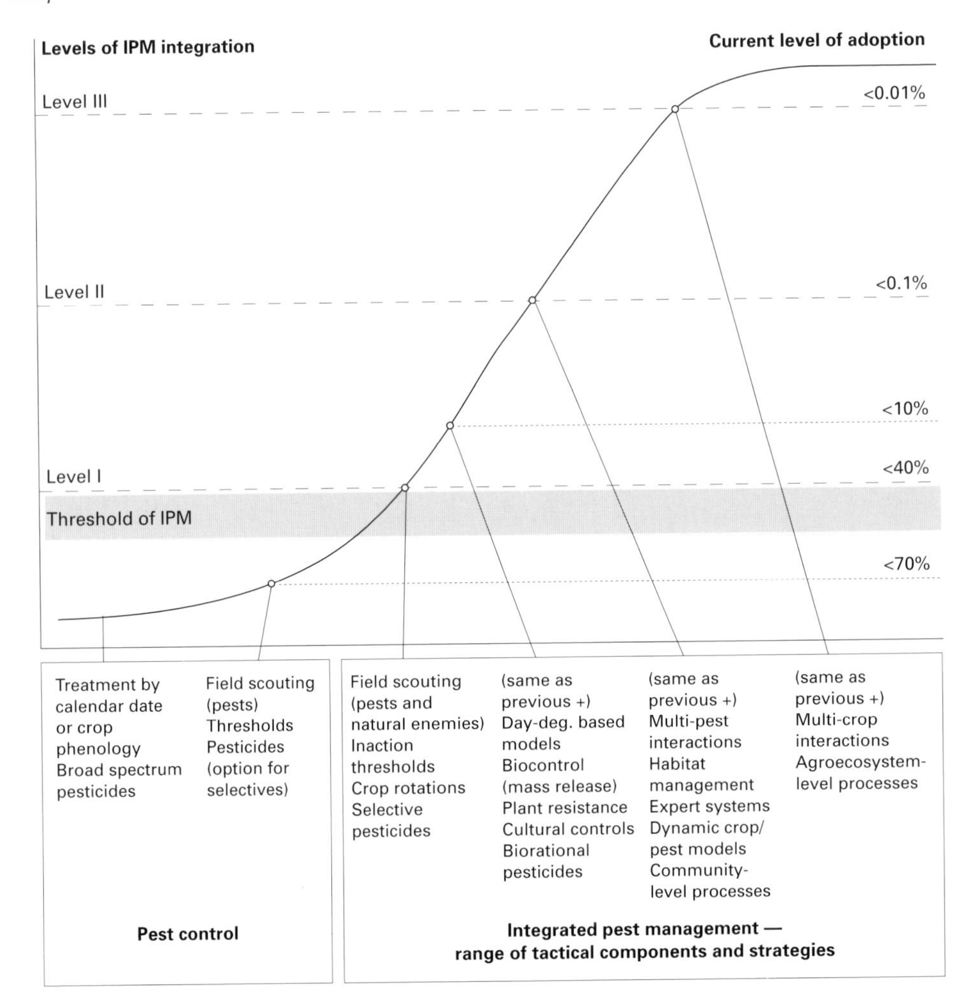

Fig. 10.31 Hypothetical continuum from conventional pest control to level III IPM, based on field and vegetable crops in the USA, illustrating that there is a minimum set of tactical components combined with a basic strategy that define the 'IPM threshold'. (From Kogan 1998. With permission, from the *Annual Review of Entomology*, ©1998, by Annual Reviews.)

narios, without which IPM cannot function properly. Table 10.22 shows some of these thresholds. The collection of such data needs some sort of capture or counting system. The use of sex-pheromone trapping has been discussed above, and other approaches require growers to examine their crops at crucial times of growth according to prescribed protocols. Such tactics do, of course, demand more care and attention from growers, but the whole essence of IPM relies upon the intimate involvement of agriculturists, horticulturists and silviculturists to promote the health and hygiene of their crops. Perhaps one of the greatest strengths of monitoring and prediction based on cost–benefit analyses is the luxurious ability to decide confidently to do nothing, even when some pests can be observed in our crops!

10.7.2 Examples of IPM in practice

The pine looper moth, Bupalus piniaria (Lepidoptera: Geometridae), in the UK

Bupalus piniaria (Plate 10.6, opposite p. 158) is widespread in Europe as a potentially serious defoliator of various species of both native and exotic pine. Yield

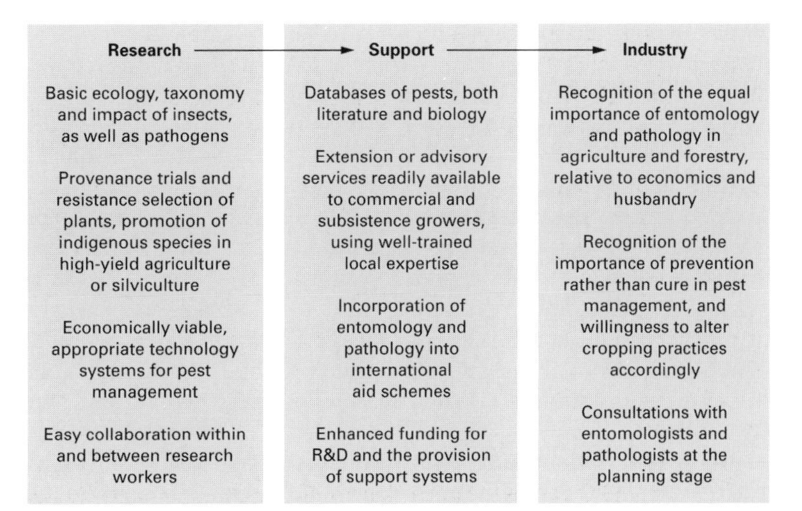

Fig. 10.32 Components required for efficient insect pest management in the future. (From Speight 1997b.)

Table 10.22 Examples of economic thresholds (critical densities) for various crop pests in the UK. (From ADAS, MAFF.)

Crop	Pest	Common name	Threshold density/infestation
Apples	*Rhopalosiphum insertum*	Apple–grass aphid	Aphids on 50% of trusses at budburst
	Dysaphis plataginea	Rosy apple aphid	Any aphids present on trusses at budburst
	Operophtera brumata	Winter moth	Larvae on 10% of trusses at budburst
Field beans	*Aphis fabae*	Black bean aphid	5% or more plants on SW headland infested (spring-sown crops only)
Peas	*Cydia nigricana*	Pea moth	10 or more moths per pheromone trap in two consecutive 2 day periods
Potato		Various aphids	Average 3–5 aphids per true leaf in a sample of 30 each of top, middle and lower leaves taken across the field
Rape	*Meligethes* spp.	Blossom beetle	15–20 adults per plant at flowering bud stage
Wheat	*Metopolophium dirhodum*	Rose grain aphid	30 or more aphids per flag leaf at flowering
	Sitobion avenae	Cereal grain aphid	5 aphids per ear at the start of flowering, with weather fine and settled

losses are usually tolerated, but the main risk from the moth is that the defoliated trees are stressed temporarily and become susceptible to attacks by the so-called pine shoot beetle, *Tomicus piniperda* (Coleoptera: Scolytidae). The girdling activities of the beetle's larvae can ring bark and kill trees that would otherwise reflush their needles after the defoliation. In addition, young adult beetles tunnel into the shoots of growing trees during maturation feeding and can cause serious stunting and deformation. As direct control of the beetle is not practical because of the concealed nature of the larvae, the avoidance of defoliation by *Bupalus* is the main defence against *Tomicus*. It is possible to recognize certain site conditions that seem to predispose pine trees to attack by *Bupalus* larvae, which may be related to soil type (Barbour 1988), or latitude and rainfall (Broekhuizen *et al.* 1993), but hazard or risk rating is not sufficiently dependable to be relied upon. Instead, the UK's Forestry Commission (now the Forest Authority) carries out routine annual surveys of the density of *Bupalus* pupae overwintering in forest soils and litter. These surveys provide data on the peak mean abundance of pupae for between 35 and 50 sites across the country, four of which are shown in Fig. 10.33. As can be seen, population densities fluctuate year by year as well as between sites, but there is some suggestion of a rough periodicity to these densities, indicating a potentially regulatory mechanism. Indeed, the parasitoid wasp *Dusona oxycanthae* (Hymenoptera: Ichneumonidae) does exert some natural biological control on late stage larvae, though not enough to prevent the occasional outbreak. The critical density (economic threshold) for *Bupalus piniaria* in the UK is estimated to be around 50 pupae/m^2 at peak (Speight & Wainhouse 1989), and if the pupal surveys discover densities greater than that in any one year in a particular site, then control may have to be carried out, but only in that localized area. Before that, however, a second round of limited monitoring may be carried out when the eggs are laid in late spring of the following year, to double check that the indications from the winter pupal survey did indeed point correctly to an impending outbreak. As Fig. 10.33 shows, pupal densities rarely greatly exceed the critical threshold, though in Tentsmuir in Scotland, aerial spraying has occasionally had to be carried out using diflubenzuron (Dimilin). Even then, it is important not to interfere with the activities of the natural enemy complex, such as it is. Finally, it is possible that even with the best survey techniques available, defoliation might occur unexpectedly, by which time it is too late to use insecticides. To protect defoliated trees from *Tomicus piniperda*, a few trees in a stand can be felled in a sanitation programme, and placed in a log pile as a trap crop. *Tomicus* adults preferentially lay eggs in these logs rather than in the stressed but standing trees, and the logs can then be destroyed before the next generation of bark beetles appear. A number of different components are represented, including monitoring, plus chemical, biological and ecological control. Pivotal to this IPM strategy is monitoring followed by prediction followed by decision making; in most years, in most places, no further management is required.

Insect pests in apple orchards in Europe

Apple orchards are plagued by a wide range of insect and mite pests, and the relative importances of the various species depend very much on the geographical location, the type of management, the crops being cultivated, and the legislation pertaining to approved pest management substances, which vary from country to country (Blommers 1994). It is hence rather difficult to give a general overview of IPM in European orchards, and thus only one or two groups of pest have been selected for this example. Table

Fig. 10.33 Densities of pupae of *Bupalus piniaria* in selected sites in the UK, showing oscillations over the years. Data taken from the Forestry Commission annual *Bupalus* survey (unpublished data).

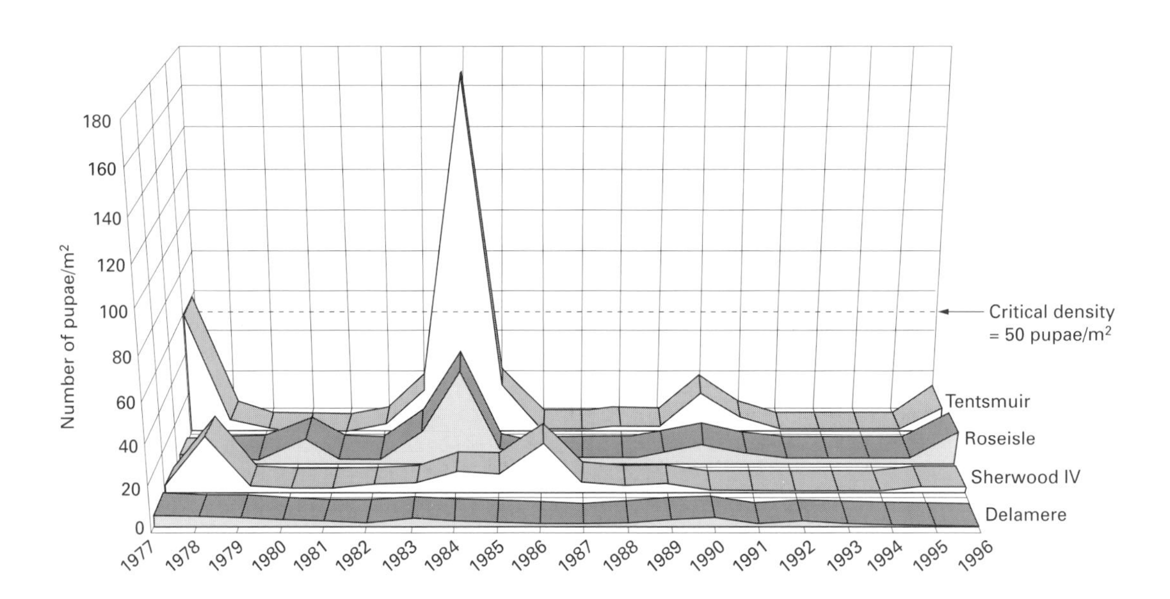

10.23 provides details of some of the tactics used against three particular pests. Some techniques work well for most pests; juvenile hormone analogues prevent the development of the pupal stage in most species, whereas the granulosis virus is specific to codling moth, and then is effective for only a short while before the young larvae enter the fruit bud. Non-insect arthropods such as mites are also notorious pests in apple orchards, and their management revolves these days around the natural control by a predatory mite released periodically into the orchard. As many pesticides that might be useful against insect pests such as the moths are also toxic to the predatory mite, IPM has to employ insecticide-resistant predators, and microbial pathogens wherever possible, which do no harm to other arthropods.

The role of advisory services in the support of apple orchard IPM is very important indeed (Blommers 1994). One form of expert advice is now commercially available as a computer package for British fruit growers (Morgan & Solomon 1993). 'PESTMAN' is an inexpensive system designed to help in rational decision making for apple and pear pest management. It is based on computer models that run on a desktop computer and simulate the development of pests in the orchard. It requires input of daily temperature data, usually provided by an on-farm automatic weather station or temperature logger, and it provides information such as start dates for flight activity and the timing of peak flight of both codling moth, *Cydia pomonella*, and also the summer fruit tortrix moth, *Adoxophyes orana* (Lepidoptera: Tortricidae). This allows growers to make informed decisions about the need to use insecticides, and if so, when. The software also deals with the sap-feeding pear psyllid, *Cacopsylla pyricola* (Hemiptera: Psyllidae), and it is able to forecast the timing of the life cycle of this pest so that the grower is able to target insecticidal treatments at the nymphal stage when predators are not abundant.

Insect pests of rice in SE Asia

The annual world production of paddy rice rises steadily but surely; statistics for the Food and Agriculture Organization of the United Nations (FAO 1998a) suggest 560×10^6 tons are produced globally, of which at least 507×10^6 tons are from Asia. Most of this production is intensive, where the crop is newly established every year by transplantation into standing water. This system favours the rice, but it also benefits certain pests and diseases by producing suitable microclimatic conditions (Lee 1992). Changes in rice cultivation (Plate 10.7, opposite p. 158) such as double cropping (two crops per year), growing high yielding but susceptible varieties that require heavy fertilizer treatment, earlier transplanting, close spacing and intensive use of pesticides have all contributed to very substantial increases in pest problems. An additional problem concerns the enormous diversity of insects that have the potential to exact damage to rice. In Taiwan, for example, more than 40 diseases and 135 species of insect attack rice (Chen & Yeh 1992). Many of these pests are by no means restricted to one country, and are distributed across very large areas of the world. Amongst the most

Table 10.23 Examples of tortricid moths as apple pests in Europe and their IPM. (From Blommers 1994.)

Pest	Code	Common name	Nature of damage
Cydia pomonella	A	Codling moth	Larvae tunnel in fruit
Archips podana	B	Apple leafroller moth	Defoliator, blossom bud feeder
Adoxophyes orana	C	Summer fruit tortrix moth	Leaf rollers, fruit borers

Control system	Time of application	Pest species code
Fenoxycarb (juvenile hormone)	Before flowering in spring	A, B, C
Bacillus thuringiensis	Summer	B, C
Granulosis virus (codling moth GV)	Several treatments during egg hatch period	A
Mating disruption pheromones	Beginning of adult flight period	A, B
Sex-attractants for monitoring	Beginning of adult flight period	A, B
Parasitoids and predators	Usually present, especially in non-insecticide treated orchards	A, B, C

serious in SE Asia are the brown plant-hopper, *Nilaparvata lugens* (Hemiptera: Delphacidae), green rice leafhopper, *Nephotettix* spp. (Hemiptera: Cicadellidae), whitebacked plant-hopper, *Sogatella furcifera* (Hemiptera: Delphacidae), rice stem borer, *Chilo suppressalis* (Lepidoptera: Pyralidae), rice leaf-folder, *Cnaphalocrocis medinalis* (Lepidoptera: Pyralidae), and armyworm, *Spodoptera* spp. (Lepidoptera: Noctuidae). Brown plant-hopper (BPH), for instance, has become increasingly severe all over SE Asia in the last decade or so. In Thailand, it has damaged between 300 000 and 400 000 ha of rice in 25 provinces (Rumakom *et al.* 1992). In Indonesia, BPH increased from 100 to 1000 times after the use of large quantities of pesticides (FAO 1998a), mainly due, it is suggested, to deleterious effects on natural enemies.

Depending on the growth stage of the crop, and the time of year, certain pests and diseases can be expected to be more serious than others. Figure 10.34 illustrates this variation in rice problems as the year and the crop varies in Vietnam (Thuy & Thieu 1992). In countries with less climatic variation, however, these temporal zonations of pest incidence are likely to be less pronounced, but the temptation to spray almost continually with insecticides is very great, as there has been until relatively recently little scope for monitoring and prediction in many rice cultivational systems (Teng 1994). None the less, many SE Asian countries such as China (Zhaohui *et al.* 1992) have managed to reduce the quantities of insecticide applied to paddy rice, with no obvious reduction in yields, and IPM systems are being introduced instead.

One of the key components of IPM in rice is the use of host plant resistance, and these days a large number of rice varieties are available that exhibit multiple resistance to insect pests and/or diseases (Heinrichs 1996). The International Rice Research Institute (IRRI), based in the Philippines, has been working on the development of plant resistance since the early 1960s, and more than 400 resistant cultivars of rice have now been identified. Many of these cultivars are now in widespread use over much of the world. In the Philippines, for example, 94% of the rice growing area is planted with modern crop varieties, whereas in China, Japan and the USA, all rice grown is of this type (Heinrichs 1996) (but see section 10.3.1).

As with most other IPM systems, rice pest management involves pest surveillance and forecasting, based on previously established economic thresholds for the major pests and diseases (Table 10.24). In Malaysia, nine pest surveillance centres distributed around the Peninsula examine 10% of the planted areas once a week for the major insect pests listed in the table, as well as a number of rice pathogens (Chin *et al.* 1992). If any of these threshold levels are exceeded, farmers are informed, and either spray chemicals of their own for small outbreaks, or take part in insecticidal treatments coordinated centrally for more widespread epidemics. In other countries, the most notorious pests have individual monitoring systems developed for them. Figure 10.35 shows a scheme formulated for brown plant-hopper in South Korea, wherein a highly organized system of monitoring and reporting is coordinated at county, city and village level. Of particular interest in this scheme is the routine consultation that goes into it between agronomists, plant pathologists and meteorologists, as well as entomologists (Choi *et al.* 1992). Chemical control is then recommended only when predeter-

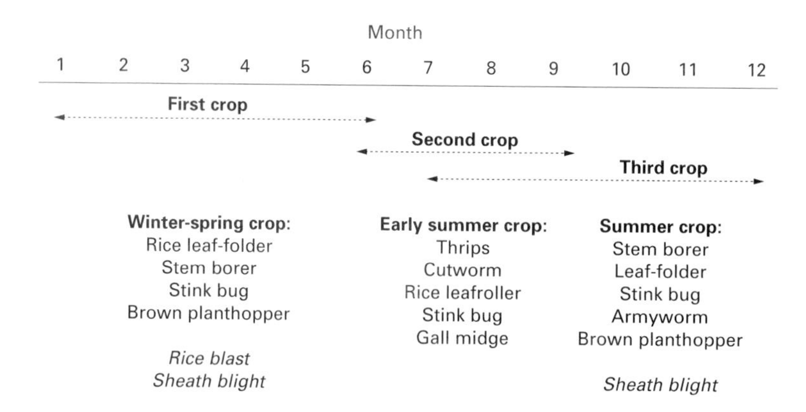

Fig. 10.34 Cropping seasons of rice in Vietnam, showing the variation in major pest species according to season (plant diseases in italics). (From Thuy & Thieu 1992.)

mined action (economic) thresholds are exceeded, and even then the chemicals used by rice growers are usually those recommended by the advisory bodies involved.

The reduction in chemical reliance for rice pest control is possible by virtue of an increase in other forms of management, which include the screening, breeding and use of multiple-resistant rice varieties, the enhancement and protection of natural enemies including microbials such as fungi, bacteria and viruses, and the adoption of new pesticide formulations and technologies such as ULV to apply them.

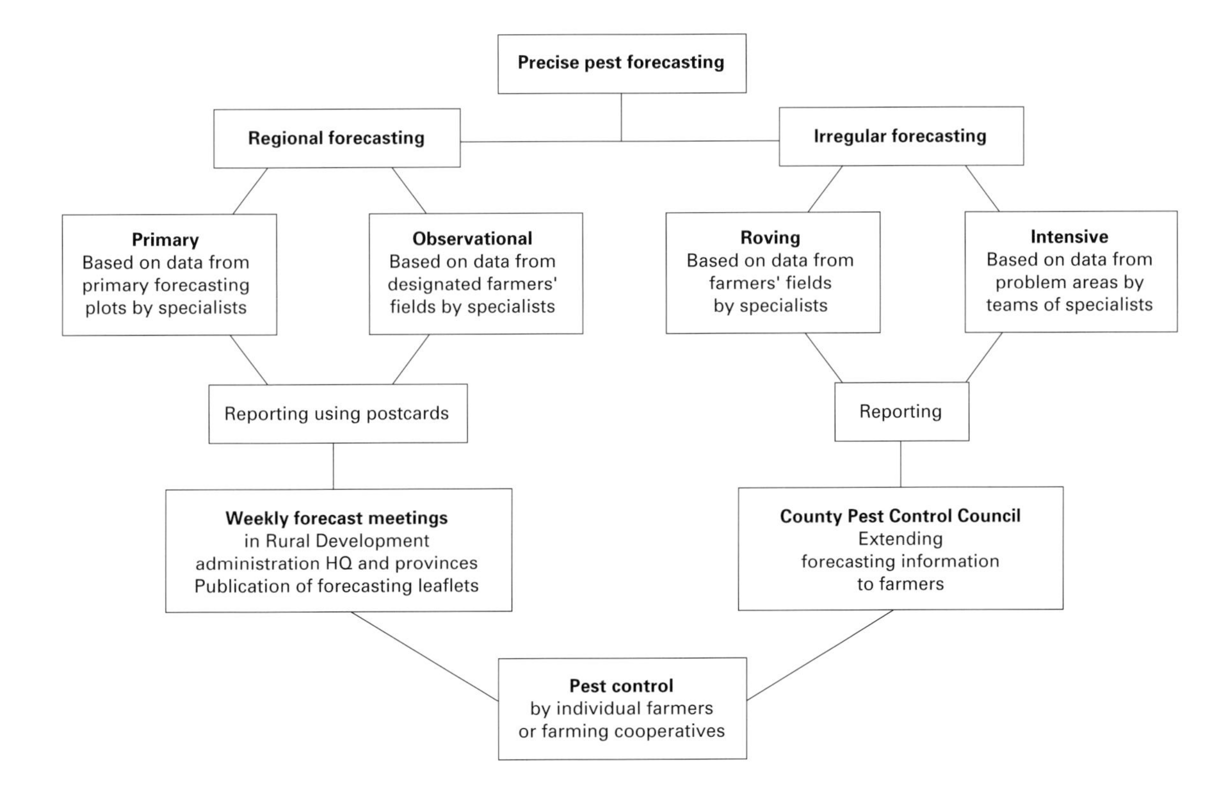

Fig. 10.35 Forecasting system for brown plant-hopper in rice crops in Korea. (Note the reliance on specialist input and efficient extension services.) (From Choi *et al.* 1992.)

Table 10.24 Major insect pests of rice in Malaysia, monitored under the pest surveillance system. (From Chin *et al.* 1992.)

Pest	Common name	Tentative economic threshold level
Nephotettix spp.	Green leaf hopper	1 adult/hill* if mosaic virus present. 50 adults/hill if no virus
Nilaparvata lugens	Brown plant-hopper	7 adults or 15 nymphs/hill
Sogatella furcifera	Whitebacked plant-hopper	7 adults or 25 nymphs/hill
Chilo spp.	Stem borer	10% infestation
Spodoptera spp.	Armyworm	10% infestation
Cnaphalocrocis medinalis	Leaf-folder	10% infestation

* A 'hill' is a single bundle of rice plants in the paddy.

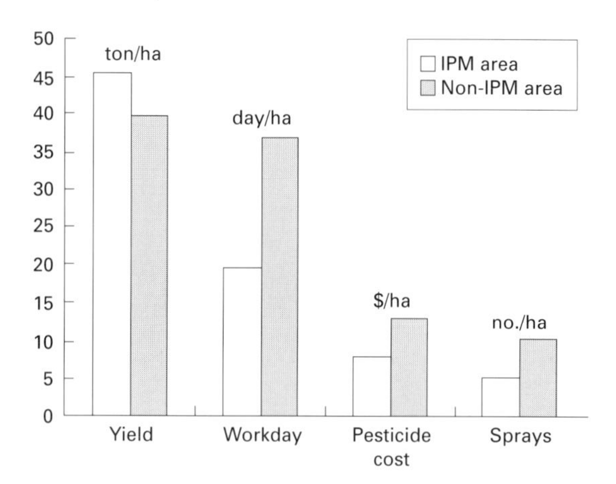

Fig. 10.36 Component average benefits in rice production in China comparing 104 IPM and non-IPM counties. (From Zhaohui *et al.* 1992.)

Modified rice husbandry systems whereby the crop is grown in as healthy and vigorous a manner as possible are also employed. The benefits of such an IPM system are many-fold. In Indonesia, FAO (1998a) has helped to train 200 000 farmers in IPM techniques for rice production, which has saved the Indonesian government around US$120 million in pesticide subsidies to farmers. In China, the adoption of IPM for rice pests and diseases has resulted not only in an increase in crop yield, but also in a reduction in the use of insecticides and a substantial decrease in the workload of farmers (Fig. 10.36) (Zhaohui *et al.* 1992).

The development of appropriate IPM tactics for a particular set of pests, crop type, country and socio-economic characteristics is under the central guidance of institutions such as the IRRI and FAO's intercountry programme on rice IPM (Teng 1994), illustrating the need for coordinated, multidisciplinary approaches to the widespread adoption of IPM tactics. IPM systems as complex and large scale as this one for rice are under constant development and fine-tuning. No pest management package can be considered permanent. The ecology of insects is no more constant than the crop types and climates in which they live and evolve—the war against insect pests will never be won, though a few battles may be chalked up to the pest manager or grower.

References

Abbott, I. (1993) Review of the ecology and control of the introduced bark beetle, *Ips grandicollis* (Eichhoff) (Coleoptera: Scolytidae) in Western Australia, 1952–90. *CalmScience*, **1** (1), 35–46.

Abel, C.A., Wilson, R.L. & Robbins, J.C. (1995) Evaluation of Peruvian maize for resistance to European corn borer (Lepidoptera: Pyralidae) leaf feeding and ovipositional preference. *Journal of Economic Entomology*, **88** (4), 1044–8.

Abrams, P. (1983) Theory of limiting similarity. *Annual Review of Ecological Syst*, **14**, 359–76.

Ackonor, J.B. & Vajime, C.K. (1995) Factors affecting *Locusta migratoria migratorioides* egg development in the Lake Chad basin outbreak area. *International Journal of Pest Management*, **41** (2), 87–96.

Adams, E.S. & Tschinkel, W.R. (1995) Effects of foundress number on brood raids and queen survival in the fire ant, *Solenopsis invicta*. *Behavioral Ecology Sociobiology*, **37**, 233–42.

Adler, F.R. & Karban, R. (1994) Defended fortresses or moving targets? another model of ducible defenses inspired by military metaphors. *American Naturalist*, **144**, 813–32.

Adler, L.S., Schmitt, J. & Bowers, M.D. (1995) Genetic variation in defensive chemistry in *Plantago lanceolata* (Plantaginaceae) and its effect on the specialist herbivore *Junonia coenia* (Nymphalidae). *Oecologia (Berlin)*, **101**, 75–85.

Agoua, H., Alley, E.S., Hougard, J.M., Akpoboua, K.L.B., Boatin, B. & Seketeli, A. (1995) Procedure of definitive cessation of larviciding in the Onchocerciasis Control Programme in West Africa: Entomological post-control studies. *Parasite*, **2** (3), 281–8.

Aguilar, J.M. & Boecklen, W.J. (1992) Patterns of herbivory in the *Quercus-grisea* × *Quercus gambelli* species complex. *Oikos*, **64**, 498–504.

Ahmad, S., Govindarajan, S., Funk, C.R. & Johnson-Cicalese, J.M. (1985) Fatality of house crickets on perennial ryegrasses (*Lolium perenne*) infected with a fungal endophyte. *Entomologia Experimentalis et Applicata*, **39**, 183–90.

Ahmad, S., Govindajaran, S., Johnson-Cicalese, J.M. & Funk, C.R. (1987) Association of a fungal endophyte in perennial ryegrass with antibiosis to larvae of the southern armyworm, *Spodoptera eridania*. *Entomologia Experimentalis et Applicata*, **43**, 287–94.

Ahmed, S.I. (1995) Investigations on the nuclear polyhedrosis of teak defoliator, *Hyblaea puera* (Cram) (Lepidoptera: Hyblaeidae). *Journal of Applied Entomology*, **119** (5), 351–4.

Agrios G.N. (1988) *Plant Pathology*, 3rd edn, p. 803. Academic Press, London.

Aide, T.M. (1992) Dry season leaf production an escape from herbivory. *Biotropica*, **24**, 532–7.

Akimoto, S. & Yamaguchi, Y. (1994) Phenotype selection on the process of gall formation of a *Tetranuera* aphid (Pemphigidae). *Journal of Animal Ecology*, **63**, 727–38.

Akinlosotu, T.A. (1973) *The role of* Diaeretiella rapae *(McIntosh) in the control of the cabbage aphid*. PhD Thesis, University of London.

Alagusunduram, K., Jayas, D.S., White, N.D.G., Muir, W.E. & Sinha, R.N. (1995) Controlling *Cryptolestes ferrugineus* (Stephens) adults in wheat stored in bolted-metal bins using elevated carbon dioxide. *Canadian Agricultural Engineering*, **37** (3), 217–23.

Alborn, H.T., Turlings, T.C.J., Jones, T.H., Stenhagen, G., Loughrin, J.H. & Tumlinson, J.H. (1997) An elicitor of plant volatiles from beet armyworm oral secretion. *Science*, **276**, 945–9.

Aleksic, I., Gliksman, I., Milanovic, D. & Tucic, N. (1993) On *r*- and *K*-selection: Evidence from the bean weevil (*Acanthoscelides obtectus*). *Zeitschrift fuer Zoologische Systematik und Evolutionsforschung*, **31** (4), 259–68.

Allan, J.D. (1995) Stream ecology. *Structure and Function of Running Waters*. Chapman & Hall, London, UK.

Allee, W.C., Emerson, O., Park, T. & Schmidt, K.P. (1949) *Principles of Animal Ecology*. Saunders, Philadelphia, PA.

Allison, A., Samuelson, G.A. & Miller, S.E. (1993) Patterns of beetle species diversity in New Guinea rainforest revealed by canopy fogging: preliminary findings. *Selbyana*, **14**, 16–20.

Allsopp, P.G., Bull, R.M. & McGill, N.G. (1991) Effect of *Antitrogus consanguineus* (Blackburn) (Coleoptera: Scarabaeidae) infestations on sugarcane yield in Australia. *Crop Protection*, **10** (3), 205–8.

Alstad, D.N. (1998) Population structure and the conundrum of local adaptation. In: *Genetic Structure and Local Adaptation in Natural Insect Populations. Effects of Ecology, Life History, and Behavior* (eds S. Mopper & S.Y. Strauss), pp. 3–21. Chapman & Hall, New York.

Alstad, D.N. & Corbin, K.W. (1990) Scale insect allozyme differentiation within and between host trees. *Ecological Ecology*, **4**, 43–56.

Alward, R.D. & Joern, A. (1993) Plasticity and overcompensation in grass responses to herbivory. *Oecologia*, **95**, 358–64.

Ambrus, A. & Csoka, G. (1992) Density estimation and swarming studies on the wintermoth, *Operophtera brumata* L. (Lep.: Geometridae) by marking and using pheromone traps. *Anzeiger fuer Schadlingskunde Pflanzenschutz Umweltschutz*, **65** (5), 88–92.

Amerasinghe, F.P. & Indrajith, N.G. (1994) Post-irrigation breeding patterns of surface water mosquitoes in the Mahaweli Project, Sri Lanka, and comparisons with preceding developmental phases. *Journal of Medical Entomology*, **31** (4), 516–23.

Andersen, A.N. (1995) Measuring more of biodiversity—genus richness as a surrogate for species richness in Australian ant faunas. *Biological Conservation*, **73**, 39–43.

Anderson, J.B. & Brower, L.P. (1996) Freeze-protection of overwintering monarch butterflies in Mexico: Critical role of the forest as a blanket and an umbrella. *Ecological Entomology*, **21** (2), 107–16.

Anderson, N.H., Steedman, R.J. & Dudley, T. (1984) Patterns of exploitation by stream invertebrates of wood debris (xylophagy). *Verh Int Version Limnol*, **22**, 1847–52.

Andersen, N.M. (1991) Marine insects: Genital morphology, phylogeny and evolution of sea skaters, genus *Halobates* (Hemiptera. Gerridae). *Zoological Journal of the Linnean Society*, **103** (1), 21–60.

Andersen, N.M. (1997) Phylogenetic tests of evolutionary scenarios: The evolution of flightlessness and wing polymorphism in insects. *Mémoires du Muséum National d'Histoire Naturelle*, **173**, 91–108.

Andersen, N.M., Farma, A., Minelli, A. & Piccoli, G. (1994) A fossil *Halobates* from the Mediterranean and the origin of sea skaters (Hemiptera: Gerridae). *Zoological Journal of the Linnean Society*, **112** (4), 479–89.

Anderson, R.M. & May, R.M. (1991) *Infectious Diseases of Humans: Dynamics and Control*. Oxford University Press, Oxford.

Andrade, A.L.S.S.D., Zicker, F., Oliveira, R.M.D., Silva, I.G.D., Silva, S.A., Andrade, S.S.D. & Martelli, C.M.T. (1995a) Evaluation of risk factors for house infestation by *Triatoma infestans* in Brazil. *American Journal of Tropical Medicine and Hygiene*, **53** (5), 443–7.

Andrade, A.L.S.S.D., Zicker, F., Silva, I.G., Souza, J.M.P. & Martelli, C.M.T. (1995b) Risk factors for *Trypanosoma cruzi* infection among children in central Brazil: a case-control study in vector control settings. *American Journal of Tropical Medicine and Hygiene*, **52** (2), 183–7.

Andrewartha, H.G. & Birch, L.C. (1954) *The Distribution and Abundance of Animals*, p. 782. University of Chicago Press, Chicago, IL.

Anonymous (1992) *Climate Change* (1992) Supplementary report to the Intergovernmental panel on climate change scientific assessment. Cambridge University Press, New York.

Anonymous (1996) No-go verandahs. *The Guardian*, 19 September 1996.

Appel, H.M. (1993) Phenolics in ecological interactions: The importance of oxidation. *Journal of Chemical Ecology*, **19**, 1521–52.

Appel, H.M. & Maines, L.W. (1995) The influence of host plant on gut conditions of gypsy moth (*Lymantria dispar*) caterpillars. *Journal of Insect Physiology*, **41**, 241–6.

Arn, H., Toth, M. & Priesner, E. (1996) Pherolist. http://www.nysaes.cornell.edu/fst/faculty/acree/pheronet/index.html.

Arocha-Pinango, C.I., de Bosch, N.B., Torres, A., *et al.* (1992) Six new cases of a caterpillar-induced bleeding syndrome. *Thrombosis and Haemostasis*, **67** (4), 402–7.

Ash, S. (1996) Evidence of arthropod–plant interactions in the Upper Triassic of the southwestern United States. *Lethaia*, **29** (3), 237–48.

Askew, R.R. (1971) *Parasitic Insects*. Heinemann, London.

Auerbach, M. (1991) Relative impact of interactions within and between trophic levels during an insect outbreak. *Ecology*, **72**, 1599–608.

Averof, M. & Cohen, S.M. (1997) Evolutionary origin of insect wings from ancestral gills. *Nature*, **385** (6617), 627–30.

Ayala, F.J. (1970) Competition, coexistence, and evolution. In: *Essays in Evolution and Genetics* (eds M.K. Hecht & W.C. Steere). Appleton-Century-Crofts, New York.

Ayala, F.J. (1971) Frequency-dependent competition. *Science*, **171**, 820–4.

Ayres, M.P., Suomela, J. & McLean, S.F. Jr. (1987) Growth performance of *Epirrita autumnata* (Lepidoptera: Geometridae) on mountain birch: trees, broods, and tree x brood interactions. *Oecologia*, **74**, 450–7.

Babuder, G., Pohleven, F. & Brelih, S. (1996) Selectivity of synthetic aggregation pheromones Linoprax© and Pheroprax© in the control of the bark beetles (Coleoptera: Scolytidae) in a timber storage yard. *Journal of Applied Entomology*, **120**, 131–6.

Baik, D.H. & Joo, C.Y. (1991) Epidemio-entomological survey of Japanese encephalitis in Korea. *Korean Journal of Parasitology*, **29** (1), 67–86.

Baker, J.E., Perez-Mendoza, J. & Beeman, R.W. (1997) Inheritance of malathion resistance in the parasitoid *Anisopteromalus calandrae* (Hymenoptera: Pteromalidae). *Journal of Economic Entomology*, **90** (2), 304–8.

Bakhvalov, S.A. & Bakhvalova, V.N. (1990) Ecology of a baculovirus in *Ocneria monacha* L. (Lepidoptera: Lymantriidae): Virus persistence in insect populations. *Ekologiya (Sverdlovsk)*, **6**, 53–9.

Bakker, K., Bagehee, S.N., Van Zwet, W.R. & Meelis, E. (1967) Host discrimination in *Pseudeucola bochei* (Hymenoptera: Cynipidae). *Entomologia Experimentalis et Applicata*, **10**, 295–311.

Balas, M.T. & Adams, E.S. (1996) Nestmate discrimination and competition in incipient colonies of fire ants. *Animal Behavior*, **51**, 49–59.

Baldwin, I.T. & Schultz, J.C. (1983) Rapid changes in tree leaf chemistry induced by damage: Evidence for communication between plants. *Science*, **221**, 277–9.

Baldwin, I.T. & Schultz, J.C. (1988) Phylogeny and the patterns of leaf phenolics in gap and forest-adapted *Piper* and *Miconia* understory shrubs. *Oecologia (Heidelberg)*, **75**, 105–9.

Balmford, A. & Long, A. (1995) Across country analyses of biodiversity congruence and current conservation effort in the tropics. *Conservation Biology*, **9**, 1539–47.

Balmford, A., Green, M.J.B. & Murray, M.G. (1996a) Using higher-taxon richness as a surrogate for species richness. 1. Regional tests. *Proceedings of the Royal Society of London, Series B: Biological Sciences*, **263**, 1267–74.

Balmford, A., Jayasuriya, A.H.M. & Green, M.J.B. (1996b) Using higher-taxon richness as a surrogate for species richness. 2. Local applications. *Proceedings of the Royal Society of London, Series B: Biological Sciences*, **263**, 1571–5.

Baltensweiler, W. (1984) The role of environment and reproduction in the population dynamics of the larch budmoth *Zieraphera diniana* Gn. (Lep.: Torticidiae). In: *Advances in Invertebrate Reproduction*, Vol. 3 (eds W. Engels, W.H. Clark, A. Fischer, P.J.W. Olive & F.F. Went), pp. 291–301. Elsevier, Amsterdam.

Baltensweiler, W. & Fischlin, A. (1988) The larch budmoth in the Alps. In: *Dynamics of Forest Insect Populations: Patterns, Causes, Implications* (ed. A.A. Berryman), pp. 331–351. Plenum, New York.

Barbour, D.A. (1985) Patterns of population fluctuation in the pine looper moth *Bupalus piniaria* L. in Britain. In: *Site Characteristics and Population Dynamics of Lepidotern and Hymenopteran Forest Pests* (eds D. Bevan & J.T. Stoakley), pp. 8–20. HMSO, London.

Barbour, D.A. (1988) The pine looper in Britain and Europe. In: *Dynamics of Forest Insect Populations. Patterns, Causes, Implications* (ed. A.A. Berryman), pp. 291–308. Plenum, New York.

Bari, A. & Qazilbash, A.A. (1995) Plague: A medieval killer in the present world: A review. *Punjab University Journal of Zoology*, **10**, 125–35.

Barlow, N.D. & Dixon, A.F.G. (1980) *Simulation of Lime Aphid Population Dynamics*. Pudoc, Wageningen.

Barnby, M.A. & Resh, V.H. (1988) Factors affecting the distribution of an endemic and a widespread species of brine fly (Diptera: Ephydridae) in a northern California (USA) thermal saline spring. *Annals of the Entomological Society of America*, **81** (3), 437–46.

Barquero, M. & Constenla, M.A. (1986) Remnants of organochlorine pesticides in human fatty tissue in Costa Rica. *Revista de Biologia Tropical*, **34** (1), 7–12.

Bartlet, E., Parson, D., Williams, I.H. & Clark, S.J. (1994) The influence of glucosinolates and sugars on feeding by the cabbage stem flea beetle, *Psylliodes chrysocephala*. *Entomologia Experimentalis et Applicata*, **73**, 77–83.

Basaglia, M., Concheri, G., Cardinali, S., Pasti-Grigsby, M.B. & Nuti, M.P. (1992) Enhanced degradation of ammonium-pretreated wheat straw by lignocellulolytic *Streptomyces* spp. *Canadian Journal of Microbiology*, **38**, 1022–5.

Basanez, M.G., Townson, H., Williams, J.R., Frontado, H., Villamizar, N.J. & Anderson, R.M. (1996) Density-dependent processes in the transmission of human onchocerciasis: Relationship between microfilarial intake and mortality of the simuliid vector. *Parasitology*, **113** (4), 331–55.

Bass, M. & Cherett, J.M. (1995) Fungal hyphae as a source of nutrients for the leaf-cutting ant *Atta sexdens*. *Physiological Entomology*, **20** (1), 1–6.

Basset, Y. (1991) The taxonomic composition of the arthropod fauna associated with an Australian rain-forest tree. *Australian Journal of Zoology*, **39**, 171–90.

Basset, Y. (1992) Host specificity of arboreal and free-living insect herbivores in rain forests. *Biological Journal of the Linnean Society*, **47** (2), 115–33.

Basset, Y., Samuelson, G.A., Allison, A. & Miller, S.E. (1996) How many species of host-specific insects feed on a species of tropical tree? *Biological Journal of the Linnean Society*, **59**, 201–16.

Bassuk, N.L., Hunter, L.D. & Howard, B.H. (1981) The apparent involvement of polyphenol oxidase and phloridzin in the production of apple rooting cofactors. *Journal of Horticultural Science*, **56**, 313–22.

Bazely, D.R., Da Myers, J.H. & Silva, K.B. (1991) The response of numbers of bramble prickles to herbivory and depressed resource availability. *Oikos*, **61**, 327–36.

Beaver, R.A. (1989) Insect–fungus relations in the bark and ambrosia beetles. In: *Insect–Fungus Interactions 14th Symposium of the Royal Entomological Society of London in Collaboration with the British Mycological Society* (eds N. Wilding, N.M. Collins, P.M. Hammond & J.F. Webber), pp. 121–43. Academic Press, London.

Beccaloni, G.W. & Gaston, K.J. (1995) Predicting the species richness of neotropical forest butterflies - Ithomiinae (Lepidoptera, Nymphalidae) as indicators. *Biological Conservation*, **71**, 77–86.

Becerra, J.X. (1997) Insects on plants: Macroevolutionary chemical trends in host use. *Science*, **276**, 253–6.

Beck, L.R., Rodriguez, M.H., Dister, S.W., *et al.* (1994) Remote sensing as a landscape epidemiological tool to identify villages at high risk for malaria transmission. *American Journal of Tropical Medicine and Hygiene*, **51** (3), 271–80.

Becker, N., Zgomba, M., Petric, D., Beck, M. & Ludwig, M. (1995) Role of larval cadavers in recycling processes of *Bacillus sphaericus*. *Journal of the American Mosquito Control Association*, **11** (3), 329–34.

Beddington, J.R., Free, C.A. & Lawton, J.H. (1975) Dynamic complexity in predator–prey models framed in difference equations. *Nature*, **225**, 58–60.

Bedford, S.E. & Usher, M.B. (1995) Distribution of arthropod species across the margins of farm woodlands. *Agriculture, Ecosystems and Environment*, **49**, 295–305.

Begon, M., Harper, J.L. & Townsend C.R. (1986) *Ecology—Individuals, Populations and Communities*. Blackwell Scientific Publications, Oxford.

Begon, M., Harper, J.L. & Townsend, C.R. (1990) *Ecology—Individuals Populations and Communities*. Blackwell Scientific Publications, Inc, Cambridge, MA.

Begon, M., Harper, J.L. & Townsend, C.R. (1996) *Ecology*, 3rd edn. Blackwell Science, Oxford.

Beier, J.C. (1998) Malaria parasite development in mosquitoes. *Annual Review of Entomology*, **43**, 519–43.

Beingolea, G.O.D. (1985) The locust *Schistocerca interrita* in northern coast of Peru during 1983. *Revista Peruana de Entomologia*, **28**, 35–40.

Bejer-Peterson, B. (1972) The nun moth, *Lymantria monacha* L., in Denmark (Lep., Lymantriidae). *Entomologiske Meddelelser*, **40**, 129–39.

Bell, G., Lechowicz, M.J., Appenzeller, A. *et al.* (1993) The spatial structure of the physical environment. *Oecologia (Heidelberg)*, **96** (1), 114–21.

Belli, G., Fortusini, A., Casati, P., Belli, L., Bianco, P.A. & Prati, S. (1994) Transmission of a grapevine leafroll associated closterovirus by the scale insect *Pulvinaria vitis* L. *Rivista di Patologia Vegetale*, **4** (3), 105–8.

Bellows, T.S., Vandriesche, R.G. & Elkinton, J.S. (1992) Life-table construction and analysis in the evaluation of natural enemies. *Annual Review of Entomology*, **37**, 587–614.

Belsky, A.J., Carson, W.P., Jensen, C.L. & Fox, G.A. (1993) Overcompensation by plants, herbivore optimization or red herring? *Evolutionary Ecology*, **7**, 109–21.

Bennett, S. & Mill, P.J. (1995) Lifetime egg production and egg mortality in the damselfly *Pyrrhosoma nymphula* (Sulzer) (Zygoptera: Coenagrionidae). *Hydrobiologia*, **310** (1), 71–8.

Bennetts, D.A. (1995) The Hadley Centre transient climate experiment. In: *Insects in a Changing Environment* (eds R. Harrington & N.E. Stork), pp. 49–58. Academic Press, London.

Bennewicz, J. (1995) Assessment of susceptibility of varieties and lines of sugar beet to feeding of aphids: Black bean aphid (*Aphis fabae* Scop.) and green peach aphid (*Myzus persicae* Sulz.). *Hodowla Roslin Aklimatyzacja i Nasiennictwo*, **39** (3), 41–72.

Benz, G. (1974) Negative feedback by competition for food and space, and by cyclic induced changes in the nutritional base as regulatory principles in the population dynamics of the larch budmoth, *Zeiraphera diniana* (Guenee) (Lep., Tortricidae). *Zeitschrift fuer Angewandte Entomologie*, **76**, 196–228.

Berenbaum, M. (1980) Adaptive significance of midgut pH in larval lepidoptera. *American Naturalist*, **115**, 138–46.

Berenbaum, M.R. (1983) Coumarins and caterpillars: a case for coevolution. *Evolution*, **37**, 163–79.

Berenbaum, M.R. (1995) Turnabout is fair play: secondary roles for primary compounds. *Journal of Chemical Ecology*, **21** (7), 925–40.

Berg, H. (1995) Modelling of DDT dynamics in Lake Kariba, a tropical man-made lake, and its implications for the control of tsetse flies. *Annales Zoologici Fennici*, **32** (3), 331–53.

Bergelson, J. & Crawley, M.J. (1992) The effects of grazers on the performance of individuals and populations of scarlet gilia, *Ipomopsis aggregata*. *Oecologia (Heidelberg)*, **90**, 435–44.

Bergelson, J. & Purrington, C.B. (1996) Surveying patterns in the cost of resistance in plants. *American Naturalist*, **148**, 536–58.

Bergivinson, D.J., Hamilton, R.I. & Arnason, J.T. (1995a) Leaf profile of maize resistance factors to European corn borer, *Ostrinia nubilalis*. *Journal of Chemical Ecology*, **21**, 343–54.

Bergivinson, D.J., Larsen, J.S. & Arnason, J.T. (1995b) Effect of light on changes in maize resistance against the European corn borer, *Ostrinia nubilalis* (Hubner). *Canadian Entomologist*, **127**, 111–22.

Bernays, E.A. & Graham, M. (1988) On the evolution of host specificity in phytophagous arthropods. *Ecology*, **69**, 886–92.

Berry, N.A., Wratten, S.D., Mcerlich, A. & Frampton, C. (1996) Abundance and diversity of beneficial arthropods in conventional and organic carrot crops in New Zealand. *New Zealand Journal of Crop and Horticultural Science*, **24** (4), 307–13.

Berryman, A.A. (1973) Population dynamics of the fir engraver, *Scolytus ventralis* (Coleoptera: Scolytidae). I. Analysis of population behavior and survival from 1964 to 1971. *Canadian Entomologist*, **105**, 1465–88.

Berryman, A.A. (1987) The theory and classification of outbreaks. In: *Insect Outbreaks* (eds P. Barbosa & J.C. Schultz), pp. 3–29. Academic Press, San Diego.

Beshers, S.N. & Traniello, J.F.A. (1994) The adaptiveness of worker demography in the attine ant *Trachymyrmex septentrionalis*. *Ecology (Tempe)*, **75** (3), 763–75.

Bevan, D. (1987) *Forest Insects. Forestry Commission Handbook 1*. HMSO, London.

Bezemer, T.M., Jones, T.H. & Knight, K.J. (1998) Long-term effects of elevated CO_2 and temperature on populations of the peach-potato aphid *Myzus persicae* and its parasitoid *Aphidus matricariae*. *Oecologia*, **116**, 128–35.

Bibby, C.J., Crosby, M.J., Heath, M.F., *et al.* (1992) *Putting Biodiversity on the Map: Global Priorities for Conservation*. ICBP, Cambridge.

Biconet (1996) Biocontrol Network—Integrated pest management solutions for a small planet. World Wide Web site http://www.usit.net/bionet.

Bignell, D.E., Eggleton, P., Nunes, L. & Thomas, K.L. (1997) Termites as mediators of carbon fluxes in tropical forests: budgets for carbon dioxide and methane emissions. In: *Forests and Insects* (eds A.D. Watt, N.E. Stork & M.D. Hunter), pp. 109–34. Chapman & Hall, London.

Bignell, D.E., Slaytor, M., Veivers, P.C., Muhlemann, R. & Leuthold, R.H. (1994) Functions of symbiotic fungus gardens in higher termites of the genus *Macrotermes*: Evidence against the acquired enzyme hypothesis. *Acta Microbiologica et Immunologica Hungarica*, **41** (4), 391–401.

Biller, A., Boppre, M., Witte, L. & Hartmann, T. (1994) Pyrrolizidine alkaloids in *Chromolaena odorata*: Chemical and chemoecological aspects. *Phytochemistry (Oxford)*, **35**, 615–9.

Birch, L.C. (1953) Experimental background to the study of the distribution and abundance of insects. I. The influence of temperature, moisture, and food on the innate capacity for increase of three grain beetles. *Ecology*, **34**, 698–711.

Birch, M.C. (1978) Chemical Communication in Pine Bark Beetles. *American Scientist*, **66**, 409–19.

Birch, M.C., Svihra, P., Paine, T.D. & Miller, J.C. (1980) Influence of chemically mediated behavior on host tree colonization by four cohabitating species of bark beetles. *Journal of Chemical Ecology*, **6**, 395–414.

Bissan, Y., Hougard, J.M., Doucoure, K., *et al.* (1995) Drastic reduction of populations of *Simulium sirbanum* (Diptera: Simuliidae) in central Sierra Leone after 5 years of larviciding operations by the Onchocerciasis Control Programme. *Annals of Tropical Medicine and Parasitology*, **89** (1), 63–72.

Blakely, N. & Dingle, H. (1978) Butterflies eliminate milkweed bugs from a Carribean island. *Oecologia*, **37**, 133–6.

Blackman, M.W. (1924) The effect of deficiency and excess in rainfall upon the hickory bark beetle. *Journal of Economic Entomology*, **17**, 460–70.

Blackman R.L. & Easthop V.F. (1984) *Aphids on the World's Crops: an Identification and Information Guide*, 466 pp. Wiley, Chichester.

Blank, R.H., Gill, G.S.C., Oslon, M.H. & Upsdell, M.P. (1995) Greedy scale (Homoptera: Diaspididae) phenology on taraire based on Julian day and degree-day accumulations. *Environmental Entomology*, **24** (6), 1569–75.

Blommers, L.H.M. (1994) Integrated pest management in European apple orchards. *Annual Review of Entomology*, **39**, 213–43.

Bockarie, M.J., Service, M.W., Toure, Y.T., Traore, S., Barnish, G. & Greenwood, B.M. (1993) The ecology and behaviour of the forest form of *Anopheles gambiae s.s. Parassitologia*, **35** (Suppl), 5–8.

Boecklen, W.J. & Spellenburg, R. (1990) Structure of herbivore communities in two oak (*Quercus* spp.) hybrid zones. *Oecologia (Heidelberg)*, **85**, 92–100.

Boiteau, G., Osborn, W.P.L. & Drew, M.E. (1997) Residual activity of imidacloprid controlling Colorado potato beetle (Coleoptera: Chrysomelidae) and three species of potato colonizing aphids (Homoptera: Aphidae). *Journal of Economic Entomology*, **90** (2), 309–19.

Bokononganta, A.H., Vanalphen, J.J.M. & Neuenschwander, P. (1996) Competition between *Gyranusodea tebygi* and *Anagyrus mangicola*, parasitoids of the mango mealybug, *Rastrococcus invadens*, Interspecific host discrimination and larval competition. *Entomologia Experimentalis et Applicata*, **79**, 179–85.

Bombick, D.W., Arlian, L.G. & Livingston, J.M. (1987) Toxicity of jet fuels to several terrestrial insects. *Archives of Environmental Contamination and Toxicology*, **16** (1), 111–18.

Bond, W.J. & Stock, W.D. (1989) The costs of leaving home: ants disperse myrmecochorous seeds to low nutrient sites. *Oecologia*, **81**, 412–17.

Bonsall, M.B. & Hassell, M.P. (1995) Identifying density-dependent processes—a comment on the regulation of winter moth. *Journal of Animal Ecology*, **64**, 781–4.

Bonsall, M.B., Jones, T.H. & Perry, J.N. (1998) Determinants of dynamics: population size, stability and persistence. *Trends in Ecology & Evolution*, **13**, 174–6.

Borchert, R. (1994) Water status and development of tropical trees during seasonal drought. *Trees*, **8**, 115–25.

Borden, J.H. (1984) Semiochemical-mediated aggregation and dispersion. In: *Insect Communication*. 12th Symposium Royal Entomological Society, London. (ed. T. Lewis). Academic Press, San Diego.

Borkent, A. (1996) Biting midges from Upper Cretaceous New Jersey amber (Ceratopogonidae: Diptera). *American Museum Novitates*, 3159, Feb. 15, 1–29.

Borror, D.J., De Long, D.M. & Triplehorn, C.A. (1981) *An Introduction to the Study of Insects*, 5th edn. Saunders College Publishing, Philadelphia.

Bottrell, D.G., Barbosa, P. & Gould, F. (1998) Manipulating natural enemies by plant variety selection and modification: a realistic strategy? *Annual Review of Entomology*, **43**, 347–67.

Boudreaux, H.B. (1987) *Arthropod Phylogeny with Special Reference to Insects*, 320 pp. Krieger, Florida.

Bourchier, R.S. & Smith, S.M. (1996) Influence of environmental conditions and parasitoid quality on field performance of *Trichogramma minutum*. *Entomologia Experimentalis et Applicata*, **80** (3), 461–8.

Bowie, M.H., Wratten, S.D. & White, A.J. (1995) Agronomy and phenology of 'companion plants' of potential for enhancement of biological control. *New Zealand Journal of Crop and Horticultural Science*, **23** (4), 423–7.

Braasch, H. (1996) Pathogenicity tests with *Bursaphelenchus mucronatus* on pine and spruce seedlings in Germany. *European Journal of Forest Pathology*, **26** (4), 205–16.

Brabin, B.J., Verhoeff, F.H., Kazembe, P., Chimsuku, L. & Broadhead, R. (1997) Antimalarial drug policy in Malawi. *Annals of Tropical Medicine and Parasitology*, **91** (Suppl. 1), S113–S115.

Bradford, M.J. & Roff, D.A. (1993) Bet hedging and the diapause strategies of the cricket *Allonemobius fasciatus*. *Ecology (Tempe)*, **74** (4), 1129–35.

Bradshaw, W.E. (1983) Estimating biomass of mosquito populations. *Environmental Entomology*, **12** (3), 779–81.

Brandao, C.R.F., Martins-Neto, R.G. & Vulcano, M.A. (1989) The earliest known fossil ant (first southern hemisphere Mesozoic record) (Hymenoptera: Formicidae: Myrmeciinae). *Psyche*, **96** (3–4), 195–208.

Brandenburg, R.L. & Herbert, D.A. (1991) Effect of timing on prophylactic treatments for southern corn rootworm (Coleoptera: Chrysomelidae) in peanut. *Journal of Economic Entomology*, **84** (6), 1894–8.

Brasier, C.M. (1990) The unexpected element: mycovirus involvement in the outcome of two recent pandemics, Dutch elm disease and chestnut blight. In: *Pests, Pathogens and Plant Communities* (eds J.J. Burdon & S.R. Leather), pp. 289–308. Blackwell Scientific Publications, Oxford.

Braun, S. & Fluckiger, W. (1984) Increased population of the aphid *Aphis pomi* at a motorway. Part 2. The effect of drought and deicing salt. *Environmental Pollution Series,* **A 36**, 261–70.

Briese, D.T. (1996) Life history of the *Onopordum capitulum* weevil *Larinus latus* (Coleoptera: Curculionidae). *Oecologia (Berlin)*, **105** (4), 454–63.

Broakhuizen, N., Evans, H.F. & Hassell, M.P. (1993) Site characteristics and the population dynamics of the pine looper moth. *Journal of Animal Ecology,* **62** (3), 511–18.

Brodmann, P.A., Wilcox, C.V. & Harrison, S. (1997) Mobile parasitoids may restrict the spatial spread of an insect outbreak. *Journal of Animal Ecology,* **66**, 65–72.

Brookes, D.L., Ritchie, S.A., van den Hurk, A.F., Fielding, J.R. & Loewenthal, M.R. (1997) *Plasmodium vivax* malaria acquired in far north Queensland. *Medical Journal of Australia*, **166** (2), 82–3.

Brower, L.P. & Brower, J.V.Z. (1964) Birds, butterflies and plant poisons: a study in ecological biochemistry. *Zoologica*, **49**, 137–59.

Brower, L.P., Nelson, C.J., Seiber, J.N., Fink, L.S. & Bond, C. (1988) Exaptation as an alternative to coevolution in the cardenolide-based chemical defense of monarch butterflies (*Danaus plexippus* L.) against avian predators. In: *Chemical Mediation of Coevolution* (ed. K.C. Spencer), pp. 447–475. Academic Press, San Diego.

Brown, J.H. & Davidson, D.W. (1977) Competition between seed-eating rodents and ants in desert ecosystems. *Science*, **196**, 800–2.

Brown, J.M., Abrahamson, W.G., Packer, R.A. & Way, P.A. (1995) The role of natural enemy escape in a gallmaker host-plant shift. *Oecologia*, **104**, 52–60.

Brown, K.S. (1991) Conservation of neotropical environments: insects as indicators. In: *The Conservation of Insects and their Habitats* (eds N.M. Collins & J.A. Thomas), pp. 350–404. Academic Press, London.

Brown, M.W. (1993) Resilience of the natural arthropod community on apple to external disturbance. *Ecological Entomology,* **18** (3), 169–83.

Brown, M.W., Nebeker, T.E. & Honea, C.R. (1987) Thinning increases loblolly pine vigour and resistance to bark beetles. *Southern Journal of Applied Forestry,* **11** (1), 28–31.

Browne, J. & Peck, S.B. (1996) The long-horned beetles of south Florida (Cerambycidae: Coleoptera): Biogeography and relationships with the Bahama Islands and Cuba. *Canadian Journal of Zoology,* **74** (12), 2154–69.

Bruin, J., Dicke, M. & Sabelis, M.W. (1992) Plants are better protected against spider-mites after exposure to volatiles from infested conspecifics. *Experienta (Basel)*, **48**, 525–9.

Bruin, J., Sabelis, M.W. & Dicke, M. (1995) Do plants tap SOS signals from their infested neighbors? *Trends in Ecology and Evolution*, **10**, 167–70.

Brust, G.E. & House, G.J. (1990) Influence of soil texture, soil moisture, organic cover, and weeds on oviposition preference of southern corn rootworm (Coleoptera: Chrysomelidae). *Environmental Entomology,* **19** (4), 966–71.

Bryant, J.P., Chapin, F.S. & Klein, D.R. (1983) Carbon/nutrient balance of boreal plants in relation to vertebrate herbivory. *Oikos*, **40**, 357–68.

Bryant, J.P., Reichardt, P.B., Clausen, T.P. & Werner, R.A. (1993) Effects of mineral nutrition on delayed inducible resistance in Alaska paper birch. *Ecology*, **74**, 2072–84.

Bryceson, K.P. (1989) The use of Landsat MSS data to determine the distribution of locust eggbeds in the Riverina region of New South Wales, Australia. *International Journal of Remote Sensing,* **10** (11), 1749–62.

Bryceson, K.P. (1990) Digitally processed satellite data as a tool in detecting potential Australian plague locust outbreak areas. *Journal of Environmental Management,* **30** (3), 191–208.

Buchner, P. (1965) *Endosymbiosis of Animals with Plant Microorganisms*. Wiley, New York.

Bulmer, M. (1994) *Theoretical Evolutionary Ecology*. Sinauer Associates, Sunderland, MA.

Bultman, T.L. & Faeth, S.H. (1985) Patterns of intra- and inter-specific association in leaf-mining insects on three oak species. *Ecological Entomology*, **10**, 121–9.

Bultman, T.L. & Faeth, S.H. (1987) Impact of irrigation and experimental drought stress on leaf-mining insects of Emory oak. *Oikos*, **48**, 5–10.

Bultman, T.L. & Ganey, D.T. (1995) Induced resistance to fall armyworm (Lepidoptera: Noctuidae) mediated by a fungal endophyte. *Environmental Entomology*, **24**, 1196–200.

Bultman, T.L., Borowicz, K.L., Schneble, R.M., Coudron, T.A. & Bush, L.P. (1997) Effect of a fungal endophyte on the growth and survival of two *Euplectrus* parasitoids. *Oikos*, **78**, 170–6.

Buntin, G.D. & Raymer, P.L. (1994) Pest status of aphids and other insects in winter canola in Georgia. *Journal of Economic Entomology,* **87** (4), 1097–104.

Buntin, G.D., Hargrove, W.L. & McCracken, D.V. (1995) Populations of foliage-inhabiting arthropods on soybean with reduced tillage and herbicide use. *Agronomy Journal,* **87** (5), 789–94.

Burkot, T.R., Graves, P.M., Paru, R., Wirtz, R.A. & Heywood, P.F. (1988) Human malaria transmission studies in the *Anopheles punctulatus* complex in Papua New Guinea: sporozoite rates, inoculation rates, and sporozoite densities. *American Journal of Tropical Medicine and Hygiene*, **39** (2), 135–44.

Burges, H.D. (1993) Foreword. In: Bacillus thuringiensis, *an Environmental Biopesticide: Theory and Practice* (eds P.F. Entwistle, J.S. Cory, M.J. Bailey & S. Higgs) pp. xv–xvii. John Wiley, Chichester, UK.

Burnett, T. (1958) Dispersal of an insect parasite over a small plot. *Canadian Entomologist*, **90**, 279–83.

Buse, A. & Good, J.E.G. (1996) Synchronization of larval

emergence in winter moth (*Operophtera brumata* L.) and budburst in pedunculate oak (*Quercus robur* L.) under simulated climate change. *Ecological Entomology*, **21** (4), 335–43.

Bush, M.B. & Whittaker, R.J. (1991) Krakatau: colonization patterns and hierarchies. *Journal of Biogeography*, **18**, 341–56.

Byers, J.A. (1993) Avoidance of competition by spruce bark beetles, *Ips typographus* and *Pityogenes chalcographus*. *Experientia*, **49**, 272–5.

Bylund, H. & Tenow, O. (1994) Long-term dynamics of leaf miners, *Eriocrania* spp, on mountain birch – Alternate year fluctuations and interaction with *Epirrita autumnata*. *Ecological Entomology*, **19**, 310–8.

Byrne, D.N., Rathman, R.J., Orum, T.V. & Palumbo, J.C. (1996) Localized migration and dispersal by the sweet potato whitefly, Bemisia tabaci. *Oecologia (Berlin)*, **105** (3), 320–8.

Cabaleiro, C. & Segura, A. (1997) Field transmission of grapevine leafroll associated virus 3 (GLRaV-3) by the mealybug *Planococcus citri*. *Plant Disease*, **81** (3), 283–7.

Cadman, C.H. (1960) Inhibition of plant virus infection by tannins. In: *Plants in Health and Disease* (ed. J.B. Pridham), pp. 101–105. Pergamon Press, London.

Campbell, B.C. & Duffy, S.S. (1979) Tomatine and parasitic wasps: potential incompatability of plant antibiosis with biological control. *Science*, **205**, 700–2.

Campbell, B.C., Bragg, T.S. & Turner, C.E. (1992) Phylogeny of symbiotic bacteria of four weevil species (Coleoptera: Curculionidae) based on analysis of 16S ribosomal DNA. *Insect Biochemistry and Molecular Biology*, **22** (5), 415–21.

Cambell, R.W. (1975) The gypsy moth and its natural enemies. *USDA Forest Service, Agricultural Information Bulletin*, 381.

Camuffo, D. & Enzi, S. (1991) Locust invasions and climatic factors from the Middle Ages to 1800. *Theoretical and Applied Climatology*, **43** (1–2), 43–74.

Cardé, R.T. & Minks, A.K. (1995) Control of moth pests by mating disruption: sucesses and constraints. *Annual Review of Entomology*, **40**, 559–86.

Carpaneto, G.M., Piattella, E. & Spampinato, M.F. (1995) Analysis of a scarab dung beetle community from an Apenninic grassland (Coleoptera, Scarabaeoidea). *Bollettino dell'Associazione Romana di Entomologia*, **50** (1–4), 45–60.

Carroll, A.L. & Quiring, D.T. (1994) Intratree variation in foliar development influences the foraging strategy of a caterpillar. *Ecology*, **75**, 1978–90.

Carroll, S.P. & Dingle, H. (1996) The biology of post-invasion events. *Biological Conservation*, **78**, 207–14.

Carroll, W.J. (1956) History of the hemlock looper, *Lambdina fiscellaria* (Guen.) (Lepidoptera: Geometridae), in Newfoundland, and notes on its biology. *Canadian Entomologist*, **88**, 587–99.

Carson, R. (1962) *Silent Spring*. Hamish Hamilton, London.

Carter, C.I., Nichols, J.F.A., Bevan, D. & Stoakley, J.T. (1985) Host plant susceptibility and choice by conifer aphids. In: *Site characterisitics and population dynamics of lepidopteran and hymenopteran forest pests*. Forestry Commission Research & Development Paper no. 135 (eds D. Bevan &. J.T Stoakley), pp. 94–9.

Carton, Y. (1978) Olfactory responses of *Cothonaspis* sp. (parasitic Hymenoptera, Cynipidae) to the food habit of its host (*Drosophila melanogaster*). *Drosophila Information Service*, **53**, 183–4.

Casher, L.E. (1996) Leaf toughness in *Quercus agrifolia* and its effects on tissue selection by first instars of *Phryganidia californica* (Lepidoptera: Dioptidae) and *Bucculatrix albertiella* (Lepidoptera. Lyonetiidae). *Annals of the Entomological Society of America*, **89**, 109–21.

Casida, J.E. & Quistad, G.B. (1998) Golden age of insecticide research: past, present or future? *Annual Review of Entomology*, **43**, 1–16.

Cassier, P., Levieux, J., Morelet, M. & Rougon, D. (1996) The mycangia of *Platypus cylindrus* Fab. & *P. oxyurus* Dufour (Coleoptera: Platypodidae): Structure and associated fungi. *Journal of Insect Physiology*, **42** (2), 171–9.

Casson, D.S. & Hodkinson, I.D. (1991) The Hemiptera (Insecta) communities of tropical rain forest in Sulawesi (Indonesia). *Zoological Journal of the Linnean Society*, **102** (3), 253–76.

Cavalieri, L.F. & Kocak, H. (1994) Chaos in biological control systems. *Journal of Theoretical Biology*, **169** (2), 179–87.

Chadenga, V. (1994) Epidemiology and control of trypanosomiasis. *Onderstepoort Journal of Veterinary Research*, **61** (4), 385–90.

Challice, J.S. & Williams, A.H. (1970) A comparative biochemical study of phenolase specificity in *Malus*, *Pyrus* and other plants. *Phytochemistry*, **9**, 1261–9.

Chang, Y.C. (1991) Integrated pest management of several forest defoliators in Taiwan. *Forest Ecology and Management*, **39**, 65–72.

Chapin, F.S. III (1980) Nutrient allocation and responses to defoliation in tundra plants. *Arctic and Alpine Research*, **12**, 553–63.

Chararas, C. (1979) *Ecophysiologie des Insectes Parasites des Forêts*. Printed by the author, Paris.

Chauhan, R. & Dahiya, B. (1994) Responses to different chickpea genotypes by *Heliocoverpa armigera* at Hisar. *Indian Journal of Plant Protection*, **22** (2), 170–2.

Chemengich, B.T. (1993) Ecological relationships between Argomyzidae feeding on leguminous plants and species-area effects in Kenya. *Insect Science and its Application*, **14**, 603–9.

Chen, C.N. & Yeh, Y. (1992) Integrated pest management in Taiwan. In: *Integrated Pest Management in the Asia–Pacific Region* (eds P.A. C. Ooi, G.S. Lim, T.H. Ho, P.L. Manalo & J. Waage), pp. 283–302. CAB International, Wallingford, UK.

Cheng, H., Yang, W., Kang, W. & Liu, C. (1995) Large-scale spraying of bednets to control mosquito vectors and malaria in Sichuan, China. *Bulletin of the World Health Organization*, **73** (3), 321–8.

Cheng, L., Baars, M.A. & Oosterhuis, S.S. (1990) *Halobates* in the Banda Sea (Indonesia): Monsoonal differences in abundance and species composition. *Bulletin of Marine Science*, **47** (2), 421–30.

Chenuil, A. & McKey, D.B. (1996) Molecular phylogenetic study of a myrmecophyte symbiosis: Did *Leonardoxa*-ant associations diversify via conspecification? *Molecular Phylogenetics and Evolution*, **6**, 270–86.

Cherrett J.M., Powell, R.J. & Stradling, D.J. (1989) The mutualism between leaf-cutting ants and their fungus. In: *Insect–Fungus Interactions 14th Symposium of the Royal Entomological Society of London in Collaboration with the British Mycological Society* (eds N. Wilding, N.M. Collins, P.M. Hammond & J.F. Webber), pp. 93–120. Academic Press, London.

Cherry, R.H. & Klein, M.G. (1997) Mortality induced by *Bacillus popilliae* in *Cyclocephala parallela* (Coleoptera: Scarabaeidae) held under simulated field temperatures. *Florida Entomologist*, **80** (2), 261–5.

Chesson, P.L. (1985) Coexistence of competitors in spatially and temporally varying environments: a look at the combined effect of different sorts of variability. *Theoretical Population Biology*, **28**, 263–87.

Chesson, P.L. & Warner, R.R. (1981) Environmental variability promotes coexistence in lottery competitive systems. *American Naturalist*, **117**, 923–43.

Chey, V.K., Holloway, J.D. & Speight, M.R. (1997) Diversity of moths in forest plantations and natural forests in Sabah. *Bulletin of Entomological Research*, **87**, 371–85.

Chey, V.K., Holloway, J.D. & Speight, M.R. (1998) Canopy knockdown of arthropods in exotic plantations and natural forest in Sabah, north-east Borneo, using insecticidal mist-blowing. *Bulletin of Entomological Research*, **88**, 15–24.

Chilcutt, C.F. & Tabashnik, B.E. (1997) Host-mediated competition between the pathogen *Bacillus thuringiensis* and the parasitoid *Cotesia plutellae* of the diamondback moth (Lepidoptera, Plutellidae).

Chin, H., Othman, Y., Loke, W.H. & Rahman, S.A. (1992) National integrated pest management in Malaysia. In: *Integrated Pest Management in the Asia–Pacific Region* (eds P.A. C. Ooi, G.S. Lim, T.H. Ho, P.L. Manalo & J. Waage), pp. 191–210. CAB International, Wallingford, UK.

Chiykowski, L.N. (1991) Vector–pathogen–host plant relationships of clover phyllody mycoplasma-like organism and the vector leafhopper *Paraphlepsius irroratus*. *Canadian Journal of Plant Pathology*, **13** (1), 11–18.

Choi *et al* (1992) IPM in Korea. In: *Integrated Pest Management in the Asia-Pacific Region* (eds P.A.C. Ooi, G.S.T.H. Lim Ho, P.L. Manalo & J. Waage). CAB International, Wallingford.

Choi, H.W., Breman, J.G., Teutsch, S.M., Liu, S., Hightower, A.W. & Sexton, J.D. (1995) The effectiveness of insecticide-impregnated bednets in reducing cases of malarial infection; a meta-analysis of published results. *American Journal of Tropical Medicine and Hygiene*, **52** (5), 377–82.

Choong, M.F. (1996) What makes a leaf tough and how this affects the patterns of *Castanopsis fissa* leaf consumption by caterpillars. *Functional Ecology*, **10**, 668–74.

Christensen, K.M., Whitman, T.G. & Keim, P. (1995) Herbivory and tree mortality across a pinton pine hybrid one. *Oecologia (Berlin)*, **101**, 29–36.

Claridge, M.F. & Evans, H.F. (1990) Species–area relationships: relevance to pest problems of British trees? In: *Population Dynamics of Forest Insects* (eds A.D. Watt, S.R. Leather, M.D. Hunter & N.A.C. Kidd), pp. 59–69. Intercept, Andover, UK.

Claridge, M.F. & Wilson, M.R. (1977) British Insects and Trees: A Study in Island Biogeography or Insect/Plant Coevolution. *American Naturalist*, 451–456.

Claridge, M.F. & Wilson, M.R. (1978) British insects and trees: a study in island biogeography or insect/plant coevolution? *American Naturalist*, **112** (984), 451–6.

Claridge, M.F. & Wilson, M.R. (1982) Insect herbivore guilds and species-area relationships—leafminers on British trees. *Ecological Entomology*, **7**, 19–30.

Claridge, M.F., Denhollander, J. & Furet, I. (1982) Adaptations of brown planthopper (*Nilaparvata lugens*) populations to rice varieties in Sri Lanka. *Entomologia Experimentalis et Applicata*, **32**, 222–6.

Clay, K. & Cheplick, G.P. (1989) Effect of ergot alkaloids from fungal endophyte-infected grasses on fall armyworm (*Spodoptera frugiperda*). *Journal of Chemical Ecology*, **15**, 169–82.

Clay, K., Hardy, T.N. & Hammond, A.M. (1985) Fungal endophytes of *Cyperus* and their effect on an insect herbivore. *American Journal of Botany*, **72**, 1284–9.

Cobb, N.S. & Whitham, T.G. (1993) Herbivore deme formation on individual trees: a test case. *Oecologia*, **94**, 496–502.

Cohen, D., Green, M., Block, C., *et al.* (1991) Reduction of transmission of shigellosis by control of houseflies (*Musca domestica*). *Lancet (North American edn)*, **337** (8748), 993–7.

Coley, P.D. (1986) Costs and benefits of defense by tannins in a neotropical tree. *Oecologia (Heidelberg)*, **70**, 238–41.

Coley, P.D. & Aide, T.M. (1991) Comparison of herbivory and plant defenses in temperate and tropical broad-leaved forests. In: Plant-animal interactions. *Evolutionary Ecology in Tropical and Temperate Regions* (eds P.W. Price, T.M. Lewinshon, G.W. Fernandes & W.W. Benson), pp. 25–49. John Wiley and Sons, Inc.

Coley, P.D., Bryant, J.P. & Chapin, F.S. (1985) Resource availability and plant antiherbivore defense. *Science*, **230**, 895–9.

Collier, K.J. (1995) Environmental factors affecting the taxonomic composition of aquatic macroinvertebrate communities in lowland waterways of Northland, New Zealand. *New Zealand Journal of Marine and Freshwater Research*, **29**, 453–65.

Collins, F.H. & Paskewitz, S.M. (1995) Malaria: current and future prospects for control. *Annual Review of Entomology*, **40**, 195–219.

Collins, N.M. & Morris, M.G. (1985) *Threatened Swallowtail Butterflies of the World*. International Union for the Conservation of Nature, Gland.

Coluzzi, M. (1994) Malaria and the afrotropical ecosystems: impact of man-made environmental changes. *Parassitologia (Rome)*, **36** (1–2), 223–7.

Colwell, R.K. & Coddington, J.A. (1994) Estimating terrestrial biodiversity through extrapolation. *Philosophical Transactions of the Royal Society of London, Series B: Biological Sciences*, **345**, 101–18.

Comins, H.N. & Hassell, M.P. (1979) The dynamics of optimally foraging predators and parasitoids. *Journal of Animal Ecology*, **48**, 335–51.

Comins, H.N., Hassell, M.P. & May, R.M. (1992) The spatial dynamics of host–parasitoid systems. *Journal of Animal Ecology*, **61**, 735–48.

Commins, H.N. & Noble, I.R. (1985) Dispersal, variability and transient niches: species coexistence in a uniformly varying environment. *American Naturalist*, **126**, 706–23.

Compton, S.G., Lawton, J.H. & Rashbrook, V.K. (1989) Regional diversity, local community structure and vacant niches: The herbivorous arthropods of bracken in South Africa. *Ecological Entomology*, **14** (4), 365–73.

Connell, J.H. (1980) Diversity and the coevolution of competitors, or the ghost of competition past. *Oikos*, **35**, 131–8.

Connell, J.H. (1983) On the prevalence and importance of interspecific competition: evidence from field experiments. *American Naturalist*, **122**, 661–96.

Connell, J.H. & Slatyer, R.O. (1977) Mechanisms of succession in natural communities and their role in community stability and organisation. *American Naturalist*, **111**, 1119–1144.

Coope, G.R. (1994) The response of insect faunas to glacial–interglacial climatic fluctuations. *Philosophical Transactions of the Royal Society of London, Series B: Biological Sciences*, **344** (1307), 19–26.

Coope, G.R. (1995) The effects of Quaternary climatic changes on insect populations: lessons from the past. In: *Insects in a Changing Environment* (eds R. Harrington & N.E. Stork), pp. 29–48. Academic Press, London.

Coppedge, B.R., Stephen, F.M. & Felton, G.W. (1995) Variation in female southern pine beetle size and lipid content in relation to fungal associates. *Canadian Entomologist*, **127** (2), 145–54.

Cornell, H.V. & Hawkins, B.A. (1995) Survival patterns and mortality sources of herbivorous insects—some demographic trends. *American Naturalist*, **145**, 563–93.

Cory, J.S., Hirst, M.L., Williams, T., *et al.* (1994) Field trial of a genetically improved baculovirus insecticide. *Nature*, **370** (6485), 138–40.

Costa, E.G. & Link, D. (1992) Damage evaluation of the rice stem stink bug, *Tibraca limbativentris* Stal. On paddy rice. *Anais da Sociedade Entomologica do Brasil*, **21** (1), 187–95.

Coulson, R.N., Pope, D.N., Gagne, J.A., Fargo, W.S. & Pulley, P.E. (1980) Impact of foraging by *Monochamus titilator* (Col. Cerambycidae) on within-tree populations of *Dendroctonus frontalis* (Col. *Scolyditae*). *Entomophaga*, **25**, 155–70.

Coulson, S.J., Hodkinson, I.D., Strathdee, A.T., *et al.* (1995) Thermal environments of Arctic soil organisms during winter. *Arctic and Alpine Research*, **27** (4), 364–70.

Craig, T.P., Itami, J.K. & Price, P.W. (1990) Intraspecific competition and facilitation by a shoot-galling sawfly. *Journal of Animal Ecology*, **59**, 147–59.

Craighead, F.C. (1925) Bark-beetle epidemics and rainfall deficiency. *Journal of Economic Entomology*, **18**, 577–86.

Crawley, M.J. (1992) Population dynamics of natural enemies and their prey. In: *Natural Enemies* (ed. M.J. Crawley), pp. 40–89. Blackwell Scientific, Oxford.

Creamer, R., Luque-Williams, M. & Howo, M. (1996) Epidemiology and incidence of beet curly top germinivirus in naturally infected weed hosts. *Plant Disease*, **80** (5), 533–5.

Cronin, J.T. & Strong, D.R. (1993) Superparasitism and mutual interference in the egg parasitoid *Anagrus delicatus* (Hymenoptera: Mymaridae). *Ecological Entomology*, **18**, 293–302.

CSIRO (1979) *The Insects of Australia*, pp. 1028. Melbourne University Press, Melbourne.

CSIRO (1991) *The Insects of Australia*, 2nd edn. Melbourne University Press, Melbourne.

Cuffney, T.F., Wallace, J.B. & Lugthart, G.J. (1990) Experimental evidence quantifying the role of benthic invertebrates in organic matter dynamics of headwater streams. *Freshwater Biology*, **23**, 28199.

Cummins, K.W. & Klug, M.J. (1979) Feeding ecology of stream invertebrates. *Annual Review of Ecology and Systematics*, **10**, 147–72.

Cummins, K.W., Petersen, R.C., Howard, F.O., Wuycheck, J.C. & Holt, V.I. (1973) The utilization of leaf litter by stream detritivores. *Ecology*, **54**, 336–45.

Cupp, E.W. & Cupp, M.S. (1997) Black fly (Diptera: Simuliidae) salivary secretions: Importance in vector competence and disease. *Journal of Medical Entomology*, **34** (2), 87–94.

Curtis, A.D. & Waller, D.A. (1995) Changes in nitrogen fixation rates in termites (Isoptera: Rhinotermitidae) maintained in the laboratory. *Annals of the Entomological Society of America*, **88** (6), 764–7.

Cushing, C.E., Minshall, G.W. & Newbold, J.D. (1993) Transport dynamics of fine particulate organic matter in two Idaho streams. *Limnology and Oceanography*, **38**, 1101–115.

D'Abrera, B. (1982) *Butterflies of the Oriental Region, Part I, Papilionidae, Pieridae and Danaidae*. Hill House, Ferny Creek.

D'Abrera, B. (1985) *Butterflies of the Oriental Region, Part II, Nymphalidae, Satyridae and Amathusidae*. Hill House, Melbourne.

D'Abrera, B. (1986) *Butterflies of the Oriental Region, Part III, Lycaenidae and Riodinidae*. Hill House, Melbourne.

D'Amico, V. & Elkinton, J.S. (1995) Rainfall effects on transmission of gypsy moth (Lepidoptera: Lymantriidae) nuclear polyhedrosis virus. *Environmental Entomology*, **24** (5), 1144–9.

Dadour, I.R. & Cook, D.F. (1996) Survival and reproduction in the scarabaeine dung beetle *Onthophagus binodis* (Coleoptera: Scarabaeidae) on dung produced by cattle on grain diets in feedlots. *Environmental Entomology*, **25**, 1026–31.

Daly, H.V., Doyen, J.T. & Purcell, A.H. (1998) *Introduction to Insect Biology and Diversity*, 2nd edn. Oxford University Press, Oxford.

Damman, H. (1993) Patterns of herbivore interaction among herbivore species. In: *Caterpillars: Ecological and Evolutionary Constraints on Foraging* (eds N.E. Stamp & T.M. Casey), pp. 132–169. Chapman & Hall, New York.

Danks, H.V. (1994) Regional diversity of insects in North America. *American Entomologist*, **40** (1), 50–5.

Danyk, T.P. & Mackauer, M. (1996) An extraserosal envelope in eggs of *Proan pequodorum* (Hymenoptera, aphidae), a parasitoid of pea aphid. *Biological Contrology*, **7**, 67–70.

Darwin, C. (1866) *The Origin of the Species*. Dent, London.

Davidson, D.W., Inouye, R.S. & Brown, J.H. (1984) Granivory in a desert ecosystem: experimental evidence for indirect facilitation of ants by rodents. *Ecology*, **65**, 1780–6.

Davidson, D.W., Longino, J.T. & Snelling, R.R. (1988) Pruning of host plant neighbors by ants: an experimental approach. *Ecology*, **69**, 801–8.

Davidson, J. & Andrewartha, H.G. (1948a) Annual trends in a natural population of *Thrips imaginis* (Thysanoptera). *Journal of Animal Ecology*, **17**, 193–9.

Davidson, J. & Andrewartha, H.G. (1948b) The influence of rainfall, evaporation and atmospheric temperature on fluctuations in the size of a natural population of *Thrips imaginis* (Thysanoptera). *Journal of Animal Ecology*, **17**, 200–22.

Davidson, R.H. & Lyon, W.F. (1979) *Insect Pests of Farm, Garden and Orchard*. John Wiley & Sons, New York.

Davies J.B. (1994) Sixty years of onchocerciasis vector control: a chronological summary with comments on eradication, reinvasion and insecticide resistance. *Annual Review of Entomology*, **39**, 23–46.

Davis, A.L.V. (1996) Habitat associations in a South African, summer rainfall, dung beetle community (Coleoptera. Carabaeidae, Aphodiidae, Staphylinidae, Histeridae, Hydrophilidae). *Pedobiologia*, **40**, 260–80.

Day, W.H. (1994) Estimating mortality caused by parasites and diseases of insects—comparisons of the dissection and rearing methods. *Environmental Entomology*, **23**, 543–50.

De Bach, P. (ed.) (1964) *Biological Control of Insect Pests*, 844 pp. Chapman & Hall, London.

De Bortoli, S.A., Marconato, J.R. & Daniel, L.A. (1994) Effects of tillage methods on some arthropod populations in soybean *Glycine max* (L.) Merrill. *Cientifica (Jaboticabal)*, **22** (1), 9–14.

De Jong, G. (1981) The evolution of dispersal pattern on the evolution of fecundity. *Netherlands Journal of Zoology*, **32**, 1–30.

De Sole, G., Baker, R., Dadzie, K.Y., *et al.* (1991) Onchocerciasis distribution and severity in five West African countries. *Bulletin of the World Health Organization*, **69** (6), 689–98.

De Vries, P.J. (1987) *The Butterflies of Costa Rica and their Natural History*. Princeton University Press, Princeton, NJ.

De Vries, P.J. (1997) *The Butterflies of Costa Rica and their Natural History*. Princeton University Press, Princeton, NJ.

De Zulueta, J. (1994) Malaria and ecosystems: from prehistory to posteradication. *Parassitologia (Rome)*, **36** (1–2), 7–15.

Debouzie, D. & Ballanger, Y. (1993) Dynamics of *Ceutorhynchus napi* in winter rape fields. *Acta Oecologica*, **14** (5), 603–18.

Delucchi, V. (1982) Parasitoids and hyperparasitoids of *Zeiraphera diniana* (Lep., Tortricidae) and their role in population control in outbreak areas. *Entomophaga*, **27**, 77–92.

Dempster, J.P. (1982) The ecology of the cinnabar moth, *Tyria jacobaeae* L (Lepidoptera, Arctiidae). *Advances in Ecological Research*, **12**, 1–36.

Dempster, J.P. (1983) The natural control of populations of butterflies and moths. *Biological Reviews of the Cambridge Philosophical Society*, **58**, 461–81.

Dempster, J.P. & Pollard, E. (1981) Fluctuations in resource availability and insect populations. *Oecologia*, **50**, 412–16.

Dempster, J.P. & McLean, I.F.G. (1998) *Insect Populations*. Chapman and Hall, London.

Den Boer, P.J. (1986) Density dependence and the stabilization of animal numbers 1. The winter moth. *Oecologia*, **69**, 507–12.

Den Boer, P.J. (1988) Density dependence and the stabilization of animal numbers 3. The winter moth reconsidered. *Oecologia*, **75**, 161–8.

Den Boer, P.J. (1991) Seeing the trees for the wood—random-walks or bounded fluctuations of population-size. *Oecologia*, **86**, 484–91.

Dennill, G.B. & Pretorius, W.L. (1995) The status of diamondback moth, *Plutella xylostella* (Linnaeus) (Lepidoptera; Plutellidae), and its parasitoids on cabbages in South Africa. *African Entomology*, **3** (1), 65–71.

Dennis, P. & Wratten, S.D. (1991) Field manipulation of populations of individual staphylinid species in cereals and their impact on aphid populations. *Ecological Entomology*, **16**, 17–24.

Dennis, P., Usher, G.B. & Watt, A.D. (1995) Lowland woodland structure and pattern and the distribution of arboreal, phytophagous arthropods. *Biodiversity and Conservation*, **4**, 728–44.

Dennis, P., Young, M.R., Howard, C.L. & Gordon, I.J. (1997) The response of epigeal beetles (Col: Carabidae, Staphylinidae) to varied grazing regimes on upland *Nardus stricta* grasslands. *Journal of Applied Ecology*, **34**, 433–43.

Denno, R.F. (1994) The evolution of dispersal polymorphisms in insects: The influence of habitats, host plants and mates. *Researches on Population Ecology (Kyoto)*, **36** (2), 127–35.

Denno, R.F. & McClure, M.S. (1983) *Variable Plants and Herbivores in Natural and Managed Systems*. Academic Press, New York.

Denno, R.F. & Roderick, G.K. (1992) Density-related dispersal in planthoppers: effects of interspecific crowding. *Ecology*, **73**, 1323–34.

Denno, R.F., McClure, M.S. & Ott, J.R. (1995) Interspecific interactions in phytophagous insects: Competition reexamined and resurrected. *Annual Review of Entomology*, **40**, 297–331.

Dent, D. (1995) *Integrated Pest Management*, 365 pp. Chapman & Hall, London.

Dethier, V.G. (1941) Chemical factors determining the choice of food plants by *Papilio* larvae. *American Naturalist*, **75**, 61–73.

Dicke, M. (1994) Local and systemic production of volatile herbivore-induced terpenoids: their role in plant–carnivore mutualism. *Journal of Plant Physiology*, **143**, 465–72.

Dicke, M. & Sabelis, M.W. (1988) Infochemical terminology: based on a cost benefit analysis rather than origin of compounds? *Functional Ecology*, **2**, 131–9.

Dicke, M., Van Baarlen, P., Wessels, R. & Djikman, H. (1993) Herbivory induces systematic production of plant volatiles that attract predators of the herbivore extraction of endogenous elicitor. *Journal of Chemical Ecology*, **19**, 581–99.

Didham, R.K., Ghazoul, J., Stork, N.E. & Davis, A.J. (1996) Insects in fragmented forests: a functional approach. *Trends in Ecology & Evolution*, **11**, 255–60.

Dietrich, C.H. & Vega, F.E. (1995) Leafhoppers (Homoptera: Cicadellidae) from Dominican amber. *Annals of the Entomological Society of America*, **88** (3), 263–70.

Dimitri, L., Gebauer, U., Loesekrug, R. & Vaupel, O. (1992) Influence of mass trapping on the population dynamics and damage-effect of bark beetles. *Journal of Applied Entomology*, **114** (1), 103–9.

Diotaiuti, L., Loiola, C.F., Falcao, P.L. & Dias, J.C.P.O. (1993) The ecology of *Triatoma sordida* in natural environments in two different regions of the state of Minas Gerais, Brazil. *Revista do Instituto de Medicina Tropical de São Paulo*, **35** (3), 237–45.

Diss, A.L., Kunkel, J.G., Montgomery, M.E. & Leonard, D.E. (1996) Effects of maternal nutrition and egg provisioning on parameters of larval hatch, survival and dispersal in the gypsy moth, *Lymantria dispar* L. *Oecologia (Berlin)*, **106** (4), 470–7.

Dixon, A.F.G., Kindlmann, P., Leps, J. & Holman, J. (1987) Why there are so few species of aphids, especially in the tropics. *American Naturalist*, **129**, 580–92.

Dohm, D.J., Romoser, W.S., Turell, M.J. & Linthicum, K.J. (1991) Impact of stressful conditions on the survival of *Culex pipiens* exposed to rift valley fever virus. *Journal of the American Mosquito Control Association*, **7** (4), 621–3.

Douglas, A.E. (1996) Reproductive failure and the free amino acid pools in pea aphids (*Acyrthosiphon pisum*) lacking symbiotic bacteria. *Journal of Insect Physiology*, **42** (3), 247–55.

Douglas, A.E. (1998) Nutritional interactions in insect-microbial symbioses: aphids and their symbiotic bacteria *Buchnera*. *Annual Review of Entomology*, **43**, 17–37.

Dover, J.W. (1991) The conservation of insects on arable farmland. In: *The Conservation of Insects and their Habitats* (eds N.M. Collins & J.A. Thomas), pp. 291–318. Academic Press, London.

Dowd, P.F. & Shen, S.K. (1990) The contribution of symbiotic yeast to toxin resistance of the cigarette beetle (*Lasioderma serricorne*). *Entomologia Experimentalis et Applicata*, **56** (3), 241–8.

Doyle, C.J. & Entwistle, P.F. (1988) Aerial application of mixed virus formulations to control joint infestations of *Panolis flammea* and *Neodiprion sertifer* on lodgepole pine. *Annals of Applied Biology*, **113** (1), 119–28.

Drake, V.A. & Farrow, R.A. (1989) The 'aerial plankton' and atmospheric convergence. *Trends in Ecology and Evolution*, **4** (12), 381–5.

Dransfield, R. (1975) *The ecology of grassland and cereal aphids*. PhD Thesis, University of London.

Dransfield, R.D., Williams, B.G. & Brightwell, R. (1991) Control of tsetse flies and trypanosomiasis: myth or reality? *Parasitology Today*, **7** (10), 287–91.

Drukker, B., Scutareanu, P. & Sabelis, M.W. (1995) Do anthocorid predators respond to synomones from *Psylla*-infested pear trees under field conditions? *Entomologia Experimentalis et Applicata*, **77**, 193–203.

Drumont, A., Gonzalez, R.D.E., Windt, N., Gregoire, J.C.D.E., Proft, M. & Seutin, E. (1992) Semiochemicals and the integrated management of *Ips typographus* (L.) (Col., Scolytidae) in Belgium. *Journal of Applied Entomology*, **114** (4), 333–7.

Dudgeon, D. & Chan, I.K.K. (1992) An experimental-study of the influence of periphytic algae on invertebrate abundance in a Hong Kong stream. *Freshwater Biology*, **27**, 53–63.

Dudley, T. & Anderson, N.H. (1982) A survey of invertebrates associated with wood debris in aquatic habitats. *Melanderia* **39**, 1–22.

Duelli, P., Studer, M., Marchand, I. & Jakob, S. (1990) Population movements of arthropods between natural and cultivated areas. *Biological Conservation*, **54** (3), 193–208.

Duffels, J.P. & van Mastrigt, H.J.G. (1991) Recognition of cicadas (Hemiptera: Cicadidae) by the Ekagi people of Irian Jaya (Indonesia), with a description of a new species of *Cosmopsaltria. Journal of Natural History,* **25** (1), 173–82.

Duffey, S.S. & Stout, M.J. (1996) Antinutritive and toxic components of plant defense against insects. *Archives of Insect Biochemistry and Physiology,* **32**, 3–37.

Dufour, D.L. (1987) Insects as food: a case study from the northwest Amazon. *American Anthropologist,* **89** (2), 383–97.

Dumas, M. & Bouteille, B. (1996) Human African trypanosomiasis. *Comptes Rendus des Séances de la Société de Biologie et de ses Filiales,* **190** (4), 395–408.

Dupont, A., Belanger, L. & Bousquet, J. (1991) Relationships between balsam fir vulnerability to spruce budworm and ecological site conditions of fir stands in central Quebec. *Canadian Journal of Forest Research,* **21**, 1752–9.

Durrett, R. & Levin, S. (1996) Spatial models for species–area curves. *Journal of Theoretical Biology,* **179** (2), 119–27.

Dwyer, G. & Elkinton, J.S. (1995) Host dispersal and the spatial spread of insect pathogens. *Ecology (Washington, DC),* **76** (4), 1262–75.

Dye, C. (1992) The analysis of parasite transmission by bloodsucking insects. *Annual Review of Entomology,* **37**, 1–20.

Dyer, L.A. & Bowers, M.D. (1996) The importance of sequestered iridoid glycosides as a defense against an ant predator. *Journal of Chemical Ecology,* **22**, 1527–39.

Dyer, M.I. & Bokhari, U.G. (1976) Plant-animal interactions: studies of the effects of grasshopper grazing on blue grama grass. *Ecology,* **57**, 762–72.

Dyer, M.I., Moon, A.M., Brown, M.R. & Crossley, D.A. Jr (1995) Grasshopper crop and midgut extract effects on plants: an example of reward feedback. *Proceedings of the National Academy of Sciences of the United States of America,* **92**, 5475–8.

Dyer, M.I., Turner, C.L. & Seastedt, T.R. (1993) Herbivory and its consequences. *Ecological Applications,* **3**, 10–6.

Eastop, V.F. (1973) Deductions from the present day host plants of aphids and related species. *Symposium of the Royal Entomological Society of London,* **6**, 157–77.

Edenius, L., Danell, K. & Bergstrom, R. (1993) Impact of herbivory and competition on compensatory growth in woody plants winter browsing by moose on scots pine. *Oikos,* **66**, 286–92.

Edmunds, G.F. Jr (1973) Ecology of black pineleaf scale (Homoptera. *Diaspididae). Environmental Entomology,* **2**, 765–77.

Edmunds, G.F. Jr & Alstad, D.N. (1978) Coevolution in insect herbivores and conifers. *Science,* **199**, 941–5.

Edson, J.L. (1985) The influences of predation and resource subdivision on the coexistence of goldenrod aphids. *Ecology,* **66**, 1736–43.

Edson, K.M., Vinson, S.B., Stoltz, D. & Summers, M.D. (1980) Virus in a parasitoid wasp: suppression of the cellular immune response in the parasitoid's host. *Science,* **211**, 582–3.

Edwards, C.A., Sunderland, K.D. & George, K.S. (1979) Studies on polyphagous predators of cereal aphids. *Journal of Applied Ecology,* **16**, 811–23.

Edwards, P.J. & Wratten, S.D. (1985) Induced plant defenses against insect grazing: Fact or artefact? *Oikos,* **44**, 70–4.

Eggleton, P. & Belshaw, R. (1992) Insect parasitoids—an evolutionary overview. *Philosophical Transactions of the Royal Society of London, Series B: Biological Sciences,* **337**, 1–20.

Eggleton, P., Williams, P.H. & Gaston, K.J. (1994) Explaining global termite diversity—productivity or history. *Biodiversity and Conservation,* **3**, 318–30.

Eggleton, P., Bignell, D.E., Sands, W.A., Mawdsley, N.A., Lawton, J.H., Wood, T.G. & Bignell, N.C. (1996) The diversity, abundance and biomass of termites under differing levels of disturbance in the Mbalmayo Forest Reserve, southern Cameroon. *Philosophical Transactions of the Royal Society of London. Series B Biological Sciences,* **351**, 51–68.

Ehrlich, P.R. & Raven, P.H. (1964) Butterflies and plants: a study in coevolution. *Evolution,* **18**, 586–608.

Eigenbrode, S.D., Moodie, S. & Castagnola, T. (1995) Predators mediate host plant resistance to a phytophagous pest in cabbage with glossy leaf wax. *Entomologia Experimentalis et Applicata,* **77**, 335–42.

Eigenbrode, S.D., Trumble, J.T., Millar, J.G. & White, K.K. (1994) Topical toxicity of tomato sesquiterpenes to the beet armyworm and the role of these comounds in resistance derived form an accession of *Lycopersicon hirsitum* f. *typicum. Journal of Agricultural and Food Chemistry,* **42**, 807–10.

Elton, C. (1927) *Animal Ecology.* Sidwick & Jackson, London.

Elton, C. (1933) *The Ecology of Animals. Methuen's Monographs on Biological Subjects,* 97 pp. Methuen, London.

Ely, S. (1993) The engineering of plants to express *Bacillus thuringiensis* δ-endotoxins. In: Bacillus thuringiensis, *an Environmental Biopesticide: Theory and Practice* (P.F. Entwistle, J.S. Cory, M.J. Bailey & S. Higgs), pp. 105–24. John Wiley, Chichester, UK.

Elzen, G.W. (1997) Changes in resistance to insecticides in tobacco budworm populations in Mississippi, 1993–5. *Southwestern Entomologist,* **22** (1), 61–72.

Embree, D.G. (1966) The role of introduced parasites in the control of the winter moth in Nova Scotia. *Canadian Entomologist,* **98**, 1159–68.

English-Loeb, G.M. (1989) Nonlinear responses of spider mites to drought-stressed host plants. *Ecological Entomology,* **14**, 45–55.

Entwistle, P.F., Adams, P.H.W., Evans, H.F., Rivers, C.F., Bird, F.T. & Burk, J.M. (1983) Epizootiology of a nuclear polyhedrosis virus (Baculoviridae) in European spruce sawfly (Gilpinia hercyniae): spread of disease from small epicentres in comparison with spread of baculovirus diseases in other hosts. *Journal of Applied Entomology,* **2**, 473–87.

Entwistle, P.F., Cory, J.S., Bailey, M.J. & Higgs, S. (eds)

(1993) Bacillus thuringiensis, *an Environmental Biopesticide: Theory and Practice*. John Wiley, Chichester, UK.

Erb, W.A., Lindquist, R.K., Flickinger, N.J. & Casey, M.L. (1994) Resistance of selected *Lycopersicon* hybrids to greenhouse whitefly (Homoptera: Aleurodidae). *Florida Entomologist*, **77** (1), 104–16.

Erhardt, A. & Thomas, J.A. (1991) Lepidoptera as indicators of change in the semi-natural grasslands of lowland and upland Europe. In: *The Conservation of Insects and their Habitats* (eds N.M. Collins & J.A. Thomas), pp. 213–36. Academic Press, London.

Erwin, T.L. (1982) Tropical forests: their richness in Coleoptera and other arthropod species. *Coleopterists Bulletin*, **36**, 74–5.

Erwin, T.L. (1983) Tropical forest canopies: the last biotic frontier. *Bulletin of the Entomological Society of America*, **30**, 14–19.

Erzinclioglu, Y.Z., Baker, J.M. & Howell, S.E. (1990) Cyclorrhaphous maggots from a hypersaline oil spill site. *Entomologist*, **109** (4), 250–5.

Escarre, J., Lepart, J. & Sentuc, J.J. (1996) Effects of simulated herbivory in three old field Compositae with different inflorescence architectures. *Oecologia (Berlin)*, **105**, 501–8.

Eubanks, M.D., Nesci, K.A., Petersen, M.K., Liu, Z. & Sanchez, H.B. (1997) The exploitation of an ant-defended host plant by a shelter-building herbivore. *Oecologia (Berlin)*, **109**, 454–60.

Evans, E.W. & England, S. (1996) Indirect interactions in biological-control of insects—pests and natural enemies in alfalfa. *Ecological Applications*, **6**, 920–30.

Evans, G. (1977) *The Life of Beetles*, p. 232. Allen & Unwin, London.

Evans, H.F. & Fielding, N.J. (1994) Integrated management of *Dendroctonus micans* in the UK. *Forest Ecology and Management*, **65**, 17–30.

Fabre, J.H. (1882) *Nouveaux Souverirs Entomologiques: Etudes sur L'instincte et les Moeurs des Insectes*. Librairie Delagrave, Paris.

Fabres, G. & Nenon, J.P. (1997) Biodiversity and biological control: The case of the cassava mealybug in Africa. *Journal of African Zoology*, **111** (1), 7–15.

Faeth, S. (1985) Host leaf selection by leaf-miners: interaction among three trophic levels. *Ecology*, **66**, 479–94.

Faeth, S. (1986) Indirect interactions between temporally separated herbivores mediated by the host plant. *Ecology*, **67**, 479–94.

Faeth, S. (1987) Community structure and folivorous insect outbreaks: the role of vertical and horizontal interactions. In: *Insect Outbreaks* (eds P. Barbosa & J.C. Schultz), pp. 135–71. Academic Press, New York.

Faeth, S.H. (1992) Interspecific and intraspecific interactions via plant responses to folivory—An experimental field-test. *Ecology*, **73**, 1802–13.

Fain, H.D. (1995) Genetic foraging for variability. *Evolutionary Theory*, **11** (1), 15–29.

FAO (1988) *An interim report on the state of the forest resources in the developing countries*. Food and Agriculture Organization, Rome.

FAO (1991) *Second interim report on the state of tropical forests by forest resources assessment 1990 project*. Tenth World Forestry Congress, Paris. Food and Agriculture Organization, Rome.

FAO (1998a) *FAOSTAT* database on rice prodcution. http: //apps.fao.org/lim500/nph-wrap.pl?Production.Crops.Primary &Domain=SUA.

FAO (1998b) The Programme against African Trypanosomiasis. Http: //www.fao.org/waicent/faoinfo/agricult/aga/agah/pd/vector.htm.

Fargette, D., Muniyappa, V., Fauquet, C., N'guessan, P. & Thouvenel, J.C. (1993) Comparative epidemiology of three tropical whitefly-transmitted gemininviruses. *Biochimie (Paris)*, **75** (7), 547–54.

Farrar, R.R., Kennedy, G.G. & Kashyap, R.K. (1992) Influence of life-history differences of 2 tachinid parasitoids of *Helicoverpa zea* (Boddie) (Lepidoptera, Noctuidae) on their interactions with glandular trichome methyl ketone-based insect resistance in tomato. *Journal of Chemical Ecology*, **18**, 499–515.

Farrell, B.D., Mitter, C. & Futuyma, D.J. (1992) Diversification at the insect–plant interface. *Bioscience*, **42**, 34–42.

Fauss, D.L. & Pierce, W.R. (1969) Stand conditions and spruce budworm damage in a western Montana forest. *Journal of Forestry*, **67**, 322–9.

Febvay, G., Bonnin, J., Rahbe, Y., Bournoville, R., Delrot, S. & Bonnemain, J.L. (1988) Resistance of different lucerne cultivars to the pea aphid *Acyrthosiphon pisum*: Influence of phloem composition on aphid fecundity. *Entomologia Experimentalis et Applicata*, **48** (2), 127–34.

Feeny, P. (1970) Seasonal changes in oak leaf tannins and nutrients as a cause of spring feeding by winter moth caterpillars. *Ecology*, **51**, 565–81.

Feeny, P. (1976) Plant apparency and chemical defense. Rec. Adv. Phytochem. 10: 1-40. In: *Coevolution* (ed J.B. Harborne), pp. 163–206. London: Academic Press.

Feinsinger, P. (1983) Coevolution and pollination., In: *Coevolution* (eds D.J. Futuyma & M. Slatkin), pp. 282–310. Sinauer Associates, Mass.

Feller, I.C. (1995) Effects of nutrient enrichment on growth and herbivory of dwarf red mangrove (*Rhizophora mangle*). *Ecological Monographs*, **65**, 477–505.

Fensham, R.J. (1994) Phytophagous insect–woody sprout interactions in tropical eucalypt forest: I. Insect herbivory. *Australian Journal of Ecology*, **19** (2), 178–88.

Ferdig, M.T., Beernsten, B.T., Spray, F.J., Li, J. & Christensen, B.M. (1993) Reproductive costs with resistance in a mosquito–filarial worm system. *American Journal of Tropical Medicine and Hygiene*, **49** (6), 756–62.

Fergusen, K.I. & Stiling, P. (1996) Non-additive effects of multiple natural enemies on aphid populations. *Oecologia (Berlin)*, **108** (2), 375–9.

Fernandes, G.W. (1994) Plant mechanical defenses against insect herbivory. *Revista Brasileira de Entomologia*, **38** (2), 421–33.

Ferrell, G.T. & Hall, R.C. (1975) Weather and tree growth associated with white fir mortality caused by fir engraver and roundheaded fir borer. *US Forest Service Research Paper PSW*, PSW, 109.

Ferriss, R.S. & Berger, H. (1993) A stochastic simulation model of epidemics of arthropod-vectored plant virus. *Phytopathology*, **83** (12), 1269–78.

Fiala, B. & Linsenmair, K.E. (1995) Distribution and abundance of plants with extrafloral nectaries in the woody flora of a lowland primary forest in Malaysia. *Biodiversity and Conservation*, **4**, 165–82.

Fiala, B. & Maschwitz, U. (1991) Extrafloral nectaries in the genus *Macaranga* (Euphorbiaceae) in Malaysia: Comparitive studies of their possible significance as predispositions for myrmecophytism. *Biological Journal of the Linnean Society*, **44**, 287–306.

Fiala, B. & Maschwitz, U. (1992) Domatia as most important adaptations in the evolution of myrmecophyes in the paleotropical tree genus *Macaranga* (Euphorbiaceae). *Plant Systematics and Evolution*, **180**, 53–64.

Fiddler, G.O., Hart, D.R., Fiddler, T.A. & Mcdonald, P.M. (1989) Thinning decreases mortality and increases growth of ponderosa pine in northeastern California (USA). *US Forest Service Research Paper, Pacific South West*, 194, I-II, 1–7.

Fiedler, K. (1996) Host-plant relationships of lycaenid butterflies: Large-scale patterns, interactions with plant chemistry, and mutualism with ants. *Entomologia Experimentalis et Applicata*, **80**, 259–67.

Fielding, N.J. & Evans, H.F. (1996) The pine wood nematode, *Bursaphelenchus xylophilus* (Steiner and Buhrer) Nickle (= *B. lignicolis* Mamimya and Kiyohara): an assessment of the current position. *Forestry (Oxford)*, **69** (1), 35–46.

Fielding, N.J., O'Keefe, T. & King, C.J. (1991) Dispersal and host-finding capability of the predatory beetle, *Rhizophagus grandis* (Coleoptera: Rhizophagidae). *Journal of Applied Entomology*, **112** (1), 89–98.

Filichkin, S.A., Brumfield, S., Filichkin, T.P. & Young, M.J. (1997) *In vitro* interactions of the aphid endosymbiotic symL chaperonin with barley yellow dwarf virus. *Journal of Virology*, **71** (1), 569–77.

Fincke, O.M. (1994) Population regulation of a tropical damselfly in the larval stage by food limitation, cannibalism, intraguild predation and habitat drying. *Oecologia (Berlin)*, **100** (1–2), 118–27.

Findlay, J.A., Li, G., Penner, P. & Miller, J.D. (1995) Novel diterpenoid insect toxins from a conifer endophyte. *Journal of Natural Products (Lloydia)*, **58**, 197–200.

Findlay, S.E.G., Meyer, J.L. & Smith, P.J. (1984) Significance of bacterial biomass in the nutrition of a freshwater isopod (*Lirceus*). *Oecologia*, **63**, 38–42.

Fineblum, W.L. & Rausher, M.D. (1995) Tradeoff between resistance and tolerance to herbivore damage in a morning glory. *Nature*, **377**, 517–20.

Finidori, Logli, V., Bagneres, A.G. & Clement, J.L. (1996) Role of plant volatiles in the search for a host by parasitoid *Diglyphus isaea* (Hymenoptera. *Eulophidae) Journal of Chemical Ecology*, **22**, 541–58.

Fink, L.S. & Brower, L.P. (1981) Birds can overcome the cardenolide defence of monarch butterflies in Mexico. *Nature*, **291**, 67–70.

Flamm, R.O., Wagner, T.L., Cook, S.P., Pulley, P.E., Coulson, R.N. & Mcardle, T.M. (1987) Host colonization by cohabitating *Dendroctonus frontalis*, *Ips avulsus*, and *I. calligraphus* (Coleoptera. *Scolytidae). Environmental Entomology*, **16**, 390–9.

Fleming, J.G.W. (1992) Polydna viruses. *Mutualists and Pathogens. Annual Review of Entomology*, **37**, 377–400.

Fletcher, M., Teklehaimanot, A., Yemane, G., Kassahun, A., Kidane, G. & Beyene, Y. (1993) Prospect for the use of larvivorus fish for malarial control in Ethiopia: search for indigenous species and evaluation of their feeding capacity for mosquito larvae. *Journal of Tropical Medicine and Hygiene*, **96** (1), 12–21.

Floate, K.D. & Whitman, T.G. (1995) Insects as traits in plant systematics: their use in discriminating between hybrid cottonwoods. *Canadian Journal of Botany*, **73**, 1–13.

Floate, K.D., Kearsley, M.J.C. & Whitham, T.G. (1993) Elevated herbivory in plant hybrid zones: *Chrysomela confluens*, *Populus* and phenological sinks. *Ecology (Tempe)*, **74**, 2056–65.

Floate, K.D., Martinsen, G.D. & Whitham, T.G. (1997) Cottonwood hybrid zones as centres of abundance for gall aphids in western North America: Importance of relative habitat size. *Journal of Animal Ecology*, **66** (2), 179–88.

Foggo, A., Speight, M.R. & Gregoire, J.C. (1994) Root disturbance of common ash, *Fraxinus excelsior* (Oleaceae), leads to reduced foliar toughness and increased feeding by a folivorous weevil, *Stereonychus fraxini* (Coleoptera, Curculionidae). *Ecological Entomology*, **19**, 344–8.

Fonseca, C.R. (1994) Herbivory and the long-lived leaves of an Amazonian ant-tree. *Journal of Ecology*, **82**, 833–42.

Fontenille, D., Lepers, J.P., Coluzzi, M., Campbell, G.H., Rakotoarivony, I. & Coulanges, P. (1992) Malaria transmission and vector biology on Sainte Marie Island, Madagascar. *Journal of Medical Entomology*, **29** (2), 197–202.

Foster, G.N. (1991) Conserving insects of aquatic and wetland habitats, with special reference to beetles. In: *The Conservation of Insects and their Habitats* (eds N.M. Collins & J.A. Thomas), pp. 238–62. Academic Press, London.

Foster, M.A., Schultz, J.C. & Hunter, M.D. (1992) Modeling gypsy-moth virus leaf chemistry interactions—implications of plant-quality for pest and pathogen dynamics. *Journal of Animal Ecology*, **61**, 509–20.

Foster, W.A. & Treherne, J.E. (1986) The ecology and behavior of a marine insect, *Halobates fijiensis* (Hemiptera

(Heteroptera): Gerridae). *Zoological Journal of the Linnean Society*, **86** (4), 391–412.

Fowler, S.V. & Lawton, J.H. (1985) Rapidly induced defenses and talking trees: The devil's advocate position. *American Naturalist*, **126**, 181–95.

Fox, J.W., Wood, D.L., Akers, R.P. & Parmeter, J.R.J.R. (1992) Survival and development of *Ips paraconfusus* Lanier (Coleoptera: Scolytidae) reared axenically and with tree-pathogenic fungi vectored by cohabiting *Dendroctonus* species. *Canadian Entomologist*, **124** (6), 1157–67.

Free, C.A., Beddington, J.R. & Lawton, J.H. (1977) On the inadequacy of simple models of mutual interference for parasitism and predation. *Journal of Animal Ecology*, **46**, 543–4.

Freitas, A.V.L. & Oliveira, P.S. (1996) Ants as selective agents on herbivore biology: Effects on the behaviour of a non-myrmecophilous butterfly. *Journal of Animal Ecology*, **65**, 205–10.

Friend, J. (1979) Phenolic substances and plant disease. In: *Biochemistry of Plant Phenolics* (eds T. Swain, J.B. Harborne & C.F. Van Sumere), pp. 557–88. Plenum Press, New York.

Fritz, R.S. & Simms, E.L., eds (1992) Plant resistance to herbivores and pathogens. *Ecology, Evolution, and Genetics* University of Chicago Press.

Fritz, R.S., Nichols-Orians, C.M. & Brunsfield, S.J. (1994) Interspecific hybridization of plants and resistance to herbivores: Hypothesis, genetics, and variable responses in a diverse herbivore community. *Oecologia (Berlin)*, **97**, 106–17.

Fritz, R.S., Sacchi, C.F. & Price, P.W. (1986) Competition versus host plant phenomenon in species composition: willow sawflies. *Ecology*, **67**, 1608–18.

Fukatsu, T. (1994) Endosymbiosis of aphids with micro-organisms: a model case of dynamic endosymbiotic evolution. *Plant Species Biology*, **9** (3), 145–54.

Funk, D.J., Futuyma, D.J., Orti, G. & Meyer, A. (1995) A history of host associations and evolutonary diversification for *Ophraella* (Coleoptera, Chrysomelidae)—new evidence from mitochondrial DNA. *Evolution*, **49**, 1008–17.

Fuxa, J.R. (1989) Seasonal occurrence of *Spodoptera frugiperda* larvae on certain host plants in Louisiana (USA). *Journal of Entomological Science*, **24** (3), 277–89.

Fuxa, J.R. & Richter, A.R. (1994) Distance and rate of spread of *Anticarsia gemmatalis* (Lepidoptera: Noctuidae) nuclear polyhedrosis virus released into soybean. *Environmental Entomology* **23** (5), 1308–16.

Fuxa, J.R., Richter, A.R. & Strother, M.S. (1993) Detection of *Anticarsia gemmatalis* nuclear polyhedrosis virus in predatory arthropods and parasitoids after viral release in Louisiana soybean. *Journal of Entomological Science*, **28** (1), 51–60.

Gagne, W.C. & Howarth, F.G. (1985) Conservation status of endemic Hawaiian Lepidoptera. In: *Proceedings of the 3rd Congress European Lepidopterologists, Cambridge 1982*, pp. 74–84. Societus Europaea Lepidopterologica, Karlsruhe.

Galil, J. & Eisikowitch, D. (1968) On the pollination ecology of *Ficus sycomorus* in East Africa. *Ecology*, **49**, 259–69.

Gange, A.C. (1995) Aphid performance in an alder (*Alnus*) hybrid zone. *Ecology*, **76**, 2074–83.

Gange, A.C. & Brown, V.K. (1989) Effects of root herbivory by an insect on a foliar-feeding species, meditated through changes in the host plant. *Oecologia*, **81**, 38–42.

Garcia, M.A. & Altieri, M.A. (1992) Explaining differences in flea beetle *Phyllotreta cruciferae* Goeze densities in simple and mixed broccoli cropping systems as a function of individual behaviour. *Entomologia Experimentalis et Applicata*, **62** (3), 201–9.

Garcia-Barros, E. (1992) Evidence for geographic variation of egg size and fecundity in a satyrine butterfly, *Hipparchia semele* (L.) (Lepidoptera, Nymphalidae–Satyrinae). *Graellsia*, **48**, 45–52.

Garnett, G.P. & Antia, R. (1994) Population biology of virus–host interactions. In: *The Evolutionary Biology of Viruses* (ed. S.S. Morse), pp. 51–73. Raven Press, New York.

Gaston, K.J. (1991) The magnitude of global insect species richness. *Conservation Biology*, **5** (3), 283–96.

Gaston, K.J. (1994) Spatial patterns of species description—how is our knowledge of the global insect fauna growing. *Biological Conservation*, **67**, 37–40.

Gaston, K.J. (1996a) Species richness: measure and measurement. In: *Biodiversity: a Biology of Numbers and Difference* (ed. K.J. Gaston), pp. 77–113. Blackwell Science, Oxford.

Gaston, K.J. (1996b) What is biodiversity? In: *Biodiversity: a Biology of Numbers and Difference* (ed. K.J. Gaston), pp. 1–9. Blackwell Science, Oxford.

Gaston, K.J. & Hudson, E. (1994) Regional patterns of diversity and estimates of global insect species richness. *Biodiversity and Conservation*, **3** (6), 493–500.

Gaston, K.J. & May, R.M. (1992) Taxonomy of taxonomists. *Nature*, **356** (6367), 281–2.

Gaston, K.J. & Williams, P.H. (1996) Spatial patterns in taxonomic diversity. In: *Biodiversity: a Biology of Numbers and Difference* (ed. K.J. Gaston), pp. 202–29. Blackwell Science, Oxford.

Gaston, K.J., Gauld, I.D. & Hanson, P. (1996) The size and composition of the hymenopteran fauna of Costa Rica. *Journal of Biogeography*, **23** (1), 105–13.

Gate, I.M., McNeill, S. & Ashmore, M.R. (1995) Effects of air-pollution on the searching behavior of an insect parasitoid. *Water, Air and Soil Pollution*, **85**, 1425–30.

Gathmann, A., Greiler, H.J. & Tscharntke, T. (1994) Trap-nesting bees and wasps colonising set-aside fields: succession and body size, management by cutting and sowing. *Oecologia*, **98** (1), 8–14.

Gaylor, E.S., Preszler, R.W. & Boecklen, W.J. (1996) Interaction between host plants, endophytic fungi, and a phytophagous insect in an oak (*Quercus grisea* × *Q. gambelii*) hydrid zone. *Oecologia (Berlin)*, **105**, 336–42.

Gehring, W.J. & Wehner, R. (1995) Heat shock protein

synthesis and thermotolerance in *Cataglyphis*, an ant from the Sahara desert. *Proceedings of the National Academy of Sciences of the USA*, **92** (7), 2994–8.

Georghiou, G.P. & Wirth, M.C. (1997) Influence of exposure to single versus multiple toxins of *Bacillus thuringiensis* subsp. *israelensis* on development of resistance in the mosquito *Culex quinquefasciatus* (Diptera: Culicidae). *Applied and Environmental Microbiology*, **63** (3), 1095–101.

Gershenzon, J. (1994) Metabolic costs of terpenoid accumulation in higher plants. *Journal of Chemical Ecology*, **20**, 1281–328.

Gibson, C.W.D. (1980) Niche use patterns among some Stenodemini (Heteroptera: Miridae) of limestone grassland, and an investigation of the possibility of interspecific competition between Notostira elongata Geoffroy and Megaloceraea recticornis Geoffroy. *Oecologia*, **47**, 352–64.

Gilbert, L.E. (1975) Ecological consequences of a coevolved mutualism between butterflies and plants. In: *Coevolution of Animals and Plants* (eds L.E. Gilbert & P.R. Raven), pp. 210–40. University of Texas Press, Austin.

Gilbert, L.E. (1977) The role of insect–plant coevolution in the organization of ecosystems. *College of international CNRS*, **265**, 399–413.

Gilbert, L.E. (1991) Biodiversity of a central American *Heliconius* community: Pattern, process, and problems. In: *Plant–Animal Interactions: Evolutionary Ecology in Tropical and Temperate Regions* (eds P.W. Price, T.M. Lewinshon, G.W. Fernandes & W.W. Benson), pp. 403–27. John Wiley and Sons, Inc, New York.

Gilbert, L.E. & Raven, P.H., eds (1975) *Coevolution of Animals and Plants*. University of Texas Press, Austin.

Giles, H.M. & Worrell, D. (1989) *Bruce-Chwatt's Essential Malariology*. Edward Arnold, London.

Gillman, M. & Hails, R. (1997) *An Introduction to Ecological Modelling: Putting Practice into Theory*. Blackwell Science, Oxford.

Gillott, C. (1995) *Entomology*, 2nd edn, 798 pp. Plenum, New York & London.

Gilpin, M.E. & Ayala, F.J. (1973) Global models of growth and competition. *Proceedings of the National Academy of Science*, **70**, 3590–3.

Gilpin, M.E. & Hanski, I. (1991) Metapopulation dynamics. *Empirical and Theoretical Investigations*. Academic Press, London.

Glinski, Z. & Jarosz, J. (1996) Immune mechanisms of vector insects in parasite destruction. *Central European Journal of Immunology*, **21** (1), 61–70.

Godfray, H.C.J. (1993) *Parasitoids: Behavioural and Evolutionary Ecology*. Princeton University Press, Princeton.

Godfrey, L.D. & Leigh, T.F. (1994) Alfalfa harvest strategy effect on *Lygus* bug (Hemiptera: Miridae) and insect predator population density: Implications for use as trap crop in cotton. *Environmental Entomology*, **23** (5), 1106–18.

Godfrey, L.D., Godfrey, K.E., Hunt, T.E. & Spomer, S.M. (1991) Natural enemies of European corn-borer *Ostrinia nubilalis* (Hübner) (Lepidoptera, Pyralidae) larvae in irrigated and drought-stressed corn. *Journal of the Kansas Entomological Society*, **64**, 279–86.

Goldstein, P.Z. (1997) How many things are there? *A reply to Oliver and Beattie, Beattie and Oliver, Oliver and Beattie, and Oliver and Beattie. Conservation Biology*, **11**, 571–4.

Gould, F. (1998) Sustainability of transgenic insecticidal cultivars — integrating pest genetics and ecology. *Annual Review of Entomology*, **43**, 701–26.

Gould, J.R., Elkinton, J.S. & Wallner, W.E. (1990) Density-dependent suppression of experimentally created gypsy-moth, *Lymantria dispar* (Lepidoptera, Lymantriidae), populations by natural enemies. *Journal of Animal Ecology*, **59**, 213–33.

Goulson, D. & Cory, J.S. (1995) Sublethal effects of baculovirus in the cabbage moth, *Mamestra brassicae*. *Biological Control*, **5** (3), 361–7.

Gowling, G.R. & van Emden, H.F. (1994) Falling aphids enhance impact of biological control by parasitoids on partially aphid-resistant plant varieties. *Annals of Applied Biology*, **125** (2), 233–42.

Graham-Bryce, I.J. (1977) Recent developments in the chemical control of agricultural pests in relation to ecological effects. In: *Ecological Effects of Pesticides* (eds F.H. Perring & K. Mellanby), pp. 47–60. Academic Press, London.

Grammatikopoulous, G. & Manetas, Y. (1994) Direct absorption of water by hairy leaves of *Phlomis fruticosa* and its contribution to drought avoidance. *Canadian Journal of Botany*, **72**, 1805–11.

Gratz, N.G. & Jany, W.C. (1994) What role for insecticides in vector control programs? *American Journal of Tropical Medicine and Hygiene*, **50** (Suppl. 6), 11–20.

Gray, S.M., Smith, D. & Altman, N. (1993) Barley yellow dwarf virus isolate-specific resistance in spring oats reduced virus accumulation and aphid transmission. *Phytopathology*, **83** (7), 716–20.

Grayson, J., Edmunds, M., Evans, E.H. & Britton, G. (1991) Carotenoids and coloration of popular hawkmoth caterpillars *Laothoe populi*. *Biological Journal of the Linnean Society*, **42**, 457–66.

Green, T.R. & Ryan, C.A. (1972) Wound-Induced Proteinase Inhibitor in Plant Leaves: a Possible Defense Mechanism Against Insects. *Science*, **175**, 776–7.

Greenblatt, J.A. & Barbosa, P. (1981) Effects of host diet on two pupal parasitoids of the gypsy moth: *Brachymeria intermedia* (Nees.) and *Coccygominus turionellae* (L.). *Journal of Applied Entomology*, **18**, 1–10.

Greenwood, B.M. (1997) What can be expected from malaria vaccines? *Annals of Tropical Medicine and Parasitology*, **91** (Suppl. 1), S9–S13.

Grégoire, J.C., Baisier, M., Drumont, A., Dahlsten, D.L., Meyer, H. & Francke, W. (1991) Volatile compounds in the larval frass of *Dendroctonus valens* and *Dendroctonus micans* (Coleoptera: Scolytidae) in relation to oviposition by the predator, *Rhizophagous grandis* (Coleoptera: Rhizophagidae). *Journal of Chemical Ecology*, **17** (10), 2003–20.

Grenier, A.M. & Nardon, P. (1994) The genetic control of ovariole number in *Sitophilus oryzae* L. (Coleoptera, Curculionidae) is temperature sensitive. *Genetics Selection Evolution (Paris)*, **26** (5), 413–30.

Grez, A.A. (1992) Species richness of herbivorous insects versus patch size of host plant: An experimental test. *Revista Chilena de Historia Natural*, **65** (1), 115–20.

Grimaldi, D. (1997) The birdflies, genus *Carnus*: species revision, generic relationships, and a fossil Meoneura in amber (Diptera: Carnidae). *American Museum Novitates*, **0**, (3190), 1–30.

Grimaldi, D., Bonwich, E., Delannoy, M. & Doberstein, S. (1994) Electron microscopic studies of mummified tissues in amber fossils. *American Museum Novitates*, **0** (3097), 1–31.

Grinnell, J. (1904) The origin and distribution of the chestnut-backed chickadee. *Auk*, **21**, 364–82.

Groombridge, B. (ed.) (1992) *Global Biodiversity. Status of the Earth's Living Resources*. Chapman and Hall, London.

Growchowska, M.J. & Ciurzynska, W. (1979) Differences between fruit-bearing and non-bearing apple spurs in activity of an enzyme system decomposing phloridzin. *Biological Plant*, **21**, 201–5.

Guderian, R.H. & Shelley, A.J. (1992) Onchocerciasis in Ecuador: the situation in 1989. *Memorias do Instituto Oswaldo Cruz, Rio de Janeiro*, **87** (3), 405–15.

Guerold, F., Vein, D., Jacquemin, G. & Pihan, J.C. (1995) The macroinvertebrate communities of streams draining a small granitic catchment exposed to acidic precipitations (Vosges Mountains, northeastern France). *Hydrobiologia*, **300-1**, 141–8.

Guillon, M. & Biache, G. (1995) IPM strategies for control of codling moth (*Cydia pomonella* L.) (Lepidoptera: Olethreutidae) interest of CmGV for long-term biological control of pest complex in orchards. *Mededelingen Faculteit Landbouwkundige en Toegepaste Biologische Wetenschappen Universiteit Gent*, **60** (3A), 695–705.

Gullan, P.J. & Cranston, P.S. (1996) *The Insects, An Outline of Entomology* (reprint). Chapman & Hall, London.

Haack, R.A. & Benjamin, D.M. (1982) The biology and ecology of the two-lined chestnut borer, *Agrilus biliniatus* (Coleoptera: Buprestidae), on oaks, (*Quercus* spp.) in Wisconsin. *Canadian Entomologist*, **114**, 385–96.

Hackstein, J.H.P. & Stumm, C.K. (1994) Methane production in terrestrial arthropods. *Proceedings of the National Academy of Sciences of the USA*, **91** (12), 5441–5.

Hafernik, J.E.J. (1992) Threats to invertebrate biodiversity: implications for conservation strategies. In: *The Theory and Practice of Nature Conservation, Preservation and Management* (eds P.L. Fielder & S.K. Jain), pp. 172–95. Chapman and Hall, London.

Haggis, M.J. (1996) Forecasting the severity of seasonal outbreaks of African armyworm, *Spodoptera exempta* (Lepidoptera: Noctuidae) in Kenya from the previous year's rainfall. *Bulletin of Entomological Research*, **86** (2), 129–36.

Haggstrom, H. & Larsson, S. (1995) Slow larval growth on a suboptimal willow results in high predation mortality in the leaf beetle *Galerucella lineola*. *Oecologia*, **104**, 308–15.

Hairston, N.G., Smith, F.E. & Slobodkin, L.B. (1960) Community structure, population control, and competition. *American Naturalist*, **44**, 421–5.

Halffter, G. & Favila, M.E. (1993) The Scarabaeinae (Insecta: Coleoptera), an animal group for analysing, inventorying and monitoring biodiversity in tropical rainforest and modified landscapes. *Biology International*, **27**, 15–21.

Hambler, C. & Speight, M.R. (1996) Extinction rates in British non-marine invertebrates since 1900. *Conservation Biology*, **10**, 892–6.

Hamilton, W.D. (1964) The genetical evolution of social behaviour. *Journal of Theoretical Biology*, **7**, 1–52.

Hamilton, W.D., Axelrod, R. & Tanese, R. (1990) Sexual reproduction as an adaptation to resist parasites (a review). *Proceedings of the National Academy of Sciences of the United States of America*, **87** (9), 3566–73.

Hammond, P.M. (1990) Insect abundance and diversity in the Dumoga-Bone National Park, North Sulawesi, with special reference to the beetle fauna of lowland rainforest in the Toraut region. In: *Insects and the Rain Forests of South East Asia (Wallacea)* (eds W.J. Knight & J.D. Holloway), pp. 197–254. Royal Entomological Society, London.

Hammond, P.M. (1992) Species inventory. In: *Global Biodiversity, Status of the Earth's Living Resources* (ed. B. Groombridge), pp. 17–39. Chapman & Hall, London.

Hammond, P.M., Stork, N.E. & Brendell, M.J.D. (1996) Tree-crown beetles in context: a comparison of canopy and other ecotone assemblages in a lowland tropical forest in Sulawesi. In: *Canopy Arthropods* (eds N.E. Stork, J. Adis & R.K. Didham), pp. 184–223. Chapman and Hall, London.

Han, E.N. & Bauce, E. (1995) Glycerol synthesis by diapausing larvae in response to the timing of low temperature exposure, and implications for overwintering survival of the spruce budworm, *Choristoneura fumiferana*. *Journal of Insect Physiology*, **41** (11), 981–5.

Han, S.S., Lee, M.H., Lim, C.Y., Yun, T.Y., Kim, J.H., Paik, J.C. & Ryoo, M.I. (1996) Cabbage worm, *Artogeia rapae*, immunodeficiency syndrome caused by virus-like particles in the parasitoid wasp. *Korean Journal of Entomology*, **26**, 279–87.

Hanhimaeki, S., Senn, J. & Haukioja, E. (1995) The convergence in growth of foliage-chewing insect species on individual mountain birch trees. *Journal of Animal Ecology*, **64**, 543–52.

Hanks, L.M. & Denno, R.F. (1993a) Natural enemies and plant water relations influence the distribution of an armored scale insect. *Ecology*, **74**, 1081–91.

Hanks, L.M. & Denno, R.F. (1993b) The role of demic adaptation in colonization and spread of scale insect populations. In: *Evolution of Insect Pests: Patterns and Variation* (eds K.C. Kim & B.A. McPheron), pp. 393–411. John Wiley, New York.

Hanks, L.M., Paine, T.D., Millar, J.G. & Hom, J.L. (1995) Variation among *Eucalyptus* species in resistance to eucalyptus long-horned borer in Southern California. *Entomologia Experimentalis et Applicata*, **74**, 185–94.

Hansen, L.O. & Somme, L. (1994) Cold hardiness of the elm bark beetle *Scolytus laevis* Chapuis, 1873 (Col., Scolytidae) and its potential as Dutch elm disease vector in the northernmost elm forests of Europe. *Journal of Applied Entomology*, **117** (5), 444–50.

Hanski, I. & Gilpin, M. (1991) Metapopulation dynamics—brief history and conceptual domain. *Biological Journal of the Linnean Society*, **42**, 3–16.

Hanski, I. & Gyllenberg, M. (1997) Uniting two general patterns in the distribution of species. *Science*, **275** (5298), 397–400.

Hanski, I. & Niemela, J. (1990) Elevational distributions of dung and carrion beetles in northern Sulawesi. In: *Insects and the Rain Forests of South East Asia (Wallacea)* (eds W.J. Knight & J.D. Holloway), pp. 145–52. Royal Entomological Society, London.

Harada, H. & Ishikawa, H. (1993) Gut microbe of aphid closely related to its intracellular symbiont. *Biosystems*, **31** (2–3), 185–91.

Harborne, J.B. (1982) *Introduction to Ecological Biochemistry*. Academic Press, London.

Harborne, J.B. (1988) Flavonoids in the environment: structure-activity relationships. In: *Plant Flavonoids in Biology & Medicine* (eds V. Cody, E. Middleton, J.B. Harborne & A. Beretz), pp. 17–27. I.I.A.R. Liss, New York, USA.

Hardy, I.C.W., Griffiths, N.T. & Godfray, H.C.J. (1992) Clutch size in a parasitoid wasp—a manipulation experiment. *Journal of Animal Ecology*, **61**, 121–9.

Harris, M.K., Chung, C.S. & Jackman, J.A. (1996) Masting and pecan interaction with insectan predehiscient nut feeders. *Environmental Entomology*, **25**, 1068–76.

Hartley, S.E. & Lawton, J.H. (1987) The effects of different types of damage on the chemistry of birch foliage and the responses of birch feeding insects. *Oecologia*, **74**, 432–7.

Hartley, S.E. & Lawton, J.H. (1990) Damage-induced changes in birch foliage: mechanisms and effects on insect herbivores, In: *Population Dynamics of Forest Insects*. (eds A.D. Watt, S.R. Leather, M.D. Hunter & N.A.C. Kidd), pp. 147–55. Intercept, Andover.

Hassell, M.P. (1971) Mutual interference between searching insect parasites. *Journal of Animal Ecology*, **40**, 473–86.

Hassell, M.P. (1976) *The Dynamics of Competition and Predation*. Edward Arnold, London.

Hassell, M.P. (1978) *The Dynamics of Arthropod Predator–Prey Systems*. Princeton University Press, Princeton, NJ.

Hassell, M.P. (1981) Arthropod predator–prey systems. In: *Theoretical Ecology: Principles and Applications* (ed. R.M. May), pp. 105–31. Blackwell Scientific Publications, Oxford.

Hassell, M.P. (1985) Insect natural enemies as regulating factors. *Journal of Animal Ecology*, **54**, 323–34.

Hassell, M.P. (1985) Parasitism in patchy environments: inverse density dependence can be stabilizing. *IMA Journal of Math Applied in Medicine and Biology*, **1**, 123–33.

Hassell, M.P. & May, R.M. (1973) Stability in insect host–parasite models. *Journal of Animal Ecology*, **42**, 693–736.

Hassell, M.P. & May, R.M. (1974) Aggregation in predators and insect parasites and its effect on stability. *Journal of Animal Ecology*, **43**, 567–94.

Hassell, M.P. & Varley, G.C. (1969) New inductive population model for insect parasites and its bearing on biological control. *Nature*, **223**, 1113–37.

Hassell, M.P., Lawton, J.H. & May, R.M. (1976) Patterns of dynamical behaviour in single species populations. *Journal of Animal Ecology*, **45**, 471–86.

Hassell, M.P., Lawton, J.H. & Beddington, J.R. (1977) Sigmoid functional responses by invertebrate predators and parasitoids. *Journal of Animal Ecology*, **46**, 249–62.

Hassell, M.P., Southwood, T.R.E. & Reader, P.M. (1987) The dynamics of the viburnum whitefly, *Aleurotrachelus jelinekii* Fraunf. A case study on population regulation. *Journal of Animal Ecology*, **56**, 283–300.

Hatcher, P.E., Paul, N.D., Ayres, P.G. & Whitaker, J.B. (1994) The effect of a foliar disease on the development of *Gastrophysa viridula* (Coleoptera, Chrysomelidae). *Ecological Entomology*, **19**, 349–60.

Haukioja, E. (1980) On the role of plant defences in the fluctuations of herbivore populations. *Oikos*, **35**, 202–13.

Haukioja, E. (1991) Cyclic fluctuations in density: Interactions between a defoliator and its host tree. *Acta Oecologica*, **12** (1), 77–88.

Haukioja, E. & Niemela, P. (1977) Retarded growth of a geometrid larva after mechanical damage to leaves of its host tree. *Annales Zoologici Fennici*, **14**, 48–52.

Hawkeswood,T.J. (1988) A survey of the leaf beetles (Coleoptera: Chrysomelidae) from the Townsville district, northern Queensland, Australia. *Giornale Italiano di Entomologia*, **4** (20), 93–112.

Hawkins, B.A. (1992) Parasitoid-host food webs and donor control. *Oikos*, **65**, 159–62.

Hawkins, B.A., Cornell, H.V. & Hochberg, M.E. (1997) Predators, parasitoids, and pathogens as mortality agents in phytophagous insect populations. *Ecology*, **78**, 2145–52.

Hawkins, B.A., Thomas, M.B. & Hochberg, M.E. (1993) Refuge theory and biological control. *Science*, **262** (5138), 1429–32.

He, E.F.G., Liu, J., Zhang, G.X.Q.U.G.M. & Yan, F.Y. (1991) Studies on the biochemical basis of resistance in sorghum to sorghum aphid *Melanaphis sacchari* (Zehntner). *Acta Entomologica Sinica*, **34** (1), 38–42.

Heath, A.C.G. (1994) Ectoparasites of livestock in New Zealand. *New Zealand Journal of Zoology*, **21** (1), 23–38.

Heie, O.E., Pettersson, J., Fuentes-Contreras, E. & Niemeyer, H.M. (1996) New records of aphids (Hemiptera: Aphidoidea) and their host plants from northern Chile. *Revista Chilena de Entomologia*, **23**, 83–7.

Heinrichs, E.A. (1994) Development of multiple pest resistant crop cultivars. *Journal of Agricultural Entomology*, **11** (3), 225–53.

Heinrichs, E.A. (1996) Management of rice insect pests. University of Minnesota—http://www.ent.agri.umn.edu/academics/classes/ipm/chapters/heinrich.htm.

Heinz, K.M. & Nelson, J.M. (1996) Interspecific interactions among natural enemies of *Bemisia* in an inundative biological-control program. *Biological Contrology*, **6**, 384–93.

Hellqvist, S. (1989) *Pachynematus pumilio* Konow (Hymenoptera: Tenthredinidae): Biology, distribution and control. *Vaxtskyddsnotiser*, **53** (3), 82–8.

Hernandez-Romero, J.C., Vera-Graziano, J., van Schoonhoven, A. & Cardona, M.C. (1984) The effects of a maize–bean association on insect pest population dynamics with emphasis on *Empasca kraemeri*. *Agrociencia*, **57**, 25–36.

Herren, H.R. (1990) Biological control as the primary option in sustainable pest management: The cassava pest project. *Mitteilungen der Schweizerischen Entomologischen Gesellschaft*, **63** (3–4), 405–14.

Herren, H.R., Neuenschwander, P., Hennessey, R.D. & Hammond, W.N.O. (1987) Introduction and dispersal of *Epidinocarsus lopezi* (Hymenoptera: Encyrtidae), an exotic parasitoid of the cassava mealybug, *Phenacoccus manihoti*, in Africa. *Agriculture, Ecosystems and Environment*, **19** (2), 131–44.

Heywood, V.H. (1995) *Global Biodiversity Assessment.* Cambridge University Press, Cambridge.

Hielkema, J.U. (1990) Satellite environmental monitoring for migrant pest forecasting by FAO: the Artemis system. *Philosophical Transactions of the Royal Society of London, Series B: Biological Sciences*, **328** (1251), 705–17.

Higashi, M., Abe, T. & Burns, T.P. (1992) Carbon–nitrogen balance and termite ecology. *Proceedings of the Royal Society of London, Series B: Biological Sciences*, **249** (1326), 303–8.

Hill, D.S. (1983) *Agricultural Insect Pests of the Tropics and Their Control*, 2nd edn. Cambridge University Press, pp. 746.

Hill, J.K., Thomas, C.D. & Lewis, O.T. (1996) Effects of habitat patch size and isolation on dispersal by *Hesperia* comma butterflies: Implications for metapopulation structure. *Journal of Animal Ecology*, **65** (6), 725–35.

Hinkle, G., Wetter, J.K., Schultz, T.R. & Sogin, M.L. (1994) Phylogeny of the attine ant fungi based on analysis of small subunit ribosomal RNA gene sequences. *Science*, **266** (5191), 1695–7.

Hjalten, J., Danell, K. & Ericson, L. (1993) Effects of simulated herbivory and intraspecific competition on the compensatory ability of birches. *Ecology*, **74**, 1136–42.

Hobbs, R. (1992) The role of corridors in conservation: solution or bandwagon? *Trends in Ecology and Evolution*, **7**, 389–92.

Hobson, K.R., Wood, D.L., Cool, L.G., White, P.R., Ohtsuka, T., Kubo, I. & Zavarin, E. (1993) Chiral specificity in responses by the bark beetle *Dendroctonus valens* to host kairomones. *Journal of Chemical Ecology*, **19**, 1837–46.

Hochberg, M.E. & Waage, J.K. (1991) A model for the biological control of *Oryctes rhinoceros* (Coleoptera: Scarabaeidae) by means of pathogens. *Journal of Applied Ecology*, **28** (2), 514–31.

Hodkinson, I.D. & Casson, D. (1991) A lesser prediction for bugs: Hemiptera (Insecta) diversity in tropical forests. *Biological Journal of the Linnean Society*, **43**, 101–9.

Hoffman, T.K. & Kolb, F.L. (1997) Effects of barley yellow dwarf virus on root and shoot growth of winter wheat seedlings grown in aeroponic culture. *Plant Disease*, **81** (5), 497–500.

Hoganson, J.W. & Ashworth, A.C. (1992) Fossil beetle evidence for climatic change 18000–10 000 years BP in south–central Chile. *Quaternary Research (Duluth)*, **37** (1), 101–16.

Hogg, I.D. & Williams, D.D. (1996) Response of stream invertebrates to a global-warming thermal regime: An ecosystem-level manipulation. *Ecology (Washington, DC)*, **77** (2), 395–407.

Hogg, J.C. & Hurd, H. (1997) The effects of natural *Plasmodium falciparum* infection on the fecundity and mortality of *Anopheles gambiae* s.l. in north east Tanzania. *Parasitology*, **114** (4), 325–31.

Hoke, R.A., Giesy, J.P., Zabik, M. & Unger, M. (1993) Toxicity of sediments and sediment pore waters from the Grand Calumet River–Indiana Harbor, Indiana, area of concern. *Ecotoxicology and Environmental Safety*, **26** (1), 86–112.

Holland, J.M., Thomas, S.R. & Hewitt, A. (1996) Some effects of polyphagous predators on an outbreak of cereal aphid (*Sitobion avenae* F.) and orange wheat blossom midge (*Sitodoplosis mosellana* Gehin). *Agriculture, Ecosystems and Environment*, **59** (3), 181–90.

Hölldobler, B. & Wilson, E.O. (1990) *The Ants*. Springer, Berlin.

Holling, C.S. (1959) Some characteristics of simple types of predation and parasitism. *Canadian Entomologist*, **91**, 385–98.

Holling, C.S. (1965) The functional response of predators to prey density and its role in mimicry and population regulation. *Memoirs of the Entomological Society of Canada*, **45**, 1–60.

Holling, C.S. (1966) The functional response of invertebrate predators to prey density. *Memoirs of the Entomological Society of Canada*, **48**, 1–86.

Holmes, J.C. (1996) Parasites as threats to biodiversity in shrinking ecosystems. *Biodiversity and Conservation*, **5**, 975–83.

Holopainen, J.K., Rikala, R., Kainulainen, P. & Oksanen, J. (1995) Resource partitioning to growth, storage and defense in nitrogen-fertilized Scots pine and susceptibility of the seedlings to the tarnished plant bug *Lygus rugulipennus*. *New Phytologist*, **131**, 521–32.

Holt, R.D. (1987) Prey communities in patchy environments. *Oikos*, **50**, 276–90.

Holt, R.D. (1990) The microevolutionary consequences of climate change. *Trends in Ecology and Evolution*, **5**, 311–15.

Holzinger, F. & Wink, M. (1996) Mediation of cardiac glycoside insensitivity in the Monarch butterfly (*Danaus plexippus*): Role of an amino acid substitution in the ouabain binding site of Na+, K+ -ATPase. *Journal of Chemical Ecology*, **22**, 1921–37.

Honda, K. (1995) Chemical basis of different oviposition by lepidopterous insects. *Archives of Insect Biochemistry and Physiology*, **30**, 1–23.

Hoover, K., Wood, D.L., Storer, A.J., Fox, J.W. & Bros, W.E. (1996) Transmission of the pitch canker fungus, *Fusarium subglutinans* F. sp. *pini*, to Monterey pine, *Pinus radiata*, by cone- and twig-infesting beetles. *Canadian Entomologist*, **128** (6), 981–94.

Horn, D.J. (1978) *Biology of Insects*. Saunders, Philadelphia, PA.

Horton, D.R., Chauvin, R.L., Hinojosa, T., Larson, D., Murphy, C. & Biever, K.D. (1997) Mechanisms of resistance to Colorado potato beetle in several potato lines and correlation with defoliation. *Entomologia Experimentalis et Applicata*, **82** (3), 239–46.

Horvitz, C.C. & Schemske, D.W. (1994) Effects of dispersers, gaps, and predators on dormancy and seedling emergence in a tropical herb. *Ecology*, **75**, 1949–58.

Hougen-Eitzman, D. & Rausher, M.D. (1994) Interactions between herbivorous insects and plant–insect coevolution. *American Naturalist*, **143**, 677–97.

Houghton, J.T., Jenkins, G.J. & Ephraums, J.J. (1990) Climate change: the IPCC scientific assessment. Cambridge University Press.

Houston, D.R. (1991) Changes in nonaggressive and aggressive subgroups of *Ophiostoma ulmi* within two populations of American elm in New England (USA). *Plant Disease*, **75** (7), 720–2.

Houston, D.R. (1994) Major new tree disease epidemics: beech bark disease. *Annual Review of Phytopathology*, **32**, 75–87.

Howarth, F.G. & Ramsay, G.W. (1991) The conservation of island insects and their habitats. In: *The Conservation of Insects and Their Habitats* (eds N.M. Collins & J.A. Thomas), pp. 71–107. Academic Press, London.

Huang, G., Vergne, E. & Gubler, D.J. (1992) Failure of dengue viruses to replicate in *Culex quinquefasciatus* (Diptera: Culicidae). *Journal of Medical Entomology*, **29** (6), 911–14.

Huang, X., Renwick, J.A.A. & Sachdev-Gupta, K. (1994) Oviposition stimulants in *Barbarea vulgaris* for *Pieris rapae* and *P. napi oleracea*: Isolation, identification and differential activity. *Journal of Chemical Ecology*, **20**, 423–38.

Huffaker, C.B. (1971) *Biological Control*, p. 511. Plenum, New York.

Hugentobler, U. & Renwick, J.A.A. (1995) Effects of plant nutrition on the balance of insect relevant cardenolides and glusinolates in *Erysimum cheiranthoides*. *Oecologia*, **102**, 95–101.

Hughes, L. & Westoby, M. (1992) Capitula on stick insect eggs and elaiosomes on seeds: convergent adaptaions for burial by ants. *Functional Ecology*, **6**, 642–8.

Hughes, P.R., Wood, H.A., Breen, J.P., Simpson, S.F., Duggan, A.J. & Dybas, J.A. (1997) Enhanced bioactivity of recombinant baculoviruses expressing insect-specific spider toxins in lepidopteran crop pests. *Journal of Invertebrate Pathology*, **69** (2), 112–18.

Hulme, M. & Gray, T. (1994) Mating disruption of Douglas-fir tussock moth (Lepidoptera: Lymantriidae) using a sprayable bead formulation of Z-6-Heneicosen-11-one. *Environmental Entomology*, **23** (5), 1097–100.

Hunter, A.F. (1993) Gypsy moth population sizes and the window of opportunity in spring. *Oikos*, **68** (3), 531–8.

Hunter, A.F. (1995) The ecology and evolution of reduced wings in forest macrolepidoptera. *Evolutionary Ecology*, **9** (3), 275–87.

Hunter, D.M. (1989) The response of Mitchell grasses (*Astrebia* spp.) and button grass (*Dactylctenium radulus*) to rainfall and their importance to the survival of the Australian plague locust, *Chortoicetes terminifera* (Walker) in the arid zone. *Australian Journal of Ecology*, **14** (4), 467–72.

Hunter, D.M. & Cosenzo, E.L. (1990) The origin of plagues and recent outbreaks of the South American locust, *Schistocerca cancellata* (Orthoptera: Acrididae) in Argentina. *Bulletin of Entomological Research*, **80** (3), 295–300.

Hunter, M.D. (1987) Opposing effects of spring defoliation on late season oak caterpillars. *Ecological Entomology*, **12**, 373–82.

Hunter, M.D. (1990) Differential susceptibility to variable plant phenology and its role in competition between two insect herbivores on oak. *Ecological Entomology*, **15**, 401–8.

Hunter, M.D. (1997) Incorporating variation in plant chemistry into a spatially-explicit ecology of phytophagous insects. In: *Forests and Insects* (eds A.D. Watt, N.E. Stork & M.D. Hunter), pp. 81–96. Chapman & Hall, London.

Hunter, M.D. (1998) Interactions between *Operophtera brumata* (Lepidoptera: Geometridae) and *Tortrix viridana* (Lepidoptera: Tortricidae) on oak: New evidence from time-series analysis. *Ecological Entomology*, in press.

Hunter, M.D. & Hull, L.A. (1993) Variation in concentrations of phloridzin and phloretin in apple foliage. *Phytochemistry*, **34**, 1251–4.

Hunter, M.D. & Price, P.W. (1992) Playing chutes and ladders—heterogeneity and the relative roles of bottom-up and top-down forces in natural communities. *Ecology*, **73**, 724–32.

Hunter, M.D. & Price, P.W. (1998) Cycles in insect populations: delayed density dependence or exogenous driving variables? *Ecological Entomology*, **23** (2), 216–22.

Hunter, M.D. & Schultz, J.C. (1995) Fertilization mitigates chemical induction and herbivore responses within damaged oak trees. *Ecology*, **76**, 1226–32.

Hunter, M.D. & Willmer, P.G. (1989) The potential for interspecific competition between two abundant defoliators on oak: Leaf damage and habitat quality. *Ecological Entomology*, **14**, 267–77.

Hunter, M.D., Watt, A.D. & Docherty, M. (1991) Outbreaks of the winter moth on Sitka spruce in Scotland are not influenced by nutrient deficiencies of trees, tree budburst, or pupal predation. *Oecologia*, **86**, 62–9.

Hunter, M.D., Ohgushi, T. & Price, P.W. (1992) *The Effects of Resource Distribution on Animal-Plant Interactions*. Academic Press, San Diego.

Hunter, M.D., Biddinger, D.J., Carlini, E.J., Mcpheron, B.A. & Hull, L.A. (1994) Effects of apple leaf allelochemistry on tufted apple bud moth (Lepidoptera: Tortricidae) resistance to azinphosmethyl. *Journal of Economic Entomology*, **87**, 1423–9.

Hunter, M.D., Malcolm, S.B. & Hartley, S.E. (1996) Population-level variation in plant secondary chemistry and the population biology of herbivores. *Chemoecology*, **7**, 45–56.

Hunter, M.D., Varley, G.C. & Gradwell, G.R. (1997) Estimating the relative roles of top-down and bottom-up forces on insect herbivore populations: a classic study revisited. *Proceedings of the National Academy of Sciences*, **94**, 9176–81.

Husband, B.C. & Barrett, S.C.H. (1996) A metapopulation perspective in plant population biology. *Journal of Ecology*, **84** (3), 461–9.

Hutchinson, G.E. (1957) Concluding remarks. *Cold Spring Harbor Symp, Quantitative Biology*, **22**, 415–27.

Hutchinson, G.E. (1959) Homage to Santa Rosalia, or why are there so many kinds of animals? *American Naturalist*, **93**, 145–59.

Hutchinson, G.E. (1978) *An Introduction to Population Ecology*, 260 pp. Yale University Press, New Haven, CT.

Hyman, P.S. & Parsons, M.S. (1992) *A Review of the Scarce and Threatened Coleoptera of Great Britain. Part 1*. The UK Joint Nature Conservation Committee, Peterborough.

Idris, A.B. & Grafius, E. (1995) Wildflowers as nectar sources for *Diadegma insulare* (Hymenoptera: Ichneumonidae), as a parasitoid of diamondback moth (Lepidoptera: Yponomeutidae*). Environmental Entomology*, **24** (6), 1726–35.

Inoue, T. & Kato, M. (1992) Inter and intra specific morphological variation in bumblebee species and competition in flower utilization In: *Effects of Resource Distribution on Animal-Plant Interactions* (ed. M.D. Hunter, T. Ohgushi and P.W. Price), pp. 393–427. Academic Press, San Diego.

Inoue, T., Murashima, K., Azuma, J.I., Sugimoto, A. & Slaytor, M. (1997) Cellulose and xylan utilisation in the lower termite *Reticulitermes speratus*. *Journal of Insect Physiology*, **43** (3), 235–42.

Irwin, M.E. & Thresh, J.M. (1990) Epidemiology of barley yellow dwarf: study in ecological complexity. *Annual Review of Phytopathology*, **28**, 393–424.

Ishihara, M. & Shimada, M. (1995) Photoperiodic induction of larval diapause and temperature-dependent termination in a wild multivoltine bruchid, *Kytorhinus sharpianus*. *Entomologia Experimentalis et Applicata*, **75** (2), 127–34.

Itioka, T. & Inoue, T. (1996) Density-dependent ant attendance and its effects on the parasitism of a honeydew-producing scale insect, *Ceroplastes rubens*. *Oecologia*, **106**, 448–54.

Itô, K. & Kobayashi, K. (1991) An outbreak of the Cryptomeria borer, *Semanotus japonicus* Lacordaire (Coleoptera: Cerambycidae) in a young Japanese cedar (*Crytpomeria japonica* D.Don) plantation I. Annual fluctuations in population size and impact on host trees. *Applied Entomology and Zoology*, **26** (1), 63–70.

Itô, Y. (1960) Ecological studies on population increase and habitat segregation among barley aphids. *Bulletin of the National Institute of Agricultural Sciences Services C*, **11**, 45–130.

IUCN (1994) *IUCN Red List Categories*. IUCN, Gland.

Ivory, M. & Speight, M.R. (1993) Pests and diseases. In: *Tropical Forestry Handbook Vol II*, (ed. L. Pancel), Section 19. Springer Verlag, Berlin.

Ives, W.G.H. (1974) Weather and outbreaks of the spruce budworm, *Choristoneura rosaceana. Information Report NOR-X*, Northern Forest Research Center. NOR-X-118.

Iwao, K. & Rausher, M.D. (1997) Evolution of plant resistance to multiple herbivores: Quantifying diffuse coevolution. *American Naturalist*, **149**, 316–35.

Iwasaki, T., Aoyagi, M., Dodo, Y. & Ishii, M. (1994) Emergence periods of overwintering generation from mantis egg case and oviposition, and longevity of adult dermestid beetle, *Thaumaglossa rufocapillata*. *Japanese Journal of Applied Entomology and Zoology*, **38** (3), 147–51.

Izquierdo, J.I. (1996) *Helicoverpa armigera* (Hübner) (Lep., Noctuidae): relationship between captures in pheromone traps and egg counts in tomato and carnation crops. *Journal of Applied Entomology*, **120**, 281–90.

Jackai, L.E.N. & Oghiakhe, S. (1989) Pod wall trichomes and resistance of two wild cowpea, *Vigna vexillata*, accessions to *Muruca testulalis* Geyer (Lepidoptera: Pyralidae) and *Clavigralla tomentosicollis* Stal (Hemiptera. *Coreidae). Entomological Bulletin of Research*, **79**, 595–605.

Jamieson-Dixon, R.W. & Wrona, F.J. (1992) Life history and production of the predatory caddisfly *Rhyacophila vao* Milne in a spring-fed stream. *Freshwater Biology*, **27** (1), 1–11.

Janzen, D.H. (1966) Coevolution of mutualism between ants and acacias in Central America. *Evolution*, **20**, 249–75.

Janzen, D.H. (1967) Interaction of the bull's horn acacia (*Acacia cornigera* L.) with an ant inhabitant (*Pseudomyrmex ferruginea* F. Smith) in eastern Mexico. *Kansas University Science Bulletin*, **47**, 315–558.

Janzen, D.H. (1968) Host plants in evolutionary and contemporary time. *American Naturalist,* **102**, 592–5.

Janzen, D.H. (1973) Host plants as islands. II. Competition in evolutionary and contemporary time. *American Naturalist,* **107**, 786–90.

Janzen, D.H. (1980) When is it coevolution? *Evolution,* **34**, 611–2.

Janzen, D.H. (1997) Wildland biodiversity management in the tropics. In: *Biodiversity II, Understanding and Protecting our Biological Resources* (eds M.L. Reaka-Kudla, D.E. Wilson & E.O. Wilson), pp. 411–31. Joseph Henry Press, Washington, DC.

Jayanthi, R., David, H. & Goud, Y.S. (1994) Physical characters of sugarcane plant in relation to infestation by *Melanaspis glomerata* (G.) and *Saccharicoccus sacchari* (Ckll.). *Journal of Entomological Research (New Delhi),* **18**, 305–14.

Jeffree, C.E. (1986) Insects and the plant surface. In: *The Cuticle, Epicuticular waxes and Trichomes of Plants, with Reference to their Structure, Functions and Evolution* (eds B.E. Juniper & R. Southwood), pp. 23–64. Edward Arnold, London.

Jeffree, C.E. & Jeffree, E.P. (1996) Redistribution of the potential geographical ranges of mistletoe and Colorado beetle in Europe in response to the temperature component of climate change. *Functional Ecology,* **10** (5), 562–77.

Jeffries, M.J. & Lawton, J.H. (1984) Enemy-free space and the structure of ecological communities. *Biological Journal of the Linnean Society,* **23** (4), 269–86.

Jenkins, E.B., Hammond, R.B., St-Martin, S.K. & Cooper, R.L. (1997) Effect of soil moisture and soybean growth stage on resistance to Mexican bean beetle (Coleoptera: Coccinellidae). *Journal of Economic Entomology,* **90** (2), 697–703.

Jermy, A.C., Long, D., Sands, M.J.S., Stork, N.E. & Winser, S. (1995) *Biodiversity Assessment, A Guide to Good Practice.* HMSO, London.

Jervis, M. & Kidd, N. (1996) *Insect Natural Enemies: Practical Approaches to their Study and Evaluation.* Chapman and Hall, London.

Jervis, M.A., Kidd, N.A.C. & Walton, M. (1992) A review of methods for determining dietary range in adult parasitoids. *Entomophaga,* **37**, 565–74.

Johnson, K.H., Vogt, K.A., Clark, H.J., Schmitz, O.J. & V ogt, D.J. (1996) Biodiversity and the productivity and stability of ecosystems. *Trends in Ecology and Evolution,* **11**, 372–7.

Johnson, S. (1995) Production and trade of tropical logs. *Tropical Forest Update,* **5**, 21–3.

Jones, C.G. & Lawton, J.H. (1991) Plant chemistry and insect species richness of British umbellifers. *Journal of Animal Ecology,* **60** (3), 767–78.

Jones, T.H. (1993) Analytical models. In: *Decision Tools for Pest Management* (eds G.A. Norton & J.D. Mumford), pp. 101–18. CAB International, London.

Jones, T.H., Cole, R.A. & Finch, S. (1988) A cabbage root fly oviposition deterrent in the frass of garden pebble moth caterpillars. *Entomologia Experimentalis et Applicata,* **49**, 277–82.

Jones, T.H., Hassell, M.P. & Pacala, S.W. (1993) Spatial heterogenicity and the population dynamics of a host–parasitoid system. *Journal of Animal Ecology,* **62**, 251–62.

Judd, G.J.R., Gardiner, M.G.T. & Thomson, D.R. (1996) Commercial trials of pheromone-mediated mating disruption with Isomate-C to control codling moth in British Columbia apple and pear orchards. *Journal of the Entomological Society of British Columbia,* **93**, 23–34.

Juniper, B. & Southwood, R. (1986) *Insects and the Plant Surface,* 360pp. Edward Arnold, London.

Jupp, P.G. & Lyons, S.F. (1987) Experimental assessment of bedbugs (*Cimex lectularius* and *Cimex hemipterus*) and mosquitoes (*Aedes aegypti formosus*) as vectors of human immunodeficiency virus. *AIDS,* **1** (3), 171–4.

Jupp, P.G. & Williamson, C. (1990) Detection of multiple blood meals in experimentally fed bedbugs (Hemiptera: Cimicidae). *Journal of the Entomological Society of Southern Africa,* **53** (2), 137–40.

Kaakeh W. & Dutcher J.D. (1993) Survival of yellow pecan aphids and black pecan aphids (Homoptera: aphididae) at different temperature regimes. *Environmental Entomology,* **22** (4), 810–7.

Kaitala, A. (1991) Phenotypic plasticity in reproductive behavior of waterstriders: Trade-offs between reproduction and longevity during food stress. *Functional Ecology,* **5** (1), 12–18.

Kambhampati, S. (1995) A phylogeny of cockroaches and related insects based on DNA sequence of mitochondrial ribosomal RNA genes. *Proceedings of the National Academy of Sciences of the USA,* **92** (6), 2017–20.

Kaneko, S. (1995) Frequent successful multiparasitism by five parasitoids attacking the scale insect *Nipponaclerda biwakoensis. Researches on Population Ecology,* **37**, 225–8.

Kang, L.L.I.H.C. & Chen, Y.L. (1989) Analyses of numerical character variations of geographical populations of *Locusta migratoria* phase *solitaria* in China. *Acta Entomologica Sinica,* **32** (4), 418–26.

Karban, R. (1989) Fine-scale adaptation of herbivorous thrips to individual host plants. *Nature,* **340**, 60–1.

Karban, R. (1993) Costs and benefits of reduced resistance and plant density for a native shrub, *Gossypium thurberi. Ecology,* **74**, 9–19.

Karban, R. & Baldwin, I.T. (1997) *Induced Responses to Herbivory.* University of Chicago Press, Chicago, IL.

Karban, R. & Strauss, S.Y. (1993) Effects of herbivores on growth and reproduction of their perennial host, *Erigeron glaucus. Ecology,* **74**, 39–46.

Kato, M. (1994) Structure, organization, and response of a species-rich parasitoid community to host leafminer population-dynamics. *Oecologia,* **97**, 17–25.

Kearsley, M.J.C. & Whitham, T.G. (1992) Guns and butter, a no cost defense against predation for *Chrysomela confluens. Oecologia (Heidelberg),* **92**, 556–62.

Keating, S.T., Hunter, M.D. & Schultz, J.C. (1990) Leaf phenolic inhibition of gypsy-moth nuclear polyhedrosis-virus—role of polyhedral inclusion body aggregation. *Journal of Chemical Ecology,* **16**, 1445–57.

Keeling, M.J. & Rand, D.A. (1995) A spatial mechanism for the evolution and maintenance of sexual reproduction. *Oikos,* **74** (3), 414–24.

Keenlyside, J. (1989) Host choice in aphids. DPhil Thesis, University of Oxford.

Kehat, M., Anshelevich, L., Harel, M. & Dunkelblum, E. (1995) Control of the codling moth (*Cydia pomonella*) in apple and pear orchards in Israel by mating disruption. *Phytoparasitica,* **23** (4), 285–96.

Kellou, R., Mahjoub, N., Benabdi, A. & Boulahya, M.S. (1990) Algerian case study and the need for permanent desert locust monitoring. *Philosophical Transactions of the Royal Society of London, Series B: Biological Sciences,* **328** (1251), 573–84.

Kendall, D.A., George, S. & Smith, B.D. (1996) Occurrence of barley yellow dwarf viruses in some common grasses (Gramineae) in south west England. *Plant Pathology (Oxford),* **45** (1), 29–37.

Kennedy, C.E.J. (1986) Attachment may be a basis for specialization in oak aphids. *Ecological Entomology,* **11**, 291–300.

Kennedy, C.E.J. & Southwood, T.R.E. (1984) The number of species of insects associated with British trees—a re-analysis. *Journal of Animal Ecology,* **53**, 455–78.

Kennedy, G.G., Farrar, R.R. & Kashyap, R.K. (1991) 2-Tridecanone glandular trichome-mediated insect resistance in tomato—effect on parasitoids and predators of *Heliothis zea*. *American Chemical Society Symposium Series,* **449**, 150–65.

Keogh, R.G. & Lawrence, T. (1987) Influence of *Acremonium lolii* presence on emergence and growth of ryegrass seedlings. *New Zealand Journal of Agricultural Research,* **30**, 507–10.

Kerslake, J.E., Kruuk, L.E.B., Hartley, S.E. & Woodin, S.J. (1996) Winter moth (*Operophtera brumata* (Lepidoptera, Geometridae)) outbreaks on Scottish heather moorlands—effects of host-plant and parasitoids on larval survival and development. *Bulletin of Entomological Research,* **86**, 155–64.

Kettle, D.S. (1984) *Medical and Veterinary Entomology*. Croom Helm, London.

Kettle, D.S. (1995) *Medical and Veterinary Entomology,* 2nd edn, 725 pp. CAB International, Wallingford, UK.

Kidd, N.A.C. & Jervis, M.A. (1996) In: *Population dynamics. Insect Natural Enemies* (eds M.A. Jervis & N.A.C. Kidd), pp. 293–374. Chapman & Hall, London.

Kidd, N.A.C., Lewis, G.B. & Howell, C.A. (1985) An association between two species of pine aphid, *Schizolachnus pineti* and *Eulachnus agilis*. *Ecological Entomology,* **10**, 427–32.

Kimberling, D.N., Scott, E.R. & Price, P.W. (1990) Testing a new hypothesis: plant vigor and *Phylloxera* distribution in wild grape in Arizona (USA). *Oecologia (Heidelberg),* **84**, 1–8.

King, C.J., Fielding, N.J. & O'Keefe, T. (1991) Observations of the life cycle and behaviour of the *Rhizophagus grandis* (Coleoptera: Rhizophagidae) in Britain (UK). *Journal of Applied Entomology,* **11** (3), 286–96.

Kingsland, S.E. (1985) *Modeling Nature: Episodes in the History of Population Ecology,* 267 pp. University of Chicago Press, Chicago, IL.

Kingsolver, J.G. & Koehl, M.A.R. (1994) Selective factors in the evolution of insect wings. *Annual Review of Entomology,* **39**, 425–51.

Kirby, P. (1992) *Habitat Management for Invertebrates: a Practical Handbook*. RSPB, Sandy.

Kirby, W. & Spence, W. (1822) *An Introduction to Entomology: or Elements of the Natural History of Insects*. Printed for Longman, Hurst, Rees, Orme and Brown, London.

Kirk, A.A. & Wallace, M.M.H. (1990) Seasonal variations in numbers, biomass and breeding patterns of dung beetles (Coloeptera. *Scarabidae)* in Southern France. *Entomophaga,* **35**, 569–82.

Kitching, R.L. (1993) Towards rapid biodiversity assessment—lessons following studies of arthropods of rainforest canopies. *Rapid Biodiversity Assessment: Proceedings of the Biodiversity Assessment Workshop, Macquarie University, 1993* (ed. Research Unit for Biodiversity and Bioresources), pp. 26–30. Macquarie University, Sydney.

Klaper, R.D. & Hunter, M.D. (1998) Genetic versus environmental effects on the phenolic chemistry of turkey oak, *Quercus laevis*. In: *Diversity and Adaptation in Oak Species* (ed. K.C. Steiner), pp. 262–8. The Pennsylvania State University Press, Pennsylvania.

Klein, B.C. (1989) Effects of forest fragmentation on dung and carrion beetle communities in central Amazonia. *Ecology,* **70**, 1715–25.

Klemm, U. & Schmutterer, H. (1993) Effects of neem preparations on *Plutella xylostella* L. and its natural enemies of the genus *Trichogramma*. *Zeitschrift fuer Pflanzenkrankheiten und Pflanzenschutz,* **100** (2), 113–28.

Kocourek, F., Havelka, J., Berankova, J. & Jarosik, V. (1994) Effect of temperature on development rate and intrinsic rate of increase of *Aphis gossypii* reared on greenhouse cucumbers. *Entomologia Experimentalis et Applicata,* **71** (1), 59–64.

Kogan, M. (1998) Integrated pest management—historical perspectives and contemporary developments. *Annual Review of Entomology,* **43**, 243–70.

Koricheva, J., Larsson, S. & Haukioja, E. (1998) Insect performance on experimentally stressed woody plants: a meta-analysis. *Annual Review of Entomology,* **43**, 195–216.

Kostal, V. & Finch, S. (1994) Influence of background on host-plant selection and subsequent oviposition by the cabbage root fly (*Delia radicum*). *Entomologia Experimentalis et Applicata,* **70** (2), 153–63.

Kotze, A.C., Sales, N. & Barchia, I.M. (1997) Diflubenzuron tolerance associated with monooxygenase activity in field strain larvae of the Australian sheep blowfly (Diptera: Calliphoridae). *Journal of Economic Entomology,* **90** (1), 15–20.

Kouki, J., Niemelä, P. & Viitasaari, M. (1994) Reversed

latitudinal gradient in species richness of sawflies (Hymenoptera, Symphyta). *Annales Zoologici Fennici*, **31** (1), 83–8.

Krassilov, V.A. & Rasnitsyn, A.P. (1996) Pollen in the guts of Permian insects: First evidence of pollenivory and its evolutionary significance. *Lethaia*, **29** (4), 369–72.

Krebs, C.J. (1989) *Ecological Methodology*, 1st edn, 654 pp. Harper Collins, New York.

Kruess, A. & Tscharntke, T. (1994) Habitat fragmentation, species loss and biological control. *Science*, **264** (5165), 1581–54.

Kumar, H. (1992) Inhibition of ovipositional responses of *Chilo partekkus* (Lepidoptera: Pyralidae) by the trichomes on the lower leaf surafce of a maize cultivar. *Journal of Economic Entomology*, **85** (5), 1736–9.

Kumar, N.S., Hewavitharanage, P. & Adikaram, N.K.B. (1995) Attack on tea by *Xyleborus fornicatus*: Inhibition of the symbiote, *Monacrosporium ambrosium*, by caffeine. *Phytochemistry (Oxford)*, **40** (4), 1113–16.

Kunin, W.E. & Lawton, J.H. (1996) Does biodiversity matter? Evaluating the case for conserving species. In: *Biodiversity: a Biology of Numbers and Difference* (ed. K.J. Gaston), pp. 283–308. Blackwell Science, Oxford.

Kyto, M., Niemela, P. & Larsson, S. (1996) Insects on trees: Population and individual response to fertilization. *Oikos*, **75**, 148–59.

Labandeira, C.C. & Sepkoski, J.J.J.R. (1993) Insect diversity in the fossil record. *Science*, **261** (5119), 310–15.

Labandeira, C.C., Beall, B.S. & Hueber, F.M. (1988) Early insect diversification: evidence from a lower Devonian bristletail from Quebec (Canada). *Science (Washington D.C.)*, **242** (4880), 913–16.

Labandeira, C.C., Dilcher, D.L., Davis, D.R. & Wagner, D.L. (1994) Ninety-seven million years of angiosperm–insect association; paleobiological insights into the meaning of co-evolution. *Proceedings of the National Academy of Sciences of the USA*, **91** (25), 12278–82.

Lack, D. (1969) The number of bird species on islands. *Bird Study*, **16**, 193–209.

Lack, D. (1976) *Island Birds*. Blackwell Scientific, Oxford.

Lamb, R.J. & Mackay, P.A. (1987) *Acyrthosiphon kondoi* influences alata production by the pea aphid, *A. pisum*. *Entomologia Experimentalis et Applicata*, **45**, 195–8.

Lambert, A.L., Mcpherson, R.M. & Espelie, K.E. (1995) Soybean host plant resistance mechanisms that alter abundance of whiteflies (Homoptera: aleyrodidae). *Environmental Entomology*, **24**, 1381–6.

Lambert, L. & Heatherly, L.G. (1991) Soil water potential: effects on soybean looper feeding on soybean leaves. *Crop Science*, **31** (6), 1625–8.

Lamberti, G.A. & Resh, V.H. (1985) Distribution of benthic algae and macroinvertebrates along a thermal stream gradient. *Hydrobiologia*, **128** (1), 13–22.

Lamy, M. (1990) Contact dermatitis (erucism) produced by processionary caterpillars (genus *Thaumetopoea*). *Journal of Applied Entomology* **110** (5), 425–37.

Lancien, J. (1991) Campaign against sleeping sickness in south-west Uganda by trapping tsetse flies. *Annales de la Societé Belge de Médecine Tropicale*, **71** (suppl 1), 35–47.

Landsberg, J. & Gillieson, D.S. (1995) Regional and local variation in insect herbivory, vegetation and soils of eucalypt associations in contrasted landscape positions along a climatic gradient. *Australian Journal of Ecology*, **20**, 299–315.

Langley, P.A. (1994) Understanding tsetse flies. *Onderstepoort Journal of Veterinary Research*, **61** (4), 361–7.

Lanza, J., Schmitt, M.A. & Awad, A.B. (1992) Comparative chemistry of elaiosomes of three species of *Trillium*. *Journal of Chemical Ecology*, **18**, 209–21.

Larsson, S. (1989) Stressful times for the plant stress-insect performance hypothesis. *Oikos*, **56**, 277–83.

Larsson, S. & Tenow, O. (1984) Areal distribution of a *Neodiprion sertifer* (Hym., Diprionidae) outbreak on Scots pine as related to stand condition. *Holarctic Ecology*, **7**, 81–90.

Latto, J. & Hassell, M.P. (1987) Do pupal predators regulate the winter moth? *Oecologia*, **74**, 153–5.

Lawson, G.L. (1994) Indigenous trees in West African forest plantations: the need for domestication by clonal techniques. In: *Tropical Trees: The Potential for Domestication and the Rebuilding of Forest Resources* (eds R.R.B. Leakey & A.C. Newton), pp. 112–23. HMSO, London.

Lawson, S.A., Furuta, K. & Katagiri, K. (1996) The effect of host tree on the natural enemy complex of *Ips typographus japonicus* Niijima (Col, Scolytidae) in Hokkaido, Japan. *Zeitschrift fuer Angewandte Entomologie*, **120**, 77–86.

Lawton, J.H. (1982) Vacant niches and unsaturated communities: a comparison of bracken herbivores at sites on two continents. *Journal of Animal Ecology*, **51**, 573–95.

Lawton, J.H. (1984) Herbivore community organization: general models and specific tests with phytophagous insects. In: *A New Ecology—Novel Approaches to Interactive Systems*, (ed P.W. Price, C.N. Slobodchikoff, and W.S. Gaud), pp. 329–52. John Wiley, New York.

Lawton, J.H. (1991) From physiology to population dynamics and communities. *Functional Ecology*, **5**, 155–61.

Lawton, J.H. (1992) There are not 10 million kinds of population-dynamics. *Oikos*, **63**, 337–8.

Lawton, J.H. (1994) What do species do in ecosystems? *Oikos*, **71**, 367–74.

Lawton, J.H. (1995) The response of insects to environmental change. In: *Insects in a Changing Environment* (eds R. Harrington & N.E. Stork), pp. 3–26. Academic Press, London.

Lawton, J.H. & Hassell, M.P. (1981) Asymmetrical composition in insects. *Nature*, **289**, 793–5.

Lawton, J.H. & Schroder, D. (1977) Effects of plant type, size of geographical range and taxonomic isolation on number of insect species associated with British plants. *Nature*, **265**, 137–40.

Lawton, J.H. & Strong, D.R. (1981) Community patterns and competition in folivorous insects. *American Naturalist*, **118**, 317–38.

Lawton, J.H., Bignell, D.E., Bolton, B., *et al.* (1998) Biodiversity inventories, indicator taxa and effects of habitat modification in tropical forest. *Nature*, **391**, 72–6.

Leather, S.R. (1986) Insect species richness of the British Rosaceae—the importance of host range, plant architecture, age of establishment, taxonomic isolation and species area relationships. *Journal of Animal Ecology*, **55**, 841–60.

Leather, S.R. (1989) Do alate aphids produce fitter offspring? The influence of maternal rearing history and morph on life-history parameters of *Rhopalosiphum padi* (L). *Functional Ecology*, **3** (2), 237–44.

Leather, S.R. & Owuor, A. (1996) The influence of natural enemies and migration on spring populations of the green spruce aphid, *Elatobium abietinum* Walker (Hom., Aphididae). *Journal of Applied Entomology*, **120** (9), 529–36.

Leather, S.R. & Walsh, P.J. (1993) Sublethal plant defenses—the paradox remains. *Oecologia*, **93**, 153–5.

Lee, S.C. (1992) Towards integrated pest management of rice in Korea. *Korean Journal of Applied Entomology*, **31** (3), 205–40.

Legner, E.F. (1995) Biological control of Diptera of medical and veterinary importance. *Journal of Vector Ecology*, **20** (1), 59–120.

Legrand, P. (1993) Management of biological control against *Dendroctonus micans* infestation of spruce in Auvergne and Limousin (Coleoptera: Scolytidae). *Revue des Sciences Naturelles d'Auvergne*, **56**, 49–57.

Leibee, G.L. & Horn, D.J. (1979) Influence of tillage in survivorship of cereal leaf beetle and its larval parasites, *Tetrastichus julis* and *Lemophagus curtus*. *Environmental Entomology*, **8** (3), 485–6.

Leimar, O. (1996) Life history plasticity: Influence of photoperiod on growth and development in the common blue butterfly. *Oikos*, **76** (2), 228–34.

Lelis, A.T. (1992) The loss of intestinal flagellates in termites exposed to the juvenile hormone analogue (JHA) methoprene. *Material und Organismen (Berlin)*, **27** (3), 171–8.

Levine, E. (1993) Effect of tillage practices and weed management on survival of stalk borer (Lepidoptera: Noctuidae) eggs and larvae. *Journal of Economic Entomology*, **86** (3), 924–8.

Li, T., Yan, S., Wang, G. *et al.* (1994) A study of controlling grassland acrididae with *Nosema locustae*. *Sichuan Daxue Xuebao (Ziran Kexueban)*, **31** (4), 556–62.

Liadouze, I., Febvay, G., Guillaud, J. & Bonnot, G. (1995) Effect of diet on the free amino acid pools of symbiotic and aposymbiotic pea aphids, *Acyrthosiphon pisum*. *Journal of Insect Physiology*, **41** (1), 33–40.

Libert, M. (1994) Biodiversity: *Rhopalocera* fauna of two hills of Yaounde area, Cameroun (Lepidoptera). *Bulletin de la Société Entomologique de France*, **99** (4), 335–55.

Lim, G.S. (1992) Integrated pest management in the Asia–Pacific context. In: *Integrated Pest Management in the Asia–Pacific Region* (eds P.A.C. Ooi, G.S. Lim, T.H. Ho, P.L. Manalo & J. Waage), pp. 1–12. CAB International, Wallingford, UK.

Lindroth, R. & Hwang, S.-Y. (1996) Clonal variation of foliar chemistry of quaking aspen (*Populus tremuloides* Michx.). *Biochemical Systematics and Ecology*, **24**, 357–64.

Linfield, M.C.J., Raubenheimer, D., Hambler, C. & Speight, M.R. (1993) Leaf miners on *Ochna ciliata* (Ochnaceae) growing on Aldabra Atoll. *Ecological Entomology*, **18**, 332–8.

Liu, C. (1990) Comparative studies on the role of *Anopheles anthropophagus* and *Anopheles sinensis* in malaria transmission in China. *Chung Hua Liu Hsing Ping Hsueh-Tsa-Chih*, **11** (6), 360–3.

Liu, H. & Beckenbach, A.T. (1992) Evolution of the mitochondrial cytochrome oxidase II gene among 10 orders of insects. *Molecular Phylogenetics and Evolution*, **1** (1), 41–52.

Liu, S., Yang, Z., Li, T., Wang, Y. & Wu, T. (1992) Studies on the mutagenicity of human body cells affected by insect viruses. *Sichuan Daxue Xuebo (Ziran Kexueban)*, **29** (4), 541–5.

Lo, P.L. (1995) Size and fecundity of soft wax scale (*Ceroplastes destructor*) and Chinese wax scale (*C. sinensis*) (Hemiptera: Coccidae) on citrus. *New Zealand Entomologist*, **18**, 63–9.

Loader, C. & Damman, H. (1991) Nitrogen content of food plants and vulnerability of *Pieris rapae* to natural enemies. *Ecology*, **72**, 1586–90.

Longo, S. (1994) The role of beekeeping within agrarian and natural ecosystems. *Ethology, Ecology and Evolution, Special Issue*, **3**, 5–9.

Lorio, P.L. Jr, Stephen, F.M. & Paine, T.D. (1995) Environment and ontogeny modify loblolly pine response to induced acute water deficits and bark beetle attack. *Forest Ecology and Management*, **73**, 97–110.

Lotka, A.J. (1925) *Elements of Physical Biology*. Williams and Wilkins, Baltimore, MD.

Lotka, A.J. (1932) The growth of mixed populations: two species competing for a common food supply. *Journal of Washington Academy of Science*, **22**, 461–9.

Louda, S.M. & Collinge, S.K. (1992) Plant resistance to insect herbivores: a field test on the environmental stress hypothesis. *Ecology*, **73**, 153–69.

Louda, S.M. & Rodman, J.E. (1983) Ecological patterns in the glucosinolate content of a native mustard, *Cardamine cordifolia*, in the Rocky Mountains. *Journal of Chemistry Ecology*, **9**, 397–421.

Louda, S.M., Farris, M.A. & Blua, M.J. (1987a) Variation in methylglucosinolate and insect damage to *Cleome serrulata* (Capparaceae) along a natural soil moisture gradient. *Journal of Chemical Ecology*, **13**, 569–81.

Louda, S.M., Huntly, N. & Dixon, P.M. (1987b) Insect herbivory across a sun/shade gradient: Response to experimentally-induced in situ plant stress. *Acta Oecologica Oecologia Generalis*, **8**, 357–64.

Louton, J., Gelhaus, J. & Bouchard, R. (1996) The aquatic macrofauna of water-filled bamboo (Poaceae:

Bambusoideae: Guadua) internodes in a Peruvian lowland tropical forest. *Biotropica*, **28** (2), 228–42.

Lovejoy, T.E., Bierregaard, R.O., Rylands, A.B., *et al.* (1986) Edge and other effects of isolation on Amazonan forest fragments. In: *Conservation Biology: The Science of Scarcity and Diversity* (ed. M.E. Soulé), pp. 257–85. Sinauer, Sunderland.

Lovejoy, T.E., Erwin, N. & Boren, S. (1997) Insect conservation. In: *Forests and Insects* (eds A.D. Watt, N.E. Stork & M.D. Hunter), pp. 395–400. Chapman and Hall, London.

Lovett, G.M. & Ruesink, A.E. (1995) Carbon and nitrogen mineralization from decomposing gypsy moth frass. *Oecologia*, **104**, 133–8.

Lowman, M.D. & Nadkarni, N.M. (1995) *Forest Canopies*. Academic Press, New York.

Lozano, C. & Campos, M. (1993) Colonisation and estimation of population size in the bark beetle *Hylesinus varius* (Coleoptera: Scolytidae), a pest of olives in southern Spain. *International Journal of Pest Management*, **39** (3), 277–80.

Luff, M.L. & Woiwod, I.P. (1995) Insects as indicators of land-use change: a European perspective, focusing on moths and ground beetles. In: *Insects in a Changing Environment* (eds R. Harrington & N.E. Stork), pp. 399–422. Academic Press, London.

Lugo, A.E. (1995) Management of tropical biodiversity. *Ecological Applications*, **5**, 956–61.

Lundberg, P. & Astrom, M. (1990) Low nutritive quality as a defense against optimally foraging herbivores. *American Naturalist*, **135**, 547–62.

Luttrell, R.G., Fitt, G.P., Ramalho, F.S. & Sugonyaev, E.S. (1994) Cotton pest management, Part 1. A worldwide perspective. *Annual Review of Entomology*, **39**, 517–26.

Lyons, D.B. (1994) Development of the arboreal stages of the pine false webworm (Hymenoptera: Pamphiliidae). *Environmental Entomology*, **23** (4), 846–54.

Lyons, S.F., Jupp, P.G. & Schoub, B.D. (1986) Survival of HIV in the common bedbug. *Lancet*, **2** (8497), 45.

Ma, K.C. & Lee, S.C. (1996) Occurrence of major rice insect pests at different transplanting times and fertilizer levels in paddy field. *Korean Journal of Applied Entomology*, **35** (2), 132–6.

MacArthur, R.H. & Levins, R. (1967) The limiting similarity, convergence, and divergence of competing species. *American Naturalist*, **101**, 377–85.

MacArthur, R.H. & Wilson, E.O. (1967) *The Theory of Island Biogeography*, p. 203. Princeton University Press, Princeton, NJ.

MacDonald, R.E. & Bishop, C.J. (1952) Phloretin: an antibacterial substance obtained from apple leaves. *Canadian Journal of Botany*, **30**, 486–9.

Mace, G.M. & Stuart, S. (1994) Draft IUCN Red List categories, version 2.2. *Species*, **21–22**, 13–24.

Mace, J.M., Boussinesq, M., Ngoumou, P., Oye, J.E., Koeranga, A. & Godin, C. (1997) Country-wide rapid epidemiological mapping of onchocerciasis (REMO) in Cameroon. *Annals of Tropical Medicine and Parasitology*, **91** (4), 379–91.

McAuslane, H.J., Johnson, F.A., Colvin, D.L. & Sojack, B. (1995) Influence of foliar pubescence on abundance and parasitism of *Bemisia agentifolia* (Homoptera: Aleyrodidae) on soybean and peanut. *Environmental Entomology*, **24** (5), 1135–43.

McClure, M.S. (1980) Competition between exotic species: scale insects on hemlock. *Ecology*, **61**, 1391–401.

McColl, A.L. & Noble, R.M. (1992) Evaluation of a rapid mass-screening technique for measuring antibiosis to *Helicoverpa* spp. in cotton cultivars. *Australian Journal of Experimental Agriculture*, **32** (8), 1127–34.

McDougall, S.J. & Mills, N.J. (1997) The influence of hosts, temperature and food sources on the longevity of *Trichogramma platneri*. *Entomologia Experimentalis et Applicata*, **83** (2), 195–203.

McGaughey, W.H. (1994) Problems of insect resistance to *Bacillus thuringiensis*. *Agriculture, Ecosystems and Environment*, **49** (1), 95–102.

McKirdy, S.J. & Jones, R.A.C. (1997) Effect of sowing time on barley yellow dwarf virus infection in wheat: Virus incidence and grain yield losses. *Australian Journal of Agricultural Research*, **48** (2), 199–206.

McLain, D.K. (1991) The r-K continuum and the relative effectiveness of sexual selection. *Oikos*, **60** (2), 263–5.

McLaughlin, A. & Mineau, P. (1995) The impact of agricultural practices on biodiversity. *Agriculture, Ecosystems and Environment*, **55** (3), 201–12.

McManus, M.L. & Giese, R.L. (1968) The Columbia timber beetle, *Corthylus columbianus*. VII. the effect of climatic integrants on historic density fluctutations. *Forest Science*, **14**, 242–53.

McMichael, A.J. & Beers, M.Y. (1994) Climate change and human population health: Global and South Australian perspectives. *Transactions of the Royal Society of South Australia*, **118** (1–2), 91–8.

McMillin, J.D. & Wagner, M.R. (1995) Season and intensity of water stress: Host-plant effects on larval survival and fecundity of *Neodiprion gilletei* (Hymenoptera. *Diprionidae)* *Environmental Entomology*, **24**, 1251–7.

McNeil, J.N. & Quiring, D.T. (1983) Evidence of an oviposition-deterring pheromone in the alfalfa blotch leafminer, *Agromyza frontella* Rondani (Diptera: agromyzidae). *Environmental Entomology*, **12**, 990–2.

McNeill, S. & Southwood, T.R.E. (1978) The role of nitrogen in the development of insect–plant relationships. In: *Biochemical Aspects of Plant and Animal Coevolution* (ed. J.B. Marborne), pp. 77–98. Academic Press, London.

McPherson, R.M., Bondari, K., Stephenson, M.G., Stevenson, R.F. & Jackson, D.M. (1993) Influence of planting date on the seasonal abundance of tobacco budworms (Lepidoptera: Noctuidae) and tobacco aphids (Homoptera: Aphididae) on Georgia flue-cured tobacco. *Journal of Entomological Society*, **28** (2), 165–7.

Madden, D. & Young, T.P. (1992) Symbiotic ants as an alternative defense against giraffe herbivory in spinescent *Acacia drepanolobium*. *Oecologia (Heidelberg)*, **91**, 235–8.

Maddox, J.V. (1987) Protozoan diseases. In: *Epizootiology of Insect Disease* (eds J.R. Fuxa & Y. Tanada), pp. 417–52. John Wiley, New York.

Maeta, Y., Sugiura, N. & Goubara, M. (1992) Patterns of offspring production and sex allocation in the small carpenter bee, *Ceratina flavipes* Smith (Hymenoptera, Xylocopinae). *Japanese Journal of Entomology*, **60** (1), 175–90.

Magurran, A.E. (1988) *Ecological Diversity and its Measurement*, p. 179. Croom Helm, London.

Malcolm, S.B. & Zalucki, M.P., eds (1993) *Biology and Conservation of the Monarch Butterfly*. Natural History Museum of Los Angeles County, Science Series 38. Los Angeles.

Malcolm, S.B. & Zalucki, M.P. (1995) Milkweed latex and cardenolide induction may resolve the lethal plant defence paradox. *Entomological Experimental Applied*, **80**, 193–6.

Malcolm, S.B., Cockrell, B.J. & Brower, L.P. (1989) Cardenolide fingerprint of monarch butterfly reared on common milkweed, *Asclepias syriaca* L. *Journal of Chemical Ecology*, **15** (3), 819–54.

Malmqvist, B., Sjostrom, P. & Frick, K. (1991) The diet of two species of *Isoperla* (Plecoptera: Perlodidae) in relation to season, site, and sympatry. *Hydrobiologia*, **213** (3), 191–204.

Mamiya, Y. (1984) The pine wood nematode. In: *Plant and Insect Nematodes* (ed. W.R. Nickle), pp. 589–626. Marcel Dekker Inc, New York.

Mann, J.A., Harrington, R., Carter, N. & Plumb, R.T. (1997) Control of aphids and barley yellow dwarf virus in spring-sown cereals. *Crop Protection*, **16** (1), 81–7.

Margules, C., Higgs, A.J. & Rafe, R.W. (1982) Modern biogeographic theory: are there any lessons for nature reserve design? *Biological Conservation*, **24**, 115–28.

Marris, G.C. & Caspard, J. (1996) The relationship between conspecific superparasitism and the outcome of *in vitro* contests staged between different larval instars of the solitary endoparasitoid *Venturia canescens*. *Behavioral Ecology Sociobiology*, **39**, 61–9.

Martin, G.J. (1995) *Ethnobotany*. Chapman & Hall, London.

Martin, M.M. (1991) The evolution of cellulose digestion in insects. *Philosophical Transactions of the Royal Society of London, Series B: Biological Sciences*, **333** (1267), 281–8.

Martin, M.M. & Martin, J.S. (1984) Surfactants: their role in preventing the precipitation of proteins by tannins in insect guts. *Oecologia*, **61**, 342–5.

Martinez-Delclos, X. (1996) The fossil record of insects. *Boletin de la Asociacion Espanola de Entomologia*, **20** (1–2), 9–30.

Maschinski, J. & Whitham, T.G. (1989) The continuum of plant responses to herbivory: The influence of plant association, nutrient availability, and timing. *American Naturalist*, **134**, 1–19.

Masters, G.J. & Brown, V.K. (1992) Plant-mediated interactions between two spatially separated insects. *Functional Ecology*, **6**, 175–9.

Mastrers, A.R., Malcolm, S.B. & Brower, L.P. (1988) Monarch butterfly (*Danaus plexippus*) thermoregulation behaviour and adaptations for overwintering in Mexico. *Ecology*, **69** (2), 458–67.

Mathias, J.K., Ratcliffe, R.H. & Hellman, J.L. (1990) Association of an endophytic fungus in perennial ryegrass and resistance to the hairy chinch bug (Hemiptera. *Lygaeidae*). *Journal of Economic Entomology*, **83**, 1640–6.

Maton, J.L. & Harrison, S. (1997) Spatial pattern formation in an insect host–parasitoid system. *Science*, **278**, 1619–21.

Matson, P.A., Hain, F.P. & Mawby, W. (1987) Indices of tree susceptibility to bark beetles vary with silvicultural treatment in a loblolly pine plantation. *Forest Ecology and Management*, **22** (1–2), 107–18.

Matthews, G.A. (1992) *Pesticide Application Methods*, 2nd edn, 403 pp. Longman, Harlow, UK.

Matthews, R.W., Flage, L.R. & Matthews, J.R. (1997) Insects as teaching tools in primary and secondary education. *Annual Review of Entomology*, **42**, 269–90.

Mattiacci, L., Dicke, M. & Posthumus, M.A. (1995) Beta-Glucosidase: an elicitor of herbivore-induced plant odor that attracts host-searching parasitic wasps. *Proceedings of the National Academy of Sciences of the United States of America*, **92**, 2036–40.

Mattson, W.J. Jr (1980) Herbivory in relation to plant nitrogen content. *Annual Review Ecology and Syst*, **11**, 119–61.

Mattson, W.J. & Addy, N.D. (1975) Phytophagous insects as regulators of forest primary production. *Science*, **190**, 515–22.

Mattson, W.J. & Haack, R.A. (1987) The role of drought stress in provoking outbreaks of phytophagous insects. In: *Insect Outbreaks* (eds P. Barbosa & J.C. Schultz), pp. 365–407. Academic Press, San Diego.

Maurer, P., Debieu, D., Malosse, C., Leroux, P. & Riba,G. (1992) Sterols and symbiosis in the leaf-cutting ant *Acromyrmex octospinosus* (Reich) (Hymenoptera, Formicidae: Attini). *Archives of Insect Biochemistry and Physiology*, **20** (1), 13–21.

Mawdsley, N.A. & Stork, N.E. (1995) Species extinctions in insects: ecological and biogeographical considerations. In: *Insects in a Changing Environment* (eds R. Harrington & N.E. Stork), pp. 321–69. Academic Press, London.

May, R.M. (1973) *Stability and Complexity in Model Ecosystems*, 235 pp. Princeton University Press, NJ.

May, R.M. (1974a) Biological populations with nonoverlapping generations: stable points, stable cycles and chaos. *Science*, **186**, 645–847.

May, R.M. (1974b) *Stability and Complexity in Model Ecosystems*. Princeton University Press, Princeton.

May, R.M. (ed.) (1978) *Theoretical Ecology, Principles and Applications*, 2nd edn. Blackwell Scientific Publications, Oxford.

May, R.M. (1988) How many species are there on earth? *Science*, **241**, 1441–9.

May, R.M. (1992) How many species inhabit the earth? *Scientific American*, **267** (4), 18–24.

May, R.M., Lawton, J.H. & Stork, N.E. (1995) Assessing extinction rates. In: *Extinction Rates* (eds J.H. Lawton & R.M. May), pp. 1–24. Oxford University Press, Oxford.

Mbogo, C.M., Snow, R.W., Khamala, C.P.M., *et al.* (1995) Relationships between *Plasmodium falciparum* transmission by vector populations and the incidence of severe disease at nine sites on the Kenya coast. *American Journal of Tropical Medicine and Hygiene*, **52** (3), 201–6.

Mehta, P.K. & Sandhu, G.S. (1992) Feeding behaviour of red pumpkin beetle, *Aulacophora fovecollis* (Lucas) on leaf extracts of different cucurbits. *Uttar Pradesh Journal of Zoology*, **12**, 87–94.

Mendis, C., Herath, P.R.J., Rajakaruna, J., *et al.* (1992) Method to estimate relative transmission efficiencies of *Anopheles* spp. (Diptera: Culicidae) in human malaria transmission. *Journal of Medical Entomology*, **29** (2), 188–96.

Mendoza-Aldana, J., Piechulek, H. & Maguire, J. (1997) Forest onchocerciasis in Cameroon: Its distribution and implications for selection of communities for control programmes. *Annals of Tropical Medicine and Parasitology*, **91** (1), 79–86.

Merlin, J., Lemaitre, O. & Gregoire, J.C. (1996) Chemical cues produced by conspecific larvae deter oviposition by the coccidophagous ladybird beetle, *Cryptolaemus montrouzieri*. *Entomologia Experimentalis et Applicata*, **79**, 147–51.

Merritt, R.W. & Cummins, K.W. (eds) (1984) *An Introduction to the Aquatic Insects of North America*. Kendall/Hunt, Dubuque.

Mesfin, T., Thottappilly, G. & Singh, S.R. (1992) Feeding behaviour of *Aphis craccivora* (Koch) on cowpea cultivars with different levels of aphid resistance. *Annals of Applied Biology*, **121** (3), 493–501.

Messing, R.H., Klungness, L.M. & Jang, E.B. (1997) Effects of wind on movement of *Diachasmimorpha longicaudata*, a parasitoid of tephritid fruit flies, in a laboratory flight tunnel. *Entomologia Experimentalis et Applicata*, **82** (2), 147–52.

Meyer, J.L. & O'Hop, J. (1983) Leaf-shredding insects as a source of dissolved organic carbon in headwater streams. *American Midland Naturalist*, **109**, 175–83.

Michelakis, S. (1973) *A study of the laboratory interactions between* Coccinella septempunctata *larvae and its prey* Myzus persicae. M.Sc. Thesis, University of London.

Miller, K., Allegretti, M.H., Johnson, N. & Jonsson, B. (1995) Measures for conservation of biodiversity and sustainable use of its components. In: *Global Biodiversity Assessment* (ed. V.H. Heywood), pp. 915–1061. Cambridge University Press, Cambridge.

Miller, L.K. (ed.) (1997) *The Baculoviruses*, p. 447. Plenum, New York.

Miller, R.F. (1991) Chitin paleoecology. *Biochemical Systematics and Ecology*, **19** (5), 401–12.

Milli, R., Koch, U.T. & de Kramer, J.J. (1997) EAG measurement of pheromone distribution in apple orchards treated for mating disruption of *Cydia pomonella*. *Entomologia Experimentalis et Applicata*, **82** (3), 289–97.

Mills, A.P., Rutter, J.F. & Rosenberg, L.J. (1996) Weather associated with spring and summer migrations of rice pests and other insects in south-eastern and eastern Asia. *Bulletin of Entomological Research*, **86** (6), 683–94.

Mills, N.J. (1994) Parasitoid guilds: defining the structure of the parasitoid communities of endopterygote insect hosts. *Environmental Entomology*, **23** (5), 1066–83.

Milne, A. (1957a) The natural control of insect populations. *Canadian Entomologist*, **89**, 193–213.

Milne, A. (1957b) Theories of natural control of insect populations. *Cold Spring Harbor Symposia on Quantitative Biology*, **22**, 253–67.

Milner, R.J. & Lutton, G.G. (1986) Dependence of *Verticilillium lecanii* (Fungi: Hyphomycetes) on high humidities for infection and sporulation using *Myzus persicae* (Homoptera: Aphididae) as host. *Environmental Entomology*, **15** (2), 380–2.

Minckley, R.L., Wcislo, W.T., Yanega, D. & Buchmann, S.L. (1994) Behavior and phenology of a specialist bee (*Dieunomia*) and sunflower (*Helianthus*) pollen availability. *Ecology*, **75**, 1406–19.

Mnzava, A.E.P., Rwegoshora, R.T., Tanner, M., Msuya, F.H., Curtis, C.F. & Irare, S.G. (1993) The effects of house spraying with DDT or lambda-cyhalothrin against *Anopheles arabiensis* on measures of malarial morbidity in children in Tanzania. *Acta Tropica*, **54** (2), 141–51.

Moellenbeck, D.J. & Quisenberry, S.S. (1992) Identification of alfalfa resistance to the three-cornered alfalfa hopper (Homoptera: Membracidae). *Journal of Economic Entomology*, **85** (5), 2027–31.

Mogi, M. (1969) Predation response of the larvae of *Harmonia axyridis* Pallas (Coccinellidae) to the different prey density. *Japanese Journal of Applied Entomology and Zoology*, **13**, 9–16.

Mogi, M. & Sembel, D.T. (1996) Predator-prey system structure in patchy and ephemeral phytotelmata—Aquatic communities in small aroid axils. *Researches on Population Ecology*, **38**, 95–103.

Mohan, K.S. & Pillai, G.B. (1993) Biological control of *Oryctes rhinoceros* (L) using an Indian isolate of *Oryctes baculovirus*. *Insect Science and its Application*, **14** (5–6), 551–8.

Moloo, S.K., Kutuza, S.B. & Desai, J. (1988) Infection rates in sterile males of *morsitans*, *palpalis* and *fusca* groups *Glossina* for pathogenic *Trypanosoma* spp. from East and West Africa. *Acta Tropica*, **45** (2), 145–52.

Monge, J.P., Dupont, P., Idi, A. & Huignard, J. (1995) The consequences of interspecific competition between *Dinarmus basalis* (rond) (Hymenoptera, Pteromalidae) and *Eupelmusvuiletti* (crw) (Hymenoptera, Pteromalidae) on the development of their host population. *Acta Oecologica International Journal of Ecologica*, **16**, 19–30.

Montgomery, M.E. & Arn, H. (1974) Feeding response of *Aphis pomi, Myzus persicae*, and *Amphorophora agathonica* to phlorizin. *Journal of Insect Physiology*, **20**, 413–21.

Moon, A.M., Dyer, M.I., Brown, M.R. & Crossley, D.A. Jr (1994) Epidermal growth factor interacts with indole-3-acetic acid and promotes Coleoptile growth. *Plant and Cell Physiology*, **35**, 1173–7.

Moore, L. & Watson, T.F. (1987) Trap crop effectiveness in community boll weevil (Coleoptera: Curculionidae) control programmes. *Journal of Entomological Science*, **25** (4), 519–25.

Moorhead, D.L. & Reynolds, J.F. (1991) A general model of litter decomposition in the Northern Chihuahuan Desert. *Ecological Modeling*, **56**, 197–220.

Mopper, S. (1996) Adaptive genetic structure in phytophagus insect populations. *Trends in Ecology and Evolution*, **11**, 235–8.

Mopper, S. & Strauss, S.Y., eds (1998) Genetic Structure and Local Adaptation in Natural Insect Populations. *Effects of Ecology, Life History, and Behavior.* Chapman & Hall, New York.

Mopper, S. & Whitham, T.G. (1992) The plant stress paradox: Effects on pinyon sawfly sex ratios and fecundity. *Ecology*, **73**, 515–25.

Mopper, S., Beck, M., Simberloff, D. & Stiling, P. (1995) Local adaptation and agents of selection in a mobile insect. *Evolution*, **49**, 810–5.

Mopper, S., Mitton, J.B., Whitham, T.G., Cobb, N.S. & Christiansen, K.M. (1991) Genetic differentiation and heterozygosity in pinyon pine associated with resistance to herbivory and environmental stress. *Evolution*, **45**, 989–99.

Moran, M.D., Rooney, T.P. & Hurd, L.E. (1996) Top-down cascade from a bitrophic predator in an old-field community. *Ecology*, **77**, 2219–27.

Moran, N.A. & Whitham, T.G. (1990) Interspecific competition between root-feeding and leaf-galling aphids mediated by host-plant resistance. *Ecology*, **71**, 1050–8.

Moran, V.C. & Southwood, T.R.E. (1982) The guild composition of arthropod communities in trees. *Journal of Animal Ecology*, **51**, 289–306.

Morgan, D. & Solomon, M.G. (1993) PEST-MAN: A forecasting system for apple and pear pests. *Bulletin OEPP (Organisation Européenne et Méditerranéenne pour la Protection des Plantes)*, **23** (4), 601–5.

Morrillo, W.L., Gabor, J.W. & Wichman, D. (1993) Mortality of the wheat stem sawfly (Hymenoptera: Cephidae) at low temperatures. *Environmental Entomology*, **22** (6), 1358–61.

Morris, W.F. (1992) The effects of natural enemies, competition and host plant water availability on an aphid population. *Oecologia*, **90**, 359–65.

Morse, S.S. (1994) The viruses of the future? Emerging viruses and evolution. In: *The Evolutionary Biology of Viruses* (ed. S.S. Morse), pp. 325–35. Raven Press, New York.

Moscardi, F. & Sosa-Gomez, D.R. (1996) Soybean in Brazil. In: *Biotechnology and Integrated Pest Management* (ed. G.J. Persley), pp. 98–112. CABI, Wallingford, UK.

Muenster-Swendsen, M. (1987) The effect of precipitation on radial increment in Norway spruce (*Picea abies* Karst.) and on the dynamics of a lepidopteran pest insect. *Journal of Applied Ecology*, **24**, 563–72.

Mulrooney, J.E., Howard, K.D., Hanks, J.E. & Jones, R.J. (1997) Application of ultra-low-volume malathion by air assisted ground sprayer for boll weevil (Coleoptera: Curculionidae) control. *Journal of Economic Entomology*, **90** (2), 639–45.

Murdoch, W.W., Luck, R.F., Swarbrick, S.L., Walde, S., Yu-D.S. & Reeve, J.D. (1995) Regulation of an insect population under biological control. *Ecology (Washington DC)*, **76** (1), 206–17.

Murdoch, W.W., Briggs, C.J. & Nisbet, R.M. (1996) Competitive displacement and biological control in parasitoids: a model. *American Naturalist*, **148** (5), 807–26.

Murray, M.D. (1995) Influences of vector biology on transmission of arboviruses and outbreaks of disease: the *Culicoides brevitarsus* model. *Veterinary Microbiology*, **46** (1–3), 91–9.

Mutinga, M.J., Kamau, C.C., Basimike, M., Mutero, C.M. & Kyai, F.M. (1992) Studies on the epidemiology of leishmaniasis in Kenya: Flight range of phlebotomine sandflies in Marigat, Baringo District. *East African Medical Journal*, **69** (1), 9–13.

Muzika, R.-M. & Pregitzer, K.S. (1992) Effect of nitrogen fertilization on leaf phenolic production of grand fir seedlings. *Trees*, **6**, 241–4.

Myers, J.H. & Williams, K.S. (1984) Does tent caterpillar attack reduce the food quality of red alder foliage? *Oecologia*, **62**, 74–9.

Myers, J.H., Savoie, A. & van Randen, E. (1998) Eradication and pest management. *Annual Review of Entomology*, **43**, 471–91.

Myers, N. (1988) Threatened biotas: 'hot spots' in tropical forests. *Environmentalist*, **8**, 187–208.

Myers, N. (1990) The biodiversity challenge: expanded hot-spots analysis. *Environmentalist*, **19**, 243–56.

Myers, N. (1997) The rich diversity of biodiversity issues. *Biodiversity II, Understanding and Protecting our Biological Resources* (eds M.L. Reaka-Kudla, D.E. Wilson & E.O. Wilson), pp. 125–38. Joseph Henry Press, Washington, DC.

Naeem, S., Thompson, L.J., Lawler, S.P., Lawton, J.H. & Woodfin, R.M. (1994) Declining biodiversity can alter the performance of ecosystems. *Nature*, **368**, 734–7.

Naeem, S., Thompson, L.J., Lawler, S.P., Lawton, J.H. &

Woodfin, R.M. (1995) Empirical evidence that declining species diversity may alter the performance of terrestrial ecosystems. *Philosophical Transactions of the Royal Society of London, Series B: Biological Sciences*, **347**, 249–62.

Nair, K.S.S. & Sudheendrakumar, V.V. (1986) The teak defoliator, *Hyblaea puera*: Defoliation dynamics and evidences of short range migration of moths. *Proceedings of the Indian Academy of Sciences, Animal Sciences*, **95** (1), 7–22.

Nakai, M. & Kunimi, Y. (1997) Granulosis-virus infection of the small tea tortrix (Lepidoptera, Tortricidae) -Effect on the development of the endoparasitoid, *Ascogaster reticulatus* (Hymenoptera, braconidae). *Biological Control*, **8** (1), 74–80.

Nakajima, S., Kitamura, T., Baba, N., Iwasa, J. & Ichikawa, T. (1995) Oleuropein, a Secoiridoid Glucoside from Olive, as a feeding stimulant to the Olive Weevil (*Dyscerus perforatus*). *Bioscience Biotechnology and Biochemistry*, **59**, 769–70.

Nakanishi, H. (1994) Myrmecochorous adaptations of *Corydalis* species (papaveraceae) in southern Japan. *Ecological Research*, **9**, 1–8.

Nauen, R. & Elbert, A. (1997) Apparent tolerance of a field-collected strain of *Myzus nicotianae* to imidacloprid due to strong antifeedant responses. *Pesticide Science*, **49** (3), 252–8.

Neal, J.W. Jr, Buta, J.G., Pitarelli, G.W., Lusby, W.R. & Bentz, J.A. (1994) Novel Sucrose Esters from *Nicotiana gossei*. Effective Biorationals Against Selected Horticultural Insect Pests. *Journal of Economic Entomology*, **87**, 1600–7.

Nealis, V.G., Oliver, D. & Tchir, D. (1996) The diapause response to photoperiod in Ontario populations of *Cotesia melanoscela* (Ratzeburg) (Hymenoptera: Braconidae). *Canadian Entomologist*, **128** (1), 41–6.

Nebeker, T.E., Schmitz, R.F., Tisdale, R.A. & Hobson, K.R. (1995) Chemical and nutritional status of dwarf mistletoe, *Amarilla roor* rot, and *Comandra* blister rust infected trees which may influence tree susceptability to bark beetle attack. *Canadian Journal of Botany*, **73**, 360–9.

Neuenschwander, P., Hammond, W.N.O., Gutierrez, A.P., *et al.* (1989) Impact assessment of the biological control of the cassava mealybug, *Phenacoccus manihoti* (Homoptera: Pseudococcidae) by the introduced parasitoid, *Epidinocarsis lopezi* (Hymenoptera: Encrytidae). *Bulletin of Entomological Research*, **79** (4), 579–94.

Neuenschwander, P., Hammond, W.N.O., Ajuonu, O., *et al.* (1990) Biological control of the cassava mealybug, *Phenacoccus manihoti* (Homoptera: Pseudococcidae) by *Epidinocarsis lopezi* (Hymenoptera: Encrytidae), in West Africa, as influenced by climate and soil. *Agriculture, Ecosystems and Environment*, **32** (1–2), 39–56.

Neupane, F.P. (1991) Effect of soybean hosts on the development of the soybean hairy caterpillar, *Spilarctia casignata* Koll. (Lepidoptera: Actiidae). *Insect Science and its Application*, **12** (1–3), 189–92.

Newton, C. & Dixon, A.F.G. (1990) Embryonic growth rate and birth weight of the offspring of apterous and alate aphids: A cost of dispersal. *Entomologia Experimentalis et Applicata*, **55** (3), 223–30.

Newton, S.F. & Newton, A.V. (1997) The effect of rainfall and habitat on abundance and diversity of birds in a fenced protected area in the central Saudi Arabian desert. *Journal of Arid Environments*, **35** (4), 715–35.

Nicholson, A.J. (1933) The balance of animal populations. *Journal of Animal Ecology*, **2**, 131–78.

Nicholson, A.J. (1957) The self-adjustment of populations to change. *Cold Spring Harbor Symposia on Quantitative Biology*, **22**, 153–72.

Nicholson, A.J. (1958) Dynamics of insect populations. *Annual Review of Entomology*, **3**, 107–36.

Nicholson, A.J. & Bailey, V.A. (1935) The balance of animal populations. *Proceedings of the Zoological Society of London, Part* 1, **3**, 551–98.

Niemala, P., Tuomi, J., Sorjonen, J., Hokkanen, T. & Neuvonen, S. (1982) The influence of host plant growth form a phenology on the life strategies of Finnish macrolepidopterous larvae. *Oikos*, **39**, 164–70.

Nilsson, S.G. & Bengtsson, J.A.S.S. (1988) Habitat diversity or area per se? Species richness of woody plants, carabid beetles and land snails on islands. *Journal of Animal Ecology*, **57** (2), 685–704.

Nishida, H. & Hayashi, N. (1996) Cretaceous coleopteran larva fed on a female fructification of extinct gymnosperm. *Journal of Plant Research*, **109** (1095), 327–30.

Nishida, R. (1994) Oviposition stimulant of a Zeryntiine swallowtail butterfly, *Luehdorfia japonica*. *Phytochemistry (Oxford)*, **36**, 873–7.

Noda, H. & Kodama, K. (1996) Phylogenetic position of yeastlike endosymbionts of anobiid beetles. *Applied and Environmental Microbiology*, **62** (1), 162–7.

Norton, A.P. & Welter, S.C. (1996) Augmentation of the egg parasitoid *Anaphes iole* (Hymenoptera: Mymaridae) for *Lygus hesperus* (Heteroptera: Miridae) management in strawberries. *Environmental Entomology*, **25** (6), 1406–14.

Noss, R.F. (1990) Indicators for monitoring biodiversity: a hierarchical approach. *Conservation Biology*, **4**, 355–64.

Noyes, J.S. (1974) *The biology of the leek moth* Acrolepia assectella (Zeller). PhD thesis, University of London.

Nuckols, M.S. & Connor, E.F. (1995) Do trees in urban or ornamental plantings receive more damage by insects than trees in natural forests? *Ecological Entomology*, **20**, 253–60.

Nylin, S. & Gotthard, K. (1998) Plasticity in life-history traits. *Annual Review of Entomology*, **43**, 63–83.

O'Connor, J.P. (1982) *Microchironomus deribae* (Freeman) (Dipt., Chironomidae): a fuel contaminant in an Irish helicopter. *Entomologist's Monthly Magazine*, **118** (Jan–April), 44.

O'Malley, M. (1997) Clinical evaluation of pesticide exposure and poisonings. *Lancet (North American edn)*, **349** (9059), 1161–6.

Oba, G. (1994) Responses of*digofera spinosa* to simulated

herbivory in a semidesert of North-West Kenya. *Acta Oecologia*, **15**, 105–17.

Oehlschlager, A.C., McDonald, R.S., Chinchilla, C.M. & Patschke, S.N. (1995) Influence of a pheromone-based mass-trapping system on the distribution of *Rhynchophorus palmarum* (Coleoptera: Curculionidae) in oil palm. *Environmental Entomology*, **24** (5), 1005–12.

Ohgushi, T. (1992) Resource limitation on insect herbivore populations. In: *Effects of Resource Distribution on Animal–Plant Interactions* (eds M.D Hunter, T. Ohgushi, & P.W Price). Academic Press, San Diego.

Ohgushi, T. & Sawada, H. (1985) Population equilibrium with respect to available food resource and its behavioural basis in an herbivorous lady beetle *Henosepilachna niponica*. *Journal of Animal Ecology*, **54**, 781–96.

Ohnesorge, B. (1994) Regulation mechanisms in the population dynamics of phytophagous pests. *Berichte ueber Landwirtschaft Sonderheft*, **0** (209), 54–67.

Okland, B. (1994) Mycetophilidae (Diptera), and insect group vulnerable to forestry practices? A comparison of clearcut, managed and semi-natural spruce forests in southern Norway. *Biodiversity and Conservation*, **3** (1), 68–85.

Okoth, J.O., Okethi, V. & Ogola, A. (1991) Control of tsetse trypanosomiasis transmission in Uganda by applications of lambda-cyhalothrin. *Medical and Veterinary Entomology*, **5** (1), 121–8.

Olaleye, O.D., Tomori, O. & Schmitz, H. (1996) Rift Valley fever in Nigeria: Infections in domestic animals. *Revue Scientifique et Technique, Office International des Epizooties*, **15** (3), 937–46.

Olatunde, G.O. & Odebiyi, J.A. (1991) Some aspects of antibiosis in cowpeas resistant to *Clavigralla tomentosicollis* tal. (Hemiptera: Coreidae) in Nigeria. *Tropical Pest Management*, **37** (3), 273–6.

Oliver, I. & Beattie, A.J. (1993) A possible method for the rapid assessment of biodiversity. *Conservation Biology*, **7**, 562–8.

Oliver, I. & Beattie, A.J. (1996a) Designing a cost-effective invertebrate survey—a test of methods for rapid assessment of biodiversity. *Ecological Applications*, **6**, 594–607.

Oliver, I. & Beattie, A.J. (1996b) Invertebrate morphospecies as surrogates for species—a case-study. *Conservation Biology*, **10**, 99–109.

Olofsson, E. (1988) Persistence and dispersal of the nuclear polyhedrosis virus of *Neodiprion sertifer* (Geoffrey) (Hymenoptera: Diprionidae) in a virus-free lodgepole pine plantation in Sweden. *Canadian Entomologist*, **120** (10), 887–92.

Opler, P.A. (1974) Oaks as evolutionary islands for leaf-mining insects. *American Scientist*, **62**, 67–73.

Opondo-Mbai, M.L. (1995) Investigation of arthropods associated with agroforestry in Machakos, Kenya. DPhil thesis, University of Oxford.

Ortega, G.M. & Oliver, C.M. (1990) Entomology of onchocerciasis in the Soconusgo area, Chiapas (Mexico): IV. Altitudinal distribution and seasonal variation of immature forms of the three simuliid species considered as vectors. *Folia Entomologica Mexicana*, vol. 79, 175–96.

Owen, D.F. (1980) *Camouflage and Mimicry*. Oxford University Press, Oxford.

Owen, D.F. & Wiegert, R.G. (1976) Do consumers maximize plant fitness? *Oikos*, **27**, 488–92.

Ozanne, C.M.P., Hambler, C., Foggo, A. & Speight, M.R. (1997) The significance of edge effects in the management of forests for invertebrate biodiversity. In: *Canopy Arthropods* (eds N.E. Stork, J. Adis & R.K. Didham), pp. 534–50. Chapman and Hall, London.

Paige, K.N. (1992) Overcompensation in response to mammalian herbivory: From mutalistic to antagonistic interactions. *Ecology*, **73**, 2076–85.

Paige, K.N. & Whitham, T.G. (1987) Overcompensation in response to mammalian herbivory: The advantage of being eaten. *American Naturalist*, **129**, 407–16.

Paine, T.D., Malinoski, M.K. & Scriven, G.T. (1990) Rating *Eucalyptus* vigor and the risk of insect infestation: Leaf surface area and sapwood: Heartwood ratio. *Canadian Journal of Forest Research*, **20**, 1485–9.

Pair, S.D., Raulston, J.R., Westbrook, J.K., Wolf, W.W. & Adams, S.D. (1991) Fall armyworm (Lepidoptera: Noctuidae) outbreak originating in the lower Rio Grande Valley, (Texas (USA) and Mexico) 1989. *Florida Entomologist*, **74** (2), 200–13.

Pankanin-Franczyk, M. & Ceryngier, P. (1995) Cereal aphids, their parasitoids and coccinellids on oats in central Poland. *Journal of Applied Entomology*, **119** (2), 107–11.

Parameswaran, S., Babu, P.C.S., Shanthi, M. & Jayaraj, S. (1993) Control of sugarcane shoot borer *Chilo infusculatus* Snellen with granulosis virus and cultural practices. *Journal of Biological Control*, **7** (2), 81–3.

Parmesan, C. (1996) Climate and species' range. *Nature*, **382**, 765–6.

Partridge, L., Barrie, B., Fowler, K. & French, V. (1994) Thermal evolution of pre-adult life-history traits in *Drosophila melanogaster*. *Journal of Evolutionary Biology*, **7**, 645–63.

Passos, L. & Ferreira, S.O. (1996) Ant dispersal of *Croton priscus* (Euphobiaceae) seeds in a tropical semideciduous forest in Southeastern Brazil. *Biotropica*, **28**, 697–700.

Pasteels, J.M. & Rowell-Rahier, M. (1991) Proximate and ultimate causes for host plant influence on chemical defense of leaf beetles (Coleoptera Chrysomelidae). *Entomologia Generalis*, **15**, 227–35.

Pasteels, J.M., Dobler, S., Rowell-Rahier, M., Ehmke, A. & Hartman, T. (1995) Distribution of autogenous and host-derived chemical defenses in *Oreina* leaf beetles (Coleoptera. Chrysomelidae). *Journal of Chemical Ecology*, **21**, 1163–79.

Payne, C.C. (1988) Pathogens for the control of insects:

where next? *Philosophical Transactions of the Royal Society of London, Series B*, **318**, 225–48.

Pearson, D.L. (1994) Selecting indicator taxa for the quantitative assessment of biodiversity. *Philosophical Transactions of the Royal Society of London, Series B: Biological Sciences*, **345**, 75–9.

Pearson, D.L. & Cassola, F. (1992) World-wide species richness patterns of tiger beetles (Coleoptera: Cicinelidae): indicator taxon for biodiversity and conservation studies. *Conservation Biology*, **6**, 376–91.

Pellegrini, G., Levre, E., Valentini, P. & Cadoni, M. (1992) Cockroaches infestation and possible contribution in the spreading of some Enterobacteria. *Igiene Moderna*, **97** (1), 19–30.

Pellmyr, O., Thompson, J.N., Brown, J.M. & Harrison, R.G. (1996) Evolution of pollination and mutualism in the yucca moth lineage. *American Naturalist*, **148**, 827–47.

Pemberton, R.W. (1988) Myrmecochory in the introduced range weed leafy spurge *Euphorbia esula* L. *American Midland Naturalist*, **119**, 431–5.

Pemberton, R.W. (1992) Fossil extrafloral nectaries, evidence for the ant-guard antiherbivore defense in an Oligocene *Populus*. *American Journal of Botany*, **79**, 1242–6.

Pemberton, R.W. & Lee, J.H. (1996) The influence of extrafloral nectaries on parasitism of an insect herbivore. *American Journal of Botany*, **83**, 1187–94.

Perlak, F.J., Deaton, R.W., Armstrong, T.A., *et al.* (1990) Insect resistant cotton plants. *Biotechnology*, **8**, 939–43.

Peterman, R.M., Clark, W.C. & Holling, C.S. (1979) The dynamics of resilience: shifting stability domains in fish and insect systems. In: *Population Dynamics* (eds R.M. Anderson, B.D. Turner & L.R. Taylor), pp. 321–41. Blackwell Scientific Publications, Oxford.

Peters, R.L. (1988) The effect of global climatic change on natural communities. In: *Biodiversity* (ed. E.O. Wilson), pp. 450–61. National Academy Press, Washington, DC.

Peters, R.L. & Darling, J.D.S. (1985) The greenhouse effect and nature reserves. *Bioscience*, **35**, 707–17.

Phelps, D.G. & Gregg, P.C. (1991) Effects of water stress on curly Mitchell grass, the common army worm and the Australian plague locust. *Australian Journal of Experimental Agriculture*, **31** (3), 325–32.

Philippe, R., Veyrunes, J.C., Mariau, D. & Bergoin, M. (1997) Biological control using entomopathogenic viruses. Application to oil palm and coconut pests. *Plantations Recherche Developpement*, **4** (1), 39–45.

Phillips, O.L. (1997) The changing ecology of tropical forests. *Biodiversity and Conservation*, **6**, 291–311.

Pielou, E.C. (1995) Biodiversity versus old-style diversity: measuring biodiversity for conservation. In: *Measuring and Monitoring Biodiversity in Tropical and Temperate Forests* (eds T.J.B. Boyle & B. Boontawee), pp. 19–46. Centre for International Forestry Research (CIFOR), Bogor, Indonesia.

Pilcher, C.D., Rice, M.E., Obrycki, J.J. & Lewis, L.C. (1997) Field and laboratory evaluations of transgenic *Bacillus thuringiensis* corn on secondary lepidopteran pests (Lepidoptera: Noctuidae). *Journal of Economic Entomology*, **90** (2), 669–78.

Podgwaite, J.D., Reardon, R.C., Walton, G.S., Venables, L. & Kolodny-Hirsch, D.M. (1992) Effects of aerially applied Gypcheck on gypsy moth (Lepidoptera: Lymantriidae) populations in Maryland woodlots. *Journal of Economic Entomology*, **85** (4), 1136–9.

Podoler, H. & Rogers, D. (1975) A new method for the identification of key factors from life-table data. *Journal of Animal Ecology*, **44**, 85–114.

Pollard, E. (1991) Changes in the flight period of the hedge brown butterfly *Pyronia tithonus* during range expansion. *Journal of Animal Ecology*, **60**, 737–48.

Pollard, E. & Yates, T.J. (1993) *Monitoring Butterflies for Ecology and Conservation*. Chapman and Hall, London.

Pollard, E., Rothery, P. & Yates, T.J. (1996) Annual growth rates in newly established populations of the butterfly *Pararge aegeria*. *Ecological Entomology*, **21** (4), 365–9.

Possingham, H.P. (1993) Impact of elevated atmospheric CO_2 on biodiversity: Mechanistic population-dynamic perspective. *Australian Journal of Botany*, **41**, 11–21.

Potting, R.P.J., Vet, L.E.M. & Dicke, M. (1995) Host microhabit location by stem-borer parasitoid *Cotesia flavpipes*: The role of herbivore volatiles and locally and systematically induced plant volatiles. *Journal of Chemical Ecology*, **21**, 525–39.

Potts, G.R. & Vickerman, G.P. (1974) Studies on the cereal ecosystem. *Advances in Ecological Research*, **8**, 107–97.

Powell, G.V.N. & Bjork, R. (1995) Implications of intratropical migration on reserve design—a case-study using *Pharomachrus mocinno*. *Conservation Biology*, **9**, 354–62.

Power, A.G. (1992) Host plant dispersion, leafhopper movement and disease transmission. *Ecological Entomology*, **17** (1), 63–8.

Prance, G.T. (1994) A comparison of the efficacy of higher taxa and species numbers in the assessment of biodiversity in the neotropics. *Philosophical Transactions of the Royal Society of London, Series B: Biological Sciences*, **345**, 89–99.

Prendergast, J.R. (1997) Species richness covariance in higher taxa: Empirical tests of the biodiversity indicator concept. *Ecography*, **20**, 210–16.

Prendergast, J.R., Quinn, R.M., Lawton, J.H., Eversham, B.C. & Gibbons, D.W. (1993) Rare species, the coincidence of diversity hotspots and conservation strategies. *Nature*, **365**, 335–7.

Prestidge, R.A. & Gallagher, R.T. (1988) Endophyte fungus confers resistance to ryegrass: argentine stem weevil larval studies. *Ecological Entomology*, **13**, 429–36.

Preszler, R.W. & Boecklen, W.J. (1994) A three-trophic level analysis of the effects of plant hybridization on a leaf-mining moth. *Oecologia (Berlin)*, **100**, 66–73.

Price, P.W. (1975) *Insect Ecology*, 514 pp. John Wiley, New York.

Price, P.W. (1989) Clonal development of coyote willow, *Salix exigua* (Salicaceae), and attack by the shoot-galling sawfly, *Euura exiguae* (Hymenoptera. *Tenthredinidae) Environmental Entomology*, **18**, 61–8.
Price, P.W. (1990) Evaluating the role of natural enemies in latent and eruptive species: New approaches in life table construction. In: *Population Dynamics of Forest Insects* (eds A.D. Watt, S.R. Leather, M.D. Hunter & N.A.C. Kidd) pp. 221–32. Intercept, Andover.
Price, P.W. (1991) The plant vigor hypothesis and herbivore attack. *Oikos*, **62**, 244–51.
Price, P.W. (1996) *Biological Evolution*. Saunders College Publishing, Philadelphia.
Price, P.W. (1997) *Insect Ecology* (3rd edn), p. 874. John Wiley, New York.
Price, P.W., Andrade, I., Pires, C., Sujii, E. & Vieira, E.M. (1995) Gradient analysis using plant architecture and insect herbivore utilization. *Environmental Entomology*, **24**, 497–505.
Price, P.W., Bouton, P., Gross, B.A., McPheron, J.N., Thompson, J.N. & Weis, A.E. (1980) Interactions among three trophic levels: influence of plants on interactions between insect herbivores and natural enemies. *Annual Review of Ecology and Systematics*, **11**, 41–65.
Price, P.W., Craig, T.P. & Hunter, M.D. (1999) Population ecology of a gall-forming sawfly, *Euura lasiolepis*, and relatives. In: *Insect Populations: in Theory and Practice* (eds J.P. Dempster & I.F.G. Mclean). Chapman & Hall, London.
Pugalenthi, P. & Livingstone, D. (1995) Cardenolides (heart poisons) in the painted grasshopper *Poecilocerus pictus* F. (Orthoptera: Pyrgomorphidae) feeding on the milkweed *Calotropis gigantea* L. *Asclepiadaceae). Journal of the New York Entomological Society*, **103**, 191–6.
Pulliam, H.R. (1988) Sources, sinks and population regulation. *American Naturalist*, **132**, 652–61.
Pumpini, C.B. & Walker, E.D. (1989) Population size of adult *Aedes triseriatus* in a scrap tireyard in northern Indiana (USA*). Journal of the American Mosquito Control Association*, **5** (2), 166–72.
Putnam, R.J. (1994) *Community Ecology*. Chapman & Hall, London.
Putman, R.J. & Wratten, S.D. (1984) *Principles of Ecology*. Croom Helm, London.
Quiring, D.T. & McNeil, J.N. (1984) Intraspecific larval competition reduces efficacy of oviposition-deterring pheromone in the alfalfa blotch leafminer, *Agromyza frontella* (Diptera: agromyzidae). *Environmental Entomology*, **13**, 675–8.
Raa, J. (1968) Polyphenols and natural resistance of apple leaves against *Venturia inaequalis*. *Netherlands Journal of Plant Pathology*, **74**, 37–45.
Rajendran, R., Rajendran, N. & Venugopalan, V.K. (1990) Effect of organochlorine pesticides on the bacterial population of a tropical estuary. *Microbios Letters*, **44** (174), 57–64.
Rajukkannu, K., Basha, A.A., Habeebullah, B., Duraisamy, P. & Balasubramanian, M. (1985) Degradation and persistence of DDT. BHC, carbaryl and malathion in soils. *Indian Journal of Environmental Health*, **27** (3), 237–43.
Raman, A. & Abrahamson, W.G. (1995) Morphometric relationships and energy allocation in the apical rosette galls of *Solidago altissima* (Asteraceae) induced by *Rhopalomyia solidaginis* (Diptera: Cecidomyiidae). *Environmental Entomology*, **24**, 635–9.
Ramesh, A., Tanabe, S., Kannan, K., Subramanian, A.N., Kumaran, P.L. & Tatsukawa, R. (1992) Characteristic trend of persistent organochlorine contamination in wildlife from a tropical agricultural watershed, South India. *Archives of Environmental Contamination and Toxicology*, **23** (1), 26–36.
Ramos-Elorduy, J., Moreno, J.M.P., Prado, E.E., Perez, M.A., Otero, J.L. & Larron de Guevara, O. (1997) Nutritional value of edible insects from the State of Oaxaca, Mexico. *Journal of Food Composition and Analysis*, **10** (2), 142–57.
Randall, M.G.M. (1982a) The dynamics of an insect population throughout its altitudinal distribution—*Coleophora alticolella* (Lepidoptera) in northern England. *Journal of Animal Ecology*, **51**, 993–1016.
Randall, M.G.M. (1982b) The ectoparasitization of *Coleophora alticolella* (Lepidoptera) in relation to its altitudinal distribution. *Ecological Entomology*, **7**, 177–85.
Rao, D.R., Reuben, R. & Nagasampagi, B.A. (1995) Development of combined use of neem (*Azadirachta indica*) and water management for the control of culicine mosquitoes in rice fields. *Medical and Veterinary Entomology*, **9** (1), 25–33.
Rathcke, B.J. (1976) Competition and Coexistence within a guild of herbivorous insects. *Ecology*, **57**, 76–87.
Rathcke, B.J. (1992) Nectar distributions, pollinator behavior and plant reproductive success. In*: Effects of Resource Distribution on Animal–Plant Interactions* (eds M.D. Hunter, T. Ohgushi and P.W. Price), pp. 113–138. Academic Press, San Diego.
Raupp, M.J. (1985) Effects of leaf toughness on mandibular wear of the leaf beetle, *Plagiodera versicolora*. *Ecological Entomology*, **10**, 73–9.
Raven, P.H., Berg, L.R. & Johnson, G.B. (1993) *Environment*. Saunders College Publishing, Philadelphia.
Redmond, C.T. & Potter, D.A. (1995) Lack of efficacy of *in vivo* and putatively *in vitro* produced *Bacillus popilliae* against field populations of Japanese beetle (Coleoptera: Scarabaeidae) grubs in Kentucky. *Journal of Economic Entomology*, **88** (4), 846–54.
Reed, D.K., Kindler, S.D. & Springer, T.L. (1992) Interactions of Russian wheat aphid, a hymenopterous parasitoid and resistant and susceptible slender wheatgrasses. *Entomologia Experimentalis et Applicata*, **64** (3), 239–46.
Reeder, J.C. & Brown, G.V. (1996) Antigenic variation and immune evasion in *Plasmodium falciparum* malaria. *Immunology and Cell Biology*, **74** (6), 546–54.
Reeve, J.D. & Murdoch, W.W. (1986) Biological control by the parasitoid *Aphytis melinus*, and population stability of

the California red scale (*Aonidiella aurantii*). *Journal Of Animal Ecology*, **55** (3), 1069–82.

Reid, W.V. (1992) How many species will there be? In: *Tropical Deforestation and Species Extinction* (eds T.C. Whitmore & J.A. Sayer), pp. 55–73. Chapman and Hall, London.

Reid, W.V. & Miller, K.R. (1989) *Keeping Options Alive: the Scientific Basis for Conserving Biodiversity*. World Resources Institute, Washington, DC.

Reisen, W.K. & Lothrop, H.D. (1995) Population ecology of *Culex tarsalis* (Diptera: Culicidae) in the Coachella Valley of California. *Journal of Medical Entomology*, **32** (4), 490–502.

Reisen, W.K., Milby, M.M. & Meyer, R.P. (1992) Population dynamics of adult *Culex* mosquitoes (Diptera: Culicidae) along the Kern River, Kern County, California, in 1990. *Journal of Medical Entomology*, **29** (3), 531–43.

Reitz, S.R. (1996) Interspecific competition between two parasitoids of *Helicoverpa zea*, *Eucelatoria bryani* and *E. rubentis*. *Entomologia Experimentalis et Applicata*, **79**, 227–34.

Rengam, S.V. (1992) IPM: The role of governments and citizens' groups. In: *Integrated Pest Management in the Asia–Pacific Region* (eds P.A.C. Ooi, G.S. Lim, T.H. Ho, P.L. Manalo & J. Waage), pp. 13–20. CAB International, Wallingford, UK.

Rhoades, D.F. (1979) Evolution of plant chemical defense against herbivores. In: *Herbivores: Their Interaction with Secondary Plant Metabolites* (eds G.A. Rosenthal and D.H. Janzen), pp. 3–54. Academic Press, New York.

Rhoades, D.F. (1983) Responses of alder and willow to attack by tent caterpilars and webworms: evidence for pheromonal sensitivity of willows. In: *Plant Resistance to Insects* (ed. P.A. Hedin), pp. 55–68. The American Chemical Society, Washington, DC.

Rhoades, D.F. (1985) Offensive-defensive interactions between herbivores and plants: Their relevance to herbivore population dynamics and community theory. *American Naturalist*, **125**, 205–38.

Rhoades, D.F. & Cates, R.G. (1976) Toward a general theory of plant antiherbivore chemistry. *Record Advances in Phytochemistry*, **10**, 168–213.

Rice, W.R. (1983) Sexual reproduction: an adaptation reducing parent-offspring contagion. *Evolution*, **37**, 1317–20.

Richards, A., Matthews, M. & Christian, P. (1998) Ecological considerations for the environmental impact evaluation of recombinant baculovirus insecticides. *Annual Review of Entomology*, **43**, 493–517.

Richardson, M.D. & Bacon, C.W. (1993) Cyclic hydroxamic acid accumulation in corn seedlings exposed to reduced water potentials before, during, and after germination. *Journal of Chemical Ecology*, **19**, 1613–24.

Ritchie, S.A., Fanning, I.D., Phillips, D.A., Standfast, H.A., McGinn, D. & Kay, B.H. (1997) Ross river virus in mosquitoes (Diptera: Culicidae) during the 1994 epidemic around Brisbane, Australia. *Journal of Medical Entomology*, **34** (2), 156–9.

Ritland, D.B. & Brower, L.P. (1993) A reassessment of the mimicry relationship among viceroys, queens and monarchs in Florida. *Natural History Museum of Los Angeles County Science Series*, **38**, 129–39.

Rivas, F., Diaz, L.A., Cardenas, V.M., *et al.* (1997) Epidemic Venezuelan equine encephalitis in La Guajira, Colombia, 1995. *Journal of Infectious Diseases*, **175** (4), 828–32.

Roach, S.H. (1981) Emergence of overwintered *Heliothis* spp. moths from three different tillage systems. *Environmental Entomology*, **10** (5), 817–18.

Robbins, R.K. & Opler, P.A. (1997) Butterfly diversity and a preliminary comparison with bird and mammal diversity. In: *Biodiversity II, Understanding and Protecting our Biological Resources* (eds M.L. Reaka-Kudla, D.E. Wilson & E.O. Wilson), pp. 69–82. Joseph Henry Press, Washington, DC.

Robert, V., Le Goff, G., Essong, J., Tchuinkam, T., Faas, B. & Verhave, J.P. (1995) Detection of falciparum malarial forms in naturally infected anophelines in Cameroon using a fluorescent anti-25-kD monoclonal antibody. *American Journal of Tropical Medicine and Hygiene*, **52** (4), 366–9.

Robertson, L.N. (1993) Population dynamics of false wireworms (*Gonocephalum macleayi, Pterohelaeus alternatus, P. darlingensis*) and development of an integrated pest management program in central Queensland field crops: A review. *Australian Journal of Experimental Agriculture*, **33** (7), 953–62.

Robison, D.J., McCowan, B.H. & Raffa, K.F. (1994) Responses of gypsy moth (Lepidoptera: Lymantriidae) and forest tent caterpillar (Lepidoptera: Lasiocampidae) to transgenic poplar, *Populus* spp., containing a *Bacillus thuringiensis* δ-endotoxin gene. *Environmental Entomology*, **23** (4), 1030–41.

Roces, F. & Hoelldobler, B. (1994) Leaf density and a trade-off between load-size selection and recruitment behavior in the ant *Atta cephalotes*. *Oecologia*, **97**, 1–8.

Roff, D.A. (1990) The evolution of flightlessness in insects. *Ecological Monographs*, **60** (4), 389–421.

Rogers, D.J. & Hassell, M.P. (1974) General models for insect parasite and predator searching behaviour: interference. *Journal of Animal Ecology*, **43**, 239–53.

Rogers, D.J. & Randolph, S.E. (1991) Mortality rates and population density of tsetse flies correlated with satellite imagery. *Nature*, **351** (6329), 739–41.

Rogers, D.J. & Williams, B.G. (1994) Tsetse distribution in Africa: seeing the wood and the trees. In: *Large-scale Ecology and Conservation Biology, 35th Symposium of the British Ecological Society* (eds P.J. Edwards, R.M. May & N.R. Webb), pp. 247–72. Blackwell Scientific, Oxford.

Rohde, K. (1992) Latitudinal gradients in species diversity: the search for the primary cause. *Oikos*, **65**, 514–27.

Roland, J. & Kaupp, W.J. (1995) Reduced transmission of forest tent caterpillar NPV at the forest edge. *Environmental Entomology*, **24**, 1175–8.

Roland, J. & Taylor, P.D. (1997) Insect parasitoid species respond to forest structure at different spatial scales. *Nature*, **386**, 710–13.

Roland, J., Denford, K.E. & Jimenez, L. (1995) Borneol as an attractant for *Cyzenis albicans*, a tachinid parasitoid of the winter moth, *Operophtera brumata* L. (Lepidoptera: Geometridae). *Canadian Entomologist*, **127**, 413–21.

Roland, J., Taylor, P. & Cooke, B. (1997) Forest structure and the spatial pattern of parasitoid attack. *Forests and Insects* (eds A.D. Watt, N.E. Stork & M.D. Hunter), pp. 97–106. Chapman and Hall, London.

Romanow, L.R. & Ambrose, J.T. (1981) Effects of solid rocket fuel exhaust of honey bee colonies. *Environmental Ecology*, **10** (5), 812–16.

Root, R.B. (1973) Organization of a plant-arthropod association in simple and diverse habitats: the fauna of collards (*Brassica oleracea*). *Ecological Monography*, **43**, 95–124.

Rosemond, A.D., Reice, S.R., Elwood, J.W. & Mulholland, P.J. (1992) The effects of stream acidity on benthic invertebrate communities in the south-eastern United States. *Freshwater Biology*, **27**, 193–209.

Rosenheim, J.A. (1998) Higher-order predators and the regulation of insect herbivore populations. *Annual Review of Entomology*, **43**, 421–47.

Rosenheim, J.A., Wilhoit, L.R. & Armer, C.A. (1993) Influence of intraguild predation among generalist insect predators on the suppression of an herbivore population. *Oecologia*, **96**, 439–49.

Rosenthal, G.A. & Berenbaum, M. (eds) (1991) *Herbivores: Their Interaction with Secondary Plant Metabolites*. Academic Press, New York.

Rosenthal, J.P. & Kotanen, P.M. (1994) Terrestrial plant tolerance to herbivory. *Trends in Ecology and Evolution*, **9**, 145–8.

Ross, D.W. & Daterman, G.E. (1994) Reduction of Douglas-fir beetle infestation of high-risk stands by antiaggregation and aggregation pheromones. *Canadian Journal of Forest Research*, **24** (11), 2184–90.

Rossiter, M.C. (1994) Maternal effects hypothesis of herbivore outbreak. *Bioscience*, **44**, 752–63.

Rossiter, M.C. (1996) Incidence and consequences of inherited environmental effects. *Annual Review of Ecology and Systematics*, **27**, 451–76.

Rossiter, M.C., Schultz, J.C. & Baldwin, I.T. (1988) Relationships among defoliation, red oak phenolics, and gypsy-moth growth and reproduction. *Ecology*, **69**, 267–77.

Roth, S.K. & Lindroth, R.L. (1995) Elevated atmospheric CO_2 effects on phytochemistry, insect performance and insect parasitoid interactions. *Global Change Biology*, **1**, 173–82.

Rothschild, M. (1973) Secondary plant substances and warning colouration in insects. *Symposium of the Royal Entomological Society London*, **6**, 59–79.

Roubik, D.W. (1992) Loose niches in tropical communities: Why are there so few bees and so many trees? In: *The Effects of Resource Distribution on Animal–Plant Interactions*. (eds M.D. Hunter, T. Ohgushi & P.W. Price). Academic Press, San Diego.

Round, P.D. (1985) *The Status and Conservation of Resident Forest Birds in Thailand*. Association for the Conservation of Wildlife, Bangkok.

Rowell-Rahier, M. & Pasteels, J.N. (1992) Third trophic level influences of plant allelochemicals. *Ecological and Evolutionary Processes*, **2**, 243–77.

Royama, T. (1970) Factors governing the hunting behaviour and selection of food by the great tit (*Parus major* L.). *Journal of Animal Ecology*, **39**, 619–68.

Ryan, J.D., Gregory, P. & Tingey, W.M. (1982) Phenolic oxidase activities in the glandular trichomes of *Solanum berthaultii* (Hawkes). *Phytochemistry*, **21**, 1885–7.

Ryan, R.B. (1990) Evaluation of biological control, introduced parasites of larch casebearer (Lepidoptera, Coleophoridae) in Oregon. *Environmental Entomology*, **19**, 1873–81.

Saayman, D. & Lambrechts, J.J.N. (1993) The possible cause of red leaf disease and its effect on Barlinka table grapes. *South African Journal for Enology and Viticulture*, **14** (2), 26–32.

Sabelis, M.W. & Afman, B.P. (1994) Synomone-induced suppression of take-off in the phytoseiid mite *Phytoseiulus persimilis* Athias-Henriot. *Experimental and Applied Acarology*, **18**, 711–21.

Sadanandane, C., Sahu, S.S., Gunasekaran, K., Jambulingam, P. & Das, P.K. (1991) Pattern of rice cultivation and anopheline breeding in Koraput district of Orissa State (India). *Journal of Communicable Diseases*, **23** (1), 59–65.

Sagers, C.L. & Coley, P.D. (1995) Benefits and costs of defense in a neotropical shrub. *Ecology*, **76**, 1835–43.

Sain, M. & Kalode, M.B. (1994) Greenhouse evaluation of rice cultivars for resistance to gall midge, *Orseolia oryzae* (Wood-Mason) and studies on the mechanism or resistance. *Insect Science and its Application*, **15** (1), 67–74.

Sait, S.M., Begon, M. & Thompson, D.J. (1994) Long-term population dynamics of the Indian meal moth *Plodia interpunctella* and its granulosis virus. *Journal of Animal Ecology*, **63** (4), 861–70.

Salt, D.E., Prince, R.C., Pickering, I.J. & Raskin, I. (1995) Mechanisms of cadmium mobility and accumulation in Indian mustard. *Plant Physiology*, **109**, 1427–33.

Salt, D.T., Fenwick, P. & Wjitaker, J.B. (1996) Interspecific herbivore interactions in a high CO_2 environment—root and shoot aphids feeding on a cardamine. *Oikos*, **77**, 326–30.

Samuel, T. & Pillai, K.K. (1989) The effect of temperature and solar radiations on volatilization, mineralization and degradation of carbon-14 DDT in soil. *Environmental Pollution*, **57** (1), 63–78.

Sanchis, V., Chaufaux, J. & Lereclus, D. (1995) The use of *Bacillus thuringiensis* in crop protection and the development of pest resistance. *Cahiers Agricultures*, **4** (6), 405–16.

Saraswathi, A. & Ranganathan, L.S. (1994) The ovicidal and larvicidal effects of dimilin on *Tabanus triceps* and *Chrysops dispar* (Diptera: Tabanidae). *Insect Science and its Application*, **15** (1), 97–100.

Savage, H.M., Niebylski, M.L., Smith, G.C., Mitchell, C.J. & Craig, G.B. (1993) Host-feeding patterns of *Aedes albopictus* (Diptera: Culicidae) at a temperate North American site. *Journal of Medical Entomology*, **30** (1), 27–34.

Sayer, J.A. & Stuart, S. (1988) Biological diversity and tropical forests. *Environmental Conservation*, **15**, 193–4.

Schaefer, A., Konrad, R., Kuhnigk, T., Kaempfer, P., Hertel, H. & Koenig, H. (1996) Hemicellulose-degrading bacteria and yeasts from the termite gut. *Journal of Applied Bacteriology*, **80** (5), 471–8.

Schellhorn, N.A. & Sork, V.L. (1997) The impact of weed diversity on insect population dynamics and crop yield in collards, *Brassica oleraceae* (Brassicaceae). *Oecologia*, **111**, 233–40.

Schmitt, T.M., Hay, M.E. & Lindquist, N. (1995) Constraints on chemically mediated coevolution: Multiple functions for seaweed secondary metabolites. *Ecology*, **76**, 107–23.

Schoener, T.W. (1983) Field experiments on interspecific competition. *American Science*, **70**, 586–95.

Schoener, T.W. (1988) On testing the MacArthur–Wilson model with data on rates. *American Naturalist*, **131** (6), 847–64.

Schoonhoven, L.M., Beerling, E.A.M., Klijnstra, J.W. & Van-Vugt, Y. (1990) Two related butterfly species avoid oviposition near each other's eggs. *Experientia (Basel)*, **46** (5), 526–8.

Schops, K., Syrett, P. & Emberson, R.M. (1996) Summer diapause in *Chrysolina hyperici* and *C. quadrigemina* (Coleoptera: Chrysomelidae) in relation to biological control of St John's wort, *Hypericum perforatum* (Clusiaceae). *Bulletin of Entomological Research*, **86** (5), 591–7.

Schowalter, T.D., Hargrove W.W. & Crossley D.A. Jr (1986) Herbivory in forested ecosystems. *Annual Review of Entomology*, **31**, 177–96.

Schulthess, F., Baumgaertner, J.U., Delucchi, V. & Gutierrez, A.P. (1991) The influence of the cassava mealybug, *Phenacoccus manihoti* Mat. Ferr.) (Homoptera: Pseudococcidae) on yield formation of cassava, *Manihot esculenta* Crantz. *Journal of Applied Entomology*, **111** (2), 155–65.

Schultz, J.C. (1992) Factoring natural enemies into plant tissue availability to herbivores. In: *Effects of Resource Distribution on Animal–Plant Interactions*, (eds M.D. Hunter, T. Ohgushi, & P.W. Price), pp. 175–97. Academic Press, San Diego.

Schultz, J.C., Nothnagle, P.J. & Baldwin, I.T. (1982) Seasonal and individual variation in leaf quality of two northern hardwoods trees species. *American Journal of Botany*, **69** (5), 753–9.

Schultz, J.C., Foster, M.A. & Montgomery, M.E. (1990) Host plant-mediated impacts of a baculovirus on gypsy moth populations. In: *Population Dynamics of Forest Insects* (eds A.D. Watt, S.R. Leather, M.D. Hunter & N.A.C. Kidd), pp. 303–13. Intercept, Andover, UK.

Schultz, J.C., Hunter, M.D. & Appel, H.M. (1992) Antimicrobial activity of polyphenols mediates plant-herbivore interactions. In: *Plant Polyphenols: Biogenesis, Chemical Properties, and Significance* (eds R.W. Hemingway & P.E. Laks), pp. 621–37. Plenum Press, New York.

Schultze-Lam, S., Ferris, F.G., Sherwood-Lollar, B. & Gerits, J.P. (1996) Ultrastructure and seasonal growth patterns of microbial mats in a temperate climate saline–alkaline lake: Goodenough Lake, British Columbia, Canada. *Canadian Journal of Microbiology*, **42** (2), 147–61.

Schumacher, M.J. & Egen, N.B. (1995) Significance of Africanised bees for public health. *Archives of Internal Medicine*, **155** (19), 2038–43.

Schuman, G.L. (1991) *Plant Diseases: Their Biology and Social Impact*, 291 pp. APS Press, St Paul, MN, USA.

Schwartz, E., Mendelson, E. & Sidi, Y. (1996) Dengue fever among travelers. *American Journal of Medicine*, **101** (5), 516–20.

Schwenke, W. (1968) Neve Hinweise Auf einer Abhaengigkeit der Vermehrung blattund nadelfressender Forstinsekten vom Zuckergehalt ihrer Nahrung. *Zeitschrift fur Angewandte Entomologie*, **61**, 365–9.

Schwenke, W. (1994) On the fundamentals of forest insect outbreaks and of counter measures. *Anzeiger Fuer Schaedlingskunde Pflanzenschulz Umweltschultz*, **67**, 120–4.

Scoble, M.J., Gaston, K.J. & Crook, A. (1995) Using taxonomic data to estimate species richness in Geometridae. *Journal of the Lepidopterists' Society*, **49** (2), 136–47.

Scott, A.C., Stephenson, J. & Chaloner, W.G. (1992) Interaction and coevolution of plants and arthropods during the Palaeozoic and Mesozoic. *Philosophical Transactions of the Royal Society of London, Series B: Biological Sciences*, **335** (1274), 129–65.

Scott, T.W., Weaver, S.C. & Mallampalli, V.L. (1994) Evolution of mosquito-borne viruses. In: *The Evolutionary Biology of Viruses* (ed. S.S. Morse), pp. 293–324. Raven Press, New York.

Seastedt, T.R., Ramundo, R.A. & Hayes, D.C. (1988) Maximization of densities of soil animals by foliage herbivory: empirical evidence, graphical and conceptual models. *Oikos*, **51**, 243–8.

Seketeli, A. & Kuzoe, F.A.S. (1994) Diurnal resting sites of *Glossina palpalis palpalis* (Robineau-Desvoidy) in a forest edge habitat of Ivory Coast. *Insect Science and its Application*, **15** (1), 75–85.

Senguttuvan, T., Gopalan, M. & Chelliah, S. (1991) Impact of resistance mechanisms in rice aginst the brown plant hopper, *Nilaparvata lugens* Stal (Homoptera: Delphacidae). *Crop Protection*, **10** (2), 125–8.

Service, M.W. (1991) Agricultural development and arthropod-borne disease: a review. *Revista de Saude Publica*, **25** (3), 165–78.

Service, M.W. (1996) *Medical Entomology for Students*, 278 pp. Chapman & Hall, London.

Setala, H., Marshall, V.G. & Trofymow, J.A. (1995) Influence of micro-habitat and macro-habitat factors on collembolan communities in douglas-fir stumps during forest succession. *Applied Soil Ecology*, **2**, 227–42.

Sexton, J.D. (1994) Impregnated bednets for malaria control: biological success and social responsibility. *American Journal of Tropical Medicine and Hygiene*, **60** (Suppl. 6), 72–81.

Sgardelis, P.S. & Usher, M.B. (1994) Responses of soil Cryptostigmata across the boundary between a farm woodland and an arable field. *Pedobiologia*, **38**, 36–49.

Shapiro, M., Robertson, J.L. & Webb, R.E. (1994) Effect of neem seed extract upon the gypsy moth (Lepidoptera: Lymantriidae) and its nuclear polyhedrosis virus. *Journal of Economic Entomology*, **87** (2), 356–60.

Sharma, H.C. & Lopez, V.F. (1993) Comparison of injury levels for sorghum head bug, *Calocoris angustatus*, on resistant and susceptible genotypes at different stages of panicle development. *Crop Protection*, **12** (4), 259–66.

Sharma, H.C. & Vidyasagar, P. (1994) Antixenosis component of resistance to sorghum midge, *Contarinia sorghicola* Coq. in Sorghum bicolor (L.) Moeanch. *Annals of Applied Biology*, **124** (3), 495–507.

Sharma, V.P. (1996) Re-emergence of malaria in India. *Indian Journal of Medical Research*, **103**, 26–45.

Sharma, Y.D., Biswas, S., Pillai, C.R., Ansari, M.A., Adak, T. & Devi, C.U. (1996) High prevalence of chloroquine resistant *Plasmodium falciparum* infection in Rajasthan epidemic. *Acta Tropica*, **62** (3), 135–41.

Shimoda, T., Takabayashi, J., Ashihara, W. & Takafuji, A. (1997) Response of predatory insect *Scolothrips takahashii* toward herbivore-induced plant volatiles under laboratory and field conditions. *Journal of Chemical Ecology*, **23**, 2033–48.

Shirt, B.D. (1987) *British Red Data Books: 2, Insects*. Nature Conservancy Council, Peterborough, UK.

Shonle, I. & Bergelson, J. (1995) Interplant communication revisited. *Ecology*, **76**, 2660–3.

Shorey, H.H. & Gerber, R.G. (1996) Use of puffers for disruption of sex pheromone communication of codling moths (Lepidoptera: Tortricidae) in walnut orchards. *Environmental Entomology*, **25** (6), 1398–400.

Short, R.A. & Maslin, P.E. (1977) Processing of leaf litter by a stream detritivore: effect on nutrient availability to collectors. *Ecology*, **58**, 935–8.

Showler, A.T. (1995) Locust (Orthoptera: Acrididae) outbreak in Africa and Asia, 1992–4: An overview. *American Entomologist*, **41** (3), 179–85.

Showler, A.T. & Potter, C.S. (1991) Synopsis of the 1986–9 desert locust (Orthoptera: Acrididae) plague and the concept of strategic control. *American Entomologist*, **37** (2), 106–10.

Siemens, D.H., Ralston, B.E. & Johnson, C.D. (1994) Alternative seed defence mechanisms in a palo verde (Fabaceae) hybrid zone: Effects on bruchid beetle abundance. *Ecological Entomology*, **19**, 381–90.

Silva-Bohorquez, I. (1987) Interspecific interactions between insects on oak trees, with special reference to defoliators and the oak aphid. DPhil Thesis, University of Oxford.

Silverman, A.L., McCray, D.C., Gordon, S.C., Morgan, W.T. & Walker, E.D. (1996) Experimental evidence against replication or dissemination of hepatitis C virus in mosquitoes (Diptera: Culicidae) using detection by reverse transcriptase polymerase chain reaction. *Journal of Medical Entomology*, **33** (3), 398–401.

Simms, E.L. & Rausher, M.D. (1989) The evolution of resistance to herbivory in *Ipomoea purpurea*: II. Natural selection by insects and costs of resistance. *Evolution*, **43**, 573–85.

Simms, E.L. & Triplett, J. (1994) Costs and benefits of plant responses to diseases. *Resistance and Tolerance Evolution*, **48**, 1973–85.

Sinsabaugh, R.L., Likens, A.E. & Benfield, E.F. (1985) Cellulose digestion and assimilation by three leaf-shredding insects. *Ecology*, **66**, 1464–71.

Sjoeib, F., Anwar, E. & Tungguldihardjo, M.S. (1994) Behaviour of DDT and DDE in Indonesian tropical environments. *Journal of Environmental Science and Health, Part B—Pesticides, Food, Contaminants and Agricultural Wastes*, **29** (1), 17–24.

Skaf, R., Popov, G.B. & Roffer, J. (1990) The desert locust: An international challenge. *Philosophical Transactions of the Royal Society of London, Series B: Biological Sciences*, **328** (1251), 525–38.

Skare, J.U., Stenersen, J., Kveseth, N. & Polder, A. (1985) Time trends of organochlorine residues in 7 sedentary marine fish species from a Norwegian fjord during the period 1972–82. *Archives of Environmental Contamination and Toxicology*, **14** (1), 33–42.

Skovgard, H. & Pats, P. (1997) Reduction of stemborer damage by intercropping maize with cowpea. *Agriculture, Ecosystems and Environment*, **62** (1), 13–19.

Slosser, J.E., Bordovsky, D.G. & Bevers, S.J. (1994) Damage and costs associated with insect management options in irrigated cotton. *Journal of Economic Entomology*, **87** (2), 436–45.

Smiley, J.T. (1985) Are chemical barriers necessary for evolution of butterfly-plant associations? *Oecologia*, **65**, 580–3.

Smith, A.M. & Ward, S.A. (1995) Temperature effects on larval and pupal development, adult emergence, and survival of the pea weevil (Coleoptera: Chrysomelidae). *Environmental Entomology*, **24** (3), 623–34.

Smith, D.A.S. & Owen, D.F. (1997) Colour genes as markers for migratory activity: The butterfly *Danaus chrysippus* in Africa. *Oikos*, **78** (1), 127–35.

Smith, J.W.J.R. & Johnson, S.J. (1989) Natural mortality of the lesser cornstalk borer (Lepidoptera: Pyralidae) in a peanut agroecosystem. *Environmental Entomology*, **18** (1), 69–77.

Smith, L.L., Lanza, J. & Smith, G.C. (1990) Amino acid concentrations in extrafloral nectar of *Impatiens sultani* increase after simulated herbivory. *Ecology*, **71**, 107–15.

Snodgrass, G.L. (1991) *Deraecoris nebulosus* (Heteroptera: Miridae): Little known predator in cotton in the Mississippi Delta (USA). *Florida Entomologist*, **74** (2), 340–4.

Southwood, T.R.E. (1961) The number of species of insect associated with various trees. *Journal of Animal Ecology*, **30**, 1–8.

Southwood, T.R.E. (1973) The insect/plant relationship—an evolutionary perspective. *Symposium of the Royal Entomological Society of London*, **6**, 3–30.

Southwood, T.R.E. (1977) Habitat, the templet for ecological strategies. *Journal of Animal Ecology*, **46**, 337–65.

Southwood, T.R.E. (1978) *Ecological Methods*, 2nd edn, 524 pp. Chapman & Hall, London.

Southwood, T.R.E. & Comins, H.N. (1976) A synoptic population model. *Journal of Animal Ecology*, **45**, 949–65.

Southwood, T.R.E., Moran, V.C. & Kennedy, C.E.J. (1982) The richness, abundance and biomass of the arthropod communities on trees. *Journal of Animal Ecology*, **51**, 635–49.

Sparks, T.H. & Carey, P.D. (1995) The responses of species to climate over 2 centuries—an analysis of the Marsham phenological record, 1736–1947. *Journal of Ecology*, **83**, 321–9.

Speight, M.R. (1983) The potential of ecosystem management for pest control. *Agriculture, Ecosystems and Environment*, **10** (2), 183–99.

Speight, M.R. (1992) The impact of leaf-feeding by nymphs of the horse chestnut scale *Pulvinaria regalis* on young host trees. *Journal of Applied Entomology*, **112**, 389–99.

Speight, M.R. (1994) Reproductive capacity of the horse chestnut scale insect, *Pulvinaria regalis* Canard (Hom., Coccidae). *Journal of Applied Entomology*, **118** (1), 59–67.

Speight, M.R. (1997a) The relationship between host tree stresses and insect attack in tropical forest plantations, and its relevance to pest management. In: *Impact of Diseases and Insect Pests in Tropical Forests, International Union of Forest Research Organizations (IUFRO) Symposium, Kerala Forest Research Institute, Peechi, India* (eds K.S.S. Nair, J.K. Sharma & R.C. Varma).

Speight, M.R. (1997b) Forest pests in the tropics: current status and future threats. In: *Insects and Trees, Royal Entomological Society Symposium on Forest and Insects* (eds A.D. Watt, M.D. Hunter & N.E. Stork).

Speight, M.R., Hails, R.S., Gilbert, M. & Foggo, A. (1998) Horse chestnut scale (*Pulvinaria regalis*) (Homoptera: Coccidae) and urban host tree environment. *Ecology (Washington D.C.)* **79** (5), 1503–13.

Speight, M.R. & Lawton, J.H. (1976) The influence of weed-cover on the mortality imposed on artificial prey by predatory ground beetles in cereal fields. *Oecologia*, **23**, 211–23.

Speight, M.R. & Wainhouse, D. (1989) *Ecology and Management of Forest Insects*, 374 pp. Oxford Science Publications, Oxford.

Speight, M.R., Kelly, P.M., Sterling, P.H. & Entwistle, P.F. (1992) Field application of a nuclear polyhedrosis virus against the Brown-tail moth, *Euproctis chrysorrhoea* (L.) (Lepidoptera, Lymantriidae). *Journal of Applied Entomology*, **113** (3), 295–306.

Spencer, K.A. (1972) *Handbooks for the Identification of British Insects*. Royal Entomological Society, London 5.

Stadler, B. & Mackauer, M. (1996) Influence of plant quality on interactions between the aphid parasitoid *Ephedrus californicus* Baker (Hymenoptera: Aphidiidae) and its host, *Acyrthosiphon pisum* (Harris) (Homoptera: Aphididae). *Canadian Entomologist*, **128** (1), 27–39.

Standley, L.J. & Sweeney, B.W. (1995) Organochlorine pesticides in stream mayflies and terrestrial vegetation of undisturbed tropical catchments exposed to long-range atmospheric transport. *Journal of the North American Benthological Society*, **14** (1), 38–49.

Stephenson, A.G. (1982) The role of extrafloral nectaries of *Catalpa speciosa* in limiting herbivory and increasing fruit production. *Ecology*, **63**, 663–9.

Sterling, P.H. & Speight, M.R. (1989) Comparative mortalities of the brown-tail moth, *Euproctis chrysorrhoea* (L.) (Lepidoptera: Lymantriidae), in south-east England. *Botanical Journal of the Linnean Society*, **101**, 69–78.

Sternlicht, M., Barzakay, I. & Tamim, M. (1990) Management of *Prays citri* in lemon orchards by mass trapping of males. *Entomologia Experimentalis et Applicata*, **55** (1), 59–68.

Stevens, M., Smith, H.G. & Hallsworth, P.B. (1994) The host range of beet yellowing viruses along common arable weed species. *Plant Pathology (Oxford)*, **43** (3), 579–88.

Stevenson, P.C., Blaney, W.M., Simmonds, M.J.S. & Wightman, J.A. (1993) The identification and characterization of resistance in wild species of *Arachis* to *Spodoptera litura* (Lepidoptera: Noctuidae). *Bulletin of Entomological Research*, **83**, 421–9.

St George, R.A. (1930) Drought affected and injured trees attractive to bark beetles. *Journal of Economic Entomology*, **23**, 825–8.

Stiling, P. (1987) The frequency of density dependence in insect host parasitoid systems. *Ecology*, **68**, 844–56.

Stiling, P. (1988) Density-dependent processes and key factors in insect populations. *Journal of Animal Ecology*, **57**, 581–93.

Stiling, P. & Rossi, A.M. (1996) Complex effects of genotype and environment on insect herbivores and their enemies. *Ecology*, **77**, 2212–18.

Stiling, P. & Rossi, A.M. (1998) Deme formation in a dispersive gall-forming midge. In: Genetic Structure and Local Adaptation in Natural Insect Populations. *Effects of Ecology, Life History, and Behavior* (eds S. Mopper & S.Y. Strauss), pp. 22–36. Chapman & Hall, New York.

Stone, C. & Bacon, P.E. (1994) Relationship among moisture stress, insect herbivory, foliar cineole content and the growth of river red gum *Eucalyptus camaldulensis*. *Journal of Applied Ecology*, **31**, 604–12.

Stork, N.E. (1987) Guild structure of arthropods from Bornean rain-forest trees. *Ecological Entomology*, **12**, 69–80.

Stork, N.E. (1988) Insect diversity—facts, fiction and speculation. *Biological Journal of the Linnean Society*, **35**, 321–37.

Stork, N.E. (1991) The composition of the arthropod fauna of Bornean lowland rain forest trees. *Journal of Tropical Ecology*, **7** (2), 161–88.

Stork, N.E. (1995) Measuring and monitoring arthropod diversity in temperate and tropical forests. In: *Measuring and Monitoring Biodiversity in Tropical and Temperate Forests* (eds T.J.B. Boyle & B. Boontawee), pp. 257–70. Centre for International Forestry Research (CIFOR), Bogor, Indonesia.

Stork, N.E. (1997) Measuring global biodiversity and its decline. In: *Biodiversity II, Understanding and Protecting our Biological Resources* (eds M.L. Reaka-Kudla, D.E. Wilson & E.O. Wilson), pp. 41–68. Joseph Henry Press, Washington, DC.

Stork, N.E. & Brendell, M.J.D. (1993) Arthropod abundance in lowland rain forest of Seram. In: *Natural History of Seram* (eds I.D. Edwards, A.A. MacDonald & J. Proctor), pp. 115–30. Intercept, Andover, UK.

Stork, N.E. & Gaston, K.J. (1990) Counting species one by one. *New Scientist*, **1729**, 43–7.

Stork, N.E. & Lyal, C.H.C. (1993) Extinction or co-extinction rates. *Nature* **366**, 307.

Stork, N.E., Adis, J. & Didham, R.K. (1997) *Canopy Arthropods*, 567 pp. Chapman & Hall, London.

Strassmann, J.E., Solis, C.R., Hughes, C.R., Goodnight, K.F. & Queller, D.C. (1997) Colony life history and demography of a swarm-founding social wasp. *Behavioral Ecology and Sociobiology*, **40** (2), 71–7.

Strathdee, A.T. & Bale, J.S. (1998) Life on the edge: insect ecology in arctic environments. *Annual Review of Entomology*, **43**, 85–106.

Strathdee, A.T., Bale, J.S., Block, W.C., Webb, N.R., Hodkinson, I.D. & Coulson, S.J. (1993) Extreme adaptive life-cycle in a high arctic aphid, *Acyrthosiphon svalbardicum*. *Ecological Entomology*, **18** (3), 254–8.

Strong, D.R. (1974) Rapid asymptotic species accumulation in phytophagous insect communities: the pests of cacao. *Science*, **185**, 1064–6.

Strong, D.R. (1982) Potential interspecific competition and host specificity: hispine beetles on *Heliconia*. *Ecological Entomology*, **7**, 217–20.

Strong, D.R., Lawton, J.H. & Southwood, T.R.E. (1984) *Insects on Plants: Community Patterns and Mechanisms*. Blackwell Scientific, Oxford.

Strong, D.R., McCoy, E.D. & Rye, J.R. (1977) Time and number of herbivore species: the pests of sugarcane. *Ecology*, **58**, 167–75.

Stuening, D. (1988) Biological–ecological investigations on the Lepidoptera of the Supralittoral Zone of the North Sea coast. *Faunistisch-Oekologische Mitteilungen, Supplement*, 7, 1–116.

Styles, C.V. & Skinner, J.D. (1996) Possible factors contributing to the exclusion of saturniid caterpillars (mopane worms) from a protected area in Botswana. *African Journal of Ecology*, **34** (3), 276–83.

Sunderland, K.D. (1988) Quantitative methods for detecting invertebrate predation occurring in the field. *Annals of Applied Biology*, **112**, 201–24.

Sunderland, K.D. & Vickerman, G.P. (1980) Aphid feeding by some polyphagous predators in relation to aphid density in cereal fields. *Journal of Applied Ecology*, **17**, 389–96.

Sunderland, K.D., Crook, N.E., Stacey, D.L. & Fuller, B.T. (1987) A study of feeding by polyphagous predators on cereal aphids using ELISA and gut dissection. *Journal of Applied Ecology*, **24**, 907–33.

Sunderland, K.D., Fraser, A.M. & Dixon, A.F.G. (1986) Field and laboratory studies on money spiders (Linyphiidae) as predators of cereal aphids. *Journal of Applied Ecology*, **23**, 433–47.

Suomi, D.A. & Akre, R.D. (1993) Biological studies of *Hemicoelus gibbicollis* (Leconte) (Coleoptera: Anobiidae), a serious structural pest along the Pacific coast: Larval and pupal stages. *Pan-Pacific Entomologist*, **69** (3), 221–35.

Sutton, S.L. & Collins, N.M. (1991) Insects and tropical forest conservation. In: *The Conservation of Insects and their Habitats* (eds N.M. Collins & J.A. Thomas), pp. 405–24. Academic Press, London.

Swetnam, T.W. & Lynch, A.M. (1993) Multicentury, regional-scale patterns of western spruce budworm outbreaks. *Ecological Monographs*, **63** (4), 399–424.

Symmons, P. (1992) Strategies to combat the desert locust. *Crop Protection*, **11** (3), 206–12.

Symmons, P.M. (1986) Locust (*Chortoicetes terminifera*) displacing winds in eastern Australia. *International Journal of Biometeorology*, **30** (1), 53–64.

Tabashnik, B.E., Finson, N. & Johnson, M.W. (1991) Managing resistance to *Bacillus thuringiensis*: Lessons from the diamondback moth (Lepidoptera: Plutellidae). *Journal of Economic Entomology*, **84** (1), 49–55.

Takabayashi, J., Dicke, M. & Posthumus, M.A. (1991) Induction of direct defense against spider-mites in uninfested lima bean leaves. *Phytochemistry (Oxford)*, **30**, 1459–62.

Takabayashi, J., Dicke, M., Takahashi, S., Posthumus, M.A. & Vanbeek, T.A. (1994) Leaf age affects composition of herbivore-induced synomones and attraction of predatory mites. *Journal of Chemical Ecology*, **20**, 373–86.

Tanabe, S., Subramanian, A., Ramesh, A., Kumaran, P.L., Miyazaki, N. & Tatsuhawa, R. (1993) Persistent organochlorine residues in dolphins from the Bay of Bengal, South India. *Marine Pollution Bulletin*, **26** (6), 311–16.

Tansley, A.G. (1935) The use and abuse of vegetational concepts and terms. *Ecology*, **16**, 284–307.

Taylor, P.S., Shields, L.J., Tauber, M.J. & Tauber, C.A. (1995)

Induction of reproductive diapause in *Empoasca fabae* (Homoptera: Cicadellidae) and its implications regarding southward migration. *Environmental Entomology*, **24** (5), 1086–95.

Tebayashi, S.I., Matsuyama, S., Suzuki, T., Kuwahara, Y., Nemoto, T. & Fujii, K. (1995) Quercimeritrin: The third oviposition stimulant of the Azuki bean weevil from the host azuki bean. *Journal of Pesticide Science*, **20**, 299–305.

Teng, P.S. (1994) Integrated pest management in rice. *Experimental Agriculture*, **30** (2), 115–37.

Tercafs, R. & Brouwir, C. (1991) Population size of Pyrenean troglobiont coleopters (*Speonomus* species) in a cave in Belgium. *International Journal of Speleology*, **20** (1–4), 23–35.

Theunisse, J. & Schelling, G. (1996) Pest and disease management by intercropping: Suppression of thrips and rust in leek. *International Journal of Pest Management*, **42** (4), 227–34.

Thiery, D., Gabel, D., Farkas, P. & Jarry, M. (1995) Egg dispersion in codling moth—influence of egg extract and of its fatty-acid constituents. *Journal of Chemical Ecology*, **21**, 2015–26.

Thirakhupt, V. & Araya, J.E. (1992) Survival and life table statistics of *Rhopalosiphum padi* (L.) and *Sitobion avenae* (F.) (Hom., Aphididae) in single or mixed colonies in laboratory wheat cultures. *Journal of Applied Entomology*, **113** (4), 368–75.

Thireau J.C. & Regniere J. (1995) Development, reproduction, voltinism and host synchrony of *Meteorus trachynotus* with its hosts *Choristoneura fumiferana* and *C. rosaceana*. *Entomologia Experimentalis et Applicata*, **76** (1), 67–82.

Thomas, A.T. & Hodkinson, I.D. (1991) Nitrogen, water and stress and the feeding efficiency of lepidopteran herbivores. *Journal of Applied Ecology*, **28**, 703–20.

Thomas, D.J., Tracey, B., Marshall, H. & Norstrom, R.J. (1992) Arctic terrestrial ecosystem contamination. *Science of the Total Environment*, **122** (1–2), 135–64.

Thomas, J.A. & Lewington, R. (1991) *The Butterflies of Britain and Ireland*. Dorling Kindersley, London.

Thompson, A.J. & Shrimpton, D.M. (1984) Weather associated with the start of mountain pine beetle outbreaks. *Canadian Journal of Forest Research*, **14**, 255–8.

Thompson, J.N. (1986) Patterns in coevolution. In: *Coevolution and Systematics* (eds A.R. Stone & D.H. Hawksworth), pp. 119–43. Clarendon Press, Oxford.

Thompson, J.N. (1989) Concepts of coevolution. *Trends in Ecology and Evolution*, **4**, 179–83.

Thompson, J.N. (1994) *The Coevolutionary Process*. University of Chicago Press, Chicago.

Thompson, J.N. (1997) Evaluating the dynamics of coevolution among geographically structured populations. *Ecology*, **78**, 1619–23.

Thompson, R.A., Quisenberry, S.S., N'guessan, F.K., Heagler, A.M. & Giesler, G. (1994) Planting date as a potential method for managing the rice water weevil (Coleoptera: Curculionidae) in water-seeded rice in southwest Louisiana. *Journal of Economic Entomology*, **87** (5), 1318–24.

Thomson, M.C., Adiamah, J.H., Connor, S.J. *et al.* (1995) Entomological evaluation of the Gambia's National Impregnated Bednet Programme. *Annals of Tropical Medicine and Parasitology*, **89** (3), 229–41.

Thomson, M.C., Connor, S.J., Quinones, M.L., Jawara, M., Todd, J. & Greenwood, B.M. (1995) Movement of *Anopheles gambiae* s.l. malaria vectors between villages in The Gambia. *Medical and Veterinary Entomology*, **9** (4), 413–19.

Thorne, B.L. (1991) Ancestral transfer of symbionts between cockroaches and termites: An alternative hypothesis. *Proceedings of the Royal Society of London, Series B: Biological Sciences*, **246** (1317), 191–6.

Thuy, N.N. & Thieu, D.V. (1992) Status of integrated pest management programme in Viet Nam. In: *Integrated Pest Management in the Asia–Pacific Region* (eds P.A.C. Ooi, G.S. Lim, T.H. Ho, P.L. Manalo & J. Waage), pp. 237–50. CAB International, Wallingford, UK.

Tilman, D. (1987) The importance of the mechanisms of interspecific competition. *American Naturalist*, **129**, 769–74.

Tilman, D. & Downing, J.A. (1994) Biodiversity and stability in grasslands. *Nature*, **367**, 363–5.

Togashi, K. (1991) Spatial pattern of pine wilt diseases caused by *Bursaphelencus xylophilus* (Nematoda: Aphelenchoididae) within a *Pinus thunbergii* stand. *Researches on Population Ecology (Kyoto)*, **33** (2), 245–56.

Torre, A.D., Merzagora, L., Powell, J.R. & Coluzzi, M. (1997) Selective introgression of paracentric inversions between two sibling species of the *Anopheles gambiae* complex. *Genetics*, **146** (1), 239–44.

Torres, J.A. & Snelling, R.R. (1997) Biogeography of Puerto Rican ants: a non-equilibrium case? *Biodiversity and Conservation*, **6**, 1103–21.

Tostowaryk, W. (1972) The effect of prey defense on the functional response of *Podisus modestus* (Hemiptera: Pentatomidae) to the densities of the sawflies *Neodiprion swainei* and *N. pratti banksianae* (Hymenoptera: Neodiprionidae). *Canadian Entomologist*, **104**, 61–9.

Traveset, A. (1990) Bruchid egg mortality on *Accacia farnesiana* caused by ants and abiotic factors. *Ecological Entomology*, **15**, 463–8.

Trimble, R.M. (1995) Mating disruption for controlling the codling moth, *Cydia Pomonella* (L.) (Lepidoptera: Tortricidae), in organic apple production in southwestern Ontario. *Canadian Entomologist*, **127** (4), 493–505.

Trivedi, J.P., Srivastava, A.P., Narain, K. & Chatterjee, R.C. (1991) The digestion of wool fibres in the alimentary system of *Anthrenus flavipes* larvae. *International Biodeterioration*, **27** (4), 327–36.

Trumble, J.T., Carson, W.G. & Kund, G.S. (1997) Economics and environmental impact of a sustainable integrated pest management program in celery. *Journal of Economic Entomology*, **90** (1), 139–46.

Trumbo, S.T. & Fernandez, A.G. (1995) Regulation of brood

size by male parents and cues employed to assess resource size by burying beetles. *Ethology, Ecology and Evolution*, **7**, 313–22.

Trutmann, P. & Graf, W. (1993) The impact of pathogens and arthropod pests on common bean production in Rwanda. *International Journal of Pest Management*, **39** (3), 328–33.

Tscharntke, T. (1992a) Cascade effects among four trophic levels: Bird predation on galls affects density-dependent parasitism. *Ecology*, **73** (5), 1689–98.

Tscharntke, T. (1992b) Coexistence, tritrophic interactions and density dependence in a species-rich parasitoid community. *Journal of Animal Ecology*, **61** (1), 59–67.

Tscharntke, T. (1992c) Fragmentation of *Phragmites* habitats, minimum viable population size, habitat suitability, and local extinction of moths, midges, flies, aphids, and birds. *Conservation Biology*, **6** (4), 530–6.

Tschen, J. & Fuchs, W.H. (1969) Wirkung von Phloridzin auf die infektion von *Phaseolus vulgaris* durch *Uromyces phaseoli*. *Naturwissenshaft*, **56**, 643–8.

Tsubaki, Y., Siva-Jothy, M.T. & Ono, T. (1994) Re-copulation and post-copulatory mate guarding increase immediate female reproductive output in the dragonfly *Nannophya pygmeae* Rambur. *Behavioral Ecology and Sociobiology*, **35**, 219–25.

Turlings, T.C.J. & Tumlinson, J.H. (1991) Do parasitoids use herbivore-induced plant-chemical defenses to locate hosts? *Florida Entomologist*, **74**, 42–50.

Turlings, T.C.J., Loughrin, J.H., Mccall, P.J., Rose, U.S.R., Lewis, W.J. & Tumlinson, J.H. (1995) How caterpillar-damaged plants protect themselves by attracting parasitic wasps. *Proceedings of the National Academy of the Sciences of the United States of America*, **92**, 4169–74.

Turner, C.L., Saestedt, T.R. & Dyer, M.I. (1993) Maximization of aboveground grassland production: The role of defoliation frequency, intensity, and history. *Ecological Applications*, **3**, 175–86.

Turnipseed, S.G. (1977) Influence of trichome variations on populations of small phytophagous insects on soybean. *Environmental Entomology*, **6**, 815–7.

Turnock, W.J., Timlick, B. & Palaniswamy, P. (1993) Species and abundance of cutworms (Noctuidae) and their parasitoids in conservation and conventional tillage fields. *Agriculture, Ecosystems and Environment*, **45** (3–4), 213–27.

Umana, V. & Constenla, M.A. (1984) Organochlorinated pesticides in human milk. *Revista de Biologia Tropical*, **32** (2), 233–40.

United Kingdom Climate Change Impacts Review Group (1996) *Review of the Potential Effects of Climate Change in the United Kingdom*. HMSO, London.

University of California (1998) UC Pest Management Guidelines: Citrus—California red scale. http://www.ipm.ucdavis.edu/PMG.

Unnithan, G.C. & Paye, S.O. (1990) Factors involved in mating, longevity, fecundity and egg fertility in the maize stem-borer, *Busseola fusca* (Fuller) (Lepidoptera, Noctuidae). *Journal of Applied Entomology*, **109** (3), 295–301.

Unruh, T.R. & Luck, R.F. (1987) Deme formation in insects: a test with the pinyon needle scale and review of other evidence. *Ecological Entomology*, **12**, 439–49.

Usher, M.B. (1995) A world of change: land-use patterns and arthropod communities. *Insects in a Changing Environment* (eds R. Harrington & N.E. Stork), pp. 372–97. Academic Press, London.

Usher, M.B. & Jefferson, R.G. (1991) Creating new and successional habitats for arthropods. In: *The Conservation of Insects and their Habitats* (eds N.M. Collins & J.A. Thomas), pp. 263–91. Academic Press, London.

Usher, M.B. & Pineda, F.D. (1991) *Biological Diversity*. Fundacion Ramon Areces, Madrid.

Uvah, I.I.L. & Coaker, T.H. (1984) Effects of mixed cropping on some insect pests of carrots and onions. *Entomologia Experimentalis et Applicata*, **36** (2), 159–68.

Vaisanen, R. & Heliovaara, K. (1994) Hot-spots of insect diversity in northern Europe. *Annales Zoologici Fennici*, **31** (1), 71–81.

Valan, G.B. & Muniyappa, V. (1992) Epidemiology of tobacco leaf curl virus in India. *Annals of Applied Entomology*, **120** (2), 257–67.

Valand, G.B. & Muniyappa, V. (1992) Epidemiology of tobacco leaf curl virus in India. *Annals of Applied Biology*, **120** (2), 257–67.

Van Alphen, J.J.M. & Jervis, M.A. (1996) Foraging behaviour. In: *Insect Natural Enemies: Practical Approaches to their Study and Evaluation* (eds M. Jervis & N. Kidd), pp. 1–62. Chapman and Hall, London.

Van Dam, N.M., De Jong, T.J., Iwasa, Y. & Kubo, T. (1996) Optimal distribution of defences: are plants smart investors? *Functional Ecology*, **10**, 128–36.

Van den Bos, J. & Rabbinge, R. (1976) *Simulation of the Fluctuations of the Grey Larch Bud Moth*. Pudoc, Wageningen.

Van Der Meijden, E., Wijn, M. & Verkaar, H.J. (1988) Defense and regrowth: alternative plant strategies in the struggle against herbivores. *Oikos*, **51**, 355–63.

Van Driesche, R.G. (1983) Meaning of percent parasitism in studies of insect parasitoids. *Environmental Entomology*, **12**, 1611–22.

Van Emden, H.F. (1995) Host plant–Aphidophaga interactions. *Agriculture, Ecosystems and Environment*, **52** (1), 3–11.

Van Frankenhuyzen, K. (1993) The challenge of *Bacillus thuringiensis*. In: Bacillus thuringiensis, *an Environmental Biopesticide: Theory and Practice* (eds P.F. Entwistle, J.S. Cory, M.J. Bailey & S. Higgs), pp. 1–35. John Wiley, Chichester, UK.

Van Lenteren, J.C., Hua, L.Z., Kamerman, J.W. & Rumei, X. (1995) The parasite–host relationship between *Encarsia formosa* (Hymenoptera: Aphelinidae) and *Trialeurodes vaporariorum* (Homoptera: Aleyrodidae): XXVI. Leaf hairs reduce the capacity of *Encarsia* to control greenhouse whitefly on cucumber. *Journal of Applied Entomology*, **119** (8), 553–9.

Van Roermund, H.J.W., Van Lenteren, J.C. & Rabbinge, R. (1996) The analysis of foraging behaviour of the whitefly parasitoid *Encarsia formosa* (Hymenoptera: Aphelinidae) in an experimental arena—a simulation study. *Journal of Insect Behaviour*, **9**, 771–97.

Vanbuskirk, J. (1993) Population consequences of larval crowding in the dragonfly *Aeshna juncea*. *Ecology*, **74**, 1950–8.

Varley, G.C. (1947) The natural control of population balance in the knapweed gall fly (*Urophora jaceana*). *Journal of Animal Ecology*, **16**, 139–87.

Varley, G.C. (1949) Population changes in German forest pests. *Journal of Animal Ecology*, **18**, 117–22.

Varley, G.C. & Gradwell, G.R. (1968) Population models for the winter moth. In: *Insect Abundance* (ed. T.R.E. Southwood), pp. 132–42. Blackwell Scientific Publications, Oxford.

Varley, G.C., Gradwell, G.R. & Hassell, M.P. (1973) *Insect Population Ecology: an Analytical Approach*, 212 pp. Blackwell Scientific Publications, Oxford.

Vasconcelos, H.L. & Casimiro, A.B. (1997) Influence of *Azteca alfari* ants on the exploitation of *Cecropia* trees by a leaf-cutting ant. *Biotropica*, **29**, 84–92.

Vats, L.K. & Handa, S. (1988) Soil litter arthropods in a deciduous forest stand at Kurukshetra (India). *Indian Journal of Forestry*, **11** (1), 13–19.

Vats, L.K. & Narula, A. (1990) Soil collemboles of forest and cropland. *Uttar Pradesh Journal of Zoology*, **10** (1), 71–5.

Vega, F.E., Barbosa, P., Kuo-sell, H.L., Fisher, D.B. & Nelsen, T.C. (1995) Effects of feeding on healthy and diseased corn plants on a vector and on a non-vector insect. *Experimentia (Basel)*, **51** (3), 293–9.

Verhulst, P.F. (1838) Notice sur la loi que la population suit dans son accroissement. *Correspondances Mathématiques et Physiques*, **10**, 113–21.

Vet, L.E.M., Wackers, F.L. & Dicke, M. (1991) How to hunt for riding hosts—The reliability-detectability problem in foraging parasitoids. *Netherlands Jounal of Zoology*, **41**, 202–13.

Vezina, A. & Peterman, R.M. (1985) Tests of the role of a nuclear polyhedrosis virus in the population dynamics of its host, Douglas-fir tussock moth *Orgyia pseudotsugata* (Lepidoptera: Lymantriidae). *Oecologia (Heidelberg)*, **67** (2), 260–6.

Visser, M.E. & Driessen, G. (1991) Indirect mutual interference in parasitoids. *Netherlands Journal of Zoology*, **41**, 214–27.

Vite, J.P. (1961) The influence of water supply on oleoresin exudation pressure and resistance to bark beetle attack in *Pinus ponderosa*. *Contributions of the Boyce Thompson Institute*, **21**, 37–66.

Volterra, V. (1926) Variations and fluctuations of the numbers of individuals in animal species living together. In: *Animal Ecology* (ed R.N. Chapman). McGraw-Hill, New York.

Vos, J.G.M. & Nurtika, N. (1995) Transplant production techniques in integrated crop management of hot pepper (*Capsicum* spp.) under tropical lowland conditions. *Crop Protection*, **14** (6), 453–9.

Vrieling, K. (1991) Cost assessment of the production of pyrrolizidine alkaloids by *Senecio jacobaea* L. II. The generative phase. *Mededelingen Van de Faculteit Landbouwwetenschappen Rijksuniversiteit Gent*, **56**, 781–8.

Waage, J.K. (1992) Biological control in the year 2000. In: *Pest Management and the Environment in the Year 2000* (eds A. A. S. A. Kadir & H.S. Barlow), pp. 329–40. CAB International, Wallingford, UK.

Wagner, D.L. & Liebherr, J.K. (1992) Flightlessness in insects. *Trends in Ecology and Evolution*, **7** (7), 216–20.

Wagner, R. (1991) The influence of the diel activity pattern of the larvae of *Sericostoma personatum* (Kirby and Spence) (Trichoptera) on organic matter distribution in stream sediments: a laboratory study. *Hydrobiologia*, **224**, 65–70.

Wainhouse, D. (1980) Dispersal of first instar larvae of the felted beech scale, *Cryptococcus fagisuga*. *Journal of Animal Ecology*, **17**, 523–32.

Wainhouse, D. & Howell, R.S. (1983) Intraspecific variation in beech scale populations and susceptibility of their host, *Fagus sylvatica*. *Ecological Entomology*, **8**, 351–9.

Wall, C., Garthwaite, D.G., Blood-Smyth, J.A. & Sherwood, A. (1987) The efficacy of sex-attractant monitoring for the pea moth, *Cydia nigricana*, in England, 1980–5. *Annals of Applied Biology*, **110**, 223–9.

Wallace, J.B. & Webster, J.R. (1996) The role of macroinvertebrates in stream ecosystem function. *Annual Review of Entomology*, **41**, 115–39.

Wallace, J.B., Webster, J.R. & Cuffney, T.F. (1982) Stream detritus dynamics: regulation by invertebrate consumers. *Oecologia*, **53**, 197–200.

Wallace, J.B., Vogel, D.S. & Cuffney, T.F. (1986) Recovery of a headwater stream from an insecticide-induced community disturbance. *Journal of the North American Benthological Society*, **5**, 115–26.

Wallace, J.B., Cuffney, T.F., Webster, J.R., Lughart, G.J., Chung, K. & Goldowitz, B.S. (1991) Export of fine organic particles from headwater streams: effects of season, extreme discharges, and invertebrate manipulation. *Limnology and Oceanography*, **36**, 670–82.

Walters, D.S., Craig, R. & Mumma, R.O. (1989a) Glandular trichome exudate is the critical factor in geranium resistance to foxglove aphid. *Entomologia Experimentalis et Applicata*, **53**, 105–10.

Walters, D.S., Grossman, H., Craig, R. & Mumma, R.O. (1989b) Geranium defensive agents. *IV Journal of Chemical Ecology*, **15**, 357–72.

Walters, D.S., Craig, R. & Mumma, R.O. (1990) Fatty acid incorporation in the biosynthesis of anacardic acids of geraniums. *Phytochemistry*, **29**, 1815–22.

Waring, G.L. & Price, P.W. (1990) Plant water stress and gall formation (Cecidomyiidae: asphondylia spp.) on creosote bush. *Ecological Entomology*, **15**, 87–96.

Warren, M.S. & Key, R.S. (1991) Woodlands: past, present and potential for insects. In: *The Conservation of Insects and their Habitats* (eds N.M. Collins & J.A. Thomas), pp. 155–211. Academic Press, London.

Warren, M.S., Thomas, C.D. & Thomas, J.A. (1984) The status of the heath fritillary butterfly, *Mellicta athalia* Rott., in Britain. *Biological Conservation*, **29**, 287–305.

Waterhouse, D.F. (1977) The biological control of dung. In: *The Insects* (eds T. Eisner & E.O. Wilson), pp. 314–22. Scientific American, W.H. Freeman, San Francisco.

Waterman, P.G. & Mole, S. (1994) *Analysis of Phenolic Plant Metabolites*. Blackwell Scientific Publications, Oxford.

Watt, A.D. (1988) Effects of stress-induced changes in plant quality and host-plant species on the population dynamics of the pine beauty moth in Scotland: partial life tables of natural and manipulated populations. *Journal of Applied Ecology*, **25**, 209–21.

Watt, A.D. (1989) The growth and survival of *Panolis flammea* larvae in the absence of predators on Scots pine and lodgepole pine. *Ecological Entomology*, **14**, 225–34.

Watt, A.D. (1990) The consequences of natural, stress-induced and damage-induced differences in tree foliage on the population dynamics of the pine beauty moth. In: *Population Dynamics of Forest Insects* (eds A.D. Watt, S.R. Leather, M.D. Hunter & N.A.C. Kidd), pp. 157–68. Intercept, Andover.

Watt, A.D. (1994) The relevance of the stress hypothesis to insects feeding on foliage. In: *Individuals, Populations and Patterns in Ecology* (eds S.R. Leather, A.D. Watt, N.J. Mills & K.F.A. Walters), pp. 63–70. Intercept, Andover.

Watt, A.D., Leather, S.R. & Evans, H.F. (1991) Outbreaks of the pine beauty moth on pine in Scotland—influence of host plant-species and site factors. *Forest Ecology and Management*, **39**, 211–21.

Watt, A.D., Carey, P.D. & Eversham, B.C. (1997a) Implications of climate change for biodiversity. In: *Biodiversity in Scotland: Status, Trends and Initiatives* (eds L.V. Flemming, A.C. Newton, J.A. Vickery & M.B. Usher), pp. 147–59. The Stationery Office, Edinburgh.

Watt, A.D., Stork, N.E., Eggleton, P., Srivastava, D., Bolton, B., Larsen, T.B. & Brendell, M.J.D. (1997b) Impact of forest loss and regeneration on insect abundance and diversity. In: *Forests and Insects* (eds A.D. Watt, N.E. Stork & M.D. Hunter), pp. 274–86. Chapman & Hall, London.

Watt, A.D., Stork, N.E., McBeath, C. & Lawson, G.L. (1997c) Impact of forest management on insect abundance and damage in a lowland tropical forest in southern Cameroon. *Journal of Applied Ecology*, **34**, 985–98.

Webb, S.E. & Shelton, A.M. (1991) A simple action threshold for timing applications of a granulosis virus to control *Pieris rapae* (Lepidoptera: Pieridae). *Entomophaga*, **36** (3), 379–90.

Webster, J.R., Wallace, J.B. & Benfield, E.F. (1995) Organic processess in streams of the eastern United States. In: *River & Stream Ecosystems* (eds C.E. Cushing, K.W. Cummins & G.W. Marshall), pp. 117–87. Elsevier Science Publishers B.V., Amsterdam.

Weidner, H. (1986) The migration routes of the European locust, *Locusta migratoria migratoria* in Europe in the year 1693 (Saltatoria, Acridiidae, Oedipodinae). *Anzeiger fuer Schaedlingskunde Pflanzenschutz Umweltschutz*, **59** (3), 41–51.

Weis, A.E. & Campbell, D.R. (1992) Plant genotype: A variable factor in insect–plant interactions. In: *Effects of Resource Distribution on Animal–Plant Interactions* (eds M.D. Hunter, T. Ohgushi and P.W. Price), pp. 75–111. Academic Press, San Diego, CA.

Wellington, W.G. (1950) Climate and spruce budworm outbreaks. *Canadian Journal of Forest Research*, **28**, 308–31.

Weseloh, R.M. (1976) Behaviour of forest insect parasitoids. In: *Perspectives in Forest Entomology* (eds J.F. Anderson & H.K. Kaya), pp. 99–110. Academic Press, New York.

West, C. (1985) Factors underlying the late seasonal appearance of the lepidopterous leaf-mining guild on oak. *Ecological Entomology*, **10**, 111–20.

Wheeler, W.M. (1910) *Ants: their Structure, Development and Behaviour*. Columbia University Press, New York.

Whiles, M.R. & Wallace, J.B. (1995) Macroinvertebrate production in a headwater stream during recovery from anthropogenic disturbance and hydrologic extremes. *Canadian Journal of Fish Aquatic Science*, **52**, 2402–22.

White, A.J., Wratten, S.D., Berry, N.A. & Weigman, U. (1995) Habitat manipulation to enhance biological control of Brassica pests by hover flies (Diptera: Syrphidae). *Journal of Economic Entomology*, **88** (5), 1171–6.

White, M.J.D. (1974) Speciation in the Australian morabine grasshoppers: the cytogenetic evidence. In: *Genetic Evidence of Speciation in Insects*. (ed. M.J.D. White), pp. 57–68. Australia and New Zealand Book Company, Sydney.

White, T.C.R. (1969) An index to measure weather-induced stress of trees associated with outbreaks of psyllids in Australia. *Ecology*, **50**, 905–9.

White, T.R.C. (1976) Weather, food and plagues of locust. *Oecologia*, **22**, 119–34.

White, T.C.R. (1984) The abundance of vertebrate herbivores in relation to the availability of nitrogen in stressed food plants. *Oecologia*, **63**, 90–105.

White, T.C.R. (1993) *The Inadequate Environment: Nitrogen and the Abundance of Animals*. Springer, Berlin.

Whitehead, L.F. & Douglas, A.E. (1993) Populations of symbiotic bacteria in the parthenogenetic pea aphid (*Acyrthosiphon pisum*) symbiosis. *Proceedings of the Royal Society of London, Series B: Biological Sciences*, **254** (1339), 29–32.

Whitford, W.G., Stinnett, K. & Anderson, J. (1988) Decomposition of roots in a Chihuahuan Desert ecosystem. *Oecologia (Heidelberg)*, **75**, 8–11.

Whitham, T.G. (1978) Habitat selection by *Pemphigus* aphids in response to resource limitation and competition. *Ecology*, **59**, 1164–76.

Whitham, T.G. (1986) Costs and benefits of territoriality: Behavioural and reproductive release by competing aphids. *Ecology*, **67**, 139–47.

Whitham, T.G. (1989) Plant hybrid zones as sinks for pests. *Science*, **23**, 1490–3.

Whitham, T.G., Morrow, P.A. & Potts, B.M. (1994) Plant hybrid zones as centers of biodiversity: The herbivore community of two endemic Tasmanian eucalypts. *Oecologia (Berlin)*, **97**, 481–90.

Whitmore, T.C. & Sayer, J.A. (1992) *Tropical Deforestation and Species Extinction*, 153 pp. Chapman & Hall, London.

WHO (1994) Bubonic plague statistics. http://www.who.org/press/1994/pr94–71.html.

WHO (1997) African Programme for Onchocerciasis Control (APOC). http://www.who.org/programmes/ocp/apoc/index.html.

Wiedenmann, R.N., Smith, J.W.J.R. & Darnell, P.O. (1992) Laboratory rearing and biology of the parasite *Cotesia flavipes* (Hymenoptera: Braconidae) using *Diatraea saccharalis* (Lepidoptera: Pyralidae) as a host. *Environmental Entomology*, **21** (5), 1160–7.

Wiley, M.J. & Warren, G.L. (1992) Territory abandonment, theft, and recycling by a lotic grazer—A foraging strategy for hard times. *Oikos*, **63**, 495–505.

Wilhoit, L.R. (1991) Modelling the population dynamics of different aphid genotypes in plant variety mixtures. *Ecological Modelling*, **55** (3–4), 257–84.

Wilkinson, T.L. & Douglas, A.E. (1995) Aphid feeding, as influenced by disruption of the symbiotic bacteria: An analysis of the pea aphid (*Acyrthosiphon pisum*). *Journal of Insect Physiology*, **41** (8), 635–40.

Williams, G. & Adam, P. (1994) A review of rainforest pollination and plant–pollinator interactions with particular reference to Australian subtropical rainforests. *Australian Zoologist*, **29** (3–4), 177–212.

Williams, K.S. & Myers, J.H. (1984) Previous herbivore attack of red alder may improve food quality for fall webworm larvae. *Oecologia*, **63**, 166–70.

Williams, M.R. (1995) An extreme-value function model of the species incidence and species: Area relations. *Ecology (Washington, DC)*, **76** (8), 2607–16.

Williams, P.H. & Gaston, K.J. (1994) Measuring more of biodiversity—can higher-taxon richness predict wholesale species richness? *Biological Conservation*, **67**, 211–17.

Williams, P.H., Humphries, C.J. & Gaston, K.J. (1994) Centers of seed-plant diversity—the family way. *Proceedings of the Royal Society of London, Series B: Biological Sciences*, **256**, 67–70.

Williamson, S.C., Detling, J.K., Dodd, J.L. & Dyer, M.I. (1989) Experimental evaluation of the grazing optimization hypothesis. *Journal of Range Management*, **42**, 149–52.

Willmer, P.G. (1990) *Invertebrate Relationships: Patterns in Animal Evolution*, 400 pp. Cambridge University Press, Cambridge.

Wilson, D. (1995) Fungal endophytes which invade insect galls: Insect pathogens, benign saprophytes, or fungal inquilines? *Oecologia (Berlin)*, **103**, 255–60.

Wilson, E.O. (1961) The nature of the taxon cycle in the Melanesian ant fauna. *American Naturalist*, **95**, 169–93.

Wilson, E.O. (1987) The little things that run the world (the importance and conservation of invertebrates). *Conservation Biology*, **1**, 344–6.

Wilson, E.O. (1997) Introduction. In: *Biodiversity II, Understanding and Protecting our Biological Resources* (eds M.L. Reaka-Kudla, D.E. Wilson & E.O. Wilson), pp. 1–3. Joseph Henry Press, Washington, DC.

Wimmer, M.J., Smith, R.R., Wellings, D.L., *et al.* (1993) Persistence of diflubenzuron on Appalachian forest leaves after aerial application of dimilin. *Journal of Agricultural and Food Chemistry*, **41** (11), 2184–90.

Winchester, N.N. (1997) Arthropods of coastal old-growth Sitka spruce forests: conservation of biodiversity with special reference to the Staphylinidae. In: *Forests and Insects* (eds A.D. Watt, N.E. Stork & M.D. Hunter), pp. 365–79. Chapman and Hall, London.

Wink, M. & Witte, L. (1991) Storage of quinolizidine alkaloids in *Macrosiphum albifrons* and *Aphis genistae* (Homoptera: aphididae). *Entomologia Generalis*, **15**, 237–54.

Wint, G.R.W. (1983) The role of alternative host plant species in the life of a polyphagous moth, *Operophatera brumata* (Lepidoptera. *Geometridae*). *Journal of Animal Ecology*, **52**, 439–50.

Wipking, W. (1995) Influences of daylength and temperature on the period of diapause and its ending process in dormant larvae of burnet moths (Lepidoptera, Zygaenidae). *Oecologia (Berlin)*, **102** (2), 202–10.

Wiseman, B.R. (1994) Plant resistance to insects in integrated pest management. *Plant Disease*, **78** (9), 927–32.

Wisnivesky-Colli, C., Schweigmann, N.J., Pietrokovsky, S., Bottazzi, V. & Rabinovich, J.E. (1997) Spatial distribution of *Triatoma guasayana* (Hemiptera: Reduviidae) in hardwood forest biotopes in Santiago del Estero, Argentina. *Journal of Medical Entomology*, **34** (2), 102–9.

Witanachchi, J.P. & Morgan, F.D. (1981) Behavior of the bark beetle, *Ips grandicollis*, during host selection. *Physiological Entomology*, **6**, 219–23.

Withers, T.M., Madie, C. & Harris, M.O. (1997) The influence of clutch size on survival and reproductive potential of Hessian fly. *Entomologia Experimentalis et Applicata*, **83** (2), 205–12.

Wood, T.G. & Sands, W.A. (1978) The role of termites in ecosystems. In: *Production Ecology of Ants and Termites* (ed. M.V. Brian). Cambridge University Press, Cambridge.

Wootton R.J. (1992) Functional morphology of insect wings. *Annual Review of Entomology*, **37**, 113–40.

World Conservation Monitoring Centre (1992) *Global*

Biodiversity: Status of the Earth's Living Resources. Chapman and Hall, London.

Worrall, W.D. & Scott, R.A. (1991) Differential reactions of Russian wheat aphid to various small-grain host plants. *Crop Science*, **31** (2), 312–14.

Wotton, R.S. (1994) *The Biology of Particles in Aquatic Systems*. Lewis, Boca Raton.

Wright, D.E., Hunter, D.M. & Symmons, P.M. (1988) Use of pasture growth indices to predict survival and development of *Chortoicetes terminifera* (Walker) (Orthoptera: Acrididae). *Journal of the Australian Entomological Society*, **27** (3), 189–92.

Wyatt, T.D. (1986) How a subsocial intertidal beetle, *Bledius spectabilis*, prevents flooding and anoxia in its burrow. *Behavioral Ecology and Sociobiology*, **19** (5), 323–32.

Xu, F.Y., Ge, M.H., Zhu, Z.C. & Zhu, K.G. (1996) Studies on resistance of pine species and Masson pine provenances to *Bursaphelenchus xylophilus* and the epidemic law of the nematode in Nanjing. *Forest Research*, **9** (5), 521–4.

Yencho, G.C. & Tingey, W.M. (1994) Glandular trichomes of *Solanum berthaultii* alter host preference of the Colorado potato beetle, *Leptintarsa decemlineata*. *Entomologia Experimentalis et Applicata*, **70**, 217–25.

Yoshida, H.A. & Parrella, M.P. (1992) Development and use of selected chrysanthemum cultivars by *Spodoptera exigua* (Lepidoptera: Noctuidae). *Journal of Economic Entomology*, **85** (6), 2377–82.

Yoshimura, T., Azumi, J.I., Tsunoda, K. & Takahashi, M. (1993) Changes of wood-attacking activity of the lower termite, *Coptotermes formosanus* Shiraki in defaunation–refaunation process of the intestinal Protozoa. *Material und Organismen (Berlin)*, **28** (2), 153–64.

Zangerl, A.R. & Berenbaum, M.R. (1990) Furanocoumarin induction in wild parsnip: genetics and population variation. *Ecology*, **71**, 1933–40.

Zangerl, A.R. & Berenbaum, M.R. (1993) Plant chemistry insect adaptations to plant chemistry and host plant utilization in patterns. *Ecology*, **74**, 47–54.

Zehnder, G.W., Sikora, E.J. & Goodman, W.R. (1995) Treatment decisions based on egg scouting for tomato fruitworm, *Helicoverpa zea* (Boddie), reduce insecticide use on tomato. *Crop Protection*, **14** (8), 683–7.

Zelazny, B., Lolong, A. & Patang, B. (1992) *Oryctes rhinoceros* (Coleoptera: Scarabaeidae) populations suppressed by a baculovirus. *Journal of Invertebrate Pathology*, **59** (1), 61–8.

Zhang, Z.Q. & McEvoy, P.B. (1995) Responses of ragwort flea beetle *Longitarsus jacobaeae* (Coleoptera: Chrysomelidae) to signals from host plants. *Bulletin of Entomological Research*, **85** (3), 437–44.

Zhaohui, Z., Yongfan, P. & Juru, L. (1992) National integrated pest management of major crops in China. In: *Integrated Pest Management in the Asia–Pacific Region* (eds P.A.C. Ooi, G.S. Lim, T.H. Ho, P.L. Manalo & J. Waage), pp. 255–66. CAB International, Wallingford, UK.

Zhou, X. & Carter, N. (1992) Effects of temperature, feeding position and crop growth stage on the population dynamics of the rose-grain aphid, *Metopolophium dirhodum* (Hemiptera: Aphididae). *Annals of Applied Biology*, **121** (1), 27–37.

Zucker, W.V. (1983) Tannins: does structure determine function? An ecological perspective. *American Naturalist*, **121**, 335–65.

Index

Page references in *italics* refer to figures and plates; those in **bold** refer to tables.